Major Biological Processes in European Tidal Estuaries

Developments in Hydrobiology 110

Series editor

H. J. Dumont

Major Biological Processes in European Tidal Estuaries

Edited by

C.H.R. Heip and P.M.J. Herman

Reprinted from Hydrobiologia, vol. 311 (1995)

Kluwer Academic Publishers
Dordrecht / Boston / London

Library of Congress Cataloging-in-Publication Data

A C.I.P. Catalogue record for this book is available from the Library of Congress.

ISBN-13: 978-94-010-6539-9 e-ISBN-13: 978-94-009-0117-9
DOI: 10.1007/978-94-009-0117-9

Published by Kluwer Academic Publishers,
P.O. Box 17, 3300 AA Dordrecht, The Netherlands.

Kluwer Academic Publishers incorporates
the publishing programmes of
D. Reidel, Martinus Nijhoff, Dr W. Junk and MTP Press.

Sold and distributed in the U.S.A. and Canada
by Kluwer Academic Publishers,
101 Philip Drive, Norwell, MA 02061, U.S.A.

In all other countries, sold and distributed
by Kluwer Academic Publishers,
P.O. Box 322, 3300 AH Dordrecht, The Netherlands.

Printed on acid-free paper

Contents

Hydrobiologia **311**, 1995.
C.H.R. Heip & P.M.J. Herman (eds), Major Biological Processes in European Tidal Estuaries.

Preface

The CEC-MAST JEEP-92 Project: Major Biological Processes in European Tidal Estuaries

Tidal estuaries in Europe serve important economic functions including transport, industry and tourism but also drainage of waste from domestic, industrial and agricultural activities. Estuaries are under heavy and increasing pressure from human activities in very large areas since they drain water-carrying nutrients, organic matter and toxic materials from the terrestrial environment and the river system to the sea. For example, in the Dutch delta area the water of more than 300,000 km^2 of land surface from seven European countries is concentrated.

Whereas estuaries serve an important role for economic activities they have important natural values as well. Tidal estuaries are characterized by high secundary production rates reflected in important biomass values for benthic and zooplankton populations and the larvae and juvenile stages of fish. They serve as overwintering or passage stations for large populations of palaearctic birds. Estuarine plants and animals are adapted to high environmental variability and possess potentially valuable genetic characteristics. Salt marshes may serve as a natural defence against the rising sea level.

The major biological processes in estuaries may be linked to the production and mineralisation of organic matter. The mineralisation of the organic waste originating from human activities and the subsequent increase of nutrients and primary production is a matter of great concern. There is evidence that increased primary production is absorbed at least partially by increased benthic metabolism and that the biomass of benthos in estuarine and coastal systems is now increasing in the long term. The eutrophication of coastal marine areas originates from land-based human activities and the effect of measures such as the reduction of phosphate and nitrate concentrations in effluents largely depends on how these substances are transferred through the estuary, which is a non-linear system and therefore not easily predictable.

Another matter for concern is the introduction of toxic material in estuarine and coastal environments which may severely change the functioning of the estuarine filter system. Although the toxic effects of single and even mixed pollutants in laboratory conditions have been studied intensively, the fate of many pollutants in the natural estuary remains unclear. Current efforts of modelling the transfer of pollutants in the food web or measuring the accumulated toxicity in higher food web levels will be more valuable if they are based on knowledge of the important structural and functional characteristics of the estuary.

The Joint European Estuarine Research Programme JEEP-92 aimed at a better understanding of effects of organic matter import and production on the ecology of the European estuaries. By integrating results from existing European research efforts and by specific research on certain topics it is hoped to achieve a better understanding of the responses within the estuarine system. In the first project JEEP-92 (1991–1993) the major research effort was devoted to selected biological processes (primary and secundary production, bacterial production, effects of organic loading on meio- and macrofauna) as they relate to nutrients and organic matter. A second important goal was the comparison of structures and processes between different estuaries. The major results of the project are now published in this volume.

CARLO HEIP

Hydrobiologia **311**: 1–7, 1995.
C. H. R. Heip & P. M. J. Herman (eds), Major Biological Processes in European Tidal Estuaries.

Major biological processes in European tidal estuaries: a synthesis of the JEEP-92 Project

Carlo Heip & Peter M. J. Herman
Netherlands Institute of Ecology, Centre for Estuarine and Coastal Research, Vierstraat 28, NL-4401 EA Yerseke, The Netherlands

Abstract

An overview is presented of the major results of the project JEEP-92 (Joint European Estuarine Project). A basic description was made of the ecological structure of the estuaries Elbe, Ems, Westerschelde, Somme, Gironde, Shannon and Tagus. The biological communities in these estuaries have been described and compared. Major macrobiological processes have been quantified: bacterial production, primpary production, zooplankton energy flow, meiofauna and macrofauna dynamics. Hydrodynamic and ecological models have been developed.

Introduction: aims of the project

The project JEEP-92 (Joint European Estuarine Project 92) was funded by the Commission of the European Communities as part of its Marine Science and Technology Programme. It focused on the macrobiological processes that influence the fate of organic matter in estuaries. Conversely, the influence of the microbiological and geochemical conditions and processes on structure and function of major macrobiological constituents in estuaries was studied. Seven major European estuaries were studied and compared: the Elbe, Ems, Westerschelde, Somme, Gironde, Shannon and Tagus.

The project was based on two approaches to study the reaction of estuarine ecosystems to changes in the input of organic material and nutrients, and the integration of both approaches through ecological modelling. In one approach, the rates and regulation of selected major biological processes were studied experimentally in the laboratory and in the field. A better knowledge of essential rates, especially in estuarine conditions, is necessary to improve the ecological modelling of these systems. The other approach was based on the idea that different patterns have already been established in different estuaries because of specific input levels of organic material and nutrients. Study of these patterns can yield insight in the possible results from the processes studied directly in the other parts of the project.

Finally, part of the project was devoted to modelling of estuarine ecosystems. The modelling part included hydrodynamical modelling of the estuaries studied, integration of transport terms for use in ecological models and formulation of ecological dynamics. The ecological modelling part of the proposal integrated the experimentally obtained information mentioned above.

As part of the project a 'Manual of sampling and analytical procedures for tidal estuaries' has been prepared (Kramer *et al.*, 1994). This 'Manual' refers to a set of methodological descriptions (referring to relevant literature), of sampling strategies (frequencies, spatial distributions), and sampling tools for the study of chemical and biological variables in water and sediments of tidal estuarine environments. A detailed description of both the sampling strategy and procedures, as well as the analytical methods, was presented. Comparisons between methods were made in terms of their optimal value within the estuarine research context.

To allow proper comparison and interpretation of data in the JEEP-92 project, a data base with corresponding data base management programme has been established in close connection to the Manual. The data are stored in a normalized relational data base. Consultation of this data base is possible from within the package PARADOX. Routines have been prepared for a number of standard consultations of the data base. For interactive queries for specific questions the highly flexible Paradox interface can be used with a minimum of training. The database and manual are available from the authors on request.

Ecological structure of European tidal estuaries

Loading of the systems

Among the estuaries studied, the Schelde has the most extreme organic loading from the river (Kromkamp *et al.*, 1995). The organic content and especially the ammonium concentration are extremely high at the upstream boundary. This high organic loading has immediate and obvious consequences for oxygen availability. The oxygen needed to oxidize the organic and ammonium load in the Schelde upstream boundary largely exceeds the maximum solubility of oxygen. It is no surprise that the water is anaerobic over a considerable part of the estuary, up to the zone where salinity reaches approximately 5 psu.

The Gironde is characterized by a relatively low nutrient loading, but extremely high concentrations of suspended particulate matter (Irigoien & Castel, 1993a). This is mostly inorganic material, with concentrations of 1–2 g l^{-1}. This high load of suspended matter is responsible for the very low values of *in situ* primary production in this estuary. Chlorophyll *a* concentrations seldom exceed 5 mg m^{-3}. Gironde, Elbe and Ems estuaries appear to be seriously loaded, but not to the point where oxygen problems become predominant. They do posses an upstream freshwater community where benthic and pelagic macro-organisms influence the system. The Ems-Dollard system is largely determined by the important cover by tidal flats; an important part of the primary production is due to microphytobenthos. The Shannon appears to be the lowest in loading of the estuaries studied.

The Somme estuary, although containing a full salinity gradient, is effectively more a marine embayment than an estuary. This is due to the macrotidal nature of the system, which is nearly completely drained at low tide.

Structure of the biological communities

The structure of the upstream, freshwater community in the Schelde estuary is very different from that in the other estuaries. Local primary production in the anoxic waters is extremely high, up to 900 gC m^{-2} y^{-1} (Kromkamp *et al.*, 1995; Soetaert *et al.*, 1994a). This is not caused by a good light climate, but rather by the nearly complete absence of herbivores, and the consequent building up of a relatively large biomass of primary producers (chlorophyll *a* peaks in the order of 100 mg m^{-3}). The food chain in this system is extremely short, with only phytoplankton algae and bacteria. Bacterial production is very high, compared with the rest of the estuary (Goosen *et al.*, 1995).

In the other estuaries in this study, herbivores are present in the freshwater and graze down the phytoplankton to lower concentrations than in the Schelde. The pelagic community is rather constant over the brackish extent (1 to 10 psu) of the estuaries. Mesozooplankton is dominated in winter, spring and early summer by *Eurytemora* sp. (Sautour & Castel, 1995), and important numbers of Mysidacea are present in the so-called hyperbenthos (Mees *et al.*, 1995). In summer a more marine community, in which *Acartia* is the dominant zooplankton genus, dominates the mesozooplankton. Although local differences do exist, it is striking how similar these pelagic communities are between estuaries. It is striking too to find back almost exactly the same community in the Schelde, but restricted here to that part of the brackish zone where oxygen is available (5 to 15 psu) (Soetaert & Van Rijswijk, 1993). Long-term observations in the Gironde, with a significant trend in river runoff and temperature, shows that *Eurytemora* sp. is found in lower numbers at a fixed station during years of higher salinity, even if this is correlated with higher chlorophyll *a* contents (Castel 1993, 1995; Ibanez *et al.*, 1993). The community succession along the salinity gradient has been compressed and partly shifted to higher salinities in the Schelde, compared with other estuaries. It seems justified to conclude that biological interactions, rather than salinity *per se*, determine this succession.

Apart from copepods, the hyperbenthos is a permanent constituent of the community of the brackish zone (Mees *et al.*, 1993a, b; Cattrijsse *et al.*, 1993). Biomass of hyperbenthos is lower than that of zooplankton (in the Westerschelde : 50 mg AFDW m^{-2}; yearly P:B is 6; yearly production is 0.3 g AFDW m^{-2} y^{-1}, Mees *et al.*, 1994), but this secondary production may be of crucial importance to fish stocks in estuaries. Gut content determination of fish in the Westerschelde have shown a high proportion of Mysidacea in the diet of many (young) fish. Although some species may be absent due to a limited biogeographical distribution range (Mees & Fockedey, 1993), in each estuary very similar, geographically isolated hyperbenthic communites can be identified (Mees *et al.*, 1995). The relative positions of these communities along the salinity-turbidity gradients are analogous. A marine community comprises all stations in the zone with high salinities, a brackish water community is situated in the middle part of the estuary, and an oligohaline community is present in the most inshore stations. In each estuary one or two transient stations are present which characterize zones where 2 communities meet. In the Westerschelde the oligohaline community is completely absent. In this system the gradient in dissolved oxygen becomes very important: from about 8 psu salinity the oxygen saturation values drop to a value which is critically low for all hyperbenthic life (Mees & Hamerlynck, 1992; Mees *et al.*, 1995).

Benthic communities show a clear relationship with suspended matter concentrations and, presumably, silt deposition. In sediments covered by turbid waters (e.g. Westerschelde brackish zone, Gironde almost down to the estuary mouth) communities dominated by deposit feeders are found in the intertidal areas. In subtidal channels, tidal current speeds and instability of the sediment become the limiting factors, leading to very poor communities (in terms of biomass, density and diversity) in all estuaries .

Big differences in biomass of macrobenthic deposit feeder communities have been found in different estuaries. In the brackish part of the estuaries, biomass can range from around 1 g AFDW m^{-2} (e.g. Gironde inner estuary) to over 10 g AFDW m^{-2} (e.g. Westerschelde brackish zone, Ems middle estuary). More 'marine' macrobenthic communities are found in the most seaward part of the estuaries. Here benthic biomass can be dominated by filter feeders, giving rise to a food chain typical for coastal marine situations. Most probably the lower suspended matter concentrations create the conditions favourable for the development of these communities. Even in the mouth of the Gironde, suspended particulate matter loads are presumably too high for a filter feeding community to develop. On the contrary the Somme, which is almost a marine embayment, has very high biomasses of benthic filter feeders. Cockle (*Cerastoderma edule*) biomass may reach 300 g AFDW m^{-2}. In high summer, oxygen consumption by this community can be so high that dissolved oxygen is depleted in the (often very shallow) water layer, causing severe anoxia problems (Desprez *et al.*, 1992). It can be hypothesized that these benthic filter feeders present food competition to the filter feeding zooplankton and hyperbenthos. Thus, they could be the cause for the decline of planktonic filter feeders towards coastal waters.

The meiobenthos in the brackish part of the estuaries has a remarkably similar biomass over the different estuaries studied (Soetaert *et al.*, 1994b, 1995). In these, mostly deposit feeder dominated sediments, the biomass and densities of nematodes are relatively low. Experiments in mesocosms have shown that dosage of intact cores with different quantity and quality of food had no effect on abundance, diversity or community structure of meiobenthos (Austen & Warwick, 1995). This result is consistent with the results of the field survey. It is further corroborated by a small-scale study in the Westerschelde where meiobenthos has been sampled with a high temporal resolution, and where concommitant samples of macrobenthos, epibenthos, POC, bacteria and diatoms have been taken. From a modelling analysis of these results it became clear that the community is mainly determined by predation and/or interference from macrofauna and epifauna. This again is consistent with the observation that meiobenthic communities are similar under different field conditions.

Major biological processes

Bacterial production

In the Scheldt estuary bacterial production was followed during an annual cycle in 1991 (Goosen *et al.*, 1995). From the mouth of the estuary to the river Rupel salinity decreased from 29 psu to 0.2 psu and the mean oxygen concentration from 100% to 12% saturation. Close to the maximum turbidity zone the oxygen concentration decreased very rapidly to a level between 0% and 20% saturation. One of the main causes is the heavy loading with ammonia and organic material, causing high microbial activity in this area. The nitrification of ammonium to nitrate consumes a large amount of oxygen. Approx. 25% of the nitrate seemed to get lost as molecular nitrogen due to anaerobic nitrate respiration (denitrification). The oxygen gradient in the water was permanent during the whole year (Goosen *et al.*, 1995).

A profile of the mean bacterial production and the mean oxygen concentration in 1991 shows an inverse relationship. The increase of the bacterial production from the marine to the freshwater part of the estuary was consistent during the whole year and coincided with the decrease of the oxygen concentration. A strong positive correlation was found between the bacterial production and the concentration of DOC.

In order to generalize the possible conclusions from these observations, longitudinal gradients along the Elbe and Schelde were sampled during 1992. Bacterial activity was measured, together with a number of variables relevant for the quantification and characterization of (particulate and dissolved) organic matter. Bacterial production in the Schelde was five times higher than in the Elbe and was amongst the highest ever reported in the literature (Goosen *et al.*, 1995).

Mineralization rates in the intertidal sediments constitute important sink terms for carbon and nitrogen in the estuary (Middelburg *et al.*, 1995a). They are estimated to be responsible for 15% and 30% of the estuarine retention of nitrogen and carbon respectively. Annually integrated nitrous oxide emissions, which appeared to be linearly related to rates of ammonification, nitrification and denitrification, showed a marked gradient along the estuary, ranging from 10 mmol N m^{-2} at the tidal freshwater endmember to −0.6 mmol N m^{-2} at the most saline stations (Middelburg *et al.*, 1995b).

Primary production

A detailed study of phytoplankton and primary production has also been conducted in the Westerschelde. In general, primary production followed the incident irradiance. Due to the high turbidity, partly caused by the high tidal currents, light attenuation was mainly caused by scattering. Absorption of light by algae is insignificant. Within a particular station, relationships between production and light were highly significant, especially in the central and eastern part of the basin where turbidity was higher and nutrients were in excess. In 1991, a period of reduced primary production was observed near the mouth. This coincided with a period of low silicate concentrations and during this period primary production was probably limited by silicate (Kromkamp *et al.*, 1995). Annual gross production in the Westerschelde in 1989 and 1991 was rather constant in the Western part of the basin and was reduced in the brackish, eastern part of the basin. This reduction was due to the higher turbidity in this area. In the very low salinity to freshwater part of the estuary the primary production increased again, due to freshwater phytoplankton (especially several *Scenedesmus* sp.). Model studies showed that due to the low ratio of euphotic depth to total mixing depth, net primary production was low (Soetaert *et al.*, 1994a). It did not exceed an average value of 40 gC m^{-2} y^{-1} on average for the whole estuary. In these turbid conditions, the microphytobenthos of intertidal areas has a relative advantage over the phytoplankton. It may be responsible for a considerable part of the total primary production in the system (De Jong & de Jonge, 1995).

Observations in a turbid part of the Gironde estuary showed that the chlorophyll concentration was strongly linked to the Suspended Particulate Matter concentration. The percentage chlorophyll-a relative to the total pigments, however, showed an opposite pattern thus indicating that primary production is low compared to allochtonous imports in various states of degradation (Irigoien & Castel, 1993).

Zooplankton energy flow

The principal objectives were to describe, understand, and predict the contribution of zooplankton in the transformation of organic matter in European tidal estuaries. More specifically, to determine the overall mass and energy budget of the zooplankton by quantifying somatic and gonadic production, food ingestion, and output processes (predation, export, recycling, respiration, excretion, dejection); to determine and parametrize the key processes for incorporation in the ecological model; and to provide calibration data for this model.

To determine whether estuarine copepod populations living in the brackish zone are indeed mainly detritivorous (as generally believed) or on the contrary are selectively feeding on phytoplankton as do many species in less turbid ecosystems, grazing experiments using a combination of methods were carried out simultaneously with adults of *Eurytemora affinis* and *Acartia tonsa* in natural water of the Westerschelde. The data suggest that both copepod species are capable of selecting live phytoplankton, but that this feeding mode is limited in *E. affinis* because it is energetically costly to perform under spring circumstances. *A. tonsa*, living in summer, is more successful in selecting phytoplankton (Tackx & Daro, 1993; Tackx *et al.*, 1995). It was attempted to relate this to the morphological structure of the species (Revis *et al.*, 1991; Castel & Feurtet, in press).

The gut evacuation rates were determined on board and in the laboratory. The gut content declined and the evacuation rate increased exponentially with temperature. The dependence of the evacuation rate on the initial gut content was clearly demonstrated. From this, the yearly chlorophyll-*a* consumption per copepod could be estimated. From simultaneous POC, chlorophyll-*a*, gut fluorescence and evacuation measurements it became clear that 3–16% of the diet of Eurytemora consists of phytoplankton (Irigoien *et al.*, 1993; Irigoien & Castel, 1995).

Gut fluorescence was measured as related to the tidal and diurnal cycles. For *Eurytemora* a correlation was found with the tidal cycle, the maxima coinciding with the high tides (lower Suspended Particulate Matter concentration, higher percentage chlorophyll-*a*). For *Acartia* species this correlation was not always present. The diurnal cycle did not influence the gut fluorescence. Conversely, in laboratory experiments *Eurytemora* produced more faecal pellets (after an adaptation period) in the dark.

Demographic parameters were established in laboratory experiments with *Eurytemora* cultured in natural (Westerschelde) water (Escaravage & Soetaert, 1993, 1995). The egg incubation time was found to be inversely related to temperature and did not differ from data in the literature. Naupliar development was slower for larger nauplii than for the first two stages at low temperatures. At 2 °C transformation to copepodites did not occur. At 5 °C development stopped before the C5-stage.

A length–weight relationship was determined and used to establish the mean stage (dry) weights. From these the instantaneous growth rates were calculated. A global P/B of 0.12 d^{-1} was found for the whole population. This ratio is remarkably constant between estuaries (Castel & Feurtet, 1992; Peitsch, 1995).

Meiofauna dynamics and food input

A comparative mesocosm experiment was carried out to determine the effects of natural foods of different quality and quantity on the structure of meiobenthic assemblages from the polluted Westerschelde and the comparatively undisturbed Gironde estuaries. The hypothesis that in the Westerschelde the structure of the meiobenthic communities will be most strongly influenced by the high nutrient and pollutant loadings but in the Gironde the availability of nutrients of different quality and quantity will be of more importance was not substantiated. There was no change in community structure in response to the treatment (different quantity and quality of food) in either of the estuarine meiobenthic communities (Austen & Warwick, 1995).

Macrofauna communities and food enrichment

A comparison of the macrofauna community structure of cores exposed for 20 weeks to different doses of food shows that there is no clear relationship between the amount of food supplied to a core and the composition of its fauna at the end of the experiment. Similarly, no relationship could be found with the type of diet which was used for feeding.

In field experiments with fertilizers, nutrient-releasing pellets were still present on treated sites after four months, but there was no significant difference in the concentration of either carbon or nitrogen between the controls and treatment sites. Application of the Osmocoat fertilizer had no effect on the numerical structure of the fauna of the estuaries (Kendall *et al.*, 1995).

Models

Biological processes in tidal estuaries are particular for physical reasons. In tidal estuaries the velocity is much larger than in non-tidal ones and so are the horizontal and vertical mixing. The zones of the estuary 'mechanically' available for the benthic communities are smaller, but the availability of food may be higher (particularly for filter-feeders).

The time step biologically meaningful for the evolution of the ecological properties (1 day) is of the order of the tidal period. Also the spatial gradients of the ecological properties are in general much smaller than the gradients associated with the tidal flow. So an ecological model can consider a much larger grid and a smaller number of boxes can be used. However, due to this choice an ecological model cannot consider the tidal flow explicitly. The traditional way to integrate a transport equation in space and time is by considering the mean values of the properties and include in a dispersion coefficient the effects of the non-linear terms of the equation. In a one-dimensional ecological model of an estuary the dispersion coefficients can be computed using only field data on salinity. In these models the residual flow across each open boundary of the model boxes is the river flow rate. It was shown for the case of the Westerschelde that with an automatic calibration routine an excellent fit to the salinity data of the estuary could be provided (Soetaert & Herman, 1995a).

There are, however, several advantages to coupling ecological and hydrodynamical modelling efforts. A transport model can simulate the distribution of a property (e.g. salinity) and so replace the continuous monitoring necessary to obtain the field data described. In this project an algorithm was developed to compute the dispersion coefficients for an ecological model using the results of a salinity transport model. The algorithm is general, allowing the computation of the coefficients independently of the form of the ecological model boxes (Neves *et al.*, 1993).

The ecological model MOSES provides formulations for basic pelagic and benthic processes, and their interaction. Processes covered are primary production, microbiological degradation of organic matter, nitrification, denitrification, zooplankton grazing, hyperbenthos grazing and migration, benthic mineralization, primary production by microphytobenthos, grazing by deposit and filter feeders, sedimentation, transport. Special care has been taken to close the mass balances; this was not obvious in view of the complicated benthic-pelagic coupling.

The ecological model is being applied to the Westerschelde. Formulations are generic enough to provide for an easy transfer to other estuaries. With the aid of the model general mass balances for the Westerschelde have been drawn for carbon (Soetaert & Herman, 1995b) and nitrogen (Soetaert & Herman, 1995c). A striking result is the extreme heterotrophy of the system: 80% of the carbon respired is imported into the system. Most import comes from upstream, but import of high quality organic matter from the sea was also shown (for low quality organic matter there is export towards the sea). A particular example is the import and subsequent mortality of marine zooplankton from the sea in the estuary (Soetaert & Herman, 1994). Locally produced carbon is relatively unimportant in the system. Net primary production is negative throughout the year in the brackish zone; in this zone the chemosynthetic primary production by nitrification exceeds the primary production by phytoplankton. In the more marine part net primary production is, on average, positive. It does never attain high values, however. The nitrogen balance of the system shows that there is very little cycling within the system. Approximately 25% of the nitrogen entering the system is lost by denitrification or burial. Most nitrogen is oxidized from ammonia to nitrate, and then leaves the system to the sea.

References

Austen, M.C. & R.M. Warwick, 1995. Effects of manipulation of food supply on estuarine meiobenthos. Hydrobiologia 311 (Dev. Hydrobiol. 110): 175–184.

Castel, J., 1993. Long-term distribution of zooplankton in the Gironde estuary and its relation with river flow and suspended matter. Proc. Workshop, Ecology of Zooplankton in European Estuaries, Arcachon, 1921 May 1992. Cah. Biol. mar. 34: 145–163.

Castel, J., 1995. Long-term changes in the population of *Eurytemora affinis* (Copepoda, Calanoida) in the Gironde estuary (1978–1992). Hydrobologia 311 (Dev. Hydrobiol. 110): 85–101.

Castel, J. & A. Feurtet, 1992. Fecundity and mortality rates of the copepod Eurytemora affinis in the Gironde estuary. In Colombo, G., I. Ferrari, V.U. Cecherelli & R. Rossi (eds), Marine Eutrophication and Population Dynamics. Olsen & Olsen: 143–149.

Castel, J. & A. Feurtet, 1995. Morphological variations in the estuarine copepod Eurytemora affinis as a response to environmental factors. Proc. 27th EMBS, Dublin (in press).

Cattrijsse, A., J. Mees & O. Hamerlynck, 1993. The hyperbenthic Amphipoda and Isopoda of the Voordelta and the Westerschelde estuary. Cah. Biol. mar. 34: 187–200.

de Jong, D.J. & V.N. de Jonge, 1995. Dynamics and distribution of microphytobenthic chlorophyll-*a* in the Western Scheldt estuary (SW Netherlands). Hydrobiologia 311 (Dev. Hydrobiol. 110): 21–30.

Desprez, M., H. Rybarczyk, J.G. Wilson, J.P. Ducrotoy, F. Sueur, R. Olivesi & B. Elkaim. 1992. Biological impact of eutrophication in the Bay of Somme and the induction and impact of anoxia. Neth. J. Sea Res. 30: 141–147.

Escaravage, V. & K. Soetaert, 1993. Estimating secondary production for the brackish Westerschelde copepod population Eurytemora affinis (Poppe) combining experimental data and field observations. Cah. Biol. mar. 34: 201–204.

Escaravage, V. & K. Soetaert, 1995. Secondary production of the brackish copepod communities and their contribution to the carbon fluxes in the Westerschelde estuary (The Netherlands). Hydrobiologia 311 (Dev. Hydrobiol. 110): 103–114.

Goosen, N.K., P. van Rijswijk & U. Brockmann, 1995. Comparison of heterotrophic bacterial production in early spring in the turbid estuaries of the Scheldt and the Elbe. Hydrobiologia 311 (Dev. Hydrobiol. 110): 31–42.

Ibanez, F., J.M. Fromentin & J. Castel., 1993. Application de la méthode des sommes cumulées à l'analyse des séries chronologiques en océanographie. C.r. Acad. Sci., Paris, Sci. de la vie, 316: 745–748.

Irigoien, X. & J. Castel, 1993. Dynamique des pigments chlorophylliens dans l'estuaire de la Gironde. Proc. III Workshop Oceanography of the Bay of Biscay, Arcachon, 7–9 April 1992. CNRS, ed.: 73–77.

Irigoien, X. & J. Castel, 1995. Feeding rates and productivity of the copepod *Acartia bifilosa* in a highly turbid estuary; the Gironde (SW France). Hydrobiologia 311 (Dev. Hydrobiol. 110): 115–125.

Irigoien, X., J. Castel & B. Sautour, 1993. In situ grazing activity of copepods in the Gironde estuary. Proc. Workshop Ecology of Zooplankton in European Estuaries, Arcachon, 19–21 May 1992. Cah. Biol. mar. 34: 225–237.

Kendall, M.A., J.T. Davey & S. Widdicombe, 1995. The response of two estuarine benthic communities to the quantity and quality of food. Hydrobiologia 311 (Dev. Hydrobiol. 110:) 207–214.

Kramer, K.J.M, U.H. Brockmann, R.M. Warwick, 1994. Tidal estuaries: manual of sampling and analytical procedures. Balkema, Rotterdam, The Netherlands, 304 pp.

Kromkamp, J., J. Peene, P. van Rijswijk, A. Sandee & N. Goosen, 1995. Nutrients, light and primary production by phytoplankton and microphytobenthos in the eutrophic, turbid Westerschelde estuary (The Netherlands). Hydrobiologia 311 (Dev. Hydrobiol. 110): 9–19.

Mees, J., Z. Abdulkerim & O. Hamerlynck, 1994. Life history, growth and production of Neomysis integer in the Westerschelde estuary (SW Netherlands). Mar. Ecol. Prog. Ser. 109: 43–57.

Mees, J., A. Cattrijsse & O. Hamerlynck, 1993a. Distribution and abundance of shallow-water hyperbenthic mysids (Crustacea, Mysidacea) and euphausiids (Crustacea, Euphausiacea) in the Voordelta and the Westerschelde, south-west Netherlands. Cah. Biol. mar. 34: 165–186.

Mees, J., A. Dewicke & O. Hamerlynck, 1993b. Seasonal composition and spatial distribution of hyperbenthic communities along estuarine gradients in the Westerschelde. Neth. J. Aquat. Ecol., 27: 359–376.

Mees, J. & N. Fockedey, 1993. First record of Synidotea laevidorsalis (Miers, 1881) (Crustacea: Isopoda) in Europe (Gironde estuary, France). Hydrobiologia 264: 61–63.

Mees, J., N. Fockedey & O. Hamerlynck, 1995. Comparative study of the hyperbenthos of three European estuaries. Hydrobiologia 311 (Dev. Hydrobiol. 110): 153–174.

Mees, J. & O. Hamerlynck, 1992. Spatial community structure of the winter hyperbenthos of the Schelde estuary, the Netherlands, and the adjacent coastal waters. Neth. J. Sea Res. 29: 357–370.

Middelburg, J.J., G. Klaver, J. Nieuwenhuize & T. Vlug, 1995a. Carbon and nitrogen cycling in intertidal sediments near Doel, Scheldt Estuary. Hydrobiologia 311 (Dev. Hydrobiol. 110): 57–69.

Middelburg, J.J., G. Klaver, J. Nieuwenhuize, R.M. Markusse, T. Vlug & F.J.W.A. van der Nat, 1995b. Nitrous oxide emissions from estuarine intertidal sediments. Hydrobiologia 311 (Dev. Hydrobiol. 110): 43–55.

Neves, R., L. Portela & J. Barata, 1993. Coupling hydrodynamical and ecological models. Dispersion coefficients. Report JEEP–92.

Peitsch, A, 1995. Production rates of *Eurytemora affinis* in the Elbe estuary, comparison of field and enclosure production estimates. Hydrobiologia 311 (Dev. Hydrobiol. 110): 127–137.

Revis, N., J. Castel & M.L.M. Tackx, 1991. Some reflections on the structure of the mandible plate of Eurytemora affinis (copepoda, calanoida). Hydrobiol. Bull. 25: 45–50.

Sautour, B. & J. Castel, 1995. Comparative spring distribution of zooplankton in three macrotidal European estuaries. Hydrobiologia 311 (Dev. Hydrobiol. 110): 139–151.

Soetaert, K. & P.M.J. Herman, 1994. One foot in the grave: zooplankton drift into the Westerschelde estuary (The Netherlands). Mar. Ecol. Prog. Ser. 105: 19–29.

Soetaert, K. & P.M.J. Herman, 1995a. Estimating estuarine residence times in the Westerschelde (The Netherlands) using a box model with fixed dispersion coefficients. Hydrobiologia 311 (Dev. Hydrobiol. 110): 215–224.

Soetaert, K. & P.M.J. Herman, 1995b. Carbon flows in the Westerschelde estuary (The Netherlands) evaluated by means of an ecosystem model (MOSES). Hydrobiologia 311 (Dev. Hydrobiol. 110): 247–266.

Soetaert, K. & P.M.J. Herman, 1995c. Nitrogen dynamics in the Westerschelde estuary (SW Netherlands) estimated by means of the ecosystem model (MOSES). Hydrobiologia 311 (Dev. Hydrobiol. 110): 225–246.

Soetaert, K., P.M.J. Herman & J. Kromkamp, 1994a. Living in the twilight: estimating net phytoplankton growth in the Westerschelde estuary (The Netherlands) by means of an ecosystem model (MOSES). J. Plankton. Res. 16: 1277–1301.

Soetaert, K. & P. van Rijswijk, 1993. Spatial and temporal patterns of the zooplankton in the Westerschelde estuary. Mar. Ecol. Progr. Ser. 97: 47–59.

Soetaert, K., M. Vincx, J. Wittoeck, M. Tulkens & D. van Gansbeke, 1994b. Spatial patterns of Westerschelde meiobenthos. Estuar. coast. Shelf Sci. 39: 367–388.

Soetaert, K., M. Vincx, J. Wittoeck & M. Tulkens, 1995. Meiobenthic distribution and nematode community structure in five European estuaries. Hydrobiologia 311 (Dev. Hydrobiol. 110): 185–206.

Tackx, M.L.M. & N. Daro, 1993. Potential influence of size dependent carbon uptake and chlorophyll content of phytoplankton cells on zooplankton grazing measurements. Cah. Biol. mar. 34: 253–260.

Tackx, M., X. Irigoien, N. Daro, J. Castel, L. Zhu, X. Zhang & J. Nijs, 1995. Copepod feeding in the Westerschelde and the Gironde. Hydrobiologia 311 (Dev. Hydrobiol. 110): 71–83.

Hydrobiologia **311**: 9–19, 1995.
C. H. R. Heip & P. M. J. Herman (eds), Major Biological Processes in European Tidal Estuaries.

Nutrients, light and primary production by phytoplankton and microphytobenthos in the eutrophic, turbid Westerschelde estuary (The Netherlands)

Jacco Kromkamp, Jan Peene, Pieter van Rijswijk, Adri Sandee & Nico Goosen
Netherlands Institute of Ecology, Centre for Estuarine and Coastal Ecology, Vierstraat 28, NL-4401 EA Yerseke, The Netherlands

Key words: Westerschelde estuary, phytoplankton, primary production, turbidity, microphytobenthos

Abstract

Abiotic factors and primary production by phytoplankton and microphytobenthos was studied in the turbid Westerschelde estuary. Because of the high turbidity and high nutrient concentrations primary production by phytoplankton is light-limited. In the inner and central parts of the estuary maximum rates of primary production were therefore measured during the summer, whereas in the more marine part spring and autumn bloom were observed. Organic loading is high, causing near anaerobic conditions upstream in the river Schelde. Because of this there were no important phytoplankton grazers in this part of the estuary and hence the grazing pressure on phytoplankton was minimal. As this reduced losses, biomass is maximal in the river Schelde, despite the very low growth rates.

On a number of occasions, primary production by benthic micro-algae on intertidal flats was studied. Comparison of their rates of primary production to phytoplankton production in the same period led to the conclusion that the contribution to total primary production by benthic algae was small. The main reason for this is that the photosynthetic activity declines rapidly after the flats emerged from the water. It is argued that CO_2-limitation could only be partially responsible for the noticed decrease in activity.

Introduction

The river Schelde and the Westerschelde estuary are vital to the economies of Belgium and The Netherlands and these systems are subjected to severe anthropogenic stress (Billen *et al.*, 1988; Hummel & Bakker, 1988).

Nutrients, microbial processes and phytoplankton composition of the Westerschelde has been studied by several authors (e.g. Billen *et al.*, 1985, 1988, Bokhorst, 1988, De Pauw, 1975). Abiotic factors were studied monthly since 1982 (Bokhorst, 1988). However, most reports on (biotic) data and processes studied thusfar are irregular and out of date (Heip, 1988). To get a more comprehensive view of important processes determining biological processes in the Westerschelde the CECE initiated the 'Westerschelde project'. For this reason Van Spaendonk *et al.* (1993) started a baseline study in 1989 on phytoplankton primary production, nutrients and turbidity. It was found that primary production was mainly limited by light and that nutrients were in excess of growth demands .

In 1991 this project was extended to the microbial foodweb and primary production by phytoplankton and bacterial production was measured in the Westerschelde estuary as well as in part of the river Schelde. The river Schelde was included because it was obvious from the results of Van Spaendonk *et al.* (1993) that more upstream important biological processes regarding the fate of dissolved nitrogen, bacterial activity and phytoplankton populations were going on. Also, the freshwater endmember had to be included for later modelling efforts (see Soetaert *et al.*, this issue).

In estuaries with large areas of intertidal flats, like the Westerschelde, benthic microalgae can contribute significantly to total primary production (Nienhuis *et al.*, 1985) and may provide in some estuaries as much as one third of total primary production (Sullivan & Moncreiff, 1988). For logistic reasons it was not possible to visit the intertidal mudflats regularly. But to get

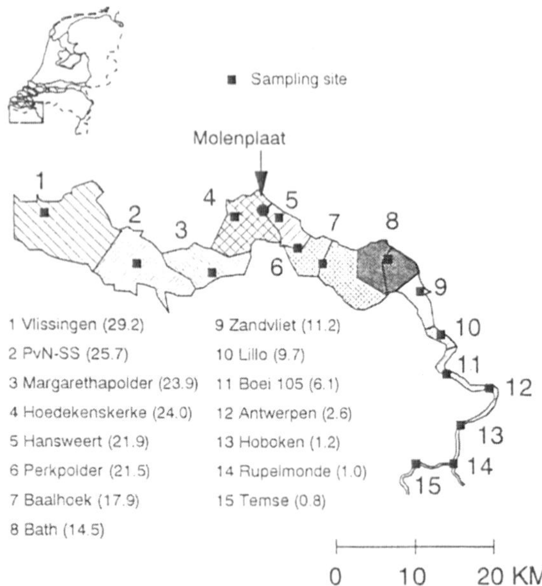

Fig. 1. The Westerschelde with sampling locations. The annual average salinity at each station is given between brackets.

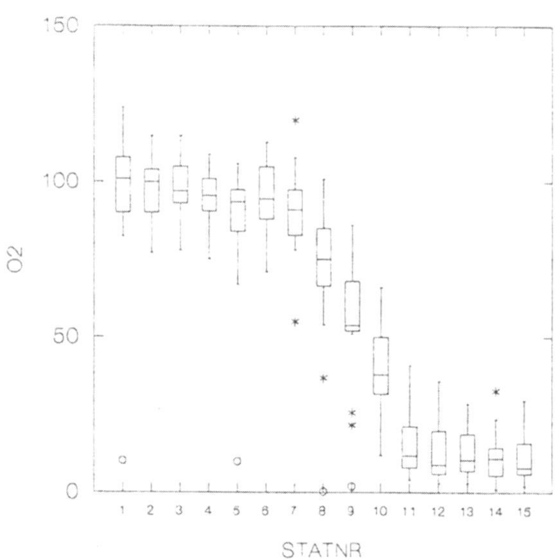

Fig. 2. Box-Whisker plot of dissolved oxygen concentration (%) at different sampling stations. The box gives the 25 and 75% percentiles and the line the annual mean. The error bars give the 95% percentiles, whereas the single points are outliers.

an estimate of the contribution by microphytobenthos to total primary production, a number of production measurements on populations of benthic algae were performed and they were compared to phytoplankton production rates in the same period.

The main aim of this study was to extend the study of nutrient concentrations and primary production at the boundary conditions beyond those investigated by Van Spaendonk *et al.* (1993) and to investigate primary production by benthic micro-algae in relation to primary production by phytoplankton .

Methods

Study site

The Westerschelde is situated in the South of the Netherlands (Fig. 1). It is the maritime zone of the river Schelde. It drains approximately 19 500 km^2 of low-lying plains, mainly Flanders (Heip, 1989). The average residence time of the Westerschelde is approx. 75 days, i.e. about 150 tides. For this reason the salinity zones in the estuary are rather stable. For more details see Van Spaendonk *et al.* (1993) The location of the sampling stations is indicated in Fig. 1.

Water sampling and physical measurements

In 1991 all stations were sampled biweekly from March to October and once a month for the rest of the year.

Temperature, pH, oxygen concentration and salinity was measured *in situ* using a CTD-probe (H_2O-datasonde coupled to a Surveyor 3, Hydrolab).

Chemical analysis

Duplicate surface samples were filtered (glass fibre filters, Schleicher & Schuell, no. 6, average pore size 1.2 μm), and filtrates were analyzed using a Technicon AA2 autoanalyzer, operated according to the manufacturers recommendations.

Seston dry weight was determined gravimetrically after drying at 70 °C.

Samples for pigment analyses of phytoplankton were filtered onto glass fibre filter (Schleicher & Schuell, no. 6). Filters and sediment samples were extracted in 90% acetone and pigments were determined using reversed phase HPLC according to Gieskes *et al.* (1988).

Primary production of phytoplankton

50 ml samples (to which 200 μl of 185 kBq ml^{-1} ^{14}C-H_2CO_3 was added) were incubated in a rotating

incubator (Vegter & De Visscher, 1984) onboard the research vessel 'Luctor' immediately after sampling at *in situ* temperatures. After incubation the samples were gently filtered onto 0.45 μm nitrocellulose filters (Schleicher & Schuell BA23). The incubation period was 2 h. Filters were placed in a HCl fume for 30 min, air dried and counted using 0.5% (w/v) PPO (2,5-diphenyloxazol, p.a. MERCK) in toluene (Baker, technical grade) in a Beckmann LSC (LS5000TD). Correction for quench took place using the shift in the Compton peak (H-number) according to the manufacturers instructions. Dissolved inorganic carbon was determined by potentiometric titration. The dark bottle values were subtracted from light bottle values in order to avoid overestimates of primary production by phytoplankton due to chemosynthetic processes.

The photosynthesis/irradiance (P/I)-curves were fitted according to Eilers & Peeters (1988):

$$P^B = \frac{E}{\mathrm{a}E^2 + \mathrm{b}E + \mathrm{c}}, \qquad (1)$$

where P^B (mgC (mg chl a)$^{-1}$ h^{-1}) is the rate of photosynthesis and E the (scalar) incubation irradiance (μE m^{-2} s^{-1}) a, b and c are fit-constants. $P^B_{\max}$ is the maximal rate of photosynthesis and α^B is the photosynthetic efficiency, i.e. the initial slope of the P/I-curve. The superscript B denotes that P/I-characteristics were calculated per mg chlorophyll a. The irradiance of the 11 compartments in the incubator was measured with a 4π sensor (Biospherical Instruments). The maximal irradiance was 895 μE m^{-2} s^{-1}.

Integral column production was calculated from the fitted P/I-curve, the attenuation coefficient and hourly incident irradiance. The latter was measured in Yerseke (close to the sampling stations) with a Kipp solarimeter fitted with a PAR (400–700 nm) sensor. Because in 1991 the instrument broke down, hourly light data were obtained from a local station (Wemeldinge) of the Dutch Meteorological Institute (KNMI). However, the sensor they used integrated radiation from 300–2200 nm. A conversion factor (0.33 ± 0.02) was calculated after comparing 500 hourly datapoints. In calculating integral column primary production, the data were corrected for basin morphology (because there are large tidal flats and shallow areas in the Westerschelde), according to Klepper (1989). Carbon turnover rates (P/B-ratio's) were calculated by dividing the column production by the C-content of the phytoplankton in the water column. The carbon content of the phytoplankton was calculated from the chlorophyll a content, assuming a C/chlorophyll a-ratio of 30. This ratio was chosen because it was the mean ratio during the period 1982–1990 from phytoplankton samples taken (and investigated microscopically) in the nearby Oosterschelde estuary (Wetsteyn & Kromkamp, 1994). In this way high C/chl-ratio's were prevented which would have occurred when the C-content was calculated from particulate organic carbon measurements.

Primary production of microphytobenthos

Rates of oxygen production were determined using custom made microelectrodes without internal reference electrode according to Revsbech & Jorgensen (1983). Steps of 0.1 mm were made using a motor driven micro-manipulator. Sediment cores were taken at the highest point on the intertidal flats as soon as they became submerged. The measurements were started approx. 15 minutes after sampling and were performed on board the ship. A 'cold light source' (Schott KL 150 B) was used to illuminate the sediment core. Light was measured with a 2π sensor (Licor LI192-SB) connected to a photonmeter (Licor LI-185B).

We did not attempt to get an accurate estimate of total microphytobenthos production, as this would include detailed measurements on several locations on an intertidal mudflat and frequent visits to several mudflats.

Underwater light

Up- and downwelling irradiance (PAR) at the fieldstations was measured with a cosine corrected quantum sensor (Licor LI-192SB) connected to a Licor LI-185B quantum meter.

The vertical attenuation of diffuse downwelling light can be described by the Lambert-Beer equation

$$E_d(z) = E_d(O)\mathrm{e}^{-k_d z}, \qquad (2)$$

where E_d(z) is the downwelling irradiance at depth z and K_d the vertical downwelling attenuation coefficient. K_u, the vertical upwelling attenuation coefficient can be calculated in the same way from the upwelling irradiance E_u.

The reflected light (R) is calculated as $R = E_u/E_d$.

For the calculation of Box-Whisker plots we used the program Systat for Windows, version 5.1.

Results and discussion

Oxyqen, salinity and nutrients

The oxygen saturation is shown in Fig. 2. As can be seen, the water remains on average fully saturated up to a salinity of about 20‰ (at station 6, Perkpolder). After this the oxygen saturation drops rapidly to 10% near Antwerpen (station 12, 2–5‰), after which it stays more or less constant. Large fluctuations occur in the brackish water and complete anoxic conditions have been found. When compared to the data gathered by Van Spaendonk *et al.* (1993) we, in general, could see no significant difference between 1989 and 1991, although the average oxygen concentrations between stations 10 and 12 were lower in 1991.

Salinity in Antwerpen remained rather constant between 1 and 5‰. Further downstream, however, large fluctuations (±8‰) occurred due to tidal movements. In the central-Western part of the basin (stations 1–5) fluctuations decreased again (not shown).

Dissolved silicate showed a more or less linear relationship with salinity above 3‰ (Fig. 3) suggesting conservative behaviour. However, at salinities below 3‰ big changes in concentration of both silicate and phosphate were observed. This is probably due to the fact that there are two freshwater endmembers, i.e. it reflects the input of phosphate and silicate of both the river Schelde and Rupel (see Fig. 1). Phosphate did not show a conservative behaviour, especially during the winter transect, where concentrations stayed more or less constant below salinities of 10–15‰. This constant concentration of phosphate suggest that the phosphate buffering method was operating during that period (Froelich, 1988).

As expected dissolved inorganic nitrogen (DIN) did not show a conservative behaviour. Winter DIN concentrations were approximately 200 μM higher at low salinities, but the differences decreased at higher salinities (Fig. 4). In winter, NH_4^+-decreased sharply from station 15 downstream, whereas in summer there was an initial rise from station 15 to 13, before the decrease started. The decrease in NH_4^+ was mainly due to nitrification of NH_4^+ to NO_3^-, as the decrease in NH_4^+ coincided with an increase in nitrate. However, the decrease in ammonium could not be totally explained by the (smaller) increase in nitrate. Most likely this difference is lost due to denitrification of nitrate to N_2. However, because the water is not completely anaerobic (see Fig. 2), most of the denitrification has to take place in either the sediments or in the particles

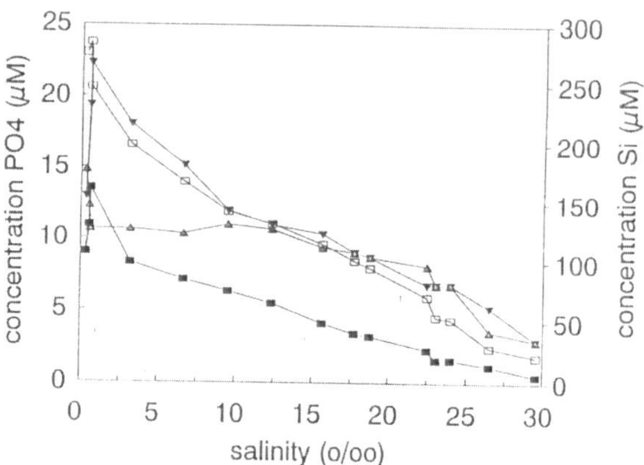

Fig. 3. Concentrations against salinity of phosphate (▼) and silicate (■) on July 31 and of phosphate (▲) and silicate (□) and December 18.

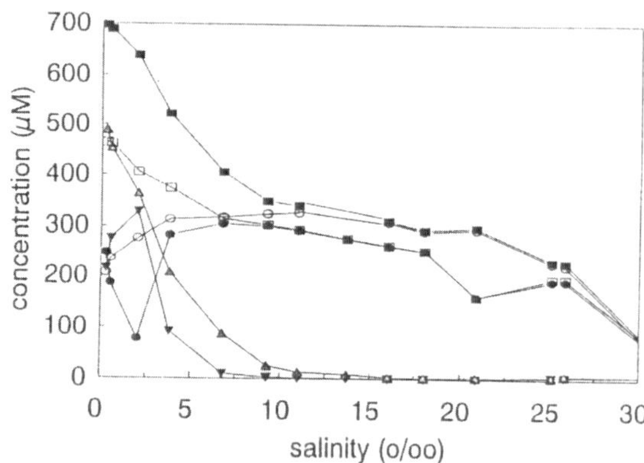

Fig. 4. Concentrations against salinity of ammonia (▼), nitrate (●) and total dissolved inorganic nitrogen (DIN, □) on July 31 and of ammonia (▲), nitrate (○) and DIN (■) on December 18.

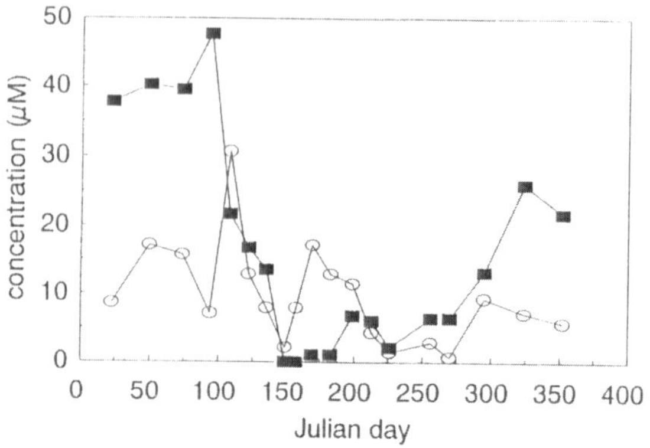

Fig. 5. Changes in silicate (■) and ammonia (○) concentrations during 1991 at station 1 (Vlissingen).

and flocculates, which will easily become anaerobic inside.

No differences in the mean summer values of the major nutrients were found when 1991 was compared to 1989 (not shown). When nutrient concentrations are lower than the half saturation concentrations for growth of marine phytoplankton (0.1–0.5 μM for phosphate, 1–5 μM for silicate and 1–2 μM for nitrogen, Gillbricht, 1988) they can limit phytoplankton growth. When the nutrient concentrations in the Westerschelde are compared to the half saturation constant mentioned above, it is very unlikely that they ever became limiting. A possible exception can be made for the most saline part of the estuary, where a silicate limitation of diatoms is possible (Fig. 5): between day 150 and 250 values were usually below 0.5 μM. Also ammonium reached potentially limiting levels, however, nitrate was always in excess of possible growth demands (>30 μM, Fig. 3). N/P-ratio's here were nearly always above the Redfield ratio of 16, indicating a relative surplus of nitrogen. Si/N-ratio's were below 1, indicating a relative shortage of silicate. From these ratio's it seemed likely that, only in the very western part of the estuary, silicate might become the first limiting nutrient, followed by phosphate. However, only silicate reached possible limiting levels. Whether this nutrient became limiting depended upon the light energy demands of the phytoplankton in relation to the ambient nutrient regime. It could explain a species shift, however: the decrease in silicate between day 100 and 150 coincided with a *Phaeocystis* bloom. On the other hand, the situation in station Vlissingen is of course very much influenced by processes going on in the adjacent North Sea.

The underwater light field

Due to the high concentrations of suspended matter (annual average at station Vlissingen was 25 mg l^{-1} and 140 mg l^{-1} at station Temse) the depth of the euphotic zone (Z_{eu} calculated at 4.6/K_d, Kirk (1983)) is small (Table 1) and the reflectance values vary around 10%. Hence, most of the light attenuation in the Westerschelde is due to scattering, increasing the vertical pathlength of incoming light. This causes a steep light gradient. Because of this high turbidity, phytoplankton spends a considerable part of the daytime in the dark, increasing respiratory losses. It is often assumed that the compensation depth (Z_c) is 6 times the euphotic zone (Cloern, 1987). When the mixing depth (Z_m) is larger than the compensation depth, respiratory losses exceed C-gains by photosynthesis and thus no net photosynthesis can take place. The average depth is close to 10 m in the Westerschelde. From this, it is calculated that if $Z_m = Z_c$, the depth of the euphotic zone must equal 1.67 m. This corresponds to a K_d of 2.76 m^{-1}. As can be seen in Fig. 6, the average K_d is only a little smaller in the western, marine part of the estuary, suggesting that net photosynthesis is only possible in this part of the basin. However, in the central and eastern part of the basin, the K_d is, on average, larger than 2.76 m^{-1}, implicating that the mixing depth is larger than the compensation depth, making net photosynthesis not possible. However, we measured significant rates of gross primary production (see below). That net photosynthesis occurred also could be concluded from the chlorophyll dynamics during the growth season (not shown). As they could not be explained by import processes, they must have been the result of local net primary production (Soetaert *et al.*, 1994). Due to physiological adaptation (see Van Spaendonk *et al.*, 1993), net photosynthesis seems possible in this highly turbid environment. The effect of the turbidity on estimates of the phytoplankton carbon balance is the subject of another paper (Kromkamp & Peene, in press).

Table 1. Depth of euphotic zone and reflectance values for some stations in the central and eastern part of the Westerschelde.

Station	k_d (m^{-1})	Z_{eu} (m)	Reflectance
Hansweert	3.20	1.44	0.10
Baalhoek	5.18	0.88	0.11
Bath	4.91	0.94	0.10
Zandvliet	4.63	1.00	0.11
Lillo	3.30	1.40	0.09
Antwerpen	2.56	1.80	0.07

Phytoplankton biomass and primary production

The average chlorophyll-*a* concentration was rather constant in the western and central part of the estuary. Below salinities of approx. 10‰ the concentration rose sharply and reached a maximum in the river Schelde (Fig. 7).

At the seaward end, at station 1 (Vlissingen), daily primary production showed the well known spring bloom followed by a summer minimum (Fig. 8). In the brackish/turbid part of the estuary, the phytoplank-

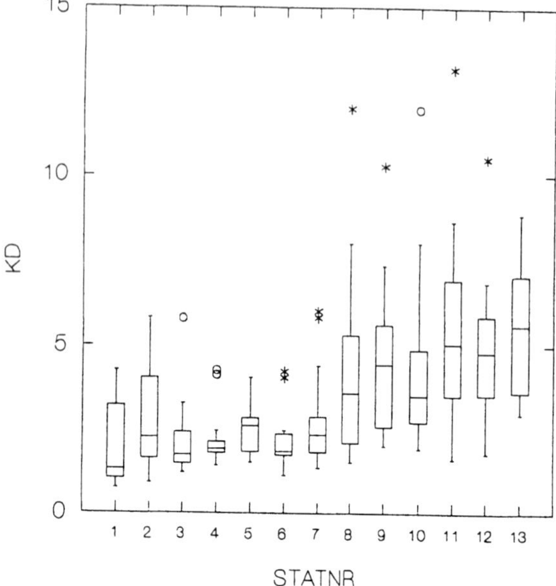

Fig. 6. Box-Whisker plot of downwelling attenuation coefficient at several sampling stations.

ton bloom occurred later in the year, and was confined to a limited period, with maximum rates occurring in the summer (Fig. 8, station Lillo as example). Also, the maximum rates of daily production were smaller than in the marine part of the estuary. These results are obviously related to the light conditions. We therefore correlated the daily production to (natural log transformed) irradiance conditions and temperature (Table 2). As irradiance conditions we took the surface irradiance on the day of measurements and the 4 and 7 day average surface irradiance preceding the day of measurements (day of measurement included). However, as the average irradiance in the watercolumn is not only a function of the surface irradiance (E_s), but also a function of the turbidity, the average irradiance in the watercolumn ($<E>$) was also calculated using Riley's formula

$$<E>=\frac{E_s(1-e^{-k_d z})}{k_d}.$$

The correlations fluctuated considerably and explained 29–85% of the primary production. Correlations were weakest with temperature. This might be expected because the primary production was light limited, i.e. the chance to absorb photons (a physical process) determined mainly the rates of photosynthesis. Contrary to what migh be expected, it could not be said that the column irradiance gave a better prediction of primary production than the surface irradiance. There was no clear cut difference between the 4 day and the 7 day averages (not shown). Daily primary production at station 1 (Vlissingen) showed the highest correlation with the light conditions.

Boynton *et al.* (1982) reviewed data on estuarine production and showed that in general single factors were poor predictors of phytoplankton primary production. Cole & Cloern (1984) developed an emperical composite parameter to estimate photic zone (net) primary production based on 24h incubations (PNP). Cole & Cloern (1987) used this model for Pudget sound, New York Bight, and South and North San Fransico Bay. As they did not found significant differences between the different estuaries the total regression they found was ($r^2=0.82$):

$$P_{N,P} = 150 + 0.73 B I_0 Z_P,$$

where B is the biomass (chl a, $\mu g\ l^{-1}$), I_0 the daily incident irradiance and z_p the depth of the photic zone. We tried to use this model for our data. The result is shown in Table 3. The regression intercept b, varied considerably, but was not significantly different from those found by Cole & Cloern (1987). The regression coefficient a, however, was lower than found by Cole & Cloern (1987). The model gave a good prediction for the primary production in the western part of the basin (stations 1–3, r^2>0.81) but gave poor prediction for the central stations 4–6. For the inner stations 7–13 the predictions varied. It is difficult to explain why for many stations the model works, whereas for others it fails. One of the reasons might be that the attenuation coefficient k_d can vary twofold over a tidal cycle which means that the calculation of the photic zone ($=4.6/k_d$) is not very accurate. Also, Cole & Cloern (1987) used 24 hour incubations, whereas we used short-term incubations. As all our samples were taken in the morning, occurrence of diel activity (which has been observed, unpublished data) might decrease the accuracy of the prediction. The fact that we used short term incubations also explains why our value of the regression coefficient (a) is lower.

Figure 9 shows the calculated annual production in 1991. Productivity rose towards the marine end. In the eastern part (5–15‰) productivity was low, due to a high turbidity. In the river Schelde productivity rose again, despite the still very high turbidity. Primary production in 1991 was very similar to that in 1989, with the exception of productivity at Antwerpen, which was higher in 1989 (Van Spaendonk *et al.*, 1993).

The increase in chlorophyl concentration in the moderate brackish and freshwater part of the study

Table 2. Correlation between daily primary production (mg C m^{-2} d^{-1})) and ^{e} log transformed environmental parameters. The (linear) regression coefficients is given (n=18, except station 14 (n=16))

Station	1	5	8	10	12	14
Temp	0.35	0.29	0.50	0.52	0.52	0.46
E_o^a	0.83	0.63	0.67	0.63	0.66	0.59
E_o (4)[b]	0.82	0.43	0.74	0.77	0.54	0.82
<E>(4)[c]	0.85	0.63	0.61	0.73	0.50	0.63

[a] Total daily surface irradiance on day of sampling.
[b] As the former, including the average daily surface irradiance of the three preceding days. [c] 4 day average column irradiance (of day of measurement and three preceding days).

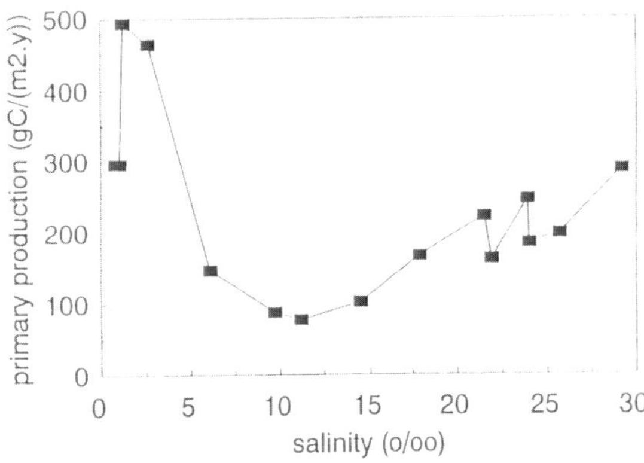

Fig. 9. Annual primary production (g C m^{-2} y^{-1}) against salinity in 1991.

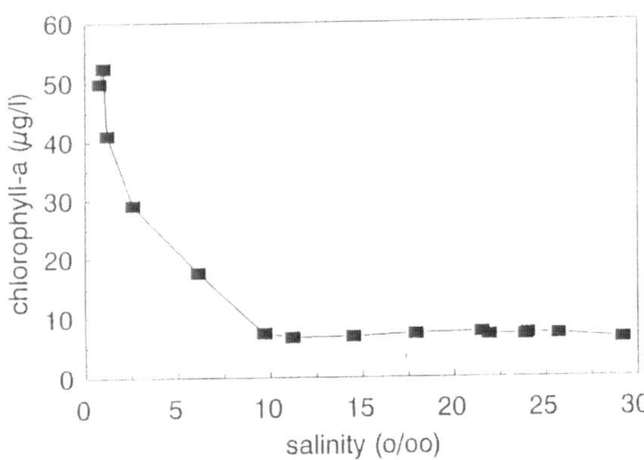

Fig. 7. Mean annual average chlorophyll *a* concentrations (μg l^{-1}) against salinity.

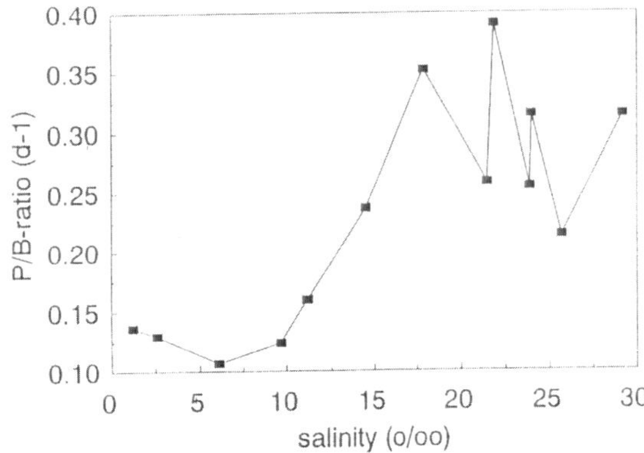

Fig. 10. Mean C-turnover rate (P/B-ratio, d^{-1}) found during the period April–September.

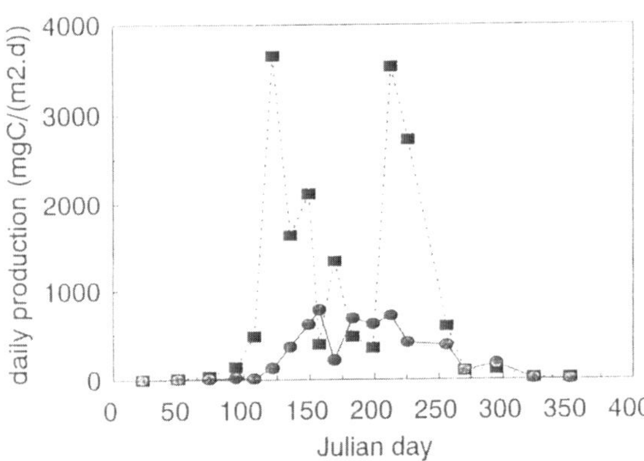

Fig. 8. Daily primary production (mg C m^{-2} d^{-1}) at the stations Vlissingen (dotted line) and Lillo (solid line). Plotted is the production calculated with the moving average irraidance of the preceding 4 days (including the day of measurements).

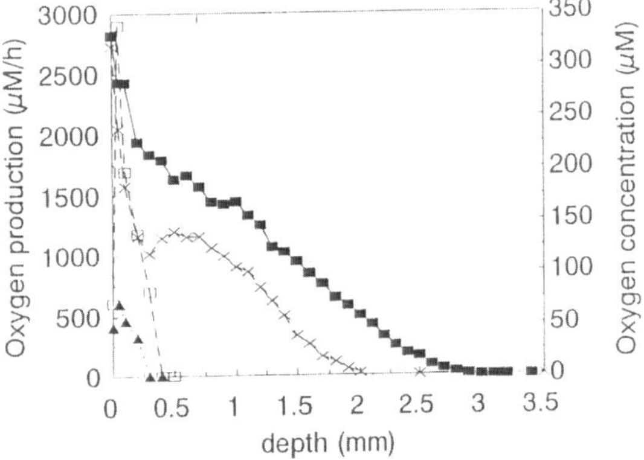

Fig. 11. Oxygen concentration (μM) and photosynthesis profiles (μM O_2 h^{-1}) measured at the Molenplaat at March 10, 1992. Solid line with solid squares: 1st core measured at 50 μE m^{-2} s^{-1}. Dashed lines with asterisks: 3d core measured 2.5 h later at identical conditions. Dashed line with open squares: benthic primary production measured at 900 μE m^{-2} s^{-1}. Dotted line: production at 200 μE m^{-2} s^{-1}.

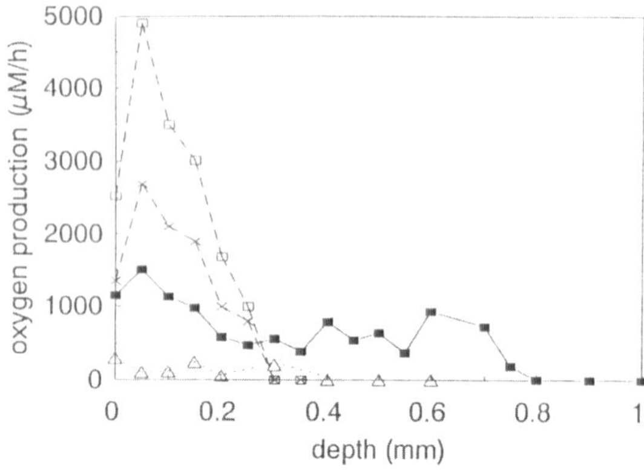

Fig. 12. Microphytobenthos primary production profiles at the Molenplaat at July 5, 1991. Filled squares and solid line: core 1 with 1500 μE m^{-2} s^{-1}. Open squares and dashed line: core 3 with 1500 μE m^{-2} s^{-1}. Dashed line and asterisks: core 1, 600 μE m^{-2} s^{-1}. Triangles with dotted line: core 3, 600 μE m^{-2} s^{-1}.

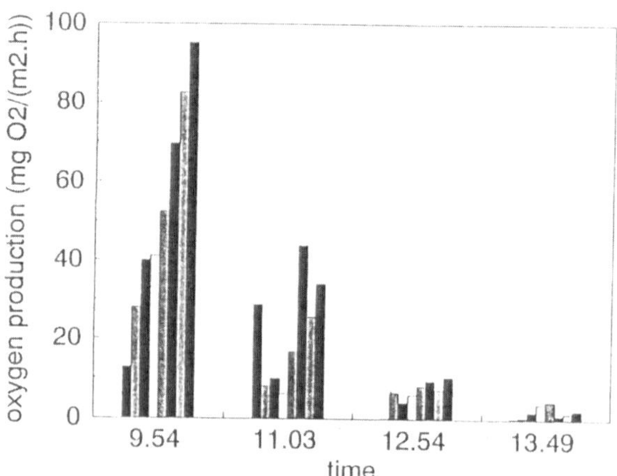

Fig. 14. Changes in benthic primary production with time at different irradiances. Each time a new core was collected. Bars from left to right: 50, 100, 300, 600, 900 and 1200 μE m^{-2} s^{-1}. Cores were taken at the Molenplaat at 28-10-91.

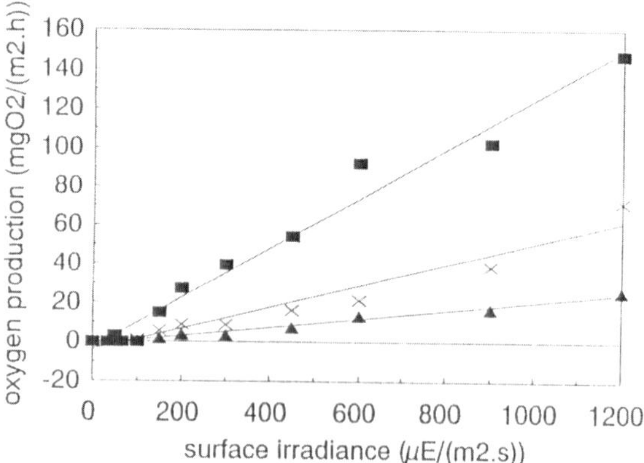

Fig. 13. Linear relationship between benthic primary production and irradiance. From top to bottom: cores taken at 11.45 h, 14.10 h and 13.20 h. The lines are the best fit through the data points (all $r^2 > 0.95$). Cores taken at March 10, 1992.

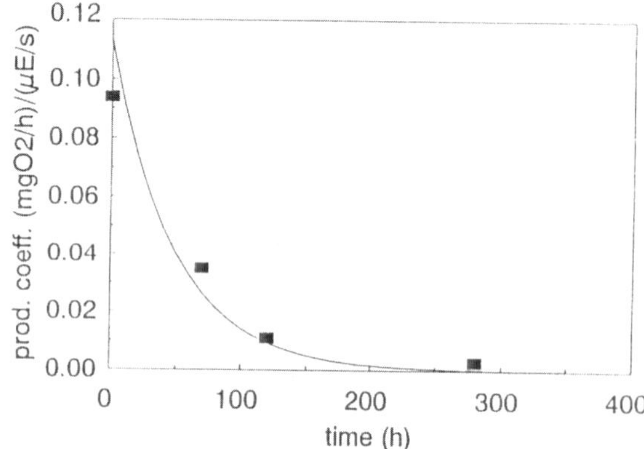

Fig. 15. Changes in the linear production coefficient with time.

area was not completely matched by an increase in production (compare Figs 7 and 9). This was caused by a decrease in C-turnover rate at lower salinities (i.e. in the P/B-ratio, Fig. 10). This decrease in P/B-ratio was caused by the increased turbidity in this part of the estuary and river. Despite the low P/B-ratio, a high plankton biomass was present in Antwerpen and more upstream. This was likely due the absence of important phytoplankton grazers. The absence of copepodes in the Antwerpen region and upstream (Soetaert & Van Rijswijk, 1993) and macrozoobenthos (Craeymeersch, pers. comm.) and hyperbenthos (Mees, pers. comm.) was most likely due to the low oxygen concentrations. Thus, the decrease in growth rates was more than compensated for by smaller loss rates, due to reduced grazing pressure.

Microphytobenthos

Anaerobic conditions in the intertidal flat sediments were nearly always found below 3 mm depth (Fig. 11). The aerobic zones became shallower with time, most likely as the result of a photosynthetic activity also reducing with time (see below).

Figure 12 show the oxygen profiles of two cores, taken 2.5 h apart, together with the photosynthetic activity in the first core. When the profiles are compared, activity is higher in the first core, and photosynthetic activity is measured at a greater depth. The decrease in activity at greater depths can be due to a decrease in activity with time (see below), but can also be caused by migration of pennate diatoms upward, as shown by Pinckney & Zingmark (1991).

When, from the production profiles, the areal production was calculated, photosynthesis was generally a linear function of irradiance, suggesting that photosynthesis is light limited (Fig. 13). This is not very surprising of course, and is due to the strong absorption of light by the sediments.

Photosynthetic activity by microphytobenthos was generally not constant with time. Whereas the maximal rate of photosynthetic activity remained more or less constant at some occasions, it normally declined with time (Fig. 14). At present it is not completely clear why photosynthetic rates decreased with time. The cores were never dry, even at low tide, and nutrients were in excess. A possible explanation can be limitation by CO_2. As photosynthesis takes up CO_2, a constant rate of photosynthesis is only possible if it is replenished by diffusion from the atmosphere and respiratory processes in the sediment below. The rate of diffusion (F) from the air into the sediments can be calculated as follows from Fick's first law of diffusion:

$$F = \frac{D(C_a - C_w)}{Z}.$$

Z is the filmthickness and decreases in proportion to the square of the wind velocity at the water surface (Emerson, 1975). Values for Z were taken between 50 and 300 μm (De Jong *et al.* 1990). E is a pH dependent chemical enhancement factor which is dependent upon the alkalinity. For $Z = 50$ μm $E = 1$, whereas for $Z = 300$ μm E ranges between 1 (at low pH) and 6 (at pH = 10) (De Jong *et al.*, 1990). D is the diffusion coefficient (taken as 1.5 10^{-9} m^2 s^{-1}, De Jong *et al.* 1990). C_a and C_w are the CO_2-concentrations in air and water respectively. From this, we calculated that the likely rate of CO_2 diffusion was 20 mg CO_2 m^{-2} h^{-1} or lower (with a possible range of 2–147 mg CO_2 m^{-2} h^{-1}). Hence, if rates of photosynthesis exceed these rates, diffusion limitation of CO_2 can occur. The rates we measured were probably too low, because the measurements were performed onboard inside the ship, where diffusion of CO_2 was mainly molecular. However, CO_2 is not only coming in from the air, but also from below due to microbial activity. CO_2-fluxes from the intertidal mudflats were measured at station Ellewoutsdijk, close to the Molenplaat. It appeared that the average annual flux from the sediments was approximately 200 mg C m^{-2} h^{-1} (Middelburg, pers. comm.). This CO_2-flux from the sediment might lessen a CO_2-limitation of microphytobenthos photosynthetic activity, but if the thin layer of algae are able to use all of the CO_2 coming from the sediments might be doubted. Onother explanation might be that the decrease in photosynthetic activity is caused by a diel rhythm. Migratory behaviour of benthic algae related to tidal stage and light cycles was observed by Pinckney & Zingmark (1991). This rhythmic behaviour even continued for a while in the abscence of tides. However, they found that activity was highest at low tide, whereas we found that activity already decreased before low tide set in. We did not investigate whether the decrease in activity with time could be attributed to some sort of internal rhythm. The pattern in benthic production we found was also different from that found on intertidal mudflats in the nearby Oosterschelde (Nienhuis *et al.*, 1985), in which the same technique was used. Nienhuis *et al.* (1985) found that productivity remained high for sustained periods, but that after pH rose to 9.4, activity dropped, which they attributed to CO_2-limitation. Unfortunately, we were not able to measure pH-microprofiles. Whether the decrease in photosynthetic activity of the microphytobenthos in the Oosterschelde could have been due to a tidal rhythm as described by Pickney & Zingmark (1991) was not investigated by Nienhuis *et al.* (1985).

Table 3. Use of a composite parameter to calculate gross primary production (PG) from biomass (B, μg chl a l^{-1}), daily incident irradiance (I_0) and photic depth (z_p, 1% of I_0) Model: P = $b + a$ (B I_0 z_p).

Station	b	a	r^2
1	119	0.53	0.81
2	317	0.68	0.81
3	123	0.50	0.83
4	231	0.38	0.32
5	252	0.22	0.36
6	310	0.47	0.36
7	112	0.50	0.69
8	44	0.45	0.73
9	84	0.45	0.62
10	32	0.72	0.72
11	153	0.50	0.48
12	226	0.51	0.48
13	275	0.32	0.77

Table 4. Primary production (mg C m^2 d) of microphytobenthos (Molenplaat) and phytoplankton (Hansweert). A PQ-ratio of 1 was assumed.

Date	Microphyto-benthos	Phytoplankton
02-10-91	9	183
28-10-91	21	48
13-12-91	9	56
10-13-92	103	16
05-07-92	80	1430

It might be clear from the above that it will be very difficult to calculate the areal productivity by benthic algae. But, in doing so we assumed that the light dependent rate of gross oxygen production was a linear function of incident irradiance (Fig. 13). We then fitted this decrease in linear activity with time as an exponential decrease as soon as the sandbank emerged due to low tide setting in (for an example see Fig. 15). Now we could calculated productivity at different times. We compared microphytobenthos production with (14 day average) daily primary production by the phytoplankton over the same period (Table 4). As can be seen, microphytobenthos production can be significant compared to phytoplankton primary production only in periods when phytoplankton production is low. As the total area of intertidal flats is approximately 25% of the total area of the Schelde estuary, it can be concluded that, in general, contribution of microphytobenthos primary production to total micro-algal primary production will be small. However, one has to realize that we compared phytoplankton gross primary production to microphytobenthos net primary production, because different techniques were used.

It is clear that the pollution of the Schelde and the Westerschelde estuary has a large impact on the biology of the systems. Due to the high organic loading near anoxic conditions prevail in the inner part of the estuary, making life for important grazers of phytoplankton impossible. This caused a relatively high, but slow growing algal biomass. In these near anoxic conditions a significant portion of the nitrogen load was lost by denitrification. It might therefore be evident that should the sewage be treated before it is released into the Schelde, the oxygen content will rise and as a consequence, the throughput of nitrate to the North Sea will increase. It is not to be expected that algal primary productivity will rise very much for the total estuary when oxygen conditions will improve in the inner part, because the turbidity will not change considerable. An increase in the waterquality of the Schelde and Westerschelde Estuary will thus, at least initially, be responsible for a higher nitrogen loading of the neighbouring North Sea.

Acknowledgement

We would like to thank the crew of the R/V Luctor for there assistance during the field campains and Jan Sinke for the pigment analyses. This is communication 726 of the NIOO-CEMO.

References

Billen, G., M. Somville, E. De Becker & P. Servais, 1985. A nitrogen budget of the Scheldt hydrographical basin. Neth. Sea. Res. 19: 223–230.

Billen, G., C. Lancelot, E. de Becker & P. Servais, 1988. Modelling microbial processes (phyto- and bacterioplankton) in the Schelde Estuary. Hydrobiol. Bull. 22: 43–55.

Bokhorst, M, 1988. Inventarisatie van een aantal a-biotische factoren in de Westerschelde. Delta Institute for Hydrobiological Research, studentenverslag D6-1988.

Cloern, J. E., 1987. Turbidity as a control on phytoplankton biomass and productivity in estuaries. Cont. Shelf Res. 7: 1367–1381.

Cole, B. E. & J. E. Cloern, 1984. Significance of biomass and light availability to phytoplankton productivity in San Francisco Bay. Mar. Ecol. Prog. Ser. 17: 15–24.

Cole, B. E. & J. E. Cloern, 1987. An emperical model for estimating phytoplankton productivity in estuaries. Mar. Ecol. Prog. Ser. 36: 299–305.

Colijn, F., 1983. Primary production in the Ems-Dollard estuary. Ph.D-thesis, University of Groningen.

De Jong, S. A., P. A. G. Hofman, A. J. J. Sandee & E. J. Wagenvoort 1990. Primary production of benthic microalgae in the Oosterschelde Estuary (S. W. Netherlands). Eindrapport BALANS, Delta Institute for Hydrobiological Research, pp. 221–223.

De Pauw, C. (1975). Bijdrage tot kennis aan het milieu en plankton in het Westerschelde estuarium. Ph.D-thesis State University of Ghent (Belgium, in Dutch).

Eilers, P. H. C. & J. C. H. Peeters, 1988. A model for the relationship between light intensity and the rate of photosynthesis in phytoplankton. Ecol. Model. 42: 199–215.

Emerson, S., 1975. Chemical enhanced CO_2 gas exchange in a eutrophic lake, a general model. Limnol. Oceanogr. 20: 743–753.

Froelich, P. N., 1988. Kinetic control of dissolved phosphate in natural rivers and estuaries: A primer on the phosphate buffer mechanism. Limnol. Oceanogr. 33: 649–668.

Gieskes, W. W. C., G. W. Kraay, A. Nontji, D. Setiapermana & Sutomo, 1988. Monsoonal alteration of a mixed and layered structure in the phytoplankton of the euphotic zone of the Banda Sea (Indonesia), a mathematical analysis of algal pigment fingerprints. Neth. J. Sea Res. 22: 435–467.

Gillbricht, M., 1988. Phytoplankton and nutrients in the Helgoland region. Helgolander Meeresunters. 22: 435–467.

Grobbelaar, J. U., 1990. Modelling phytoplankton productivity in turbid waters with small euphotic to mixing ratios. J. Plankton Res. 12: 923–931.

Heip, C., 1988. Biota and abiotic environment in the Westerschelde estuary. Hydrobiol. Bull. 22: 31–34.

Heip, C., 1989. The ecology of the estuaries of Rhine, Meuse and Scheldt in The Netherlands. Scient. Mar. 53: 457–463.

Hummel, H. & C. Bakker, 1988. Introduction into the Schelde Symposium. Hydrobiol. Bull. 22: 5.

Kirk, J. T. O., 1983. Light and photosynthesis in aquatic ecosystems. Cambridge University Press.

Klepper, O. (1988). A model of carbon flows in relation to macrobenthic food supply in the Oosterschelde estuary (S.W. Netherlands). Ph.D-thesis, University of Wageningen, The Netherlands.

Kromkamp, J & J. Peene. On the net growth of phytoplankton in the turbid, eutrophic Westerschelde Estuary (The Netherlands). Mar. Ecol. Prog. Ser. (in press).

Nienhuis, P. H., E. A. M. J. Daemen, S. A. De Jong & P. A. G. De Jong, 1985. Biomass and production of microphytobenthos. Progress Report 1985, Delta Institute for Hydrobiological Research.

Pinckey, J. & R. G. Zingmark (1991). Effects of tidal stage and sun angles on intertidal benthic microalgal productivity. Mar. Ecol. Prog. Ser. 76: 81–89.

Revsbech, N. P & B. B. Jorgensen, 1983. Photosynthesis of benthic microflora measured with high spatial resolution by the oxygen microprofile method: capabiltities and limitations of the method. Limnol. Oceanogr. 28: 1062–1074.

Soetaert, K., P. M. J. Herman & J. Kromkamp, 1994. Living in the twilight: estimating net phytoplankton growth in the Westerschelde estuary (the Netherlands) by means of an global ecosystem model (MOSES). J. Plankton Res. 16: 1277–1301.

Soetaert, K & P. Van Rijswijk, 1993. Spatial and temporal changes of the zooplankton in the Westerschelde estuary. Mar. Ecol. Prog. Ser. 97: 47–59.

Soetaert, K & P. M. J. Herman, 1995. Carbon flows in the Westerschelde estuary (The Netherlands) evaluated by means of an ecosystem model (MOSES). Hydrobiologia 311 (Dev. Hydrobiol. 110): 247–266.

Sullivan, M. & C. Moncreiff, 1988. Primary production of edaphic algal communities in a Mississippi salt marsh. J. Phycol. 24: 49–58.

Van Spaendonk, A., J. Kromkamp & P. De Visscher, 1993. Primary production of phytoplankton in the turbid, coastal plain estuary De Westerschelde (The Netherlands). Neth. J. Sea Res. 31: 267–279.

Vegter, F. & P. R. M. De Visscher, 1984. Phytoplankton primary production in brackish lake Grevelingen (S.W. Netherlands) during 1976-1981. Neth. J. Sea Res. 18: 246–259.

Wetsteyn, L. P. M. J. & J. Kromkamp, 1994. Turbidity, nutrients and phytoplankton primary production in the Oosterschelde (The Netherlands) before, during and after a large-scale coastal engineering project (1980–1990). Hydrobiologia 282/283 (Dev. Hydrobiol. 97): 61–78.

Hydrobiologia **311**: 21–30, 1995.
C. H. R. Heip & P. M. J. Herman (eds), Major Biological Processes in European Tidal Estuaries.

Dynamics and distribution of microphytobenthic chlorophyll-*a* in the Western Scheldt estuary (SW Netherlands)

D. J. de Jong[1] & V. N. de Jonge[2]
National Institute for Coastal and Marine Management/RIKZ, Directorate-General for Public Works, Ministry of Transport, Public Works and Water Management
[1] *P.O. Box 8039, NL-4330 EA Middelburg, The Netherlands*
[2] *P.O. Box 207, NL-9750 AE Haren, The Netherlands*

Key words: microphytobenthos, chlorophyll-*a*, primary production, annual cycle, depth distribution, hydrodynamic energy, sediment, elevation

Abstract

The temporal dynamics and spatial distribution of microphytobenthic chlorophyll-*a* in the layer 0–1 cm were determined in the Western Scheldt estuary over the period 1991–1992. Connections between the annually averaged benthic chlorophyll-*a* and station elevation and sediment composition (as a measure of the hydrodynamic energy caused by currents and waves) were also examined.

Microphytobenthic chlorophyll-*a* showed one main peak in early summer and a smaller peak in autumn. The mean chlorophyll-*a* concentration of 113 mg Chl-*a* m^{-2} in the upper centimeter is of the same order of magnitude as in other estuarine areas. The average annual primary production of the microphytobenthos has been estimated at 136 g C m^{-2} y^{-1}. The primary production of sediment inhabiting microalgae is at least 17% of the total primary production in the estuary.

Considerable differences in annually averaged chlorophyll-*a* emerges between the stations. These differences are related mainly to the interaction between station elevation and clay content of the sediment.

Introduction

Microphytobenthos plays an important role as a primary producer in the carbon cycle of the estuarine foodchain (Baretta & Ruardij, 1988; Klepper, 1989; De Jonge & van Beusekom, 1992). Although the primary production on the tidal flats generally comprises only 15–20% of the total primary production in an estuary it is of particular importance to groups of animals, such as sediment feeders (macrozoobenthos and meiobenthos)(e.g. Asmus, 1982a; Admiraal *et al.*, 1983), birds (e.g. Shelduck, *Tadorna tadorna*) (Meininger & Snoek, 1992) and fish (e.g. Grey mullet, *Mugil ramada*).

Being dependent on light energy for their growth, the active and mobile cells are confined to the uppermost few millimeters of the sediment (e.g. De Jonge & Colijn, 1994). They are, however, also found in substantial numbers down to depths of 10 cm and more in the sediment, due to hydrodynamic energy (De Jonge, 1992), bioturbation (Cadée, 1976) and active migration (Cadée & Hegeman, 1974). Generally about 25% of the biomass present in the 0–10 cm layer can be found in the layer of 0–1 cm (Cadée & Hegeman, 1974; De Jong *et al.*, 1994; De Jonge & Colijn, 1994).

Owing to hydrodynamic energy (tidal currents, waves), part of the microphytobenthos may be resuspended into the watercolumn, thus becoming temporarily part of the 'phytoplankton' (e.g. Baillie & Welsh, 1980; De Jonge, 1985), after which they can be redeposited. De Jonge & van Beusekom (1992) found a significant correlation between windspeed and the resuspended fraction of microphytobenthos in the Ems-estuary (Fig. 1). In the Eastern Scheldt (Fig. 1) De Jong *et al.* (1994) found that a decrease in current and wave energy, and consequently watercolumn turbidity, resulted in a significant increase (1.7×) in biomass of

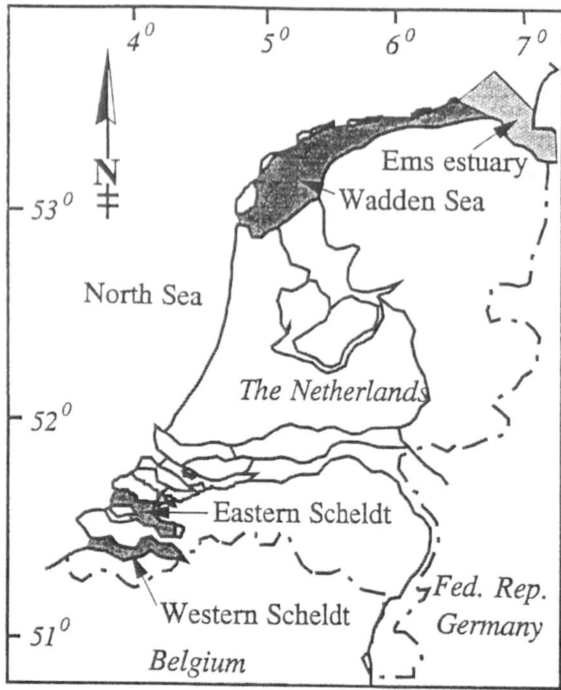

Fig. 1. Map showing the Western Scheldt and some of the other areas mentioned.

the microphytobenthos. As in the latter area the pelagic primary production has decreased in the same period, the share of the microphytobenthos in the total primary production has increased from 16% to 30%.

As a consequence of intensive dredging activities in the Western Scheldt (Figs 1, 2), the level of hydrodynamic energy may have increased in recent decades. This is thought to have caused an impoverishment of the natural habitats, due to steepening of the slopes of many intertidal areas and a strong increase in hydrodynamic energy in many areas. This has resulted in increased sediment movement and possibly the export of benthic microalgae from the intertidal areas to the channels. Moreover, it is thought that water turbidity might have increased due to the higher hydrodynamic energy. In the near future a further increase in dredging activities will occur due to an additional deepening of the main shipping channel. This may lead to further increase in hydrodynamic energy and further impoverishment of the habitats. One aspect that might change as a consequence of the new developments is the biomass and primary production of both the microphytobenthos and phytoplankton. Due to the expected increase in sediment movement and increase in turbidity, both benthic and pelagic primary production may decrease, but the first possibly less than the latter, as benthic algae benefit from exposure during the day.

Up till now, no quantitative data were available regarding microphytobenthos in the Western Scheldt. Accordingly, in 1989 a research programme has been started which has been fully extended from 1991 onwards. The first aim of this programme was to measure the annual cycle of benthic chlorophyll-*a* and to estimate the benthic primary production. A second aim was to investigate possible relations between chlorophyll-*a* and some abiotic factors. This paper presents the results of this programme for the period 1991–1992.

Material and methods

Research area

The Scheldt estuary is situated in the NW of Belgium and the SW of the Netherlands. This research programme was confined to the Dutch part of it, the Western Scheldt (Fig. 2), extending from the Dutch–Belgium border to the town of Flushing over a distance of approx. 55 km. The total area covers approx. 310 km^2, of which approx. 63 km^2 is intertidal area and approx. 32 km^2 is salt marsh. The mean tidal range increases from 3.8 m near Flushing to 4.7 m near the border. A chlorinity gradient is present from about 6‰ near the border to 18‰ near Flushing (Fig. 2). The intertidal area can be divided in mud flats (situated along the dikes, generally more sheltered from currents and waves) and sand flats (sites entirely surrounded by tidal channels, generally more exposed). The sediment of the intertidal areas varies from sandy to clayey, the precise distribution depending on the local hydrodynamic energy.

In the Western Scheldt 7 sand flats and 11 mud flats have been selected, evenly distributed between the Dutch–Belgian border and Flushing (Fig. 2). In each area 1–2 (sometimes 3 or 4) transects were situated, generally from about Mean Low Water (MLW) upwards. In each transect 3–6 stations have been marked out with poles. In this way 109 permanent stations have been established, covering all types of sand and mud flats and the entire chlorinity range in the Western Scheldt.

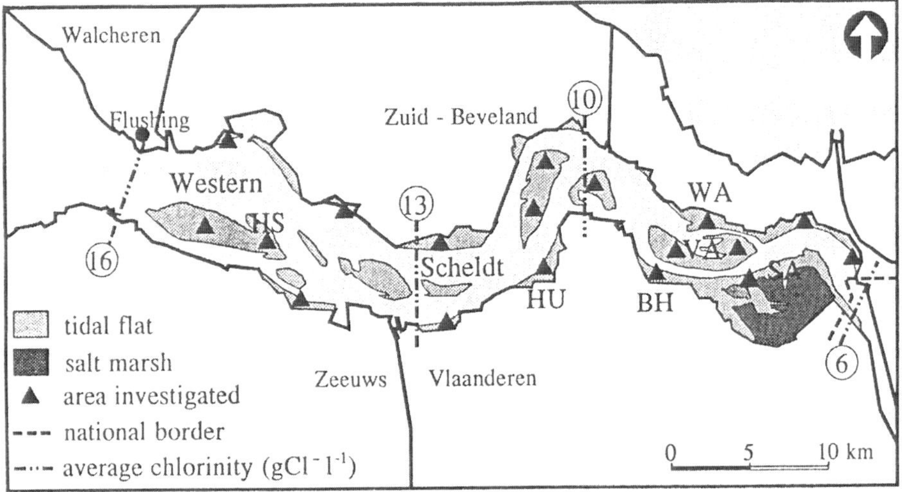

Fig. 2. Western Scheldt with the intertidal areas investigated. (HU: Hulst, BH: Baalhoek, SA: Saeftinge-east, WA: Waarde-east (all mud flats), HS: Hoge Springer, VA: Valkenisse-west (all sand flats))

Chlorophyll-a

Chlorophyll-*a* samples have been taken with a perspex corer with an inner diameter of 2.3 cm. The top 1 cm of the sediment at the selected stations was sampled monthly in 1991 and 1992. In addition to this programme in March, June, September and December 1992 at 10 stations the sediment was sampled up to a depth of 10 cm in 10 subsequent layers of 1 cm, to investigate the depth-distribution of chlorophyll-*a*. Each sample consists of 5 randomly chosen subsamples, from either the upper 1 cm of the sediment or the subsequent layers of the upper 10 cm. The mixed samples were frozen (−20 °C) before further processing in the laboratory. Chlorophyll-*a* was analysed by HPLC (Daemen, 1986), resulting in figures in μg chlorophyll-*a* g^{-1} dry sediment ($\mu g\ g^{-1}$). Conversion from μg chlorophyll-*a* g^{-1} dry sediment to mg chlorophyll-*a* m^{-2} was done by multiplication with 15.5, based on an average bulk-density of 1.55 g cm^{-3}. This figure was derived from measurements in the Eastern Scheldt, a nearby tidal area, as no data from the Western Scheldt were available.

Primary production

The primary production was estimated on the basis of an equation that enables annual primary production to be calculated as a function of the mean annual biomass (De Jong *et al.*, 1994). The authors in question based their formula on measurements of Cadée & Hegeman (1977) and Colijn & de Jonge (1984) in the western Dutch Wadden Sea and the Ems estuary respectively:

$$P = 1.13B + 8.23 \qquad (R^2 = 0.92),$$

in which P = gross primary production in g C $m^{-2}\ y^{-1}$, B = avarage biomass in mg chlorophyll-*a* $m^{-2}\ y^{-1}$ in the layer 0–1 cm. For conversion from chlorophyll-*a* to carbon a carbon/chlorophyll-*a* ratio of 40 was used, based on data from the Eastern Scheldt (unpublished data E. A. M. J. Daemen); a comparable value was found for the Ems estuary by De Jonge (1980).

Abiotic factors

The elevation of each station relative to mean sea level was either measured by a theodolite or estimated from echosounding maps from 1991 or 1992. The latter, less accurate, method had to be used sometimes as no exact measurements are as yet available for all stations.

At each station soil samples of the layer 0–2 cm and 0–10 cm were taken in autumn 1992. The sampling was carried out by taking two subsamples with a corer with a diameter of 4.3 cm, which were mixed in the field. The samples were examined for clay content (particles < 40 μm) and median grain-size by applying the Malvern-method (Weiner, 1984). As a very good correlation was found between the data of the two layers (both median grain size and %-clay), only the data for the layer 0–2 cm are presented.

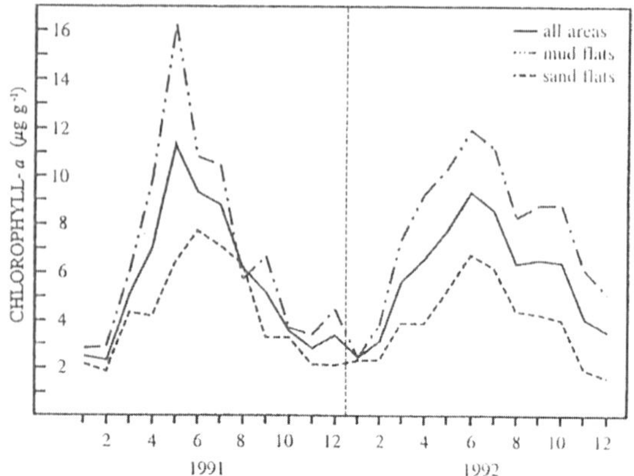

Fig. 3. Monthly averaged chlorophyll-*a* concentrations in the sediment (layer 0–1 cm) in 1991 and 1992, for all stations and for mud flats and sand flats separately.

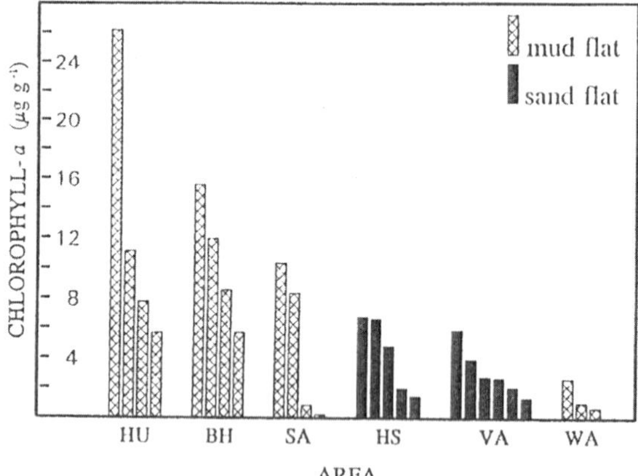

Fig. 4. Annually averaged chlorophyll-*a* concentration in the sediment (layer 0–1 cm) per station in 6 transects. Per transect the individual stations are arranged according to their elevation: from left to right is from high to low elevation on the flat. (HU, BH, etc. see Fig. 2)

Statistics

With regard to the data presented hereafter we would point out that no Standard Deviations are presented. As a clear and rather large seasonal course at each station exists (Fig. 3) on the one hand and large differences between the stations individually (Fig. 4) on the other it does not make sense to provide S.D.'s with the 'overall' averages presented. These S.D.'s will be rather high, although the actual differences at a particular station in a particular month over a number of years may be small. Hence a S.D. should be calculated in stages: first for each station and month separately for the years investigated; subsequently these averages and S.D.'s per station/month can be collected in the form of derived averages and S.D.'s, e.g. for all stations in a certain month, for one station during the years, or for all stations during the years measured. However, in this paper the results of two years of measurements are presented, giving two measurements per month per station, which is too few for these calculations.

Results

Chlorophyll-a

The seasonal variation in the chlorophyl-*a* concentration in the upper 1 cm is presented in Fig. 3 as monthly averaged values for all stations together, as well as for the mud flats and sand flats separately. The curves show one large peak in early summer (May–July). Moreover, either a very small peak or a temporary break in the post-summer decrease can be observed. Except for the May peak in 1991, the differences between the two years are small.

Considering the annually averaged chlorophyll-*a* concentration per station in 6 selected transects (Fig. 4), it is clear that large differences in chlorophyll-*a* levels exist between the individual stations (range from approx. 0.3 to 27 $\mu g\ g^{-1}$ or 4.5 to 420 $mg\ m^{-2}$) as well as between the transects as a whole.

The mean annual chlorophyll-*a* concentration of the total intertidal area is calculated at 7.3 μg chlorophyll-*a* g^{-1} dry sediment, or 113.1 $mg\ m^{-2}$, in the top 1 cm (Table 1). For a total area of intertidal flats of 63.3 km^2 this means the presence of 7.2 tonnes chlorophyll-*a* or 286 tonnes of organic carbon in the upper 1 cm. Application of the 'primary production equation' gives an average annual primary production of approx. 136 $g\ C\ m^{-2}\ y^{-1}$ or 8611 tonnes of organic carbon per annum.

The relative depth distribution of the chlorophyll-*a* in the 0–10 cm layer is presented in Fig. 5 as an average over all stations and sampling dates. The curve shows a steep decrease in concentration over the first centimeter followed by a more gradual decrease deeper in the sediment. The highest chlorophyll-*a* is clearly present in the upper 1 cm. About 25% (range 20–35%) of the total chlorophyll-*a* determined in the upper 10 cm is found here. When considering the stations individually, roughly two groups can be distinguished. One shows a gradual decrease in chlorophyll-*a* over the entire sediment column sampled during the whole year (Fig. 6: VA), while the other shows a strong decline

Table 1. Annually averaged chlorophyll-*a* concentration and primary production of microphytobenthos in the Western Scheldt (sediment layer 0–1 cm); data for Eastern Scheldt (De Jong *et al.*, 1993), western Dutch Wadden Sea (Cadée & Hegeman, 1977) and Ems estuary (Colijn & De Jonge, 1984) are added. (Eastern Scheldt: <1985 = former situation without storm surge barrier, >1985 = present situation with storm surge barrier)

	Western Scheldt	Eastern Scheldt		Wadden Sea west	Ems estuary
		<1985	>1985		
Intertidal area (ha)	6330	9300	9300		
Average chlor.-*a*					
μg chl-*a* g^{-1}	7.3	8.1	13.3		
mg chl-*a* m^{-2}	113.1	125.0	207.5	35–120	23–120
Total chlor.-*a*					
ton chl-*a*	7.2	11.7	19.0		
ton Carbon	286	469	760		
Primary production					
g C m^{-2} y^{-1}	136.0	149.5	242.5	40–180	60–100(250)
ton C y^{-1}	8611	14045	22265		

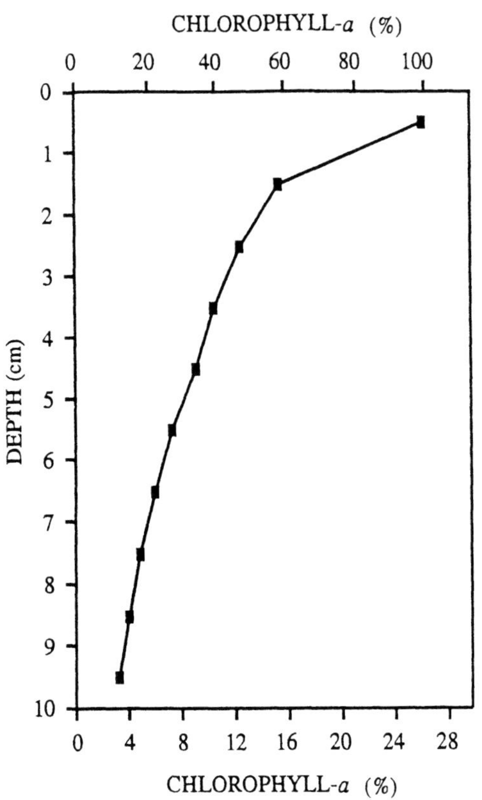

Fig. 5. Depth gradient of the benthic chlorophyll-*a* in the top 10 cm of the sediment; upper X-axis: relative to the chlorophyll-*a* content present in the 0–1 cm layer, lower X-axis: relative to the chlorophyll-*a* content present in the entire layer of 10 cm.

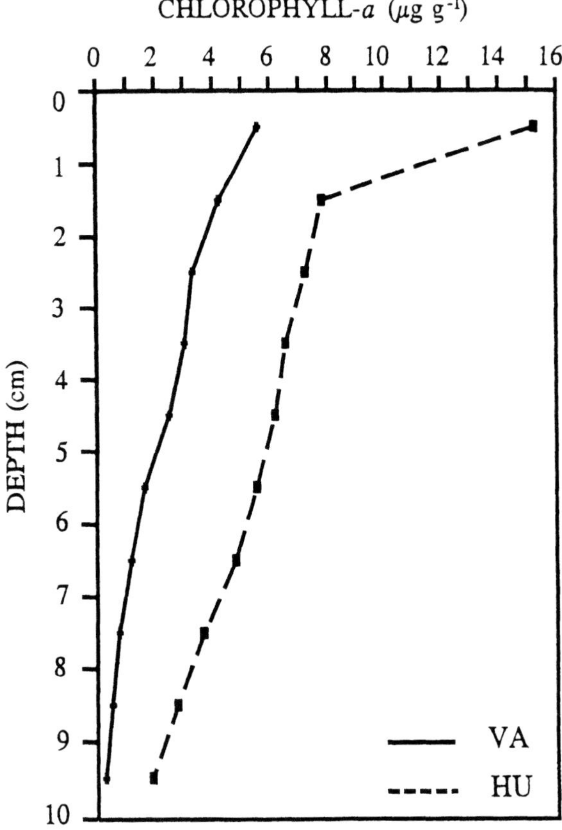

Fig. 6. Depth gradient of the benthic chlorophyll-*a* in the top 10 cm of the sediment for two stations seperately: VA = Valkenisse and HU = Hulst. (see Fig 2)

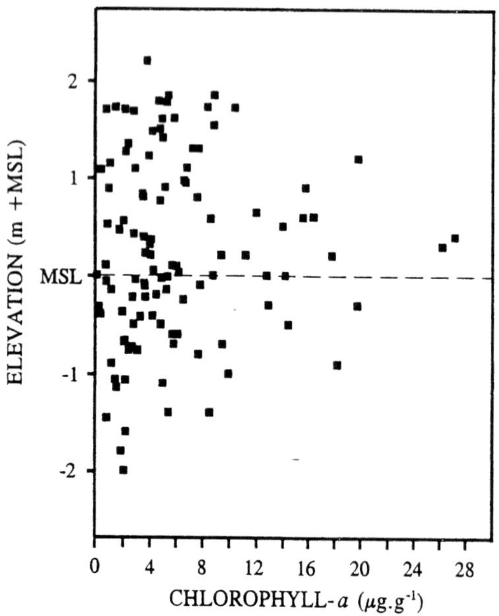

Fig. 7. Relation between elevation according to Mean Sea Level (MSL) and the annually averaged chlorophyll-*a* concentration per station (layer 0–1 cm).

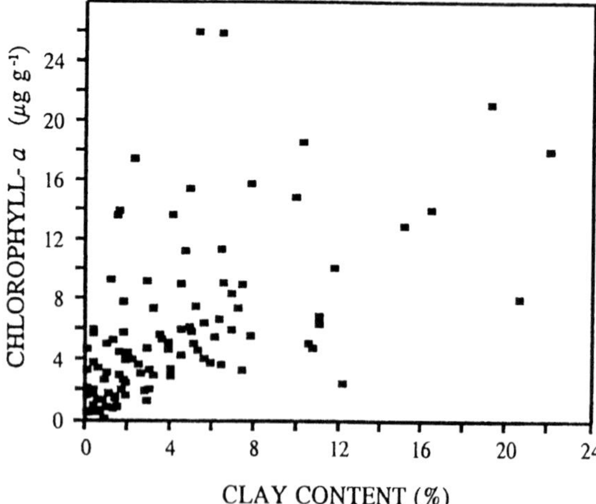

Fig. 8. Relation between clay content of the sediment (layer 0–2 cm) and the annually averaged chlorophyll-*a* concentration per station (layer 0–1 cm). ($Y = 0.65X + 3.35$; $R^2 = 0.30$)

in chlorophyll-*a* over the top 1 or 2 cm during spring and/or summer (Fig. 6: HU).

Abiotic factors

The annually averaged chlorophyll-*a* concentration per station (upper 1 cm) as a function of the station elevation is given in Fig. 7. This figure does not illustrate an over-all relation between the two parameters. In Fig. 4 the annually averaged values per station are given, ordinated per transect from high to low in the tidal range: per transect from left bar to right bar is from high to low elevation. Interestingly, these data illustrate that per transect there is a clear correlation between elevation and mean annual chlorophyll-*a*. However, the level of average chlorophyll-*a* may differ considerably between the areas.

The annually averaged chlorophyll-*a* concentration per station as a function of the clay content in that station (layer 0–2 cm) is presented in Fig. 8. This figure shows a very weak positive correlation ($Y = 0.65X + 3.35$; $R^2 = 0.30$) between both parameters, due to the wide range in chlorophyll-*a* concentrations.

Discussion

Chlorophyll-a

The chlorophyll-*a* in the top layers of estuarine sediments is derived mainly from benthic diatoms (Admiraal *et al.*, 1988; De Jonge, 1992). For the Western Scheldt this was confirmed by microscopical observations by Sabbe & Vyverman (1991). The seasonal variation in benthic chlorophyll-*a* (Fig. 3) is similar for the two years investigated, both in shape and in level. Shape and level are comparable to the data published for other areas like the Eastern Scheldt before 1985 (De Jong *et al.*, 1994), the Wadden Sea (Cadée & Hegeman, 1977) and the Ems estuary (Colijn & De Jonge, 1984). The large differences in chlorophyll-*a* levels between the areas in the Western Scheldt (Fig. 4) may be ascribed to the differences in hydrodynamic energy. Generally, mud flats represent more quiet and sand flats more turbulent hydrodynamic conditions (cf. also De Jonge, 1992); however, exceptions occur, e.g. the Waarde mud flat (Fig. 4), where the average biomass is low because of a very turbulent environment due to the presence of a flood channel at the lower side of the mud flat. Because the differences in hydrodynamic energy between the different arcas are larger during summer than during (the stormy) winter, the differences in chlorophyll-*a* values between both types of areas are also larger in summer.

The start of the increase in chlorophyll-*a* in spring is generally ascribed to an increase in light and temperature. The end of the bloom in early summer may, however, be ascribed to different factors such as a shortage of nutrients in the water and/or deficiency of inorganic carbon in the sediment (Admiraal, 1977; Admiraal *et al.*, 1982; Colijn & De Jonge, 1984) and/or grazing (Asmus, 1982b). The Western Scheldt is very eutrophic (Van Spaendonk *et al.*, 1993; Kromkamp *et al.*, 1994), thus a shortage in nutrients is not very likely to happen. Increased grazing, however, by macrozoobenthos (e.g. *Hydrobia* spec, *Corophium* spec) might exceed the primary production in estuarine areas (Cadée, 1980; Asmus, 1982b; Morrisey, 1988a,b). This may cause the end of the blooming period. In the Western Scheldt *Hydrobia ulvae* is not a very common species, while *Corophium volutator* is often abundantly present (Meire *et al.*, 1991). Thus, in the Western Scheldt intensive grazing pressure may be exerted by the latter species.

Estimation of the primary production of microphytobenthos in this study is based on the relationship with annually averaged chlorophyll-*a* values (see Methods). The production estimation presented in this paper is therefore, strongly dependent on the amount of chlorophyll-*a*. Because the chlorophyll-*a* concentrations in the Western Scheldt are comparable to those measured in Wadden Sea and Ems Estuary, the calculated primary production values are also in the same range as the values published for the Wadden Sea (Cadée & Hegeman, 1977; Cadée, 1980; Asmus, 1982b), the Ems estuary (Colijn & De Jonge, 1984) and the Eastern Scheldt before 1985 (De Jong *et al.*, 1994) (Table 1).

This benthic primary production can be compared to the phytoplankton and total primary production in the Western Scheldt. The phytoplankton primary production amounts to approx. 210 and 85 g C m^{-2} y^{-1} for the western and eastern part of the Western Scheldt resp. (Kromkamp *et al.*, 1994), or 42,300 tonnes organic carbon y^{-1} in the entire Western Scheldt (based on a water area at Mean Sea Level of approx. 178 and 58 km^2 in Western Scheldt west and east resp.). By comparison, to this the share of the benthic primary production on the tidal flats in the total primary production in the Western Scheldt is approx. 17%.

However, from previous studies (De Jonge & Van Beusekom, 1992) we also know that resuspended microphytobenthos may substantially contribute to the total primary production in the water column. De Jonge (1992) and De Jonge & Van Beusekom (1992) mention values of approx. 25–30% of the total benthic chlorophyll-*a* being present in the watercolumn in the Ems estuary. How important the contribution of resuspended microphytobenthos in the Western Scheldt is to the total chlorophyll-*a* concentration and the primary production in the water column is as yet unknown, but it might be significant, or even higher than in the Ems estuary in view of the estuary's high hydrodynamic energy.

Moreover, due to the high water turbidity and the substantial depth of the vertical mixing zone, it may be assumed that a substantial part of the phytoplankton primary production is respired by the phytoplankton during the long dark periods (see also Van Spaendonk *et al.*, 1993, and Kromkamp *et al.*, 1994). As a result, the final share of the microphytobenthic primary production in the total primary production in the Western Scheldt will be much higher than the afore mentioned 17%, and might even rise to 100% in some very turbid and dynamic parts of the estuary.

The decrease in chlorophyll-*a* with depth is similar to the situation found in other areas (Cadée & Hegeman, 1977; Cadée, 1980; De Jonge & Colijn, 1994; De Jong *et al.*, 1994). The difference in decrease between the two groups of stations (steep versus gradual decline in the toplayer during spring/summer, Fig. 6) may be the result of different hydrodynamic conditions at the sampling stations. The stations with a steep decrease in the toplayer are situated in sheltered areas (both on mud and sand flats), where tidal current and waves exert hardly any movement on the uppermost layer, especially in summer, and there is no turbation of the sediment. Consequently, the chlorophyll-*a* produced remains mostly in the toplayer of the sediment. Due to winterstorms, the sediment in these areas is reworked over greater depth, and transport between tidal flats and channels will also be greater. Another phenomenon observed in these sheltered areas during the field surveys was a strong increase in clay content in the upper layer during summer, which decreases again in the winter period. The stations with a gradual decrease in chlorophyll-*a* with depth during the whole year are situated in exposed areas. As a result, both the sediment and the chlorophyll-*a* produced is continuously reworked over a greater depth, and export of chlorophyll-*a* will be greater. Consequently, the average level of chlorophyll-*a* in the top-layer is lower in

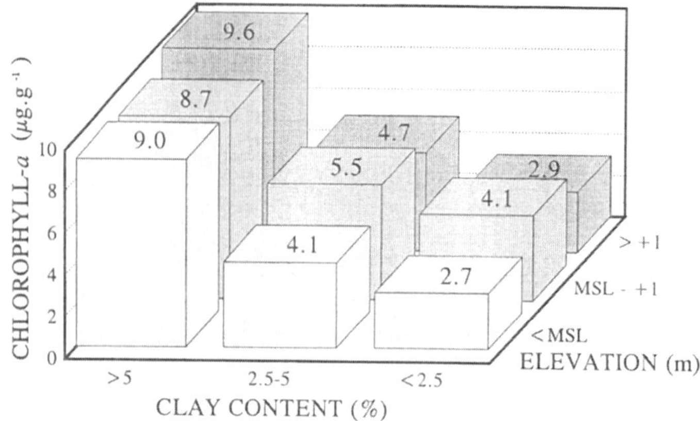

Fig. 9. Annually averaged chlorophyll-*a* concentration (layer 0–1 cm) for the stations per combination of elevation class/clay content class. (see text)

these more turbulent areas than in the more sheltered stations.

Abiotic factors

The data in Fig. 4 indicate a relationship between station elevation and annually averaged chlorophyll-*a* level within a certain transect. This picture does not emerge when these data are plotted in an X-Y diagram for all stations in the estuary (Fig. 7). At any elevation a broad range in chlorophyll-*a* values occurs. However, the maximum values at each elevation in Fig. 7 suggest the presence of an optimum curve in which the highest maximum chlorophyll-*a* values are reached around or just above Mean Sea Level. Colijn & Nienhuis (1977 (cf. Figs 2 and 8)) and Colijn & De Jonge (1994) found a positive correlation between elevation and chlorophyll-*a* in the eastern Wadden Sea and Ems estuary respectively. For the Eastern Scheldt De Jong *et al.* (1994) described an increasing chlorophyll-*a* concentration with increasing elevation in the situation before the construction of a storm surge barrier, whereas after the construction of this barrier this correlation was reversed, an increasing average chlorophyll-*a* concentration with decreasing elevation. They ascribed this reversal to the strong increase in water transparency in the area which may have led to increased primary production by microphytobenthos during submergence.

Obviously, other parameters may also have a significant influence on the average chlorophyll-*a*. One important aspect might be hydrodynamic energy (De Jonge, 1992; De Jonge & Van Beusekom, 1992); in more sheltered areas a higher biomass may occur as reworking of the sediment and resuspension and lateral transport are of minor importance. As hydrodynamic energy is a parameter which is difficult to define and measure, its result, as reflected in the sediment composition, is generally used. Sediment with a relative low clay content indicates an area with high hydrodynamic energy, because the small clay particles cannot settle, and sediment with a relative high clay content reflects an area with low hydrodynamic energy. A negative correlation between hydrodynamic energy and chlorophyll-*a*, will therefore be apparent as a positive relation between clay content of the sediment and microphytobenthic chlorophyll-*a*. Colijn & Dijkema (1981) mentioned this possibility, but were not able to demonstrate it. Figure 8 shows the plot of average chlorophyll-*a* concentration against clay content for each station. Despite the wide variation in values, this figure suggests a weak positive correlation between the two parameters in the Western Scheldt.

This implies that besides elevation clay content, representing the hydrodynamic energy level, is also important. The wide range in chlorophyll-*a* values that occur per individual parameter, obscuring clear monoparametric relations, might be caused by an interaction between both parameters. Therefore, both elevation and clay content need to be considered together. In doing so, three classes per parameter have been distinguished: elevation: <0 m, $0 \leq e \leq 1$ m and >1 m (relative to Mean Sea Level) and clay content: $<2.5\%$, $2.5 \leq c \leq 5\%$ and $>5\%$ clay. The stations were classified accordingly and the annually averaged chlorophyll-*a* concentration was calculated per group. The result is presented in Fig. 9. This figure shows an increase in annually averaged chlorophyll-*a* concentration with clay content for all elevation classes, as might be expected. It also indicates an optimum in average chlorophyll-*a* concentration with elevation at the lower and medium clay content class, while this optimum is absent at the high clay content class. The features related to elevation might be explained by the fact that at the low elevation level the light period is too short for a high net primary production and biomass, while at the upper elevation level desiccation might occur, hampering the primary production. In the highest clay content class this desiccation might not occur, as the soil is able to retain water better as the result of the high clay content. The unexpected high concentration for the combination low elevation - high clay content might be the result of a virtual absence of lateral transport in these very quiet areas.

The conclusion to be drawn from Fig. 9 might be that in a very dynamic and turbid estuary like the Western Scheldt, elevation and hydrodynamic energy and their interaction may be dominant factors that determine the level of the average chlorophyll-*a*. The effect of these factors on the chlorophyll-*a* content may vary: sometimes they are mutually enhancing and sometimes they weaken each other. The role of these parameters will be elaborated when more data are available.

The approach of a combined relation of elevation and clay content (or hydrodynamic energy) with annually averaged chlorophyll-*a* concentration seems to offer a suitable opportunity to calculate the total amount of chlorophyll-*a* in the estuary and consequently to estimate the primary production of microphytobenthos, while taking into account the variation in different sediment types and elevation zones. This approach also offers an opportunity to calculate the consequences of changes in morphology and hydrodynamics due to anthropogenic activities on microphytobenthic chlorophyll-*a* concentration and primary production and consequently the effects on food availability for the ecosystem.

References

Admiraal, W., 1977. Experiments with mixed populations of benthic estuarine diatoms in laboratory microecosystems. Bot. mar. 20: 479–485.

Admiraal, W., H. Peletier & H. Zomer, 1982. Observations and experiments on the population dynamics of epipelic diatoms from an estuarine mud flat. Estuar. coast. mar. Shelf Sci. 14: 471–487.

Admiraal, W., L. A. Bouwman, L. Hoekstra & K. Romeyn, 1983. Qualitative and quantitative interactions between microphytobenthos and herbivorous meiofauna on a brackish intertidal mud flat. Int. Revue ges. Hydrobiol. 68: 175–191.

Admiraal, W., M. A. van Arkel, J. W. Baretta, F. Colijn, W. Ebenhöh, V. N. de Jonge, A. Kop, P. Ruardij & H. G. J. Schröder, 1988. The construction of the benthic submodel. In: J. Baretta & P. Ruardij (eds), Tidal flat estuaries. Simulation and analysis of the Ems estuary. Ecological Studies 71, Springer-Verlag, Heidelberg: 105–152.

Asmus, H., 1982a. Field measurements on respiration and secondary production of a benthic community in the northern Wadden Sea. Neth. J. Sea Res. 16: 403–413.

Asmus, R., 1982b. Field measurements on seasonal variation of the activity of primary producers on a sandy tidal flat in the northern Wadden Sea. Neth. J. Sea Res. 16: 389–402.

Baillie, P. W. & B. L. Welsh, 1980. The effect of tidal resuspension on the distribution of intertidal epipelic algae in an estuary. Estuar. coast. Mar. Sci. 10: 165–180.

Baretta, J. & P. Ruardij (eds), 1988. Tidal flat estuaries. Simulation and analysis of the Ems Estuary. Ecological Studies 71, Springer-Verlag, Heidelberg, 353 pp.

Cadée, G. C., 1976. Sediment reworking by *Arenicola marina* on tidal flats in the Dutch Wadden Sea. Neth. J. Sea Res. 10: 440–460.

Cadée, G. C., 1980. Reappraisal of the production and import of organic carbon in the western Wadden Sea. Neth. J. Sea Res. 14: 305–322.

Cadée, G. C. & J. Hegeman, 1974. Primary production of the benthic microflora living on tidal flats in the Dutch Wadden Sea. Neth. J. Sea Res. 8: 260–291.

Cadée, G. C. & J. Hegeman, 1977. Distribution of primary production of the benthic microflora and accumulation of organic matter on a tidal flat area, Balgzand, Dutch Wadden Sea. Neth. J. Sea Res. 11: 24–41.

Colijn, F. & V. N. de Jonge, 1984. Primary production of the microphytobenthos in the Ems-Dollard estuary. Mar. ecol. Prog. Ser. 14: 185–196.

Colijn, F. & K. S. Dijkema, 1981. Species composition of benthic diatoms and distribution of chlorophyll-*a* on an intertidal flat in the Dutch Wadden Sea. Mar. ecol. Prog. Ser. 4: 9–21.

Colijn, F. & H. Nienhuis, 1977. The intertidal microphytobenthos of the 'Hohe Weg' shallows in the German Wadden Sea. Forschungsstelle Norderney, Jahresbericht 1977, 29: 149–174.

Daemen, E. A. M. J., 1986. Comparison of methods for the determination of chlorophyll in estuarine sediments. Neth. J. Sea Res. 20: 21–28.

De Jong, D. J., P. H. Nienhuis & B. J. Kater, 1994. Microphytobenthos in the Oosterschelde estuary (the Netherlands), 1981–1990; consequences of a changed tidal regime. Hydrobiologia 282/283 (Dev. Hydrobiol. 97): 183–195.

De Jonge, V. N., 1980. Fluctuations in the organic carbon to chlorophyll-*a* ratios for estuarine benthic diatom populations. Mar. ecol. Prog. Ser. 2: 345–353.

De Jonge, V. N., 1985. The occurrence of 'epipsammic' diatom populations: a result of interaction between physical sorting of sediment and certain properties of diatom species. Estuar. coast. Shelf Sci., 21: 607–622.

De Jonge, V. N., 1992. Physical processes and dynamics of microphytobenthos in the Ems estuary (the Netherlands). Thesis, State University of Groningen, 176 pp.

De Jonge, V. N. & J. E. E. van Beusekom, 1992. Wind and tide induced resuspension of sediment and microphytobenthos in the Ems estuary. In De Jonge, V. N., 1992. Physical processes and dynamics of microphytobenthos in the Ems estuary (the Netherlands). Thesis, State University of Groningen: 139–155.

De Jonge, V. N. & F. Colijn, 1994. Dynamics of microphytobenthos biomass in the Ems estuary measured as chlorophyll-*a* and carbon. Mar. Ecol. Progr. Ser. 104: 185–196.

Klepper, O., 1989. A model of carbon flows in relation to macrobentic food supply in the Oosterschelde estuary (SW Netherlands). Thesis, Agricultural University of Wageningen (ISBN 90-9002944-3).

Kromkamp, J., J. Peene, P. van Rijswijk, A. Sandee & N. Goosen, 1995. Nutrients, light and primary production by phytoplankton and microphytobenthos in the eutrophic, turbid Westerschelde estuary (The Netherlands). Hydrobiologia 311 (Dev. Hydrobiol. 110): 9–19.

Meininger, P. L. & H. Snoek, 1992. Non-breeding Shelduck (*Tadorna tadorna*) in the SW Netherlands: effects of habitat changes on distribution, numbers, moulting sites and food. Wildfowl 43: 139–151.

Meire, P. M., J. J. Seys, T. J Ysebaert & J. Coosen, 1991. A comparison of the macrobenthic distribution and community structure between two estuaries in SW Netherlands. In: Estuaries and coasts, spatial and temporal intercomparisons; M. Elliott & J-P Ducrotoy (eds). Olsen & Olsen, Fredensborg Denmark: 221–230.

Morrisey, D. J., 1988a. Differences in effects of grazing by deposit-feeders *Hydrobia ulvae* (Pennant)(Gastropoda: Prosobranchia) and *Corophium arenarium* Crawford (Amphipoda) on sediment microalgal populations. I. Qualitative differences. J. Exp. Mar. Biol. Ecol. 118: 33–42.

Morrisey, D. J., 1988b. Differences in effects of grazing by deposit-feeders *Hydrobia ulvae* (Pennant)(Gastropoda: Prosobranchia) and *Corophium arenarium* Crawford (Amphipoda) on sediment microalgal populations. II. Quantitative effects. J. Exp. Mar. Biol. Ecol. 118: 43–53.

Sabbe, K. & W. Vyverman, 1991. Distribution of benthic diatom assemblages in the Westerschelde (Zeeland, the Netherlands). Belg. J. Bot. 124: 91–101.

Van Spaendonk, J. C. M., J. C. Kromkamp & P. R. M. de Visser, 1993. Primary production of phytoplankton in the turbid coastal plain estuary, the Westerschelde (the Netherlands). Neth. J. Sea Res. 31: 267–279.

Weiner, B. B., 1984. Particle and droplet sizing using Fraunhofer diffraction. In H. G. Barth (ed.), 'Modern methods of particle size analysis'. Chapter 5: 135–172. Vol. 73 in 'Chemical analysis: a series of monographs on analytical chemistry and its applications', P. J. Elving & J. D. Winefordner (eds).

Hydrobiologia **311**: 31–42, 1995.
C. H. R. Heip & P. M. J. Herman (eds), Major Biological Processes in European Tidal Estuaries.

Comparison of heterotrophic bacterial production rates in early spring in the turbid estuaries of the Scheldt and the Elbe

Nico K. Goosen[1], Pieter van Rijswijk[1] & Uwe Brockmann[2]
[1]*Netherlands Institute of Ecology, Centre for Estuarine and Coastal Ecology, Vierstraat 28, NL-4401 EA Yerseke, The Netherlands*
[2]*Institute for Biogeochemistry and Marine Chemistry, University of Hamburg, Centre for Marine and Climate Research, Martin-Luther-King-Platz 6, D-20146 Hamburg, Germany*

Key words: heterotrophic bacterial production, estuaries, organic matter, oxygen, maximum turbidity zone, vertical distribution, pollution

Abstract

In spring bacterial production rates were estimated by tritiated thymidine incorporation in the turbid estuaries of the rivers Scheldt and Elbe. Bacterial production rates in the Scheldt were 5 times higher than in the Elbe. In the Scheldt bacterial production rates correlated better with the DOC concentration than in the Elbe. Organic matter concentrations in the marine part of the estuaries were the same while in the brackish part concentrations in the Scheldt were much more higher. In the Scheldt, but not in the Elbe, oxygen depletion occurred in the maximum turbidity zone caused by bacterial growth and respiration. The water in the Scheldt was well-mixed while in the turbidity maximum of the Elbe salinity and bacterial production was higher near the bottom than at the surface. Nutrient concentrations in the Scheldt were higher than in the Elbe. Bacterial production rate values in the Scheldt are among the highest reported in the literature. The relatively high bacterial production rates in both estuaries are caused by a high load of waste water. Comparison of bacterial growth rates and water residence time suggests an intensive grazing by probably protozoa. Production rates showed a tidal dynamic. In the Elbe high current velocities caused resuspension of sediment and increased bacterial production rates near the bottom. The high production rates in the turbidity maximum and freshwater part of both estuaries show that a large amount of organic matter is degraded in this region.

Introduction

Degradation of organic matter by planktonic bacteria is an important process controlling water quality, especially in aquatic ecosystems which receive high amounts of allochthonous organic matter and nutrients. The input of large quantities of allochthonous organic matter exerts a strong influence on the carbon flow in the microbial food web. An excess of organic matter may cause an increase in bacterial production and biomass (Coffin & Sharp, 1987; Painchaud & Therriault, 1989) which may lead to turbid water with low oxygen concentrations.

In the Scheldt estuary microbial processes have been studied, specially nitrogen transformations (i.e. Somville *et al.*, 1982; Billen *et al.*, 1985, 1986). In the Elbe estuary a number of studies on microbial activities have been done (i.e. Caspers, 1981; Rheinheimer, 1959, 1960, Gocke & Rheinheimer, 1988; Kerner, 1990). Data on bacterial production rates and biomass in estuaries in The Netherlands and Germany are scarce. Laanbroek *et al.* (1985) and Laanbroek & Verplanke (1986a, 1986b) have reported on the bacterial production rates in the Oosterschelde basin which is not an estuary but a tidal basin. Admiraal *et al.* (1985) have reported on bacterial productions rates in the Ems-Dollard which is a turbid estuary like the Scheldt and the Elbe, with well-mixed water. Until now no data are available on bacterial production rates and biomass in the Scheldt- and the Elbe estuary.

As bacteria are the most important organisms in connection with organic matter degradation and

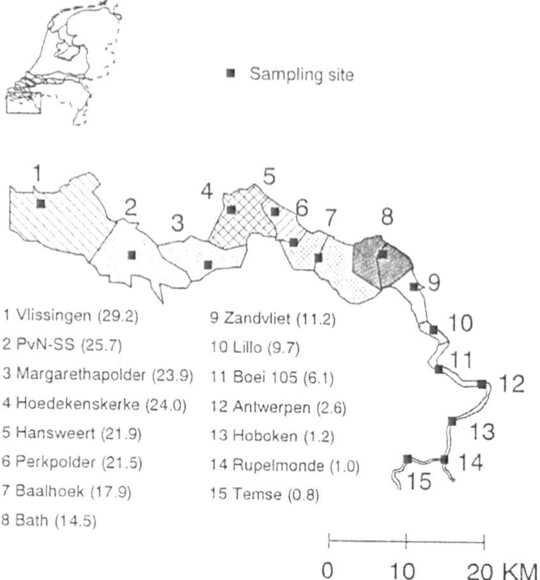

Fig. 1. Map of the Scheldt estuary in the south-western part of the Netherlands. Sampling sites and names, with salinity (g l^{-1}) between brackets, are given.

because the food webs in the Scheldt and the Elbe are dominated by heterotrophic processes we measured heterotrophic bacterial production rates and biomass in both estuaries. The data obtained are important for a better understanding of the ecological functioning of both estuaries. In this paper we describe a study on bacterial production rate and biomass in early spring along the salinity gradient and during a tidal cycle at anchor stations in and close to the turbidity zone in both estuaries. Production rates and biomass are compared together with chemical and physical parameters.

Materials and methods

Description of sampled estuaries

The turbid estuaries of the rivers Scheldt (SW-Netherlands) and Elbe (NW-Germany) are examples of heterotrophic ecosystems which are under permanent stress caused by a high load of urban, industrial and agricultural waste. Both estuaries are characterised by tidal flats and permanent dredging.

The Scheldt estuary (Fig. 1) is the last true estuary in the Dutch delta region. It draines 19500 km^2 (Heip, 1989). The average residence time of the water is about 75 days causing a gradual salinity gradient. Maximum tidal currents are about 1 m s^{-1}. The maritime part is well-mixed, while mixing in the central part is less strong. Close to Antwerpen a flocculation zone exists.

The drainage area of the Elbe (Fig. 2) includes 146500 km^2. The average residence time of the water is 13 days with maximum tidal currents of 2–3 m s^{-1} (ARGE, 1984; Duwe, 1990). The estuary is partially mixed with a steep salinity gradient mainly located close to Brunsbüttel (km 695).

Sampling and chemical analysis

In the Scheldt estuary (Netherlands-Belgium, Fig. 1) a longitudinal profile during ebb-tide was sampled (15 stations) on 17–18 April 1991 between the city of Vlissingen at the mouth of the estuary and the City of Temse in Belgium. A tidal cycle was sampled during an anchor station on 15 April 1992 at station Bath in the brackish part of the estuary. In the Elbe estuary (Germany, Fig. 2) both a longitudinal profile (8 stations), from the city of Hamburg to the German Bight, and an anchor station at km 698, near Brunsbüttel, were sampled during cruise no. 122 of RV 'Valdivia' between 27 March and 1 April 1992. The anchor station in the Scheldt was, for logistic reasons, located at the beginning of the turbidity zone while the anchor station in the Elbe was located in the turbidity maximum (Figs 6a, 6b).

At each station depth profiles of conductivity (C), temperature (T), depth (D), pH and oxygen concentration were measured *in situ* using a CTD-probe (H2O-datasonde coupled to a Surveyor 3, Hydrolab). Samples for chemical analysis were filtered through glass fiber filters (GF/C, Whatman) immediately on board. Nutrients were measured on autoanalysers (Skalar and Bran & Lubbe). Samples for measurement of dissolved organic carbon (DOC) were filtered through glass fiber filters (Schleicher & Schüll no. 6) which were pretreated at 350 °C. DOC was measured with an autoanalyser (Skalar) using UV destruction and photometric detection. Glass fiber filters (GF/C, Whatman) for sampling of suspended matter were treated with acetone and dried at 250 °C. After sampling the preweighed filters were dried at 60 °C and mass increase was measured for determination of suspended matter (SPM). On the same filters particulate organic carbon (POC) was determined with a nitrogen/carbon analyser (model NA 1500, Carlo Erba) as described by Nieuwenhuize *et al.* (1994).

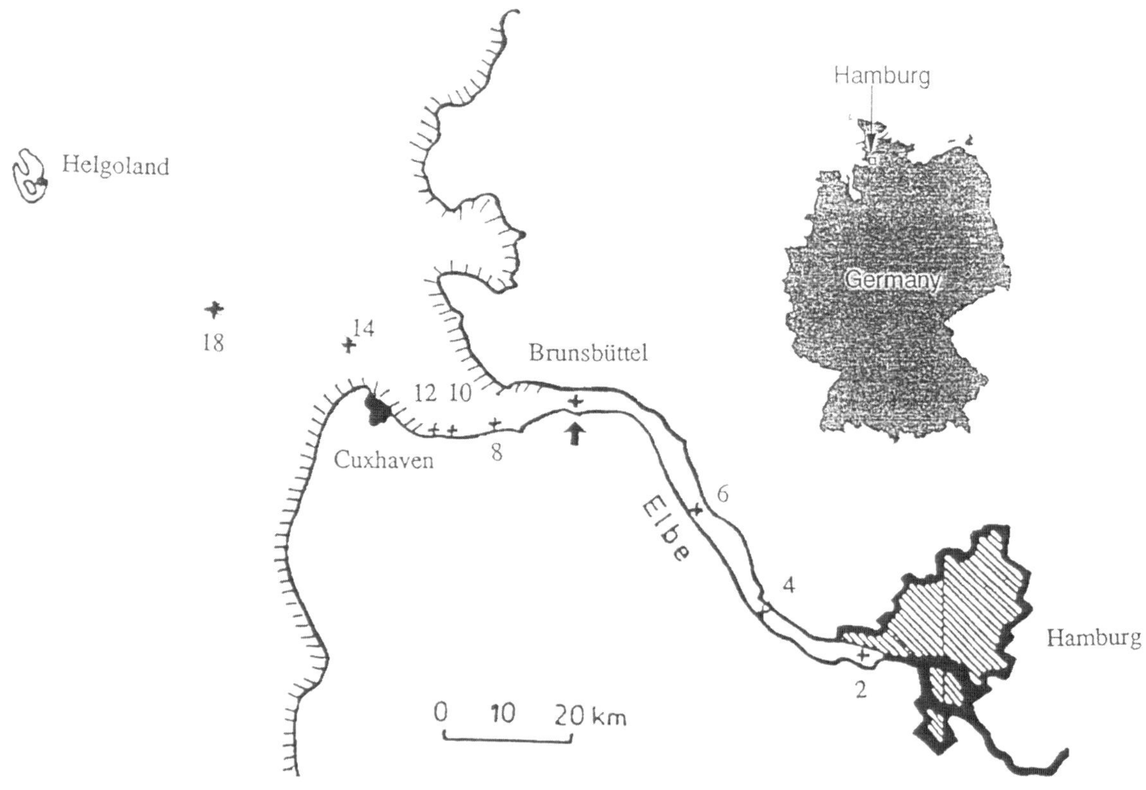

Fig. 2. Map of the Elbe estuary in the north-western part of Germany. Sampling sites are given by the black crosses and the corresponding numbers of the sample stations. For the sampling sites data for km distance and salinity are as follows (station: km, salinity (g l^{-1})): 2: 630, 0.21; 4: 650, 0.34; 6: 670, 0.39; 8: 710, 2.9; 10: 716, 9.1; 12: 718, 14.3; 14: 737, 23.3; 18: 754, 26.3. The arrow indicates the anchor station at km 695.

Bacterial numbers and biomass

Water samples for enumeration of bacteria were preserved with 1.5% glutaraldehyde (final concentration). Bacterial numbers were determined by epifluorescence microscopy after staining with 4′6-diamidino-2-phenylindole (DAPI), following the procedure of Porter & Feig (1980). For calculation of bacterial biomass a conversion factor of 20 fg Carbon $cell^{-1}$ was assumed (Ducklow & Carlson, 1992).

Bacterial production estimates

Bacterial production rates were estimated by measuring the incorporation of tritiated thymidine ([methyl-^{3}H]thymidine, 2.92–3.18 TBq $mmole^{-1}$, Amersham Ltd.) into cold trichloroacetic acid(TCA)-insoluble material (Fuhrman & Azam, 1982). Incubations, in triplicate and at *in situ* temperature, were performed immediately on board. To 5 ml water samples tritiated thymidine was added giving a final concentration of 19 nM. Preliminary tests showed that in the high productive, brackish part of the Scheldt estuary this concentration saturated thymidine uptake (data not shown). Blanc incubations were prefixed with 1.5% formaldehyde (final concentration). After 45 minutes

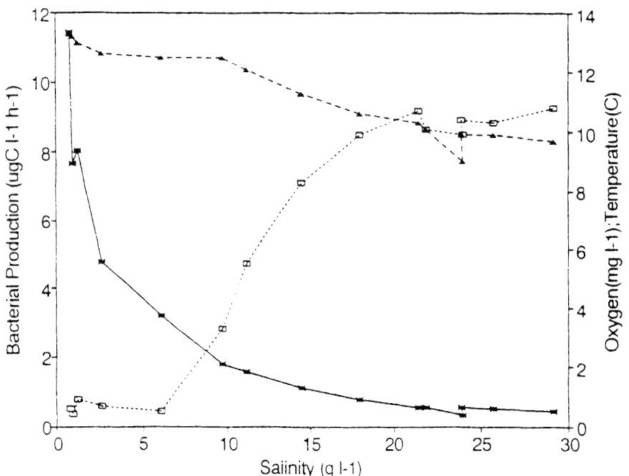

Fig. 3A. Longitudinal profile of bacterial production (-*-), oxygen concentration (...□...) and temperature (–▲–) in the surface water of the Scheldt estuary (a) and the Elbe estuary (b).

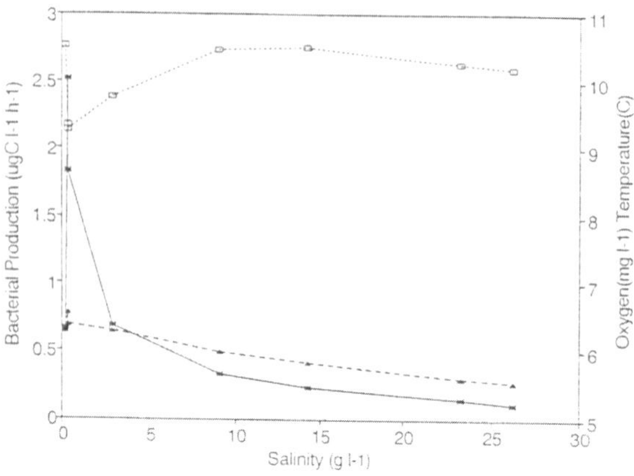

Fig. 3B.

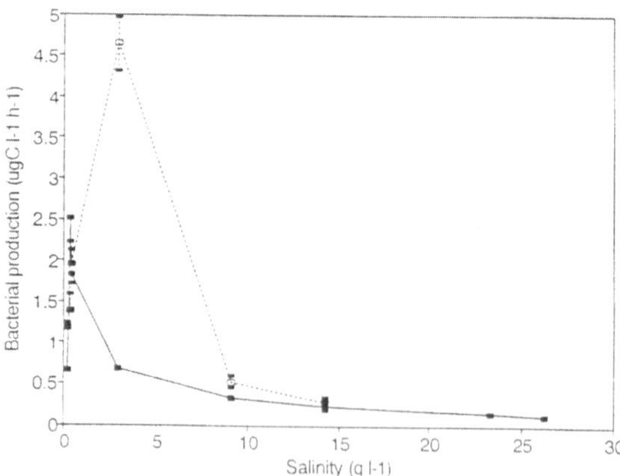

Fig. 4. Longitudinal profile of bacterial production (±SD) in the surface water (-■-) and 1.5–2 m above the sediment (...□...) in the Elbe estuary.

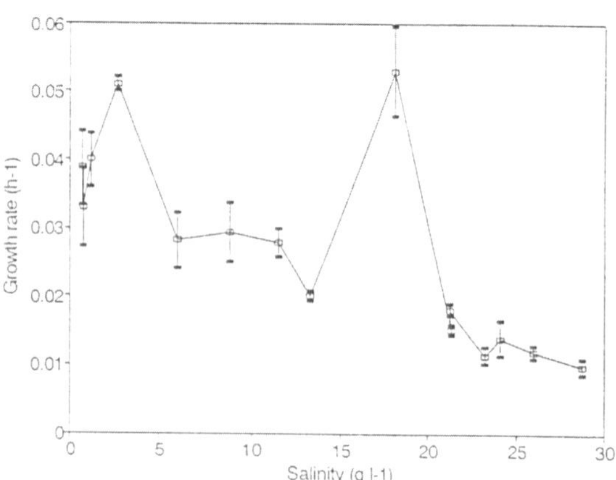

Fig. 5. Bacterial growth rates (±SD) along the salinity gradient in the surface water of the Scheldt estuary.

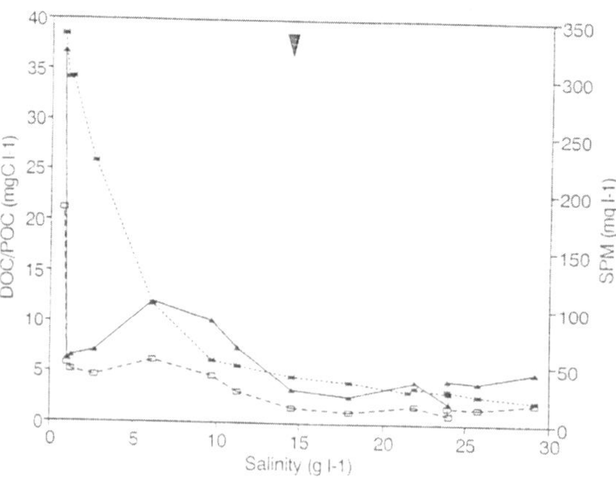

Fig. 6A. Longitudinal profile of suspended matter (SPM, -▲-), dissolved- (DOC, ...∗...) and particulate (POC, –□–) organic carbon in the surface water of the Scheldt estuary (a) and the Elbe estuary (b). Arrows indicate the location of the anchor stations. In both the Scheldt and the Elbe this station was sampled during ebb-tide.

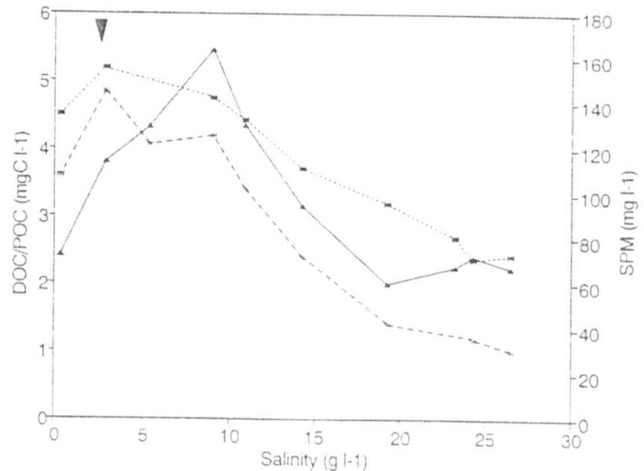

Fig. 6B.

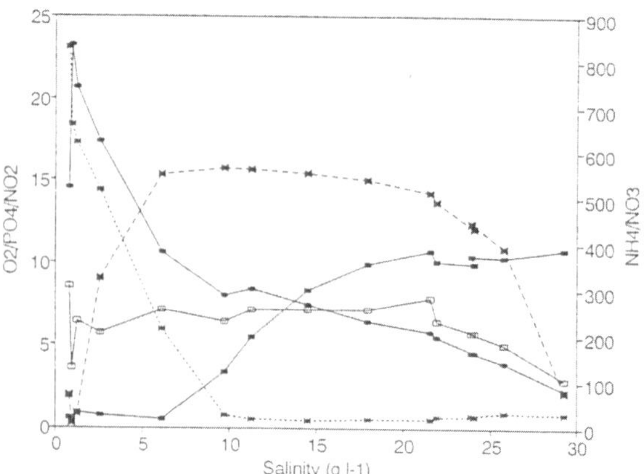

Fig. 7A. Longitudinal profile of concentrations of nitrate (- - × - - μM), nitrite (-□-, μM), ammonium (...∗..., μM), phosphate (- - + - -, μM) and oxygen (-■-, mg l^{-1}) in the surface water ammonium of the Scheldt estuary (a) and the Elbe estuary (b).

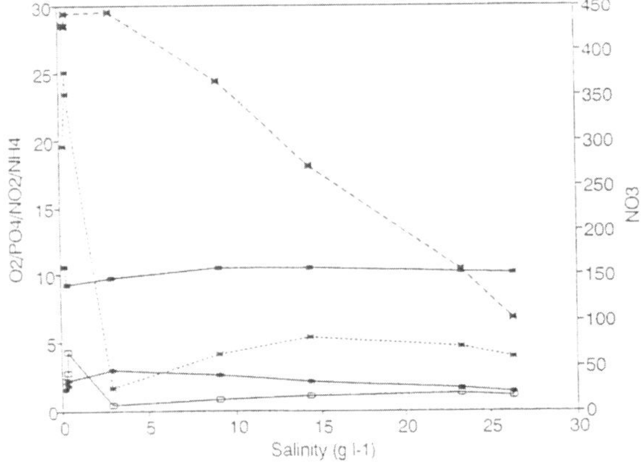

Fig. 7B.

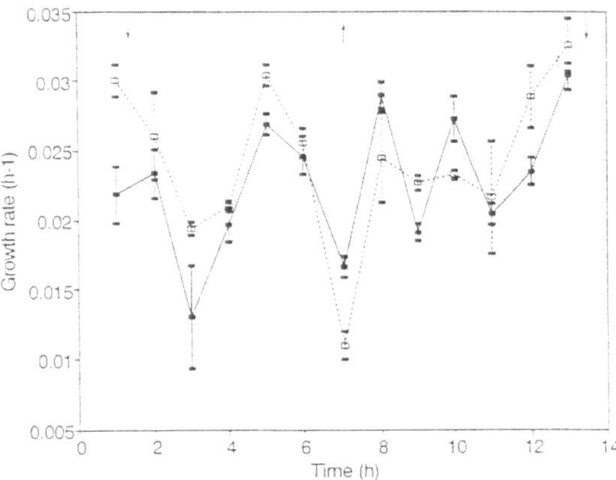

Fig. 8C.

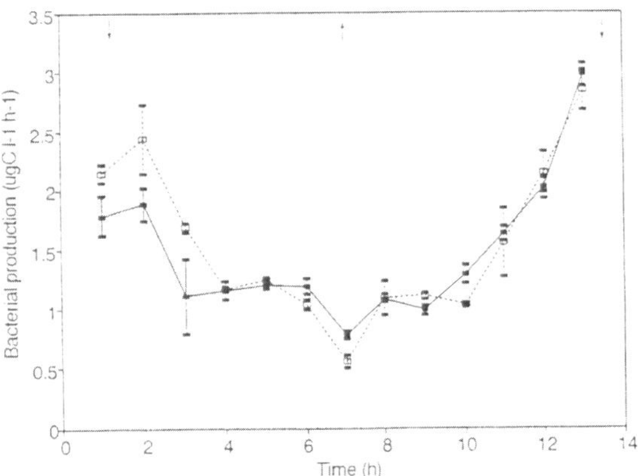

Fig. 8A. Tidal cycle of a: bacterial production (±SD); b: biomass; c: growth rate (±SD) and d: suspended matter (SPM) in the surface water (-■-) and 1.5–2 m above the bottom (...□...) during an anchor station at Bath (Scheldt estuary). Time 0 corresponds to 8.00 (h.min) on 15 April 1992. ↓ = low tide, ↑ = high tide. The differences between water at the surface and near the bottom are, except for two samples, not significant.

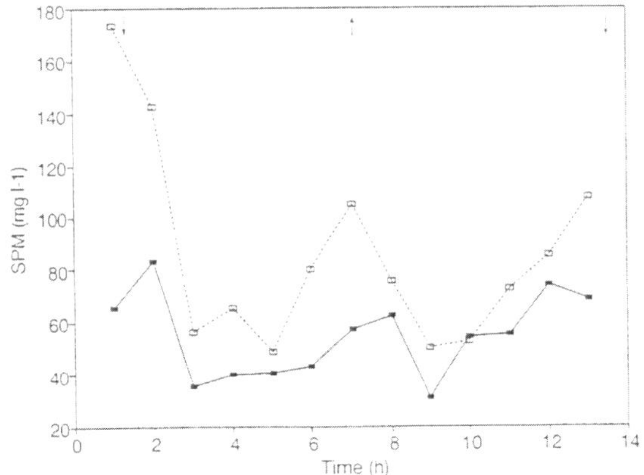

Fig. 8D.

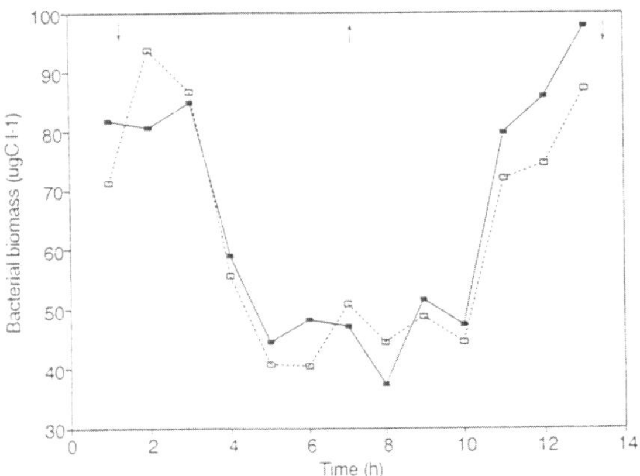

Fig. 8B.

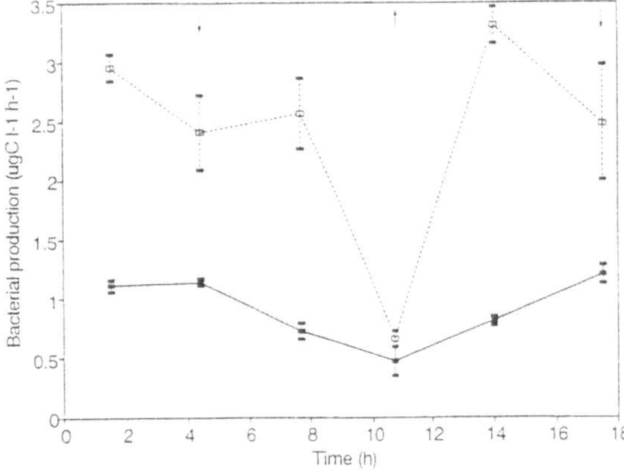

Fig. 9A. Tidal cycle of a: bacterial production (±SD); b: biomass; and c: growth rate (±SD) in the surface water (-■-) and 1.5–2 m above the bottom (...□..) during an anchor station at km 695 near Brunsbüttel (Elbe estuary). Time 0 corresponds to 12.00 (h.min) on 28 March 1992. ↓ = low tide, ↑ = high tide. All differences between water at the surface and near the bottom are significant ($P<0.05$) except for the sample at 10.75 h.

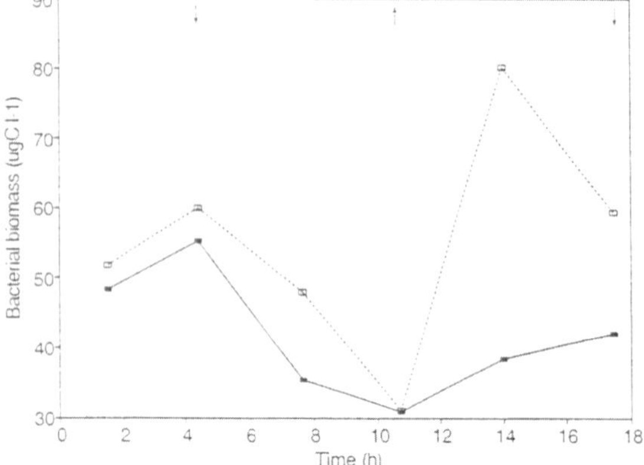

Fig. 9B.

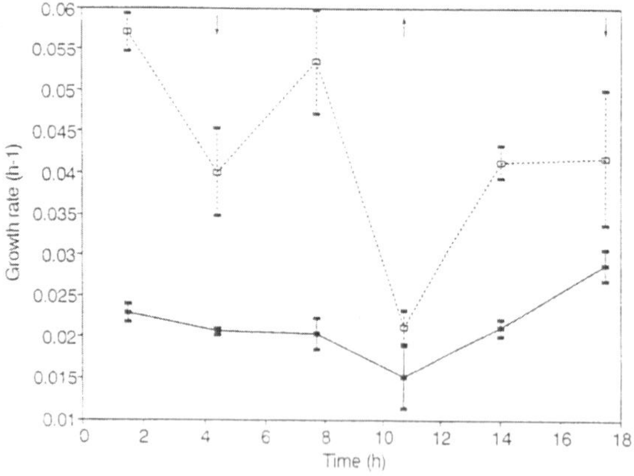

Fig. 9C.

incubations were stopped by adding 1.5% formaldehyde (final concentration) to the experimental vials. Cold TCA precipitates on 0.2 μm pore size cellulose nitrate filters were obtained as described by Ellenbroek & Cappenberg (1991). Filters were dissolved in Filter-Count (Packard) and radioactivity was measured with a Beckman TD5000 liquid scintillation counter. Quenching was accounted for by automatic external standardisation. Bacterial production rates were calculated by using the conversion factors 2×10^{18} cells mole^{-1} and 20 fg carbon cell^{-1} (Ducklow & Carlson, 1992).

Statistical analysis

Interactions between parameters were statistically analysed by two way ANOVA. The significance of differences in bacterial production rates and growth rates was analysed by Tukey multiple pairwise comparison. Differences were considered to be significant when $P<0.05$.

Results

Longitudinal profile

In both the Scheldt and the Elbe estuary bacterial production rates in the surface water decreased with increasing salinity (Figs 3a, 3b). Although the patterns of bacterial production rates along the salinity gradient are comparable in both estuaries the absolute values in the Scheldt estuary (0.3–11.4 μg C l^{-1} h^{-1}) were 5 times as high as in the Elbe estuary (0.1–2.5 μg C l^{-1} h^{-1}). In the Scheldt a steep decrease in the oxygen concentration was observed from the brackish to the freshwater part of the estuary. Below a salinity of 6 g l^{-1} a permanent low oxygen concentration was observed with a minimum of 0.4 mg l^{-1} at the station where the river Rupel enters the Scheldt (Fig. 1). Figure 3a shows an inverse correlation between bacterial production rate and the oxygen concentration in the Scheldt. In the Elbe estuary the oxygen concentration was high (Fig. 3b). In the low salinity region there was a slight decrease but the concentration was always above 9 mg l^{-1}. In both estuaries a weak positive correlation was observed between production rate and temperature. During the longitudinal profile in the Elbe estuary bacterial production rates were also measured at half depth and 1.5 m above the sediment. No difference was observed between production rates in the surface water and at half depth. At the station with a salinity of 2.9 g l^{-1} bacterial production rate in the water 1.5 m above the sediment was significantly ($P<0.05$) higher than in the surface water (Fig. 4). This station was located in the middle of the turbidity maximum.

For the Scheldt estuary bacterial growth rates along the salinity profile were calculated from the measured bacterial production rates and numbers (Fig. 5). Except for two stations, Antwerpen (0.051 h^{-1}) and Baalhoek (0.053 h^{-1}), the growth rates decreased from the most upstream station, Temse (0.039 h^{-1}), to the most marine station at the North Sea, Vlissingen (0.01 h^{-1}). For the Elbe estuary bacterial growth rates could not be calculated because the bacterial counts were performed after staining with acridine orange which gave unreliable results, due to the high turbidity of the water causing a high background label. Comparison of bacterial counts at the tidal cycle in the Scheldt estuary (data not shown) showed that in the surface water counts with acridine orange were a factor 4 to 28 higher than counts with DAPI. For the water near the bottom this factor was between 21 and 50.

Generally the suspended matter (SPM) concentration in the Elbe was higher than in the Scheldt except for the most upstream station in the Scheldt estuary where the SPM concentration was 322 mg l^{-1} (Figs 6a, b). The figures show a maximum turbidity zone between a salinity of 5 and 10 g l^{-1}. The POC concentrations in both estuaries are in the same range except for the low salinity region where POC concentrations in the Scheldt estuary were higher than in the Elbe estuary (Figs 6a, 6b).

The DOC concentrations in the marine part of the both estuaries were in the same range (Figs 6a, 6b). However, below a salinity of 10 g l^{-1} DOC concentrations in the Scheldt estuary were higher, up to 38.5 mg l^{-1}. The DOC concentrations in the Elbe showed a conservative behaviour above a salinity of 5 g l^{-1}. In the Scheldt the DOC concentrations behaved non-conservative with a steep decrease between a salinity of 1 and 10 g l^{-1}.

Figures 7a and 7b show the concentrations of some nutrients in both estuaries. In general the concentrations of all nutrients shown are higher in the Scheldt than in the Elbe. In the Scheldt for all nutrients measured a non-conservative behaviour was found. At low salinities and low oxygen concentrations a strong increase of the nitrate concentration and at the same time a strong decrease of the ammonium concentration was found. Above a salinity of 22 g l^{-1} the nitrate concentration showed a sudden decrease, at the same time the nitrite concentration also decreased. In the Elbe also a decrease of the ammonium concentration seemed to occur and a corresponding small increase of the nitrate concentration. However, the changes in the concentrations at low salinity (<1 g l^{-1}) are very small and probably not significant. Finally, the nitrate concentration showed a conservative decrease towards the German Bight.

Tidal cycle

In the Scheldt estuary a tidal cycle was sampled at an anchor station near Bath (Fig. 1). At this station frequently a small turbidity maximum exists besides the main turbidity maximum located more upstream in the estuary, close to Antwerpen. The anchor station in the Elbe (Fig. 2) was located in the maximum turbidity zone. Samples were taken from the surface water, at half depth and 1.5–2 m above the sediment. Both bacterial production rate and chemical and physical parameters were the same in the surface water and at half depth, therefore only the data for the surface and the deep water will be shown.

In the Scheldt bacterial production rates and biomass in the surface water and near the bottom were not significantly different except for two samples. Production rates showed a tidal cycle with low values between 0.6 and 0.8 μg C l^{-1} h^{-1} at high tide and high values between 1.8 and 3.0 μg C l^{-1} h^{-1} at low tide (Fig. 8a). Bacterial biomass showed the same cycle with low values, 37.5–40.5 μg C l^{-1}, at high tide and high values, 71.3–97.9 μg C l^{-1}, at low tide (Fig. 8b). From the measured biomass and production rate the growth rate of the bacteria can be calculated. Growth rates fluctuated between 0.011 and 0.033 h^{-1} (Fig. 8c). The salinity, temperature, pH and oxygen concentration showed a clear tidal cycle with again no difference between water at the surface and near bottom. Values for low and high tide are given in Table 1. A positive correlation existed between bacterial production rate and the temperature while a negative correlation was found between bacterial production rate and the other parameters. Concentrations of DOC (3.8–5.2 mg C l^{-1}) did not show a tidal cycle and no difference between surface and bottom. Concentrations of SPM (Fig. 8d) and POC did not show a strong tidal cycle either. However, in general the concentrations were highest at low tide and higher near the bottom than at the surface.

In the Elbe bacterial production rates in the water 1.5 m above the bottom were always, except for one sample at 10.75 h, significantly (P<0.05) higher than in the surface water (Fig. 9a). As in the Scheldt production rates showed a tidal cycle. During two subsequent tidal cycles at the same anchor station on 30 and 31 March (data not shown) the correlation between production rate and the tidal cycle was not as strong as on 28–29 March. However, with some exceptions the bacterial production rate near the bottom was always higher than in the surface water. In the surface water the production rate varied between 0.46 μg C l^{-1} h^{-1} at high tide and 1.13–1.20 μg C l^{-1} h^{-1} at low tide. In the water near the bottom the production rate varied between 0.65 μg C l^{-1} h^{-1} at high tide and 2.40–2.48 μg C l^{-1} h^{-1} at low tide. Bacterial biomass showed a comparable tidal fluctuation with again highest values in the deeper water (Fig. 9b). The biomass varied between 30.8 and 42–55.2 μg C l^{-1} at the surface and between 31 and 59.4–60 μg C l^{-1} near the bot-

Table 1. Values for salinity (g l^{-1}), temperature (°C), pH and oxygen (mg l^{-1}) and low tide (LT) and high tide (HT) during a tidal cycle at the anchor stations at Bath (Scheldt) and km 695 (Elbe) in spring 1992.

	Bath			km 695		
	LT	HT	Average	LT	HT	Average
Surface						
Salinity	9.7	17.1	13.1	0.4	1.4	0.8
Temperature	10.5	9.4	9.9	5.9	6.4	6.2
pH	6.9	7.2	7.1	7.8	7.6	7.7
O_2	5.5	8.9	7.4	9.6	10.1	9.8
Bottom						
Salinity	10.5	18.4	13.9	0.4	2.9	1.5
Temperature	10.4	9.3	9.8	5.8	6.3	6.2
pH	7.0	7.3	7.1	7.5	7.8	7.7
O_2	5.9	8.8	7.5	9.6	10.3	9.8

Table 2. Comparison of bacterial production estimates (pmole thymidine l^{-1} h^{-1}) in spring in the marine part, the maximum turbidity zone (MTZ) and the freshwater part of some European estuaries.

Estuary	Marine part	MTZ	River	Reference
Scheldt	8–28	39–200	200–290	This study
Elbe	3–13	17–120	16–63	This study
Loire	5–23	66–76	56–102	Relexans *et al.* (1988)
Ems-Dollard	20–40	—	—	Admiraal *et al.* (1985)

tom, at high and low tide respectively. Near the bottom production rates peaked at mid-tide, when maximum current velocities were reached, with a highest value of 3.31 μg C l^{-1} h^{-1}. Biomass reached a maximum value of 80.2 μg C l^{-1}. Bacterial growth rates in the surface water were between 0.015 and 0.029 h^{-1}. In the water near the bottom growth rates were significantly ($P<0.05$) higher, between 0.021 and 0.057 h^{-1}, except for one sample. Growth rates in the water near the bottom peaked at mid-tide and were higher at low tide than at high tide (Fig. 9c). In the surface water a gradual decrease and increase of the growth rate occurred between low and high tide. Salinity, pH and oxygen concentration more or less followed the tidal cycle with high values at high tide and low values at low tide (Table 1). The salinity was always higher near the bottom than at the surface. The temperature did not show a correlation with the tidal cycle and gradually decreased from 6.4 to 5.9 °C and 6.3 to 5.8 °C at the surface and near the bottom respectively (Table 1).

Discussion

The data in this paper show that in early spring there is a pronounced difference in both microbiological activity and in hydrography between the Scheldt and the Elbe estuary. The hydrographical difference between the Scheldt and the Elbe is illustrated by the position of the maximum turbidity zone at a salinity around 5–10 g l^{-1} (Figs 6a, 6b). In the Scheldt the turbidity zone extends from Bath to Antwerpen with a maximum close to Antwerpen which lies 70 km upstream of the mouth of the estuary. In the Elbe the turbidity maximum is positioned near Brunsbüttel, 40 km from the mouth of the estuary. The different position of the turbidity maximum is a consequence of different rates of freshwater discharge. In the Scheldt the mean freshwater discharge is 104 m^3 s^{-1} (Heip, 1989). In March 1992 during cruise no. 122 of RV 'VALDIVIA' an actual discharge of 1600 m^3 s^{-1} was measured in the Elbe corresponding to a residence time of 10 days. The difference in freshwater discharge in the two estuaries causes a different salinity gradient. In the Scheldt the salinity gradient is gradual while in the Elbe a steep salinity gradient exists and sea water is not entering the estuary as far as in the Scheldt. Furthermore the water in the Scheldt is well-mixed over a long distance, illustrated by the absence of a depth gradient for bacterial production rate and biomass during the tidal cycle (Figs 8a, 8b). In the turbidity maximum of the Elbe production rates near the bottom were higher than in the surface water (Fig. 4). A depth gradient was also observed for the salinity and turbidity (data not shown). These data show that the water column is only partially mixed.

Measurement of thymidine incorporation is a widely used method for estimating bacterial production rates. The accuracy of the method is affected by several processes (Riemann & Bell, 1990) from which isotope dilution and experimental conditions are very important. In this study the effect of isotope dilution has been eliminated by determination of the saturating concentration for thymidine incorporation in the most productive region of the Scheldt estuary. In the experimental procedures the *in situ* temperature was maintained. *In situ* oxygen concentrations could not be maintained which may have lead to aeration of samples with low oxygen concentrations. This may affect the thymidine incorporation rate, however, the effect of aeration is probably low (Ducklow, 1993). The best way to calculate production estimates from thymidine incorporation rates is to experimentally determine conversion factors. In this study we did not determine conversion factors so we used conversion factors given in the literature: 20 fg C per cell and 2×10^{18} cells per mole thymidine incorporated. We used these factors because in a recent review on bacterial production in estuaries by Ducklow & Shiah (1993) the same factors were chosen, thus giving the opportunity for a better comparison between our study and studies in other estuaries. However, the choice for these factors has some consequences. In the Scheldt estuary bacteria in the freshwater part are larger than in the marine part. The use of one carbon conversion factor for the whole estuary may result in an overestimation in the marine part and an underestimation in the freshwater part. The thymidine conversion factor may slightly overestimate production estimates, specially in the marine part of the both estuaries described here.

As Figs 3a and 3b show the bacterial production rate in the Scheldt is five times higher than in the Elbe. The massive growth of bacteria is caused by a high load of organic (Fig. 6a) and nutrient (Fig. 7a) waste in the freshwater and brackish part of the estuary. An important source of pollution is the river Rupel which loads the estuary with untreated waste of the city of Brussel. As Fig. 6a illustrates the organic matter, DOC and POC, in the scheldt was degraded very fast in the brackish part of the estuary. In the Elbe (Fig. 6b) the degradation of organic matter was not so fast and showed an almost linear dilution towards the sea. These data may indicate that the organic matter in the Scheldt is more labile than in the Elbe. This is sustained by comparison of the bacterial production rates in the marine part of the both estuaries. Although concentrations of organic matter in this part were in the same range, the bacterial production rates in the Scheldt are higher. However, an other explanation for these differences may be the difference in temperature in the both estuaries. In the Scheldt the temperature was twice as high as in the Elbe which may lead to higher bacterial activities in the Scheldt and thus a more rapid degradation of organic material. The figures 3a and 6a show a good correlation ($r^2 = 0.96$) between the bacterial production rate and the DOC concentration in the Scheldt. In the Elbe this correlation ($r^2 = 0.8$) was not as strong as in the Scheldt (Figs 3b, 6b). From the relative high bacterial production rates in the brackish part of both estuaries it can be concluded that high amounts of organic matter are degraded in this part including the maximum turbidity zone. The high production rates in the brackish part of the estuaries, as compared to the marine part, can probably not, or only for a very small part, be explained by a higher temperature as the difference in temperature along the longitudinal profile was only 1 and 3 °C for the Elbe and Scheldt respectively, for high production rates bacteria consume high amounts of oxygen leading to lowered oxygen concentrations in the water and eventually to oxygen depletion in the maximum tubidity zone in the Scheldt (Fig. 3a). The measured low oxygen concentrations in the brackish part of the Scheldt is not an incidental event but exists during the whole year. In the Elbe only a weak oxygen minimum was observed in spring. A same situation was found in the Loire estuary where oxygen depletion in the maximum turbidity zone only occurred in summer (Relexans *et al.*, 1988).

A comparison of estimates of bacterial production rates between the Scheldt and Elbe and some other european estuaries in regions with a temperate climate, Ems-Dollard (Admiraal *et al.*, 1985) and Loire (Relexans *et al.*, 1988), is given in Table 2. In the marine part the highest bacterial production rates are found in the Ems-Dollard and the lowest in the Elbe. The high values in the Ems-Dollard may partly be due to an intensive exchange with the sediment of the shallow Wadden Sea. Moreover, for the Ems-Dollard de Jonge (1992) has shown the importance of primary production by microphytobenthos in the total primary production in the estuary. Thus microphytobenthos may be an important source of organic carbon for bacterial growth. In the maximum turbidity zone and the fluvio- estuarine region highest bacterial production rates are found in the Scheldt. Ducklow & Carlson (1992: Table 1) reviewed bacterial production rates in estuaries. Comparison with our data shows that bacterial production rates in the Scheldt are among the highest values reported, certainly when it is considered that our values are from early spring. In summer bacterial production rates are even higher (N. K. Goosen, unpublished data). Values of bacterial production rates in the Elbe are at the average level given by Ducklow & Carlson (1992). It can be expected that in summer these values will also be higher.

In general the growth rates in the Scheldt (Fig. 5) are between 0.01 and 0.04 h^{-1} corresponding to a population doubling time of 1 to 4 days. With a residence time of the water of 75 days this means that a high number of bacteria should disappear either via natural death or sedimentation or grazing. Grazing by heterotrophic flagellates and ciliates, from which it has been shown that they can control bacterial populations (Sherr & Sherr, 1987; Kuosa & Kivi, 1989; Vaqué *et al.*, 1992), is very likely because we have observed a high abundance of heterotrophic protists in the Scheldt estuary (data not shown). The high growth rates for the two stations Baalhoek and Antwerpen can be explained by relatively low bacterial numbers which may have been caused by counting errors. The growth rates in the Scheldt and Elbe are in the range reported for other aquatic ecosystems (Billen *et al.*, 1988, 1990). Recently Servais & Garnier (1993) presented growth rates for small (0.01 to 0.05 h^{-1}) and large (0.04 to 0.18 h^{-1}) bacteria in the river Seine. At most stations sampled in the Scheldt estuary the growth rates for bacteria are in the same range as for the small bacteria in the Seine. Growth rates for bacteria in the surface water of the Elbe are in the same range as in the Scheldt. During the tidal cycle in the Elbe bacterial growth rates in the water near the bottom were mostly in the range given for the larger bacteria in the Seine.

The steep decrease of ammonium and the increase of nitrate in the brackish part of the Scheldt (Fig. 7a) can be explained by nitrification, a process which consumes high amounts of oxygen. The high nitrification activity in the oxygen limited zone of the Scheldt estuary seems surprising. However, beside heterotrophic bacteria this oxygen consuming nitrification process itself is partly responsible for the lowered oxygen levels in the water. The coexistence of the two bacterial population at low oxygen levels is interesting and subject to further investigations. The increased concentration of nitrate and the low oxygen concentration in this region favour denitrification activity. A high rate of pelagic denitrification, most likely in the numerous flocs which may be anoxic, can be expected in this part of the Scheldt estuary. The importance of denitrification in the nitrogen budget of the Scheldt estuary has been reported by Billen *et al.* (1985). The high concentration of nitrite in the Scheldt as compared to the Elbe is a direct consequence of the high nitrification and denitrification activity. Also in the Elbe nitrification might be an important process in the brackish part of the estuary (Fig. 7b).

The data from the anchor stations (Figs 8, 9) illustrate that in both the Scheldt and the Elbe bacterial production rate, biomass and growth rate show a tidal dynamic with in general high values at low tide and low values at high tide. A comparable tidal fluctuation of bacterial production rate and biomass has been described for the Oosterschelde basin by Laanbroek & Verplanke (1986a). The water in the Scheldt was completely mixed from the North Sea up to the beginning of the turbidity maximum at station Bath. In the Elbe, at least in the turbidity maximum (Fig. 4), the water was only partly mixed. A continuous salinity gradient was present with a higher salinity near the bottom than at the surface. The figures 8 and 9 show that the higher maximal tidal current in the Elbe (2–3 m s^{-1}) than in the Scheldt (1 m s^{-1}) has an important influence on the microbiological activity in the water. In the Elbe bacterial production rate, biomass and growth rate in the deeper water layers showed maximum values at mid-tide, when current velocities are maximal. During high current velocities resuspension of sediment takes place, also indicated by an increased turbidity (data not shown), causing higher bacterial numbers and production rates in the water near the bottom. By resuspension the sediment–water exchange will increase which may result in higher concentrations of organic matter and nutrients and thus stimulation of bacterial production rates. Increased bacterial production rates at mid-tide were also found during the two other tidal cycles on 30 and 31 March 1992 at the same anchor station in the Elbe (data not shown). A strong effect of the current velocities as in the Elbe was not found in the Scheldt.

The growth rates during the tidal cycle in the Scheldt (Fig. 8c) and the Elbe (Fig. 9c) are in the same range. However, the anchor station in the Scheldt (average salinity: 13.1 g l^{-1}) was situated at a more haline part than in the Elbe (average salinity: 0.8 and 1.5 g l^{-1} at the surface and the bottom respectively). A comparable sampling site in the Scheldt would be the maximum turbidity zone near Antwerpen where growth rates are expected to be higher than at Bath (Fig. 5). The average growth rate during the tidal cycle at station Bath in the Scheldt was 0.023 h^{-1}. This agreed well with the value found at Bath during the longitudinal profile, 0.02 h^{-1}.

The results of this study show that hydrodynamical conditions in the Scheldt and the Elbe are different. Bacterial production rates in the Scheldt estuary are much higher than in the Elbe estuary and the values observed in spring are relatively high compared with other estuaries. The data show that there is a high load of pollution in the Scheldt as compared to the Elbe. Making a comparison with primary production and microzooplankton grazing activity remains an important challenge in order to determine biogeochemical fluxes in both estuaries.

Acknowledgments

The authors are grateful to the crew of the research vessels 'Luctor' and 'Valdivia' for bringing us safely to the sampling stations and for assistance during sampling. Valuable help with statistical analysis was given by Dr P. M. J. Herman.

Publication No. 733 Netherlands Institute of Ecology, Centre for Estuarine and Coastal Ecology, Yerseke, The Netherlands.

References

Admiraal, W., J. Beukema & F. B. van Es, 1985. Seasonal fluctuations in the biomass and metabolic activity of bacterioplankton and phytoplankton in a well-mixed estuary: the Ems-Dollard (Wadden Sea). J. Plankton Res. 7: 877–890.

ARGE Elbe, 1984. Gewässeroekologische studie der Elbe. Hamburg, 98 pp.

Billen, G., P. Servais & S. Becquevort, 1990. Dynamics of bacterioplankton in oligotrophic and eutrophic aquatic environments: Bottom-up or top-down control? Hydrobiologia 207: 37–42.

Billen, G., P. Servais & A. Fontigny, 1988. Growth and mortality in bacterial population dynamics of aquatic ecosystems. Arch. Hydrobiol. Beih. 31: 173–183.

Billen, G., C. Lancelot, E. De Becker & P. Servais, 1986. The terrestrial interface: Modelling nitrogen transformations during its transfer through the Schelde river system and its estuarine zone. In J. C. J. Nihoul (ed.), Marine Interfaces Ecohydrodynamics. Elsevier Oceanography series, 42: 429–452.

Billen, G., M. Somville, E. De Becker & P. Servais, 1985. A nitrogen budget of the Scheldt hydrographical basin. Neth. J. Sea Res. 19: 223–230.

Caspers, H., 1981. Seasonal effects on the nitrogen cycle in the freshwater section of the Elbe estuary. Verh. int. Ver. Limnol. 21: 866–870.

Coffin, R. B. & J. H. Sharp, 1987. Microbial trophodynamics in the Delaware estuary. Mar. Ecol. Prog. Ser. 41: 253–266.

De Jonge, V. N., 1992. Physical processes and dynamics of microphytobenthos in the Ems estuary (The Netherlands). Thesis, University of Groningen, 176 pp.

Ducklow, H. W., 1993. Bacterioplankton distributions and production in the northwestern Indian Ocean and Gulf of Oman, September 1986. Deep-Sea Res. II 40: 753–771.

Ducklow, H. W. & C. A. Carlson, 1992. Oceanic bacterial production. In K. C. Marshall (ed.), Advances in Microbial Ecology. Vol. 12. Plenum Press, New York-London: 113–181.

Ducklow, H. W. & F.-K. Shiah, 1993. Bacterial production in estuaries. In T. E. Ford (ed.), Aquatic Microbiology, an ecological approach. Blackwell Scientific Publications, Boston: 261–287.

Duwe, K. C., 1990. Hydrography and numerical modelling of Elbe and Shannon estuaries. In H. Kausch, J. G. Wilson & H. Barth (eds), Biogeochemical cycles in two major european estuaries: The ShanElbe Project. Water Pollution Res. Rep. 15, CEC Environment and Waste Recycling, Brussels: 9–23.

Ellenbroek, F. M. & T. E. Cappenberg, 1991. DNA synthesis and tritiated thymidine incorporation by heterotrophic freshwater bacteria in continuous culture. Appl. envir. Microbiol. 57: 1675–1682.

Fuhrman, J. A. & F. Azam, 1982. Thymidine incorporation as a measure of heterotrophic bacterioplankton production in marine surface waters: evaluation and field results. Mar. Biol. 66: 109–120.

Gocke, K. & G. Rheinheimer, 1988. Microbial investigations in rivers VII. Seasonal variation of bacterial numbers and activity in eutrophied rivers of northern Germany. Arch. Hydrobiol. 112: 197–219.

Heip, C., 1989. The ecology of the estuaries of Rhine, Meuse and Scheldt in the Netherlands. In J. D. Ros (ed.), Topics in marine biology. Scient. Mar. 53: 457–463.

Kerner, M., 1990. Seasonal variation in the structure of the microbial community capable of nitrate respiration in an intertidal mud flat sediment of the Elbe estuary. Arch. Hydrobiol. (Suppl.) 75: 273–280.

Kuosa, H. & K. Kivi, 1989. Bacteria and heterotrophic flagellates in the pelagic carbon cycle in the northern Baltic Sea. Mar. Ecol. Prog. Ser. 53: 93–100.

Laanbroek, H. J. & J. C. Verplanke, 1986a. Tidal variations in bacterial biomass, productivity and oxygen uptake rates in a shallow channel in the Oosterschelde basin, The Netherlands. Mar. Ecol. Prog. Ser. 29: 1–5.

Laanbroek, H. J. & J. C. Verplanke, 1986b. Seasonal changes in percentages of attached bacteria enumerated in a tidal and a stagnant coastal basin: relation to bacterioplankton productivity. FEMS Microb. Ecol. 38: 87–98.

Laanbroek, H. J., J. C. Verplanke, P. R. M. de Visscher & R. de Vuyst, 1985. Distribution of phyto- and bacterioplankton growth and biomass parameters, dissolved inorganic nutrients and free amino acids during a spring bloom in the Oosterschelde basin, The Netherlands. Mar. Ecol. Prog. Ser. 25: 1–11.

Nieuwenhuize, J., Y. E. M. Maas & J. J. Middelburg, 1994. Rapid analysis of organic carbon and nitrogen in particulate materials. Mar. Chem. 45: 217–224.

Painchaud, J. & J. C. Therriault, 1989. Relationship between bacteria, phytoplankton and particulate organic carbon in the upper St. Lawrence estuary. Mar. Ecol. Prog. Ser. 56: 301–311.

Porter, K. G. & Y. S. Feig, 1980. The use of DAPI for identifying and counting aquatic microflora. Limnol. Oceanogr. 25: 943–948.

Relexans, J. C., M. Meybeck, G. Billen, M. Brugeaille, H. Etcheber & M. Somville, 1988. Algal and microbial processes involved in particulate organic matter dynamics in the Loire estuary. Estuar. coast. shelf Sci. 27: 625–644.

Rheinheimer, G., 1959. Mikrobiologische untersuchungen über den stickstoffhaushalt der Elbe. Arch. Mikrobiol. 34: 358–373.

Rheinheimer, G., 1960. Der jahresrythmus der bacterienkeimzahl in der Elbe zwischen Schnackeburg und Hamburg. Arch. Mikrobiol. 35: 34–43.

Riemann, B. & R. T. Bell, 1990. Advances in estimating bacterial biomass and growth in aquatic systems. Arch. Hydrobiol. 25: 385–402.

Servais, P. & J. Garnier, 1993. Contribution of heterotrophic bacterial production to the budget of the river Seine (France). Microb. Ecol. 25: 19–33.

Sherr, E. B. & B. F. Sherr, 1987. High rates of consumption of bacteria by pelagic ciliates. Nature 325: 710–711.

Somville, M., G. Billen & J. Smitz, 1982. An ecophysiological model of nitrification in the Scheldt estuary. Math. Model. 3: 523–533.

Vaqué, D., M. L. Pace, S. Findlay & D. Lints, 1992. Fate of bacterial production in a heterotrophic ecosystem: grazing by protists and metazoans in the Hudson estuary. Mar. Ecol. Prog. Ser. 89: 155–163.

Hydrobiologia **311**: 43–55, 1995.
C. H. R. Heip & P. M. J. Herman (eds), Major Biological Processes in European Tidal Estuaries.

Nitrous oxide emissions from estuarine intertidal sediments

Jack J. Middelburg[1], Gerard Klaver[2], Joop Nieuwenhuize[1], Rinus M. Markusse[1], Tom Vlug[1] & F. Jaco W. A. van der Nat[1]
[1]*Netherlands Institute of Ecology, Centre for Estuarine and Coastal Ecology, Vierstraat 28, NL-4401 EA Yerseke, The Netherlands*
[2]*Rijks Geologische Dienst, P.O. Box 157, NL-2000 AD Haarlem, The Netherlands*

Key words: nitrous oxide, nitrogen cycling, estuarine sediments, biogeochemistry, eutrophication, Scheldt Estuary

Abstract

From September 1990 through December 1991 nitrous oxide flux measurements were made at 9 intertidal mud flat sites in the Scheldt Estuary. Nitrous oxide release rates were highly variable both between sites and over time at any one site. Annual nitrous oxide fluxes vary from about 10 mmol N m^{-2} at the tidal fresh-water end-member site to almost zero at the most saline stations. Along the estuarine gradient, annual nitrous oxide fluxes are significantly correlated with sedimentary organic carbon and nitrogen concentrations, ammonium fluxes and annual nitrogen turn-over rates, that are estimated using mass-balance considerations. Nitrous oxide fluxes seem to respond linearly to an increasing nitrogen load, with one out of each 17 000 atoms nitrogen entering estuaries being emitted as nitrous oxide.

Introduction

Recent ice-core data indicate that present-day atmospheric nitrous oxide (N_2O) concentrations are significantly higher than those in the past 45 000 year (Leuenberger & Siegenthaler, 1992). At present atmospheric nitrous oxide concentrations are increasing at a rate of 0.3% per year (Schlesinger, 1991). Increasing atmospheric nitrous oxide concentrations are of concern since nitrous oxide is a greenhouse trace gas that, molecule for molecule, is about 200 times more effective than carbon dioxide. Moreover, in the stratosphere nitrous oxide is oxidized to nitric oxide, a major sink for ozone (Cicerone, 1987). Recognition of the role of nitrous oxide in both the stratospheric ozone and tropospheric heat budget has initiated and stimulated research to quantify nitrous oxide sources and sinks.

Nitrous oxide has both biogenic and anthropogenic sources (Schlesinger, 1991). Major biogenic sources are the ocean (2 Tg N yr^{-1}) and natural soils (5.7 Tg N yr^{-1}). The important anthropogenic sources are fertilized agriculture and land-use change (both 0.7 Tg N yr^{-1}) and biomass burning (2 Tg N yr^{-1}). Studies of nitrous oxide emissions from fertilized soils far outnumber those of nitrous oxide emissions from natural soils (e.g. see overview by Bouwman, 1990). Similarly, terrestrial sites have generally received more attention than aquatic sites.

Nitrous oxide emissions are determined by the interplay between nitrous oxide production, consumption and transport processes. Nitrous oxide can be produced by microbial processes in three ways: as a true intermediate species during denitrification; and as sidereaction products during both nitrification and dissimilatory reduction of nitrate to ammonium (e.g. Seitzinger, 1988). Microbial consumption of nitrous oxide occurs only in the process of denitrification. There is, unfortunately, not much known about nitrous oxide transport. By analogy to other biogenic trace gases, nitrous oxide is probably transported from the sediment to the overlying water or atmosphere by diffusion, gas bubble ebullition following stripping of nitrous oxide from sediment or plant supported transport (e.g. Martens & Chanton, 1989; Chanton & Dacey, 1991). Accordingly, any factor that affects either directly or indirectly nitrous oxide production, consumption or transport may affect nitrous oxide emission rates.

In this manuscript we present data from a seasonal study of nitrous oxide release from 9 intertidal sites in the Scheldt Estuary. In order to obtain a first-order understanding of the factors determining nitrous oxide fluxes, annual emission rates will be compared to nitrogen stocks and annual nitrogen turnover rates based on mass-balance considerations.

Scheldt Estuary

The river Scheldt (total length of 330 km) flows through France, Belgium and the Netherlands before it decharges into the North Sea (Fig. 1). The tidal amplitude varies between 4 m near Ellewoutsdijk, 5 m near Antwerpen, and to 2 m near Ghent. In the Scheldt estuary tidal exchange (about 100 000 $m^3\ s^{-1}$) is more important than fresh-water discharge (about 100 m^3/sec). Vertical salinity stratification is therefore either absent (downstream) or minor (upstream), but there is a pronounced horizontal salinity gradient which is subject to tidal and seasonal variations. The residence time for dissolved substances is about 50 to 70 days (Soetaert & Herman, 1994). Along its course, the Scheldt is heavily polluted by the discharge of important cities, active industrial areas and intensive stock-farming. In the tidal fresh-water part of the estuary, mineralization of anthropogenic organic matter inputs results in suboxic to anoxic conditions and high total dissolved nitrogen (upto 600 μM) and phosphate (upto 20 μM) concentrations. Upon estuarine mixing, dissolved oxygen concentrations increase and nutrient concentrations generally decrease due to mixing with sea water, reaeration and biogeochemical transformations (Soetaert & Herman, 1995: Boderie *et al.*, 1993; Goossen *et al.*, 1995).

In the coastal zone of Belgium and the Netherlands nitrogen is the primary limiting nutrient (Peeters & Peperzak, 1990). In the Scheldt Estuary the throughput of riverine nitrogen has therefore received considerable attention (Billen, 1975; Billen *et al.*, 1985; Soetaert & Herman, 1995; Middelburg *et al.*, 1995). Each year about 64×10^9 g of N enters the Scheldt Estuary, about 40×10^9 g of N from the Scheldt river and about 24×10^9 g of N from direct agricultural, domestic and industrial inputs (Soetaert & Herman, 1995). Meanwhile, the amount of nitrogen discharged into the North Sea has been estimated to be 49×10^9 g of N yr^{-1}. Hence, each year about 15×10^9 g of N is removed through denitrification, sedimentation and burial. Independent estimates for the annual amount of N removed through sedimentation and burial vary from 1 to 3×10^9 g of N (Billen *et al.*, 1985; van Eck *et al.*, 1991; Middelburg *et al.*, 1995). Accordingly, it seems that about 25% of the total nitrogen entering the estuary does not reach the North Sea due to denitrification. Denitrification may occur both in the suboxic/anoxic water column and in the sediments. About 14 (Middelburg *et al.*, 1995) to 30% (Soetaert & Herman, 1995) of the denitrification may take place in intertidal sediments.

Material and methods

Sampling sites

Nine intertidal mud flat stations were selected along the Scheldt Estuary, largely on basis of the salinity and the accessibility (Fig. 1; Table 1). During the measurement period extensive dredging activities resulted in the removal of muddy material from the intertidal mud flat at station Burcht. Results for station Doel have been presented before and are discussed in detail by Middelburg *et al.* (1995).

Flux measurements

A single flux measurement on at least two adjacent plots was made at each of the 9 sites at roughly four weeks intervals over a sampling period from September 1990 through December 1991. Flux measurements were made systematically at low tide and within a single week for each estuarine profile. Wooden walkboards were constructed and used to reduce the disturbance caused by repeated visits and to limit spatial variability. The release of nitrous oxide was measured by monitoring accumulation of the gas beneath chambers placed over the sediment surface. The chambers have a 0.11 m^2 base, a volume of 45 l and are made of non-transparent polypropylene. When necessary, a sun shield was placed over the chamber to maintain the chamber temperature and relative humidity within 25% of their ambient values.

Nitrous oxide, as well as carbon dioxide and methane, concentrations were measured by circulating chamber air through Teflon tubes between chamber and the gas monitor. The measurement principle of the multi-gas monitor used, a Brüel & Kjaer type 1302, is based on the photoacoustic infra-red detection method. Briefly, after thorough flushing of the Teflon tubes and analysis cell, the air sample is hermetically scaled in

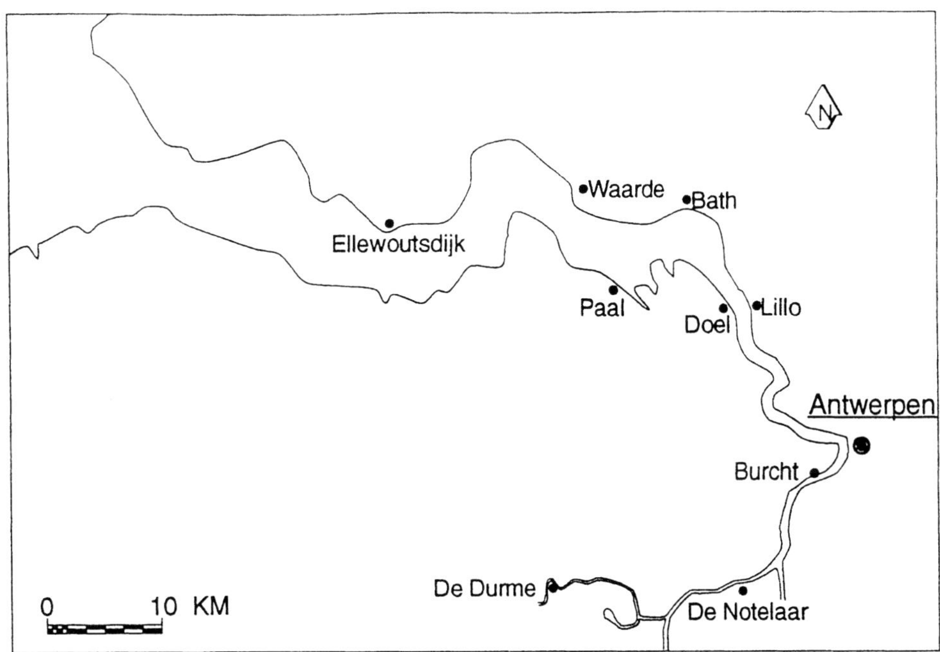

Fig. 1. Map of the Scheldt Estuary showing sampling locations.

Table 1. Site description

Station	Chlorinity ($g\,l^{-1}$)	POC (wt%)	TN (wt%)	Annual N_2O-flux ($mmol\,N\,m^{-2}\,yr^{-1}$)	Molar C/N ratio
DURME	0.4	4.58	0.42	10.1	13.9
NOTELAAR	0.9	3.06	0.24	4.0	13.7
BURCHT	2.6	1.74	0.11	3.8	17.6
LILLO	5.6	2.35	0.16	6.4	15.7
DOEL	6.3	2.60	0.18	6.6	17.5
PAAL	8.7	1.72	0.12	2.8	17.2
BATH	7.9	1.52	0.11	2.9	16.0
WAARDE	8.7	1.39	0.09	−0.6	20.2
ELLEWOUTSDIJK	14.7	1.90	0.09	0.5	21.9

the analysis cell. The light emitted by a pulsating infrared light source and purified by a narrow-band optical filter is selectively adsorbed. The temperature of the gas increases and decreases in response to the pulsating light transmitter, and this causes an equivalent increase and decrease of the pressure of the gas in the closed cell. Two ultrasensitive microphones mounted in the cell are used to measure this pressure wave, which is directly proportional to the concentration of the gas. Various gases can be measured in the same sample using different filters. The response time for a sequential measurement of nitrous oxide, carbon dioxide and methane is about 90 sec. The detection limit is 0.05 ppmv N_2O and the response is linear upto 5 ppmv of nitrous oxide. The reproducibility at ambient nitrous oxide levels is about 15%. Due to the high stability of the system calibration procedures are necessary only once each three months. The accuracy and precision of the system are controlled each measurement period by analysing ambient nitrous oxide concentrations before starting flux measurements. The monitor compensates for interferences caused by temperature and pressure fluctuations and the presence of water vapour, and it is very well adapted for field work since it is easy portable (8 kg) and has the ability to store measurements of a few days into its background memory. At high carbon dioxide concentrations, a background cross-correction was applied to obtain reliable nitrous oxide concentrations. Nitrous oxide fluxes are calculated by regression analysis from the recorded change in concentration over time. Nitrous oxide fluxes significant at $p<0.01$ are normally obtained as low as 5 mmol N_2O m^{-2} yr^{-1}, within 30 minutes. Nitrous oxide fluxes with a significance of $p>0.05$ are considered to be not significant. Since reported nitrous oxide fluxes from marine environments range from about − 1.5 to +50 mmol m^{-2} yr^{-1} (e.g. Kieskamp *et al.*, 1991), our flux measurement technique may be restricted to high flux environments such as the Scheldt Estuary. The flux detection limit can be lowered by increasing the duration of measurement, but this may cause deviations from ambient conditions.

Sediment and pore water

The sediment chlorinity was determined by potentiometric titration using $AgNO_3$. The organic carbon and nitrogen contents of the sediments were determined using a Carlo-Erba CN-analyzer following a recently developed *in situ* HCl acidification procedure (Nieuwenhuize *et al.* 1994). Pore-water samples from intertidal sediments were obtained by pressure filtration of sediments following the collection of cores by hand at low tide. No measures were take to control the conditions during squeezing, but the pore-water extraction procedure was always completed within two hours. Measurements of dissolved ammonium and nitrate concentrations were made using standard colorimetric methods on a Skalar autoanalyzer.

Mass balance calculations

At each site the rate of ammonification has been calculated from the rate of organic carbon mineralization and the C/N ratio of surficial sediments. Carbon mineralization rates are based on measured gaseous carbon dioxide and methane fluxes, which will be discussed in detail elsewhere. Nitrification rates are calculated from the difference in ammonification and flux of ammonium out of the sediments. The diffusive flux of ammonium has been estimated using Fick's first law of diffusion, measured porosities, diffusivities estimated from molecular diffusion coefficients (Li & Gregory, 1974) that are corrected for tortuosity and temperature effects (e.g. Iversen & Jørgensen, 1993), and gradients based on linear regressions of the upper few data points. The rate of denitrification has been estimated as the sum of nitrification rates and nitrate fluxes into the sediment. The diffusive flux of nitrate has been calculated using a linear gradient and Fick's first law as outlined above. This mass-balance approach is only valid if steady-state conditions apply; i.e. temporal variations in storage of pore-water components can be neglected. It is for this reason that annual rates are presented only.

Results

The estuarine distribution of dissolved ammonium and nitrate during spring, summer, autumn and winter are shown in Fig. 2. During spring high concentrations of riverine ammonium enter the estuary and a well pronounced mid-estuarine nitrate maximum is present. In contrast, during autumn relatively low concentrations of ammonium enter the estuary upstream and nitrate concentration are relatively low. Summer and winter conditions are characterized by relatively moderate riverine ammonium inputs and moderate midestuarine nitrate maxima. Nitrite concentrations are in general low and form an insignificant fraction of nitrate (usually <5%). The most important processes affecting dissolved inorganic nitrogen concentrations in the Scheldt estuary have been recognized as nitrification, denitrification, biological uptake and organic nitrogen mineralization (Billen, 1975; Billen *et al.*, 1985; van Eck *et al.*, 1991; Soetaert & Herman, 1995; Boderie *et al.*, 1993; Goossen *et al.*, 1995).

The annual average sedimental chlorinity, organic carbon and nitrogen contents are listed in Table 1. Organic carbon and nitrogen contents, as well as molar carbon-nitrogen ratios, generally decrease going downstream. Molar carbon/nitrogen ratios vary from 13.7 at tidal fresh-water stations to 21.9 at the most saline station.

Vertical profiles of ammonium in pore-water of intertidal mud flats followed normal trends of increasing concentration with depth in sediments (Fig. 3). Dissolved nitrate concentrations (data not shown) are only measurable in the top few samples and are at least one order of magnitude lower than ammonium. In sediments, ammonium is formed by the decomposition of organic nitrogen compounds and may build up in pore waters if it is not oxidized by nitrifiers (e.g. Berner, 1980). The ammonium distribution in pore water over time at any one site shows a distinct seasonal cycle. Ammonium concentrations at depth are in general higher during late summer and fall in comparison

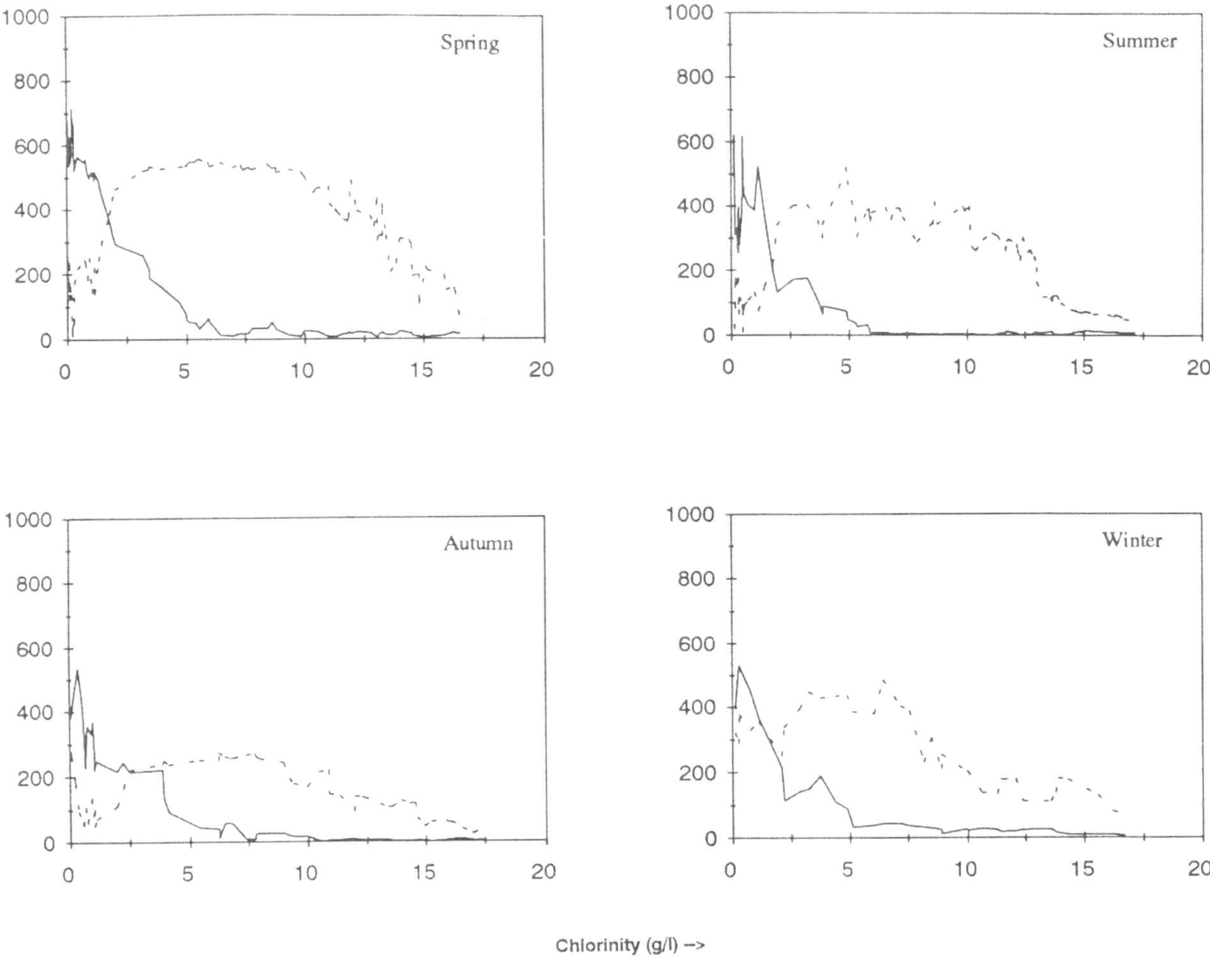

Fig. 2. The distribution of ammonium (mol m^{-3}) and nitrate (mol m^{-3}) versus chlorinity (g l^{-1}) in Scheldt water during spring, summer, autumn and winter of 1991. In order to emphasize the major trends, data from various cruises are presented as three-point moving averages.

to those during late winter and early spring. Marked seasonal variations in pore-water ammonium concentrations have been reported for various estuarine and coastal sediments (Aller, 1980; Klump & Martens, 1989; Watson *et al.*, 1985; Kemp *et al.*, 1990). The seasonality of pore-water ammonium concentrations has been attributed to temperature dependent organic matter decomposition rates and seasonal varying bioturbation or mixing rates (Aller, 1980; Klump & Martens, 1989; Middelburg *et al.*, 1995).

Emissions of nitrous oxide from intertidal sediments in the Scheldt Estuary are presented in Fig. 4. The flux data are highly variable, both spatially and temporally, and range from -25 to 75 mmol N m^{-2} yr^{-1}. Positive values indicate fluxes from the sediment to the atmosphere. Nitrous oxide fluxes at tidal freshwater stations (i.e. Durme and Notelaar) are in general higher than those from the most saline stations. Nitrous oxide fluxes during summer were higher at some sites, but not at all sites. Estimates for annual nitrous oxide emission rates were calculated for all sites by integration under the curves that connect averages of replicate measurements (Table 1).

Discussion

Nitrous oxide emission rates are highly variable, both spatially and temporarily (Fig. 4). This seems to be a general characteristic of biogenic trace gas emission rate data bases (e.g. Smith *et al.*, 1983; Bartlett *et al.*, 1993; Bouwman, 1990; Middelburg *et al.*, 1995). This

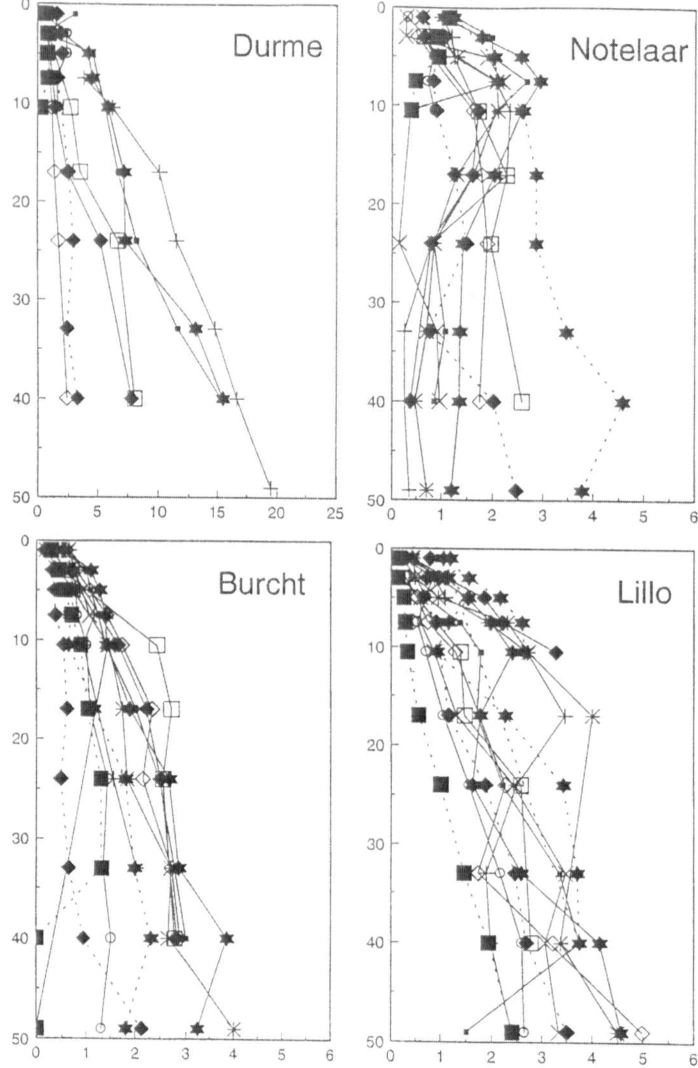

Fig. 3. Interstitial water ammonium concentrations (mol m^{-3}) versus depth (cm). Notice the different concentrations scales.

variability is related to the complexity and heterogeneity of processes involved in the production, consumption and transport of nitrogen compounds in aquatic systems. Temporal variability in nitrous oxide fluxes could be related to seasonal variations in nitrification, denitrification, nitrate-ammonification, bioturbation and gaseous exchange rates, and nitrate, nitrite and ammonium availability (Fig. 2), or short term variations in periods of light and tidal exposure. Careful examination of our data revealed not a single factor, or any combination of measured variables, that could adequately explain the time varying nitrous oxide fluxes at a single site. In order to obtain a first-order understanding of the factors determining nitrous oxide exchange between estuarine sediments and the atmosphere, our discussion will be based primarily on annual emission rates. Integration over a year will not only remove most of the temporal variability, but it also allows us to relate spatial trends in nitrous oxide fluxes to spatial trends in nitrogen biogeochemistry.

Nitrous oxide exchange along an estuarine gradient

Along an estuarine gradient significant changes in biogeochemical processes, sediment characteristics and dissolved and particulate concentrations may take place. In the Scheldt Estuary, there are considerable

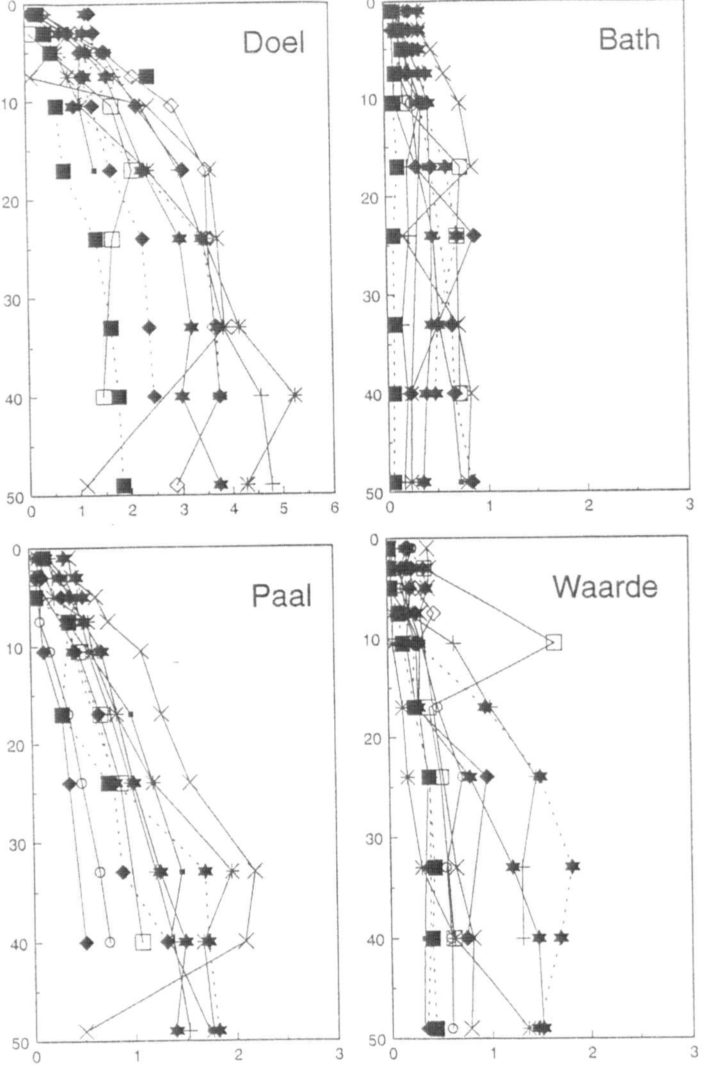

Fig. 3A.

changes in nitrogen stocks, exchanges and turnover rates along the estuarine gradient (Fig. 5).

The tidal fresh-water end-member of the estuary is characterized by high dissolved ammonium, moderate dissolved nitrate and low dissolved oxygen concentrations (Fig. 2), and high sedimentary organic carbon and nitrogen contents (Table 1; Fig. 5). Rates of sedimentary ammonification based on organic matter mineralization rates and the sedimentary C/N ratio are extremely high (upto 25 mol m^{-2} yr^{-1}). Related and directly coupled to these high rates of ammonium production, there are high fluxes of ammonium out of the sediment (upto 0.7 mol m^{-2} yr^{-1}; Fig. 5). The majority of ammonium generated is however oxidized to nitrate. Despite these high rates of nitrification, pore-water nitrate concentrations are neglible. Moreover, nitrate diffuses from the water column into the sediment. The nitrate produced by nitrifiers and that diffusing into the sediment consequently support high rates of denitrification.

The intermediate brackish sites in the Scheldt Estuary (Lillo and Doel) are characterized by low dissolved ammonium, high dissolved nitrate and moderate dissolved oxygen concentrations (Fig. 2). Sedimentary organic carbon and nitrogen concentrations are moderate (Table 1; Fig. 5). Although rates of ammonification at these sites are almost an order of magnitude lower than those at the tidal fresh-water sites, ammonium fluxes out of the sediments are comparable. Coupled

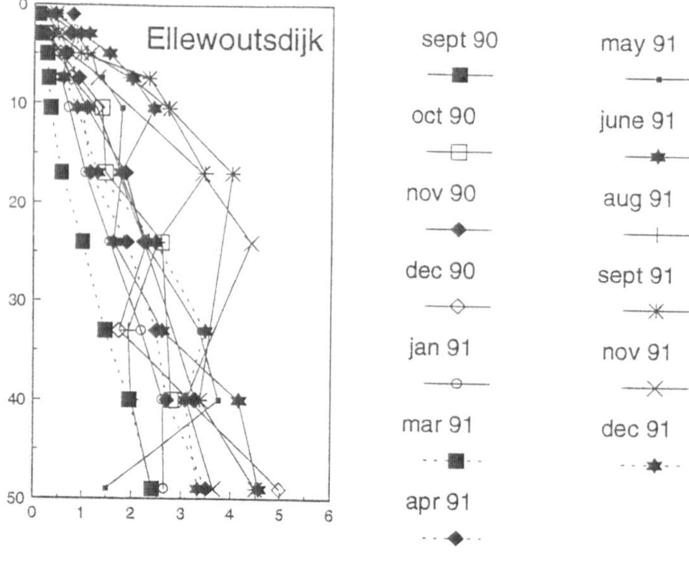

Fig. 3B.

to these lower ammonification rates, there are also lower rates of nitrification and denitrification (Fig. 5). However, the nitrate flux into the sediment and consequently bottom water nitrate supported rates of denitrification are relatively high.

The intermediate saline sites (Paal and Bath) are in the reach of low dissolved ammonium, moderate to high dissolved nitrate and high dissolved oxygen bottom waters (Fig. 2). The sedimentary concentrations of organic carbon and nitrogen are rather low (Table 1; Fig. 5). At these intermediate saline sites, rates of ammonification, nitrification and denitrification are lowest (Fig. 5). Fluxes of ammonium are intermediate between those of intermediate brackish sites and fully marine sites. Due to moderate to high dissolved nitrate concentrations, there is a significant flux of nitrate into the sediments.

The most saline stations (Waarde and Ellewoutsdijk) are characterized by low dissolved ammonium, moderate nitrate, and high dissolved oxygen concentrations (Fig. 2), moderate sedimentary organic carbon contents and low nitrogen concentrations. Rates of ammonification, nitrification and denitrification are slighty higher than those observed at the intermediate saline stations, but significant lower than at the brackish stations. The flux of ammonium out of the sediments at these sites is extremely low compared to other sites in the estuary.

Nitrous oxide fluxes from the tidal fresh-water endmember site are very high (about 10 mmol N $m^{-2} yr^{-1}$) and subsequently decrease going downstream to stations Notelaar and Burcht. Further downstream nitrous oxide emission rates initially increase, with maximum rates (about 6.5 mmol N m^{-2} yr^{-1}) at intermediate brackish sites, before they decrease to very low or even negative values at the saline stations Waarde and Ellewoutsdijk (Figs 4 and 5).

Influxes of nitrous oxide are not a feature unique to tidal flats in the marine part of the Scheldt Estuary, but have also been observed in other coastal environments. Jensen *et al.* (1984) were the first to report nitrous oxide uptake for brackish estuarine subtidal sediments in Limfjorden, Denmark. Nitrous oxide influxes were also observed for sediments from intertidal flats in San Francisco Bay (Miller *et al.*, 1986) and the western Wadden Sea (Kieskamp *et al.*, 1991) and for shelf sediments from the North Sea (van Raaphorst *et al.*, 1992). These influxes have been related to the utilization of nitrous oxide as a terminal electron acceptor for organic matter degradation in the absence of nitrate (Kieskamp *et al.*, 1991) or for oxidation of reduced sulphur compounds (e.g. Dalsgaard & Bak, 1992). Nitrous oxide consumption by anoxic estuarine and marine sediments is consistent with their in general high denitrification activity (Seitzinger, 1988; Devol, 1991; Devol & Christensen, 1993) and low pore-water nitrate concentrations.

Changes in estuarine nitrogen biogeochemistry that result in higher nitrous oxide production or consumption rates are important in the context of nitrous oxide

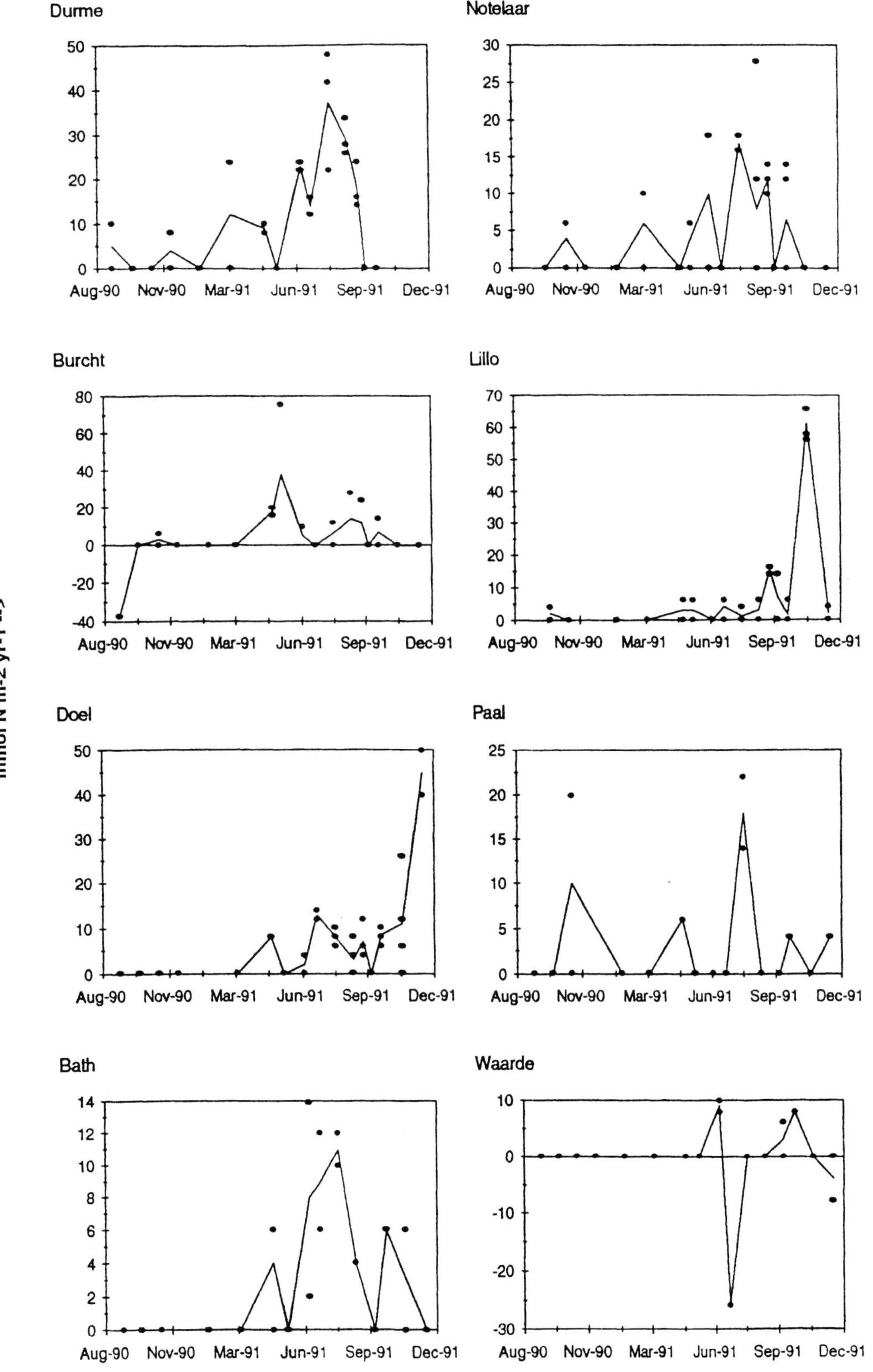

Fig. 4. Emission rates of nitrous oxide (mmol N m^{-2} yr^{-1}) versus date. The solid lines connect averages of replicate measurements.

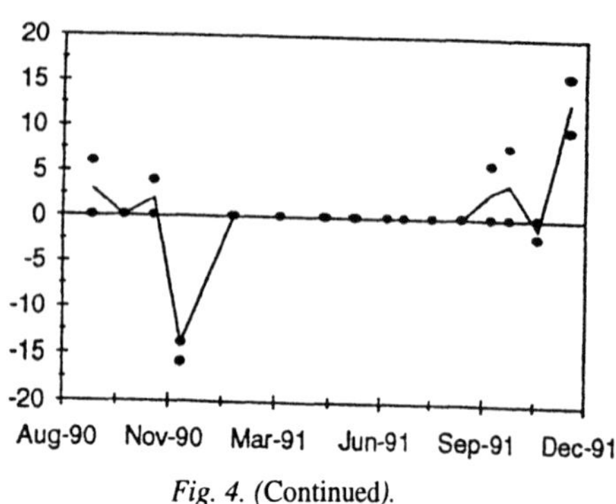

Fig. 4. (Continued).

exchange between estuarine sediments and the atmosphere. Nitrous oxide can be produced during nitrification (ammonium oxidation) or denitrification and nitrate-ammonification (nitrate reduction), but it is consumed only in the process of denitrification. The spatial distribution of the processes involved in nitrous oxide formation and consumption are very similar due to the very tight coupling between ammonification, nitrification and denitrification processes in these intertidal sediments. On the basis of nitrification and denitrification rate data, it is therefore not possible to unravel the production and consumption pathways of nitrous oxide unambiguously. For instance, the high nitrous oxide emission rate at the tidal fresh-water station (Durme) could be related to high rates of nitrification or denitrification (Fig. 5). Similarly, the mid-estuarine maximum in nitrous oxide emission rates could be due to midestuarine maxima in nitrification and denitrification. Despite the similarity of estuarine profiles of nitrogen cycling processes, there are significant differences in relative magnitudes of gradients. Rates of sedimentary ammonification, nitrification and denitrification differ by almost an order of magnitude between the tidal fresh-water station Durme and intermediate brackish stations Doel and Lillo, whereas fluxes of ammonium, nitrate and nitrous oxide at station Durme are within a factor of two from those at stations Doel and Lillo. This might indicate that exchange processes can not respond linearly to changes in production or consumption rates.

Ecosystem approach to predict nitrous oxide emission rates

Our detailed understanding of the factors regulating nitrous oxide production via nitrification and denitrification at the cellular level (e.g. Tiedje, 1988) has not yet provided a simple concept that can be applied at the ecosystem level. In order to determine and refine global nitrous oxide budgets, it is necessary to understand processes that regulate nitrous oxide emission on a ecosystem basis (Matson & Vitousek, 1990). In a cross-system comparison, Matson & Vitousek (1987) have reported a significant relationship between soil fertility and nitrous oxide fluxes in tropical forests. Although this ecosystem approach has been applied successfully to processes regulating soil emission of nitrous oxide in terrestrial systems (Matson & Vitousek, 1987, 1990), it has not yet been applied to aquatic, i.e. more open and dynamic, systems. The data presented in Fig. 5 indicate that it may apply to estuarine systems as well. Nitrous oxide emission rates correlate significantly with sedimentary organic carbon ($r=0.84$) and nitrogen ($r=0.83$) contents, diffusive ammonium fluxes ($r=0.90$) and rates of ammonification ($r=0.71$), nitrification ($r=0.71$) and denitrification ($r=0.72$).

The gradient in nitrogen biogeochemistry in the Scheldt Estuary also provides us with an opportunity to investigate the impact of enhanced nitrogen loading on nitrous oxide fluxes. To this end the total nitrogen load at our stations was estimated using the compartimentalisation of intertidal areas and average input data from Soetaert & Herman (1995). Nitrogen loadings, estimated as the sum of all dissolved and particulate nitrogen components, range from about 930 mol N m^{-2} yr^{-1} at the tidal fresh-water stations to about 27 mol N m^{-2} yr^{-1} at our most saline station Ellewoutsdijk. Nitrous oxide fluxes increase linearly with an increasing nitrogen load (Fig. 6). If we exclude the Burcht station, since these results may have been affected by dredging activities, our data can be described by a single straight line with a coefficient of determination of 0.998 ($N=6$):

$$N_2O = -0.0017\,(0.0009) + 5.81(0.12) \times 10^{-5} N - \text{load},$$

where nitrous oxide fluxes and nitrogen loads have as units mol N m^{-2} yr^{-1}. Inclusion of data from station Burcht reduces the coefficient of determination to 0.81 and reduces the slope to 5.1 (1.1) $\times$ 10^{-5}. Similarly, exclusion of data from the tidal fresh-water end-member with its high nitrogen loading lowers the coefficient of determination and slope to 0.89 and 6 (1.2) $\times$ 10^{-5}, respectively. It is surprising that the data fit so well since low-oxygenated, ammonium-rich fresh-water sites as well as oxygenated, nitrate-rich marine sites are included. This strong linear relationship reinforces the ecosystem approach to understand the factors controlling nitrous oxide emission rates.

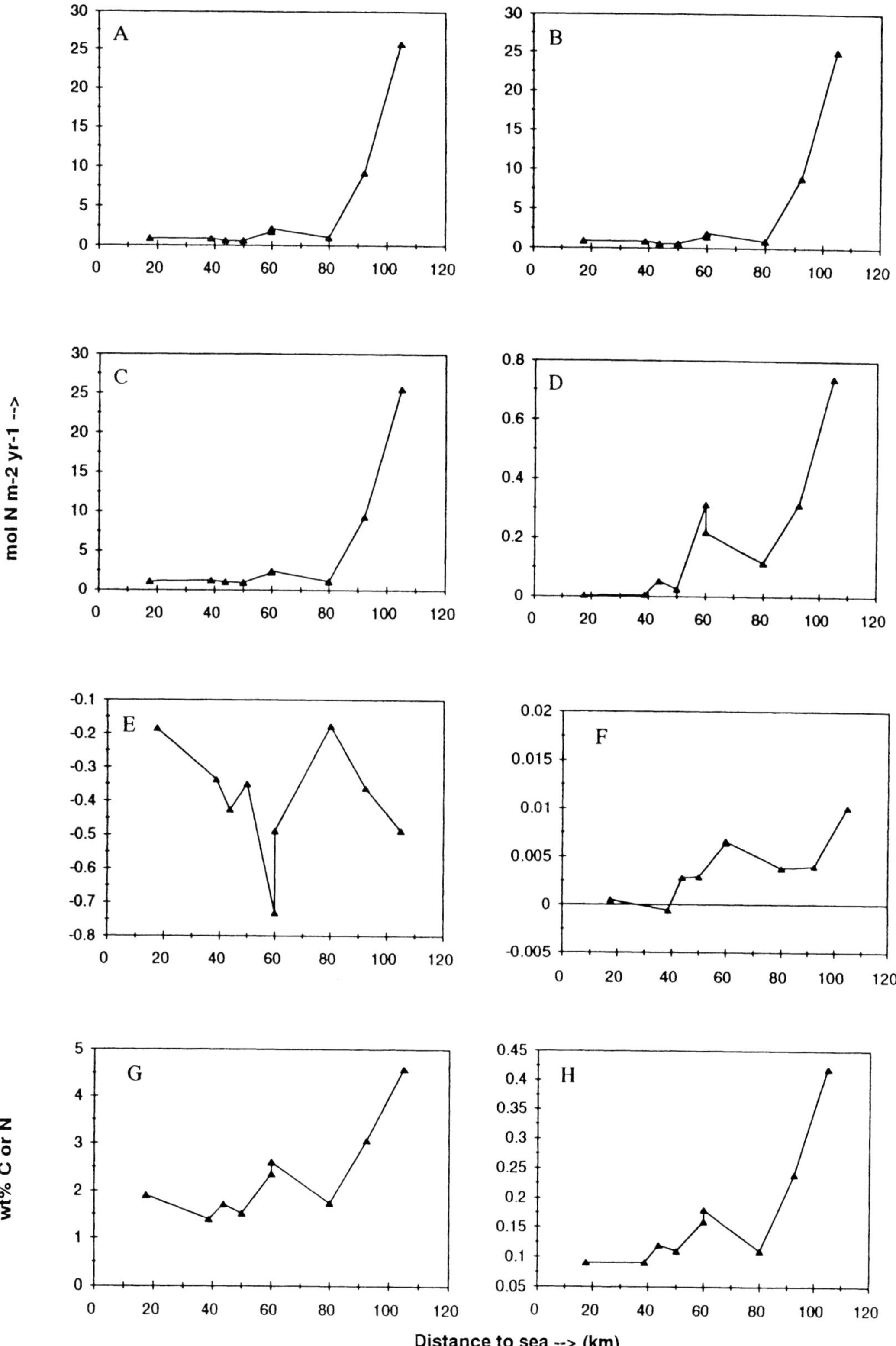

Fig. 5. Nitrogen stocks and turnover rates, and annual integrated nitrous oxide emission rates versus distance to the North Sea. (a) Ammonification (mol N m^{-2} yr^{-1}); (b) Nitrification (mol N m^{-2} yr^{-1});(c) Denitrification (mol N m^{-2} yr^{-1}); (d) diffusive ammonium flux (mol N m^{-2} yr^{-1}); (e) diffusive nitrate flux (mol N m^{-2} yr^{-1}); (f) nitrous oxide flux (mol N m^{-2} yr^{-1}); (g) sedimentary organic carbon (wt%); (h) total sedimentary nitrogen (wt%).

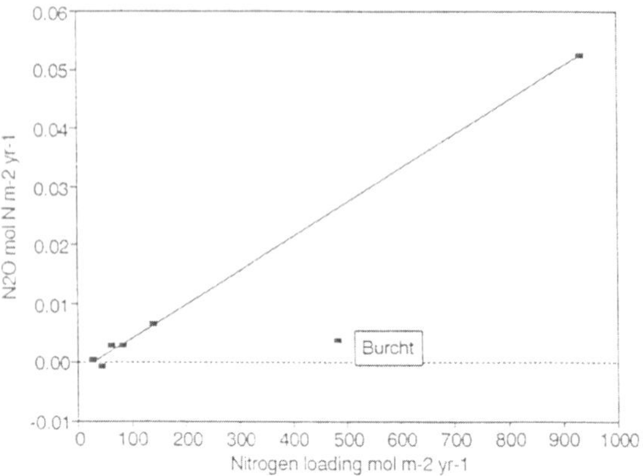

Fig. 6. Relation between nitrogen load (mol N m^{-2} yr^{-1}) and annual integrated nitrous oxide emission rates (mol N m^{-2} yr^{-1}). The straight line represents the best fit ($r^2 = 0.998$) corresponding to a slope of 5.81 (0.12) $\times 10^{-5}$ and an intercept of -0.0017 (0.0009). Data refer to model compartments (Soetaert & Herman, 1995) rather than stations. Stations Durme, Notelaar, Burcht, Lillo, Doel, Bath, Paal, Waarde and Ellewoutsdijk relate to compartments 1, 1, 2, 5, 5, 6, 7, 9 and 11 of Soetart & Herman (1995), respectively.

This relationship can also be used to constrain the global nitrous oxide flux from estuarine sediments and its enhancement due to elevated anthropogenic nitrogen loadings. The global riverine nitrogen flux is about 43×10^{12} g N yr^{-1} of which 16% (7×10^{12} g N yr^{-1}) is due to human pollution (Schlesinger, 1991). This riverine nitrogen input to estuaries relates to a global nitrous oxide source of 2.5×10^{9} g N yr^{-1}, which is rather unimportant compared to other nitrous oxide sources which total 11.1×10^{12} g N yr^{-1} (Schlesinger, 1991). Moreover, the enhancement of nitrous oxide fluxes from estuarine sediments (0.4×10^{9} g N yr^{-1}) due to elevated riverine nitrogen loadings is insignificant in a global context and measures to control it should therefore have a low priority in governmental environmental programs.

Acknowledgments

Drs Owens and Zwolsman are thanked for their constructive reviews. A. Wielemaker executed part of the data handling.

References

Aller, R. C., 1980. Diagenetic processes near the sediment–water interface of Long Island Sound. 1. Decomposition and nutrient element geochemistry (S,N,P). Adv. Geophys. 22: 238–350.

Bartlett, K. B & R. C. Harris, 1993. Review and assessment of methane emissions from wetlands. Chemosphere, 126: 261–320.

Berner, R. A., 1980. Early diagenesis: A theoretical approach. Princeton University Press, Princeton: 1–241.

Billen, G., 1975. Nitrification in the Scheldt Estuary (Belgium and Netherlands). Estuar. coastal mar. Sci. 3: 79–89.

Billen, G., M. Somville, E. de Becker & P. Servais, 1985. A nitrogen budget of the Scheldt hydrographical basin. Neth. J. Sea Res. 19: 223–230.

Boderie, P. M. A., J. J. G. Zwolsman, G. T. M. van Eck & C. H. van der Weijden, 1993. Nutrient biogeochemistry in the water column (N,P,Si) and pore-water (N) of sandy sediment of the Scheldt Estuary (SW Netherlands). Neth. J. aquat. Ecol. 27: 309–318.

Bouwman, A. E., 1990, (ed.). Soils and the greenhous effect. John Wiley & Sons.

Chanton, J. P. & J. W. H. Dacey, 1991. Effects of vegetation on methane flux, reservoirs, and carbon isotopic composition. In Trace Gas Emissions by Plants, 1st edn. Academic Press, Inc: 65–92.

Cicerone, R. J., 1987. Changes in stratospheric ozone. Science 237: 35–42.

Dalsgaard, T. & F. Bak, 1992. Effect of acetylene on nitrous oxide reduction and sulfide oxidation in batch and gradient cultures of *Thiobacillus denitrificans*. Appl. envir. Microbiol. 58: 1601–1608.

Devol, A. H., 1991. Direct measurement of nitrogen gas fluxes from continental shelf sediments. Nature 349: 319–321.

Devol, A. H. & J. P. Christensen, 1993. Benthic fluxes and nitrogen cycling in sediments of the continental margin of the eastern North Pacific J. mar. Res. 51: 345–372.

Goossen, N. K., P. van Rijswijk & U. Brockmann, 1995. Comparison of heterotrophic bacterial production rates in early spring in the turbid estuaries of the Scheldt and the Elbe. Hydrobiologia 311 (Dev. Hydrobiol. 110): 31–42.

Iversen, N. & B. B. Jørgensen, 1993. Diffusion coefficients of sulphate and methane in marine sediments: influence of porosity. Geochim. Cosmochim. Ac. 57: 571–578.

Jensen, H. B., K. S. Jørgensen & J. Sørensen, 1984. Diurnal variation of nitrogen cycling in coastal, marine sediments. II Nitrous oxide emission. Mar. Biol. 83: 177–183.

Kemp, W. M., P. Sampou, J. Caffrey, M. Mayer, K. Henriksen & W. R. Boynton, 1990. Ammonium recycling versus denitrification in Chesapeake Bay sediments. Limnol. Oceanogr. 35: 1543–1563.

Kieskamp, W. M., L. Lohse, E. Epping & W. Helder, 1991. Seasonal variation in denitrification rates and nitrous oxide fluxes in intertidal sediments of the western Wadden Sea. Mar. Ecol. Progr. Ser. 72: 145–151.

Klump, J. V. & C. S. Martens, 1989. The seasonality of nutrient regeneration in an organic-rich coastal sediment: Kinetic modelling of changing pore-water nutrient and sulfate distributions. Limnol. Oceanogr. 34: 559–577.

Leuenberger, M. & U. Siegenthaler, 1992. Ice-age atmospheric concentration of nitrous oxide from an Antarctic ice core. Nature 360: 449–451.

Li, Y.-H. & S. Gregory, 1974. Diffusion of ions in seawater and in deep-sea sediments. Geochim. Cosmochim. Acta 38: 703–714.

Martens, C. S. & J. P. Chanton. 1989. Radon as a tracer of biogenic gas equilibration and transport from methane-saturated sediments. J. Geophys. Res. 4: 3451–3459.

Matson, P. A. & P. M. Vitousek, 1987. Cross-system comparison of soil nitrogen transformations and nitrous oxide flux in tropical forest ecosystems. Global Biogeoch. Cycles 1: 163–170.

Matson, P. A. & P. M. Vitousek, 1990. Ecosystem approach to a global nitrous oxide budget. Bioscience 40: 667–672.

Middelburg, J. J., G. Klaver, J. Nieuwenhuize & T. Vlug, 1995. Carbon and nitrogen cycling in intertidal sediments near Doel, Scheldt Estuary. Hydrobiologia 311 (Dev. Hydrobiol. 110): 57–69.

Miller, L. G., Oremland, R. S. & S. Paulsen, 1986. Measurement of nitrous oxide reductase activity in aquatic sediments. Appl. envir. Microb. 51: 18–24.

Nieuwenhuize, J., Y. E. M. Maas & J. J. Middelburg, 1994. Rapid analysis of organic carbon and nitrogen in marine particles. Mar. Chem. 45: 217–224.

Peeters, J. C. H. & L. Peperzak, 1990. Nutrient limitation in the North Sea: A bioassay approach. Neth. J. Sea Res. 26: 61–73.

Schlesinger, W. H., 1991. Biogeochemistry. An analysis of global change. Academic Press Inc.

Seitzinger, S., 1988. Denitrification in freshwater and coastal marine ecosystems: ecological and geochemical significance. Limnol. Oceanogr. 33: 702–724.

Smith, C. J., R. D. deLaune & W. H. Patrick, Jr., 1983. Nitrous oxide emission from Gulf Coast wetlands. Geochim. Cosmochim. Acta 47: 1805–1814.

Soetaert, K & P. M. J. Herman, 1995. Nitrogen dynamics in the Westerschelde estuary (SW Netherlands) estimated by means of the ecosystem model MOSES. Hydrobiologia 311 (Dev. Hydrobiol. 110): 225–246.

Tiedje, J. M., 1988. Ecology of denitrification and dissimilatory nitrate reduction to ammonium. In: Zehnder, A. (ed.) Biology of anaerobic microorganisms. Wiley, New York: 179–244.

Van Eck, G. T. M., N. de Pauw, M. van den Langenbergh & G. Verreet, 1991. Emissies, gehalten, gedrag en effecten van (micro)verontreinigingen in het stroomgebied van de Schelde en Schelde-estuarium. Water 60: 164–181 (in Dutch).

Van Raaphorst, W., H. T. Kloosterhuis, E. M. Berghuis, A. J. M. Gieles, J. F. P. Malschaert & G. J. Van Noort, 1992. Nitrogen cycling in two types of sediments of the southern North Sea (Frisian front, broad fourteens): Field data and mesocosm results. Neth. J. Sea Res. 28: 293–316.

Watson, P. G., P. E. Frickers & C. M. Goodchild, 1985. Spatial and seasonal variations in the chemistry of sediment interstitial waters in the Tamar estuary. Estuar. coast. Shelf. Sci. 21: 105–119.

Hydrobiologia **311**: 57–69, 1995.
C. H. R. Heip & P. M. J. Herman (eds), *Major Biological Processes in European Tidal Estuaries.*

Carbon and nitrogen cycling in intertidal sediments near Doel, Scheldt Estuary

Jack J. Middelburg[1], Gerard Klaver[2], Joop Nieuwenhuize[1] & Tom Vlug[1]
[1]*Netherlands Institute of Ecology, Centre for Estuarine and Coastal Ecology, Vierstraat 28, NL-4401 EA Yerseke, The Netherlands*
[2]*Rijks Geologische Dienst, P.O. Box 157, NL-2000 AD Haarlem, The Netherlands*

Key words: Carbon cycling, nitrogen cycling, estuarine sediments, methane, nitrous oxide, mineralization, carbon dioxide

Abstract

Carbon and nitrogen cycling in intertidal mud flat sediments in the Scheldt Estuary was studied using measurements of carbon dioxide, methane and nitrous oxide emission rates and pore-water profiles of ΣCO_2, ammonium and nitrate. A comparison between chamber measured carbon dioxide fluxes and those based on ΣCO_2 pore-water gradients using Fick's First law indicates that apparent diffusion coefficients are 2 to 28 times higher than bulk sediment diffusion coefficients based on molecular diffusion. Seasonal changes in gaseous carbon fluxes or ΣCO_2 pore water concentrations cannot be used directly, or in a simple way, to determine seasonal rates of mineralization, because of marked seasonal changes in pore-water storage and exchange parameters.

The annual amount of carbon delivered to the sediment is 42 mol m^{-2}, of which about 42% becomes buried, the remaining being emitted as methane (7%) or carbon dioxide (50%). Each year about 2.6 mol N m^{-2} of particulate nitrogen reaches the sediment; 1.1 mol m^{-2} is buried and 1.6 mol m^{-2} is mineralized to ammonium. Only 0.42 mol m^{-2} yr^{-1} of the ammonium produced escapes from the sediments, the remaining being first nitrified (1.2 mol m^{-2} yr^{-1}) and then denitrified (1.7 mol m^{-2} yr^{-1}). Simple calculations indicate that intertidal sediments may account for about 14% and 30% of the total estuarine retention of nitrogen and carbon, respectively.

Introduction

Depending on net organic metabolism coastal systems may be a sink (autotrophy) or source (heterotrophy) of atmospheric carbon dioxide (Wollast, 1991; Smith & Hollibaugh, 1993). The balance between the transfer of nutrients and organic carbon from terrestrial systems to the coastal ocean determines whether coastal environments are net autotroph (Billen *et al.*, 1991) or heterotroph (Smith & Hollibaugh, 1993).

Human activities have led both to increased nutrient fluxes and increased organic carbon loadings. On the one hand, euthrophication can lead to enhanced new production, hence increased preservation of organic carbon in sediments and uptake of atmospheric carbon dioxide. On the other hand, enhanced carbon loadings may results in net mineralization of organic matter, hence emission of carbon dioxide, and possibly methane. In order to study human impact on coastal systems it is therefore important to quantify riverine fluxes of nutrients and organic carbon and their modification in estuarine environments.

There are two possibilities for estuaries to reduce the transfer of riverine material to the coastal ocean: by burial or by gaseous emission. Sediment burial constitutes an important removal pathway for suspended matter, particulate organic carbon, and dissolved components following fixation, sorption or precipitation (Berner & Berner, 1987). Particulate and dissolved phases can be removed from estuaries by gas exchange following biogeochemical transformations such as mineralization of particulate organic matter to carbon dioxide and denitrification of nitrate to dinitrogen and nitrous oxide (Billen *et al.*, 1991; Seitzinger, 1988).

This capacity of estuaries to retain riverine material is subject to seasonal changes in river discharge and temperature. In this study we report on the seasonality of carbon and nitrogen recycling in intertidal mud flat sediments near Doel in the Scheldt Estuary. Contemporaneous measurements of gaseous carbon dioxide, methane and nitrous oxide fluxes, as well as pore-water profiles of ΣCO_2 and ammonium are presented. The results are interpreted using mass-balance approaches to address the temporal variability of organic matter mineralization and nitrogen cycling rates. These data will also be used to derive an annual budget of organic carbon and nitrogen in estuarine intertidal sediments.

Material and methods

Study site

The river Scheldt with a total length of 330 km flows through France, Belgium and the Netherlands before it debouches into the North Sea. Along its course, the Scheldt is heavily polluted by the discharge of important cities, active industrial areas and intensive stock-farming. In the Scheldt estuary tidal exchange (about 100000 m^3 sec^{-1}) is more important than fresh-water discharge (about 100 m^3 sec^{-1}). Vertical salinity stratification is therefore either absent (downstream) or minor (upstream). The intertidal mud flat near Doel is located midestuary, just south of the Dutch-Belgian border (Fig. 1). In this part of the estuary the tidal range is about 4.5 m and surface water chlorinities vary from 2 to 8 g l^{-1}. Using ^{210}Pb isotope data, Wartel (1993) estimated an accumulation rate of 0.8±0.2 cm yr^{-1} for sediments at the intertidal mud flat near Doel. The porosity of the sediments ranges from 0.5 to 0.8, without any significant depth trend. Silt-clay sized particles (<63 μ) constitute about 80% of the sediment. All flux measurements were made in a restricted area (about 10 m^2), that is situated high in the intertidal zone and being exposed to air at least 14 hours per day.

Gaseous flux measurements

Over a sampling period from September 1990 through December 1991 replicate measurements of carbon dioxide, methane and nitrous oxide fluxes were made at low tide. The release of these gases was measured by monitoring accumulation of the gas beneath chambers placed over the sediment surface. The chambers have a 0.11 m^2 base, a volume of 45 l and are made of non-transparent polypropylene to exclude any phototrophic activity. When necessary, a sun shield was placed over the chamber to maintain the chamber temperature and relative humidity within 25% of their ambient values.

Carbon dioxide, methane and nitrous oxide, concentrations were measured by circulating chamber air through Teflon tubes between chamber and the gas monitor. The measurement principle of the multi-gas monitor used, a Brüel & Kjaer type 1302, is based on the photoacoustic infra-red detection method. Briefly, after thorough flushing of the Teflon tubes and analysis cell, the air sample is hermetically sealed in the analysis cell. The light emitted by a pulsating infra-red light source and purified by a narrow-band optical filter is selectively adsorbed. The temperature of the gas increases and decreases in response to the pulsating light transmitter, and this causes an equivalent increase and decrease of the pressure of the gas in the closed cell. Two ultrasensitive microphones mounted in the cell are used to measure this pressure wave, which is directly proportional to the concentration of the gas. Various gases can be measured in the same sample using different filters. The response time for a sequential measurement of nitrous oxide, carbon dioxide and methane is about 90 sec. Detection limits for carbon dioxide, methane and nitroux oxide are 3, 0.1 and 0.05 ppmv, respectively, and their response is linear upto several thousands of ppmv for carbon dioxide and methane, and 5 ppmv for nitrous oxide. The reproducibilities at ambient levels are about 1, 10 and 15%, respectively. Fluxes are calculated by regression analysis from the recorded change in concentration over time; fluxes with a significance of $p>0.05$ are considered to be nihil. Fluxes significant at $p<0.01$ are normally obtained as low as 0.05 mol CH_4, 1.2 mol CO_2 m^{-2} yr^{-1} and 5 mmol N_2O m^{-2} yr^{-1}, within 30 minutes.

Sediment and pore-water data

Pore-water samples from intertidal muds were obtained by pressure filtration of sediments following the collection of cores by hand at low tide. Coinciding sections of three replicate cores were combined to reduce spatial heterogeneity and treated as one sample. Measurements of pore-water ammonium and nitrate concentrations were made using standard colorimetric methods on a Skalar autoanalyzer. ΣCO_2 was deter-

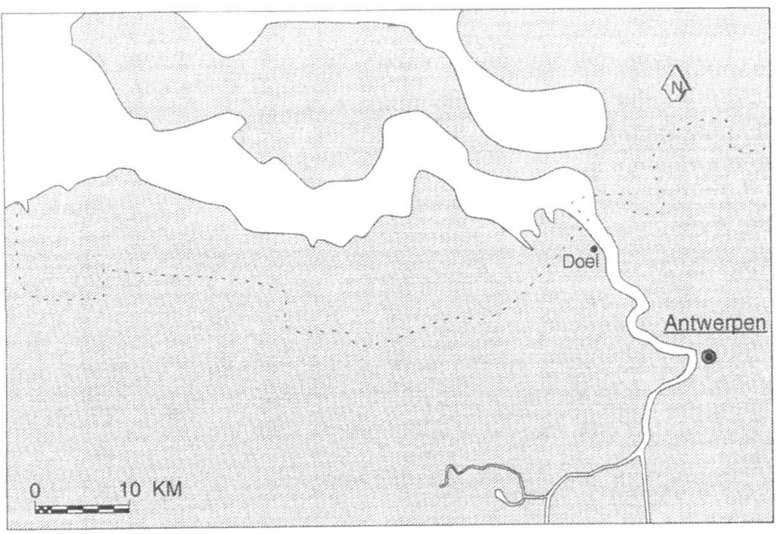

Fig. 1. Map of the Scheldt Estuary showing sampling location.

mined by potentiometric titration. The organic carbon and nitrogen contents of the sediments were determined using a Carlo-Erba CN-analyzer following a recently developed *in situ* HCl acidification prodecure (Nieuwenhuize *et al.*, 1994).

Results

Gas flux measurements

Fluxes of carbon dioxide, methane and nitrous oxide from sediments to the atmosphere during low tide are presented in Fig. 2. Replicate carbon dioxide flux measurements show less variability than those of nitrous oxide and in particular methane. Moreover, carbon dioxide fluxes seem to track closely temperature (Fig. 2), whereas methane and nitroux oxide fluxes do not. This high variability in methane and nitrous oxide fluxes is a common observation in studies of biogenic trace gas emissions (e.g. Bouwman, 1990) and results from the complexity and heterogeneity of processes controlling their production, consumption and transport. Annual emission rates calculated by integration under the curves connecting averages of replicate flux measurements are 21.1±4.7 and 3.1±1.5 mol C m^{-2} and 6.6±2.8 mmol N m^{-2} for carbon dioxide, methane and nitrous oxide respectively.

Pore water data

Depth profiles of total dissolved inorganic carbon and ammonium followed normal trends of increasing concentrations with depth in sediments (Figs 3 and 4). Dissolved nitrate concentrations are only measurable in the top few centimeters of the sediments and are at least one order of magnitude lower than interstitial water ammonium levels. Concentration versus depth profiles of ΣCO_2 and ammonium were fit to the equation:

$$C_z = (C_0 - C_\infty)\exp(-bz) + C_\infty, \qquad (1)$$

where z is the depth below the sediment–water interface (m), C is the concentration (Mm^{-3}), C_0 is the concentration at the sediment–water interface (Mm^{-3}), C_∞ is the asymptotic concentration at depth (Mm^{-3}) and b is an attenuation coefficient (m-1) equal to k/ω, where k is a first order rate constant (yr^{-1}) and ω is the sediment accumulation rate (m yr^{-1}). The best fits presented in Figs 3 and 4 were obtained using a matrix minimization procedure in which C_∞ and b were allowed to float.

Discussion

Carbon dioxide exchange

During the last few decades the exchange of material between sediments and water or air has aroused major

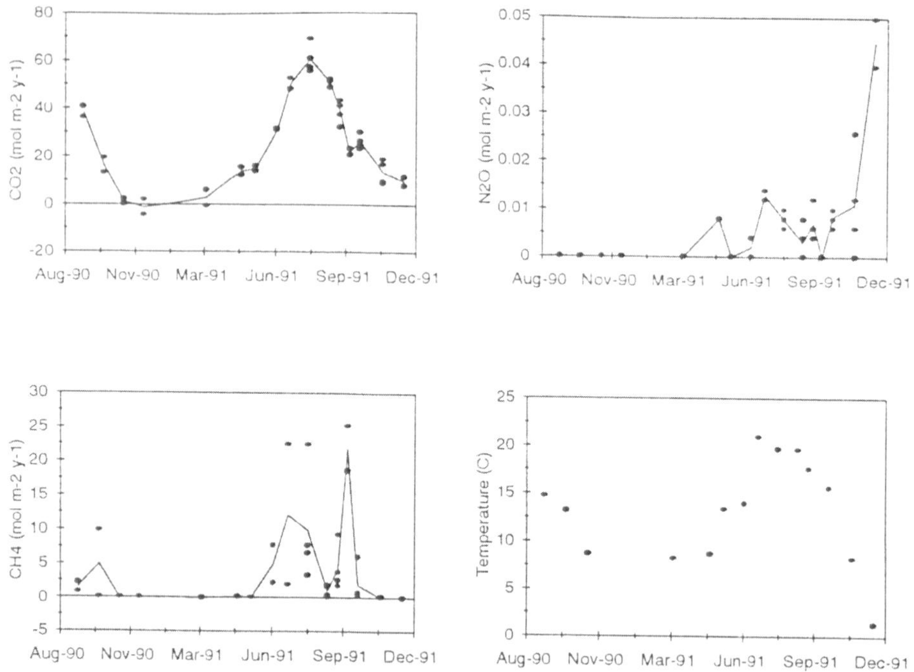

Fig. 2. Emission rates of carbon dioxide, methane and nitrous oxide and temperature versus date of sampling.

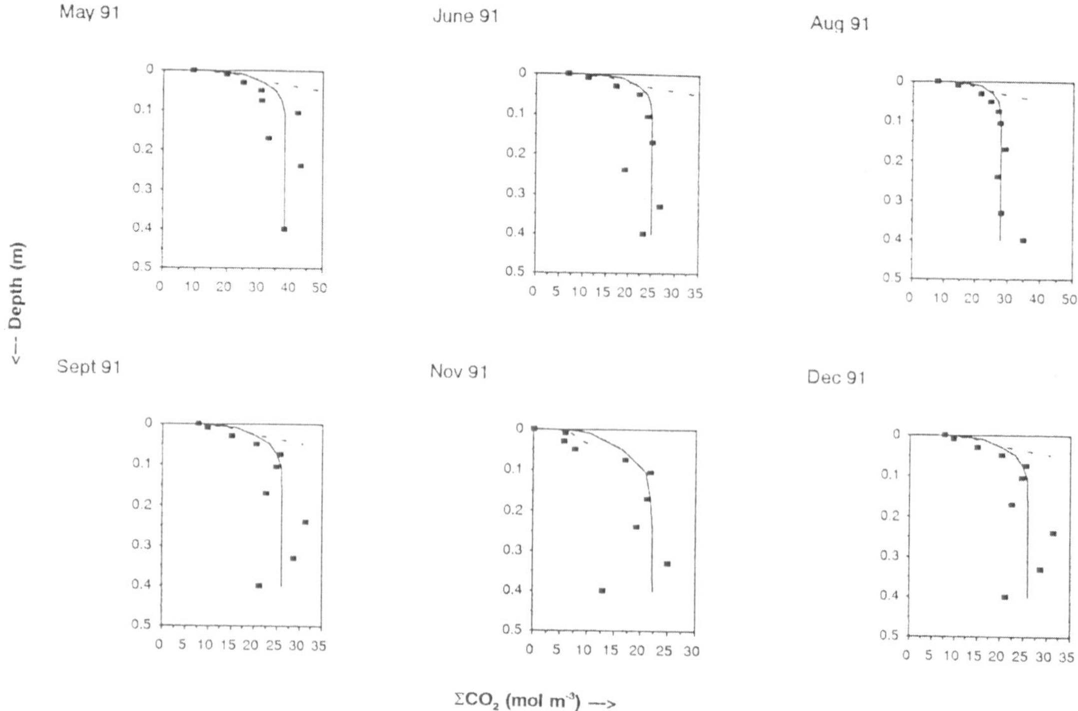

Fig. 3. Interstitial water ΣCO_2 (mol m^{-3}) versus depth (m). The solid and dashed lines represent the best fit to equation 1 and linear regression of the upper few data points respectively.

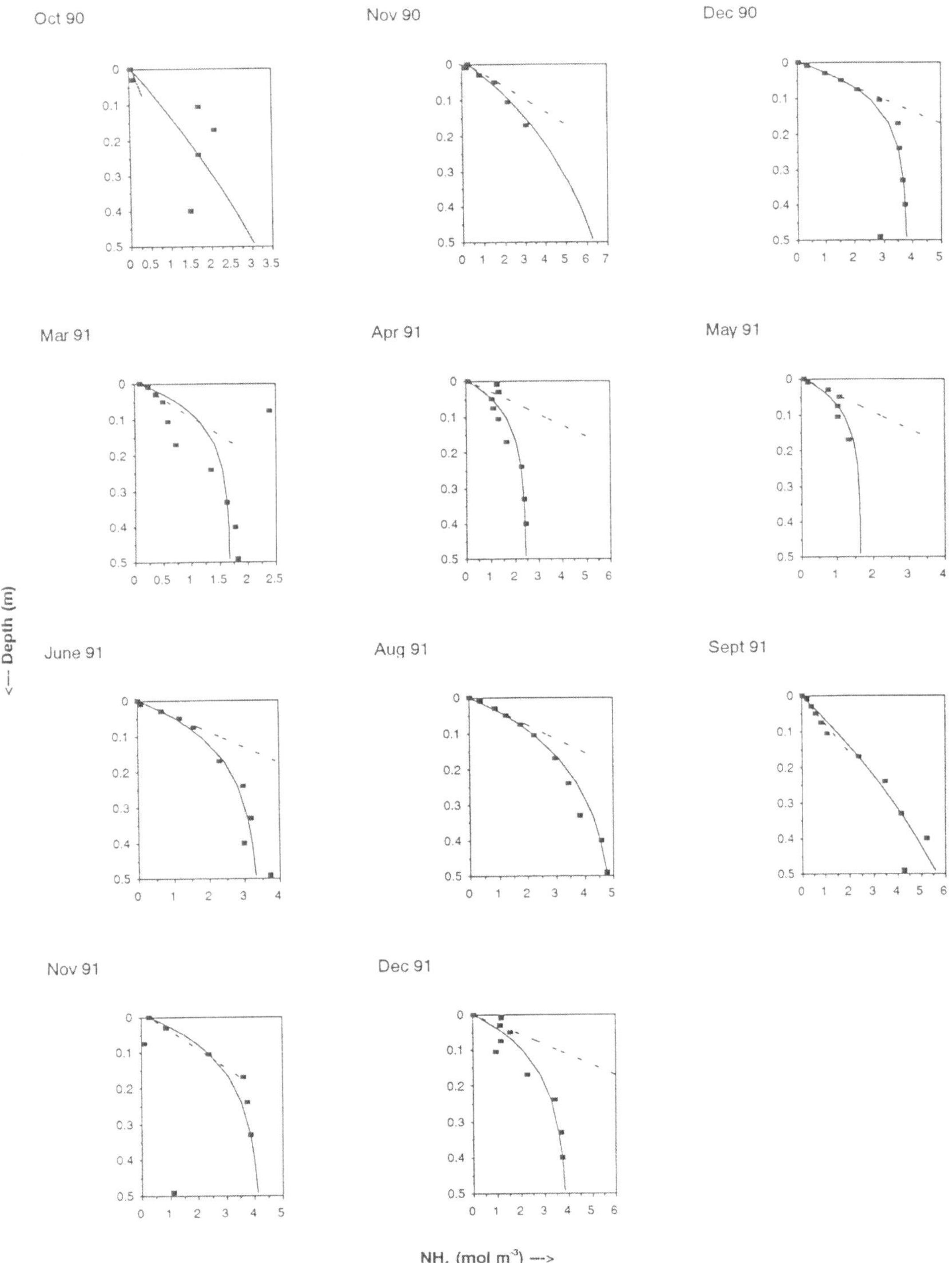

Fig. 4. Interstitial water ammonium (mol m^{-3}) versus depth (m). The solid and dashed lines represent the best fit to equation 1 and linear regression of the upper few data points respectively.

interest because of its importance for biogeochemical cycles. Fluxes from aquatic sediments can be determined either by direct *in situ* measurements using benthic chambers, or by flux calculations applying Fick's first law of diffusion. The data presented in this study allow us to apply both methods. Flux calculations will be based on Fick's first law of diffusion modified for sediments (Li & Gregory, 1974; Berner, 1980):

$$J = \phi D_S \left(\frac{\delta C}{\delta Z}\right)_{z=0}, \qquad (2)$$

where J is the flux from the sediment (mol m^{-2} yr^{-2}), ϕ is the porosity of the sediment (0.6), D_s is the whole sediment diffusion coefficient (m^2 yr^{-1}) and $(\delta C/\delta z)z = 0$ is the concentration gradient at zero depth. The whole sediment diffusion coefficient (D_s) is calculated from the formula (Iversen & Jørgensen, 1993)

$$D_S = \frac{D_0}{1 + n(1-\phi)}, \qquad (3)$$

where D_0 is the free-solution diffusion coefficient, and $n = 3$. The free-solution diffusion coefficient for HCO_3^- is obtained from Li and Gregory (1974) after correcting to observed temperatures using the Stokes-Einstein relationship. ΣCO_2 is assumed to diffuse as HCO_3^- since this is by far the dominant species in estuarine sediments. The pore-water concentration gradient at depth zero was estimated in two ways, namely by linear regression of the upper few data points versus depth (regression gradient), and from the first derivative of equation 1 at depth zero (fit gradient):

$$\frac{\delta C}{\delta Z} = -b(C_0 - C_\infty)\exp(-bz). \qquad (4)$$

ΣCO_2 fluxes based on equation (2) and either a regression gradient or fit gradient are significantly smaller than those measured in situ at low tide (Table 1).

This difference can be related to a number of factors. Firstly, due to inadequate sampling resolution near the sediment surface pore-water gradients may be underestimated, in particular those based on linear regression. Secondly, net mineralization processes at the sediment surface (i.e. mineralization minus carbon fixation) are recorded by the carbon dioxide flux measurement technique, but do not show up in the pore water composition. Thirdly, molecular diffusion in the form of HCO_3^- may not be the only transport mode of ΣCO_2. Fourthly, estimates based on linear regression can be inaccurate since they are very sensitive to the proper choice of ΣCO_2 bottom water concentrations, which are not defined during exposure. Fifthly, the pore-water profile may not have been adapted to air exposure at the time of measurement.

Under the assumption of no net mineralization at the sediment surface, a diffusive enhancement factor (E) can be calculated from the ratio between observed and predicted fluxes:

$$E = \frac{J_{\text{obs}}}{\phi D_S \frac{\delta C}{\delta Z}}. \qquad (5)$$

Calculated diffusion enhancement factors vary from 2 to 16 for gradients based on equation 4 and 4 to 28 for gradients obtained by linear regression (Table 1). These empirical diffusion enhancement factors include: (1) the effects of movement of mobile fauna and multidimensional diffusive geometry related to presence of burrows (e.g. Berner, 1980, Aller, 1980; Rutgers van der Loef *et al.*, 1984; Helder & Andersen, 1987), (2) enhanced diffusion associated with methane ebullition and the presence of bubble tubes (Martens & Klump, 1980), (3) the effects of physical disturbance by waves and tidal currents (VanderBorght *et al.*, 1977a, b) and (4) any effect due to the change from solution to the gas phase. Diffusive exchange may be enhanced by bubble structures, bioturbation and physical disturbances with factors upto 3 (Martens & Klump, 1980), 10 (Rutgers van der Loef *et al.*, 1984; Helder & Anderson, 1987) and 100 (VanderBorght *et al.*, 1977a, b) respectively.

Seasonal carbon cycling

Mineralization of organic matter in estuarine and marine sediments proceeds through the terminal electron acceptor sequence of oxygen, nitrate, MnO_2, FeOOH, sulfate and carbon dioxide. Benthic mineralization rates have traditionally been measured as oxygen uptake by the sediments. This oxygen uptake includes the oxygen consumption due to aerobic organic matter decomposition, benthic fauna respiration and oxidation of reduced components such as ammonium, methane and sulfide generated during anaerobic mineralization of organic matter (e.g. Mackin & Swider, 1989; Di Toro *et al.*, 1990). Since the majority of organic matter mineralization in coastal environments occurs anaerobically (Martens & Klump, 1984, Mackin & Swider, 1989; Middelburg *et al.*, 1993), oxygen consumption rates can be a quantitative measure of organ-

Table 1. Carbon stocks, fluxes and mineralization rates.

CO_2	May 91	June 91	Aug 91	Sept 91	Nov 91	Dec 91
C_0 (mol/m^3)[a]	3.20	3.20	3.20	3.20	3.20	3.20
C_∞ (mol/m^3)[a]	37.90	25.20	27.94	25.95	22.32	23.55
b (m^{-1})[a]	32.98	40.98	46.12	28.18	15.55	7.76
J_{obs} (mol/m^2 y)[b]	14.73	31.26	50.67	30.14	13.29	9.76
J_{fit} (mol/m^2 y)[c]	6.76	5.41	7.80	4.22	1.50	0.62
$J_{regr.}$ (mol/m^2 y)[d]	3.68	2.50	3.42	2.41	0.47	0.43
E_{fit}[e]	2.18	5.78	6.50	7.14	8.85	15.75
E_{regr}[f]	4.01	12.49	14.83	12.50	28.41	22.59
I (mol/m^2)[g]	10.51	7.09	7.89	7.14	5.83	5.38
ΔI (mol/m^2 y)[h]		−41.11	4.83	−8.97	−7.91	−5.29
$\sum R_{mol}$ (mol/m^2 y)[i]	11.54	9.19	13.19	7.22	2.66	1.17
$\sum R_{enh}$ (mol/m^2 y)[j]	24.83	52.27	84.65	50.41	22.29	16.06
production (mol/m^2 y)[k]		−15.64	48.75	38.48	20.39	6.27

[a] Fit parameters (equation 1)
[b] Measured CO_2 flux
[c] Flux calculated from fit (equations 2 and 4)
[d] Flux calculated from linear regression (equation 2)
[e] Enhancement factor based on J_{fit} (equation 5)
[f] Enhancement factor based on $J_{regression}$ (equation 5)
[g] Inventory per month (equation 6)
[h] Inventory change (equation 7)
[i] Depth integrated rate based on molecular diffusion (equation 10)
[j] Depth integrated rate based on enhanced exchange (equation 10)
[k] Production based on gaseous carbon fluxes and inventory changes

ic matter mineralization rates only, if re-oxidation of reduced nitrogen, manganese, iron, sulfur and carbon compounds is complete. The presence of substantial amounts of solid-phase reduced sulfur (e.g pyrite) in most coastal sediments and significant methane and dinitrogen fluxes from estuarine sediments, indicates that this is generally not the case. Moreover, this re-oxidation efficiency depends on a variety of factors (Chanton *et al.*, 1987; Middelburg, 1991) and may vary seasonally (e.g. Roden & Tuttle, 1993).

An alternative method to determine benthic mineralization rates is the measurement of ΣCO_2 production rates. These may, however, deviate from true mineralization rates if there is significant dissolution or precipitation of carbonate minerals, assimilation of carbon dioxide or emission of methane from the sediment (e.g. Anderson *et al.*, 1986). ΣCO_2 production rates can be obtained from the release of carbon dioxide from the sediment (Hargrave & Philips, 1981; Anderson *et al.*, 1986), from modelling ΣCO_2 pore-water profiles (McNichol *et al.*, 1988) or from anoxic incubation experiments (Mackin & Swider, 1989). Carbon mineralization rates in the intertidal sediments near Doel will be based on the first two methods and combinations thereof, under the assumption that carbonate mineral dissolution and precipitation, and carbon dioxide assimilation can be neglected. The latter assumption is reasonable while using dark chambers for flux measurements.

If steady-state conditions apply to a sediment system, then the sum of all reactions in the sediment should be balanced by the flux out of the sediment and the carbon dioxide flux method yields a reliable organic matter mineralization rate. However, if production rates exceed emission rates, there is build up in solution of ΣCO_2. Conversely, if emission rates are higher than production rates, the pore-water inventory will decrease. Under such non-steady-state conditions carbon dioxide fluxes will reflect ΣCO_2 production rates only if corrections are applied for storage changes.

The pore-water inventory (I) of ΣCO_2 has been calculated using

$$I = \phi\left[\frac{(C_0 - C_\infty)}{-b}[\exp(-0.5b) - 1] + 0.5C_\infty\right], \quad (6)$$

which is obtained by integrating equation 1 to a depth of 0.5 m. ΣCO_2 inventories range from 5.4 to 10.5 mol

m^{-2} (Table 1). Storage changes between two sampling intervals can be calculated by dividing the inventory change with the time period between two successive samplings:

$$\Delta I = \frac{I_{\text{current}} - I_{\text{previous}}}{\Delta t}. \quad (7)$$

Increases and decreases in ΣCO_2 storage appear to be significant compared to carbon dioxide fluxes (Table 1). Organic matter mineralization rates based on the sum of gaseous carbon fluxes and storage changes of ΣCO_2 are presented in Table 1.

Organic matter mineralization rates have also been determined by modelling ΣCO_2 pore-water profiles. At steady-state the concentration of ΣCO_2 is determined by diffusive and advective transport and net production or consumption (*R*) (Berner, 1980):

$$D_S \frac{\delta^2 C}{\delta Z^2} - \omega \frac{\delta C}{\delta Z} + R = 0, \quad (8)$$

where ω is the sediment accumulation rate or net water flow.

By combining the first and second derivative of equation 1 with equation 8, the following expression for the nett rate of production (*R*) is obtained:

$$R(z) = (C_0 - C_\infty)[Db^2 + \omega b]\exp(-bz). \quad (9)$$

Depth integrated ΣCO_2 dioxide production rates can then be calculated from:

$$\sum R = (C_0 - C_\infty)(Db + \omega)[1 - \exp(-0.5b)]. \quad (10)$$

Depth integrated mineralization based on ΣCO_2 profiles and molecular diffusion (D_s) and enhanced diffusion ($E \times D_s$) are listed in Table 1. On the basis of attenuation coefficients reported in Table 1, we estimate that more than 98% of the depth integrated mineralization occurs in the top 0.5 m.

Organic matter mineralization rates based on modelling pore-water ΣCO_2 profiles using both molecular and enhanced diffusion, gaseous carbon fluxes and gaseous carbon fluxes together with storage changes are compared in Fig. 5. Mineralization rates based on gaseous carbon fluxes are in general intermediate between those based on pore-water profiles. Mineralization rates based on pore-water ΣCO_2 profiles may underestimate sedimentary mineralization rates if substantial decomposition occurs at the sediment surface and if methane emission rates contribute significantly

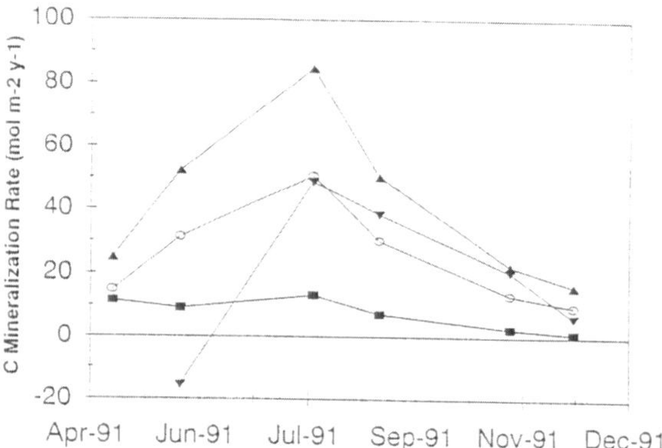

Fig. 5. Comparison between carbon mineralization rates based on gaseous carbon fluxes (circles), the sum of gaseous carbon fluxes and inventory changes (inverted solid triangle), modelling pore-water profiles using molecular diffusivities (solid square) or enhanced exchange (solid triangle).

to the total carbon flux out of the sediment. Moreover, depth integrated mineralization rates based on molecular diffusion represent a lower estimate since transport processes are underestimated, whereas those based on enhanced diffusion represent maximum values since enhanced diffusion strongly decreases with depth. Mineralization rates based on gaseous carbon dioxide and methane fluxes may under or overestimate mineralization rates depending on the magnitude of storage changes. Theoretically, the combined gaseous flux and storage change method should provide the best estimate, but the difference in time scale between measurements of gaseous fluxes (about 30 minutes) and net storage changes (one or two months) introduces uncertainties. Negative values may occur if carbon dioxide fluxes in between sampling intervals are higher than those during sampling. Accordingly, it appears that the determination of seasonal mineralization rates using changes in inorganic carbon pools or fluxes is complicated by seasonal changes in exchange parameters and pore-water storage.

Carbon mineralization rates based on inorganic carbon fluxes can also be compared to those based on sulfate reduction rate measurements made at the nearby intertidal Ballastplaat (Panutrakul, 1993). Using radiotracer incubations, she obtained depth integrated sulfate reduction rates of 12.4, 80.9 and 11.5 mmol m^{-2} d^{-1} in April, July and December 1991, respectively. These sulfate reduction rates relate to carbon consumption rates of 9.1, 59.1 and 8.4 mol C m^{-2} yr^{-1} respectively, very similar to the actually measured carbon dioxide fluxes at the intertidal flat near Doel: 13.7±2.3, 60.6±6.0 and 9.8±2.5 mol C m^{-2}

Table 2. Nitrogen cycling.

	Oct 90	Nov 90	Dec 90	Mar 91	Apr 91	May 91	Jun 91	Aug 91	Sep 91	Nov 91	Dec 91	$\int(\text{var})_{\text{year}}$
C_0 (mol/m^3)[a]	0.04	0.28	0.03	0.09	0.09	0.09	0.01	0.00	0.00	0.28	0.03	
C_∞ (mol/m^3)[a]	8.05	8.25	3.80	1.68	2.45	1.65	3.42	5.11	11.25	4.22	3.96	
b (m^{-1})[a]	0.96	2.52	10.71	10.50	10.53	12.75	7.28	5.52	1.40	7.29	7.24	
$\sum R_{mol}$ (mol/m^2 y)[a]	0.12	0.37	0.66	0.39	0.57	0.45	0.59	0.74	0.28	0.59	0.48	
J_{fit} (mol/m^2 y)[c]	0.10	0.22	0.36	0.22	0.32	0.26	0.33	0.42	0.23	0.32	0.25	
J_{regr} (mol/m^2 y)[d]	0.03	0.25	0.28	0.11	0.36	0.27	0.30	0.39	0.18	0.21	0.31	
Production (mol/m^2 y)[e]	0.08	0.28	0.43	0.24	0.42	0.33	0.40	0.52	0.23	0.37	0.35	0.42
Ammonification (mol/m^2 y)[f]	1.34	0.07	−0.09	0.16	0.88	0.94	2.31	3.29	2.76	0.85	0.62	1.57
Nitrification (mol/m^2 y)[g]	1.26	−0.21	−0.52	−0.08	0.46	0.61	1.90	2.78	2.53	0.48	0.28	1.15
$J(NO_3)$ (mol/m^2 y)[h]	−0.31	−0.25	−0.27	−0.54	−0.58	−0.67	−0.54	−0.34	− 0.37	−0.25	−0.22	−0.50
$J(N_2O)$ (mol N/m^2 y)[i]	0.0000	0.0000	0.0000	0.0000	0.0160	0.0000	0.0040	0.0060	0.0070	0.0220	0.0000	0.0065
$J(N_2)$ (mol N/m^2 y)[j]	1.57	0.04	−0.25	0.45	1.03	1.28	2.44	3.11	2.89	0.72	0.49	1.65
Denitrification (mol/m^2 y)[k]	1.57	0.04	−0.25	0.45	1.04	1.28	2.44	3.12	2.90	0.73	0.49	1.65

[a] Fit parameters (equation 1)
[b] Depth integrated net ammonium production (equation 11)
[c] Ammonium flux calculated from fit (equation 2 and 4)
[d] Ammonium flux calculated from linear regression (equation 2)
[e] Net ammonium production (average of b,c,d)
[f] Calculated from gaseous carbon fluxes and C/N ratio
[g] Nitrification = ammonification - net ammonium production
[h] Nitrate flux (equation 2)
[i] Measured nitrous oxide flux
[j] Dinitrogen flux = denitrification - nitrous oxide flux
[k] Denitrification = nitrification + nitrate flux

yr^{-1} during the same months. This close correspondence both in magnitude and seasonality indicates not only that sulfate reduction is the major organic matter decomposition pathway, but also that gaseous carbon fluxes provide a reliable measure for organic carbon mineralization rates. Moreover, recent flux measurements starting immediately following air exposure till subsequent flooding revealed no systematic variation of carbon dioxide fluxes over the 9 hour period of air exposure. In other words, gaseous fluxes measured *in situ* at low tide may be extrapolated to longer time scales.

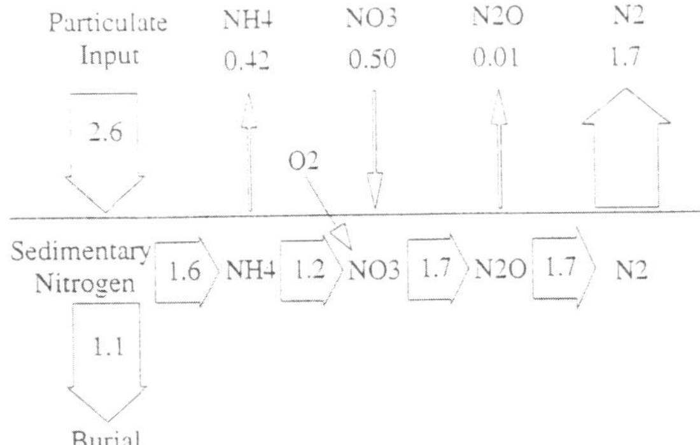

Fig. 6. Proposed nitrogen cycle in intertidal sediments near Doel, Scheldt Estuary. Nitrogen fluxes and transformation rates are given in mol N m^{-2} yr^{-1}.

Nitrogen cycling rates

Coastal sediments, and in particular estuarine sediments, have been suggested as an important sink in the nitrogen cycle (Billen *et al.*, 1985; Seitzinger, 1988; Devol, 1991; Devol & Christensen, 1993). Estuarine sediments can remove nitrogen either by burial in deeper sediment layers or by denitrification of nitrate to gaseous dinitrogen or nitrous oxide. The nitrate consumed during denitrification is either replaced by nitrate diffusing into the sediments from the bottom water, or produced within sediments by oxidation of ammonium (nitrification). The relative importance of bottom-water nitrate and nitrification as a nitrate source for denitrification depends on a number of factors (VanderBorght *et al.*, 1977b) and may vary seasonally (Kemp *et al.*, 1990; van Raaphorst *et al.*, 1992). The seasonality of nitrogen cycling in intertidal estuarine sediments will be modelled on the basis of mass-balance considerations using the nitrogen cycling scheme shown in Fig. 6. Our approach is based on depth-integrated rates, since this allows us to use measured fluxes and circumvents some of the problems related to the bioturbated and disturbed nature of the sediments.

The rate of ammonium production by mineralization of organic nitrogen compounds is calculated from the rate of organic carbon mineralization using a C/N molar ratio of 15.7, which is based on the composition of surficial sediments. The sum of carbon dioxide and methane fluxes are taken as a measure of organic carbon mineralization rates. Ammonium production rates calculated in this way may underestimate true ammonium production rates, if the organic matter being decomposed has a lower C/N ratio than sedimentary organic matter (Middelburg, 1991). Stoichiometric modelling of pore-water ΣCO_2 and ammonium concentrations to determine the C/N ratio of decomposing organic matter (Berner, 1980) yielded inconclusive results, probably due to the disturbed nature of the sediments.

The net rate of ammonification (i.e. ammonification + nitrate-ammonification – ammonium oxidation) can be estimated from pore-water profiles in two ways. One, the diffusive flux of ammonium is assumed to reflect net ammonification, i.e. changes in ammonium storage are neglected. The diffusive flux of ammonium has been calculated using equation 2 with both a gradient based on linear regression and a gradient obtained from equation 4 (Table 2). The diffusion coefficient of ammonium is taken from Li & Gregory (1974) and has been corrected for tortuosity and temperature following the approach outlined above. The second method to determine net ammonification rates is based on

$$\sum R = (C_0 - C_\infty)[Db + \omega(1+K)][1 - \exp(-0.5b)]. \tag{11}$$

This equation is similar to equation 10, but the exchange of dissolved ammonium with sedimentary solid phases is also taken into account. Net ammonification rates presented in Table 2 are based on a linear sorption coefficient (K) of 1.2 (Mackin & Aller, 1984; Oenema, 1988; Middelburg, 1991). Measured surface water ammonium concentration are used for C_0 (Table 2). Net ammonification rates estimated from equation 11 are somewhat higher than those based on ammonium fluxes, but otherwise very similar. Please notice the lack of clear seasonal cycle.

Net ammonification rates based on either method are always much lower than ammonification rates related to organic nitrogen mineralization (Table 2). This discrepancy indicates that most of the ammonium produced in the sediments is rapidly consumed by oxidation. Nitrification rates calculated from the difference in ammonification and the average of our three estimates of net ammonification vary from −0.5 to 2.8 mol m^{-2} yr^{-1} (Table 2). Negative nitrification rates are calculation artefacts related to our neglect of pore-water storage. Despite these high nitrification rates, pore-water nitrate concentrations are extremely low, which suggests high rates of denitrification.

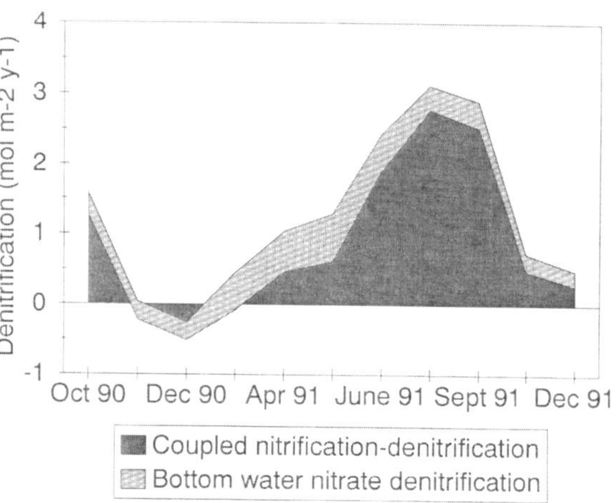

Fig. 7. Seasonal distribution of calculated denitrification rates. Nitrification supported and bottom-water nitrate supported denitrification rates are presented by black and hatched areas, respectively.

The rate of denitrification has been estimated as the sum of nitrification rates and nitrate fluxes into the sediment. The diffusive flux of nitrate has been calculated using Fick's First law (equation 2), measured bottom-water nitrate concentrations, a linear gradient and free solution diffusion coefficient reported by Li & Gregory (1974) that is corrected for tortuosity and temperature effects. Estimated denitrification rates show a pronounced seasonal pattern (Fig. 7). In April rates of denitrification are low (about 1.0 mol m^{-2} yr^{-1}) and about 40% of the nitrate consumed by denitrification is supported by nitrification, the remainder being supplied by diffusion from the overlying water. In August, rates of denitrification are high (3.1 mol m^{-2} yr^{-1}) and about 90% of the nitrate consumed has been produced by oxidation of ammonium (Table 2, Fig. 7).

In this mass-balance approach we have excluded nitrate-ammonification which may (Jørgensen & Sørensen, 1985) or may not (Rysgaard *et al.*, 1993) be important in estuarine sediments. If nitrate-ammonification is significant in these sediments, our mass-balance has led to an underestimation of nitrification rates, since any nitrate reduced to ammonium must be oxidized again. Estimated rates of denitrification represent the balance between ammonification and nitrate influx on the one hand and ammonium efflux on the other hand and are consequently not affected by nitrate ammonification.

Annual carbon and nitrogen budgets

Having established the seasonal rates of carbon and nitrogen cycling, annual budgets of carbon and nitrogen cycling can be obtained by integration. While seasonal changes in pore-water storage may complicate seasonal carbon and nitrogen budgets, these problems do not arise on an annual basis since there is no long-term storage of either ΣCO_2 or ammonium in the pore water. The net production of these compounds is therefore balanced by their fluxes out of the sediments.

The burial rate of organic carbon and nitrogen has been determined from

$$C_{\mathrm{buried}} = \rho_s \omega (1 - \phi) C_{\mathrm{org}}, \qquad (12a)$$

$$N_{\mathrm{buried}} = \rho_s \omega (1 - \phi) N, \qquad (12b)$$

where C_{buried} and N_{buried} represent the organic carbon and nitrogen burial rates (mol m^{-2} yr^{-1}), ρ is the dry density of the sediment (2550 kg m^{-3}), C_{org} is the organic carbon content (2.6 wt%) and N is the total nitrogen content (0.18 wt%). The annual rate of particulate organic carbon and nitrogen supply to the sediments can be calculated from the sum of their burial and annual integrated mineralization rates.

The annual carbon budget for intertidal estuarine sediments near Doel is shown in Fig. 8. Each year about 42±7.6 mol m^{-2} carbon is delivered to the sediment. About 42% (17.7±5.9 mol m^{-2} yr^{-1}) of the incoming carbon becomes buried, and vice versa, 58% (24.2±4.9 mol m^{-2} yr^{-1}) of the carbon supply is being mineralized. This annual carbon mineralization rate estimate is based on the sum of annual integrated gaseous carbon dioxide fluxes (21.2±4.7 mol m^{-2}) and methane emission rates (3.1±1.5 mol m^{-2}) and corresponds closely with the annual organic carbon mineralization rate through sulfate reduction at the nearby Ballasplaat intertidal station (18.8 mol C m^{-2}; Panutrakul, 1993). Only 13% of the carbon being mineralized leaves the sediment as methane. It should be realized that this does not imply that only 13% of the organic carbon is mineralized through methanogenesis, since the majority of the methane produced may have been oxidized before it could escape from the sediments.

The estimated annual nitrogen budget is presented in Fig. 6 and Table 2. Each year about 2.6 mol of particulate nitrogen per m^{-2} is delivered to the sediments. About 40% (1.1±0.3 mol m^{-2} yr^{-1}) of the incoming nitrogen becomes buried, the remaining 60% being mineralized to ammonium (1.6 mol m^{-2} yr^{-1}). About 27% (0.42 mol m^{-2} yr^{-1}) of the ammonium produced escapes from the sediment and 73% is being nitrified (1.2 mol m^{-2} yr^{-1}). Each year about 1.7 mol m^{-2} nitrate is denitrified of which 30% (0.50 mol m^{-2}) is being supplied by diffusion from the bottom water and 70% through nitrification. The majority of the nitrate denitrified escapes as dinitrogen, since the annual emission rate of nitrous oxide is only 0.006 mol N m^{-2}.

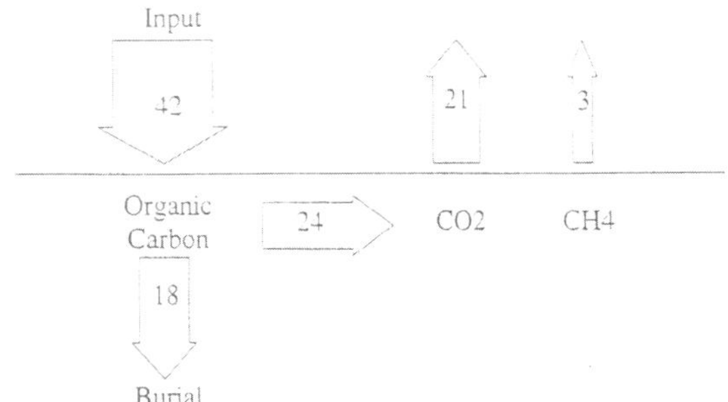

Fig. 8. Annual carbon budget of intertidal sediments near Doel, Scheldt Estuary. Carbon fluxes are given in mol C m^{-2} yr^{-1}.

Finally, the annual carbon and nitrogen budgets presented in Figs 6 and 8 can be used to estimate the contribution of intertidal sediments to the total estuarine retention of these components. Implicit in our approach is the assumption that intertidal flat sediments near Doel can be considered representative of the total area of intertidal sediments in the Scheldt estuary (59×10^6 m^2; Soetaert & Herman, 1995a). On the basis of an annual denitrification rate of about 23 g N m^{-2} and a N burial rate of 14 g N m^{-2}, it seems that each year about 1360 and 826 ton N are lost due to denitrification and burial in intertidal sediments, respectively. Accordingly, intertidal sediments account for the removal of 2200 ton N per year, which is about 14% of the total estimated estuarine loss of nitrogen (16000 ton N yr^{-1}; Soetaert & Herman, 1995a). The remainder of the nitrogen loss must take place in subtidal sediments, and more likely in the watercolumn (Billen *et al.*, 1985). Similar calculations for carbon indicate that intertidal sediments remove 12500 ton C yr^{-1} by burial and 17100 ton C yr^{-1} by remineralization to carbon dioxide and methane. Intertidal sediments therefore represent a net sink of about 29700 ton C yr^{-1}, which is about 30% of the total estimated estuarine retention (98000 ton C yr^{-1}; Soetaert & Herman, 1995b). These very simple calculations provide first-order estimates only, but they compare favourably with global ecosystem estimates (Soetaert & Herman, 1995a, b): 14 versus 30% and 30 versus 20% for nitrogen and carbon, respectively. Accordingly, intertidal sediments are a significant component in carbon and nitrogen cycles in estuarine ecosystems.

References

Aller, R. C., 1980. Diagenetic processes near the sediment-water interface of Long Island Sound. 1. Decomposition and nutrient element geochemistry (S,N,P). Adv. Geophys. 22: 238–350.

Anderson, L. G., P. O. J. Hall, M. M. L. van der Loeff, B. Sundby & S. F. G. Westerlund, 1986. Benthic respiration measured by total carbonate production. Limnol. Oceanogr. 31: 319–329.

Berner, E. K. & R. A. Berner, 1987. The Global Water Cycle, Chemistry and Environment. Prentice-Hall Inc, New Jersey.

Berner, R. A., 1980. Early diagenesis: A theoretical approach. Princeton University Press, Princeton: 1–241.

Billen, G., M. Somville, E. de Becker & P. Servais, 1985. A nitrogen budget of the Scheldt hydrographical basin. Netherl. J. Sea Res 19: 223–230.

Billen, G., C. Lancelot & M. Meybeck, 1991. N, P, and Si retention along the aquatic continuum from land to ocean. In Mantoura, R. F. C., J.-M. Martin & R. Wollast (eds), Ocean Margin Processes in Global Change. John Wiley & Sons Ltd; N, New York: 19–44.

Bouwman, A. E. (ed.) 1990. Soils and the greenhous effect. John Wiley & Sons.

Chanton, J. P., C. S. Martens & M. B. Goldhaber, 1987. Biogeochemical cycling in an organic-rich coastal marine basin. 7. Sulfur mass balance, oxygen uptake and sulfide retention. Geochim. Cosmochim. Acta 51: 1187–1199.

Devol, A. H., 1991. Direct measurement of nitrogen gas fluxes from continental shelf sediments. Nature 349: 319–321.

Devol, A. H. & J. P. Christensen, 1993. Benthic fluxes and nitrogen cycling in sediments of the continental margin of the eastern North Pacific. J. Mar. Res. 51: 345–372.

Di Toro, D. M., P. R. Paquin, K. Subburamu & D. A. Gruber, 1990. Sediment oxygen demand model: Methane and ammonia oxidation. J. envir. Eng. 116: 945–986.

Hargrave, B. T. & G. A. Philips, 1981. Annual in situ carbon dioxide and oxygen flux across a subtidal marine sediments. Estuar. coast mar. Shelf Sci. 12: 725–737.

Helder, W. & F. O. Andersen, 1987. An experimental approach to quantify biologically mediated dissolved silica transport at the sediment-water interface. Mar. Ecol. Prog. Ser. 39: 305–311.

Iversen, N. & B. B. Jørgensen, 1993. Diffusion coefficients of sulphate and methane in marine sediments: influence of porosity. Geochim. Cosmochim. Acta 57: 571–578.

Jørgensen, B. B. & J. Sørensen, 1985. Seasonal cycles of O_2, NO_3^-, and SO_4^{2-} reduction in estuarine sediments: the significance of a NO_3^- reduction maximum in spring. Mar. Ecol Progr. Ser. 48: 147–154.

Kemp, W. M., P. Sampou, J. Caffrey, M. Mayer, K. Henriksen & W. R. Boynton, 1990. Ammonium recycling versus denitrification in Chesapeake Bay sediments. Limnol. Oceanogr. 35: 1543–1563.

Li, Y.-H. & S. Gregory, 1974. Diffusion of ions in seawater and in deep-sea sediments. Geochim. Cosmochim. Acta. 38: 703–714.

Mackin, J. E. & R. C. Aller, 1984. Ammonium adsorption in marine sediments. Limnol. Oceanogr. 29: 250–257.

Mackin, J. E. & K. T. Swider, 1989. Organic matter decomposition pathways and oxygen consumption in coastal marine sediments. J. mar. Res. 47: 681–716.

Martens, C. S. & J. V. Klump, 1980. Biogeochemical cycling in an organic-rich coastal marine basin. 1. Methane sediment-water exchange processes. Geochim. Cosmochim. Acta 44: 471–490.

Martens, C. S. & J. V. Klump, 1984. Biochemical cycling in an organic-rich coastal marine basin 4. An organic carbon budget for sediments dominated by sulfate reduction and methanogenesis. Geochim. Cosmochim. Acta 48: 1987–2004.

McNichol, A. P., C. Lee & E. R. M. Druffel, 1988. Carbon cycling in coastal sediments: 1. A quantitative estimate of remineralization of organic carbon in the sediments of Buzzards Bay, MA. Geochim. Cosmochim. Acta 52: 1531–1543.

Middelburg, J. J., 1991. Organic carbon, sulphur and iron in recent semi-euxinic sediments in Kau Bay. Geochim. Cosmochim. Acta 55: 815–828.

Middelburg, J. J., T. Vlug & F. J. W. A. van der Nat, 1993. Organic matter mineralization in marine systems. Global Planetary Change 8: 47–58.

Nieuwenhuize, J., Y. E. M. Maas & J. J. Middelburg, 1994. Rapid analysis of organic carbon and nitrogen in marine particles. Mar. Chem., 45: 217–224.

Oenema, O., 1988. Early diagenesis in recent fine-grained sediments in the Eastern Scheldt. Ph. D. Thesis, Utrecht University, 222 pp.

Panutrakul, S., 1993. Impact of sulfur cycle on the mobilization of heavy metals in intertidal flat's sediments. Ph.D. thesis, Free Univ. of Brussels, 222 pp.

Roden, E. E. & J. H. Tuttle, 1993. Inorganic sulfur cycling in mid and lower Chesapeake Bay sediments. Mar. Ecol. Prog. Ser. 93: 101–118.

Rutgers van der Loeff, M. M., L. G. Anderson, P. O. Hall, A. Iverfeldt, A. B. Josefson, B. Sundby & S. F. G. Westerlund, 1984. The asphyxiation technique: An approach to distinguish between molecular diffusion and biologically mediated transport at the sediment-water interface. Limnol. Oceanogr. 29: 675–686.

Rysgaard, S., N. Risgaard-Petersen, L. P. Nielsen & N. P. Revsbech, 1993. Nitrification and denitrification in lake and estuarine sediments measured by the ^{15}N dilution technique and isotope paring. Appl. envir. Microbiol. 57: 2093–2098.

Seitzinger, S., 1988. Denitrification in freshwater and coastal marine ecosystems: ecological and geochemical significance. Limnol. Oceanogr. 33: 702–724.

Smith, S. V. & J. T. Hollibaugh, 1993. Coastal metabolism and the oceanic organic carbon balance. Rev. geophysics 31: 75–89.

Soetaert, K & P. M. J. Herman, 1995a. Nitrogen dynamics in the Westerschelde estuary (SW Netherlands) estimated by means of the ecosystem model MOSES. Hydrobiologia 311 (Dev. Hydrobiol. 110): 225–246.

Soetaert, K & P. M. J. Herman, 1995b. Carbon flows in the Westerschelde estuary (The Netherlands) evaluated by means of an ecosystem model (MOSES). Hydrobiologia 311 (Dev. Hydrobiol. 110): 247–266.

Vanderborght, J.-P., R. Wollast & G. Billen, 1977a. Kinetic model of diagenesis in disturbed sediments. Part 1. Mass transfer properties and silica diagenesis. Limnol. Oceanogr. 22: 787–793.

Vanderborght, J.-P., R. Wollast & G. Billen, 1977b. Kinetic models of diagenesis in disturbed sediments. Part 2. Nitrogen diagenesis. Limnol. Oceanogr. 22: 794–803.

Van Raaphorst, W., H. T. Kloosterhuis, E. M. Berghuis, A. J. M. Gieles, J. F. P. Malschaert & G. J. Van Noort, 1992. Nitrogen cycling in two types of sediments of the southern North

Sea (Frisian front, broad fourteens): Field data and mesocosm results. Neth. J. Sea Res. 28: 293–316.

Wartel, S, 1993. Mud layers and cyclic sedimentation patterns in the estuary of the Schelde (Belgium–The Netherlands). In Progress in Belgian Oceanographic Research.

Wollast, R., 1991. The Coastal Organic Carbon Cycle: Fluxes, Sources, and Sinks. In Mantoura, R. F. C., J.-M. Martin & R. Wollast (eds), Ocean Margin Processes in Global Change. John Wiley & Sons Ltd, New York: 365–381.

Hydrobiologia **311**: 71–83, 1995.
C. H. R. Heip & P. M. J. Herman (eds), Major Biological Processes in European Tidal Estuaries.

Copepod feeding in the Westerschelde and the Gironde

M. Tackx[1], X. Irigoien[2], N. Daro[1], J. Castel[2], L. Zhu[1], X. Zhang[3] & J. Nijs[1]
[1]*Ecology Laboratory, Free University of Brussels, Pleinlaan 2, B-1050 Brussels, Belgium*
[2]*Laboratory of Biological Oceanography, University of Bordeaux I, Rue du Professeur Jolyet 2F, F-33120 Arcachon, France*
[3]*Netherlands Institute for Ecological Research, Centre for Estuarine and Marine Ecology, Vierstraat 28, NL-4401 EA Yerseke, The Netherlands*
Present address: Department of Marine Sciences, College of Liberal Arts and Sciences, 1084 Shennecosset Road, Groton, Connecticut 06340-6097, USA

Key words: Eurytemora affinis, Acartia tonsa, estuaries, selectivity of feeding

Introduction

As in most European estuaries, the mesoplankton of the Westerschelde and the Gironde is dominated by calanoid copepods. In the Westerschelde, *Eurytemora affinis* dominates the brackish water area between station 7 and 10 in winter and spring while *Acartia tonsa* is dominant in summer (Fig. 1a). Maximum abundance of *A. tonsa* is located slightly more seawards than that of *E. affinis*. *Acartia bifilosa* occurs in winter and spring in lower abundances than *A. tonsa* and its maximum abundance is located between the brackish and the marine area (station 4–6) (Soetaert & Van Rijswijk, 1992).

In the Gironde, *E. affinis* dominates in winter and spring between station J and K, with its maximum around station K (Fig. 1b). The second important species in the Gironde is *Acartia bifilosa* with a maximum abundance in summer, located around station E and J. In the Gironde, *A. tonsa* occurs in the same area as *A. bifilosa* only in late summer and early autumn (Castel, 1985). For a detailed description of the distribution of zooplankton species in the Westerschelde and the Gironde, the reader is referred to Soetaert & Van Rijswijk (1992) and Castel (1985).

The role of the copepod populations in the foodweb of the Westerschelde and the Gironde was studied as part of the project entitled 'Major Biological Processes in European Tidal Estuaries'. Heinle & Flemer (1975) calculated that primary production in the Patuxent estuary (USA) is insufficient to sustain its abundant population of *E. affinis*. Heinle *et al.* (1977) demonstrated that *E. affinis* survives well on mixtures of detrital food with abundant microbiota and hypothesized that, in natural circumstances, *E. affinis* feeds to a substantial degree on detritus and associated microbiota. Although Heinle's hypothesis does not necessarily implicate unselective feeding, estuarine copepods have since been considered as being mainly detritivorous (Hummel *et al.*, 1988). Little attention has been given to feeding on other components such as e.g. live phytoplankton under natural conditions.

Copepods were long believed to be filter feeders, whose ingestion of a given food item was strongly dependant on the prevailing concentration of the item in the water. Presently, it is known that many calanoid copepod species can select certain (preferred) food items, and collect these from much larger volumes of water than would be possible by filter feeding (Paffenhöfer *et al.* 1982; Paffenhöfer & Van Sant, 1985; Price 1988; Paffenhöfer & Lewis, 1990). Prey detection by zooplankters can be by physical contact, remote mechanoreception, chemoreception or vision. Whether ingestion follows is the result of a sequence of complex mechanisms, involving pursuit, capture, rejection or ingestion (Price, 1988).

Taking this present knowledge on calanoid feeding strategies into account, feeding studies on estuarine

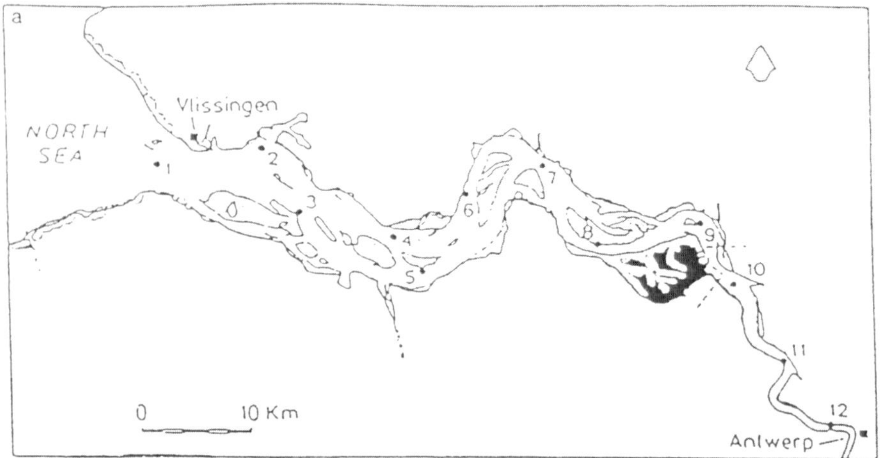

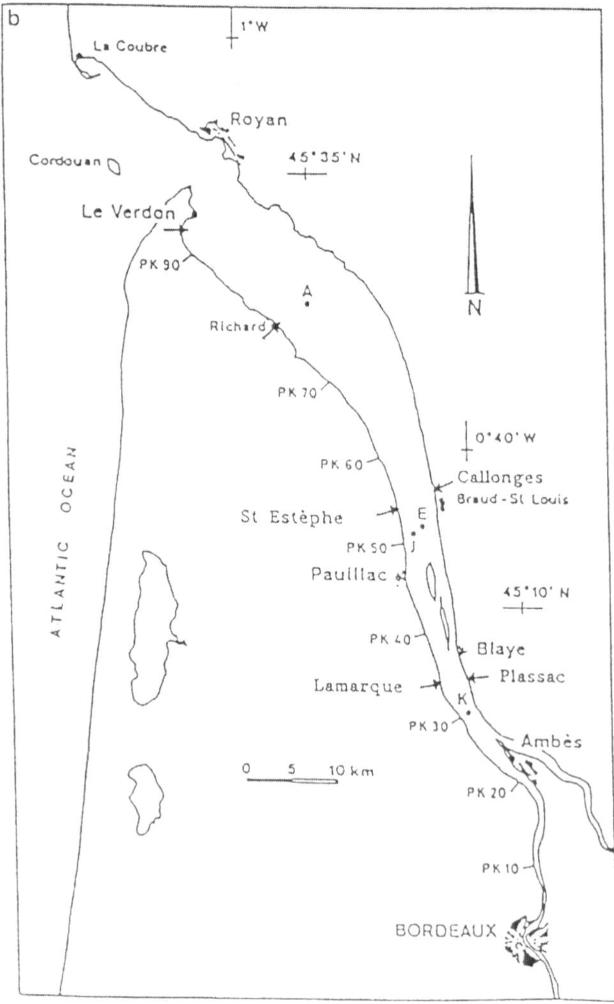

Fig. 1. Location of sampling stations in the Westerschelde (a) and the Gironde (b).

copepods should consider their predation impact on various particulate matter components separately, not only on the totality of particulate matter.

To outline the importance of this quantification to the modelling of their role in the carbon flow in the estuarine brackish water zone, let us consider the implications of unselective feeding on one hand and selective feeding on the other.

I. Feeding is totally unselective. The copepod populations feed on all available food items in proportion to their numerical abundance. In other words, clearance rate on all particles is the same. Consequently, the copepods feed mainly on detritus, a non limiting food source in these areas. The predation impact (measured as clearance rate) of the copepod population on its main food source, the detritus pool, is negligible. Because of the non selectivity of feeding, clearance rates on phytoplankton are also low (the same as on detritus), and likely to be unimportant in controlling phytoplankton populations. Such a feeding mode implies that copepods are fully adapted to the prevailing food composition, and that the predominance of detritus in their diet does not hamper development of the populations. Consequently carbon flow through the copepod population is a relatively simple function of the population's biomass and age structure, with little impact on the other compartments of the ecosystem. Copepod population development can then be considered mainly a function of physical conditions and its importance in the food chain lies mainly with its function as prey for higher trophic levels.

II. Copepods feed selectively on one or more of the following (preferred) particulate matter components: live phytoplankton, freshly formed phytoplankton detritus and microzooplankton. Bulk detritus is only eaten to the degree that it is unavoidably ingested, and/or as a supplemental food in case the preferred food components are too scarce to allow energetically beneficient feeding on them. The preferred food items can be collected in disproportion to their numerical abundance (relative to detritus) because of the animal's perceptive performance and oriented capture responses (Price, 1988; Paffenhöfer & Lewis, 1990). In this case predation by copepod populations could have an important impact on phytoplankton biomass in the brackish water zones of estuaries. Autochthonous primary production in most estuaries being low (Heinle *et al.*, 1975; CNEXO, 1977; Kromkamp, pers. comm.), it is likely that detritus indeed serves as complementary food. Roman (1984) showed that addition of detritus to a diet of *Thalassiosira weisflogii* decreased instantaneous mortality rates of *A. tonsa* and increased its growth rate. If feeding is to any extent selective however, modelling of the copepods role in the food chain requires answers to the following questions: which are the factors that determine the clearance rate exerted by the copepods on the various particulate matter components? Can these factors be quantified? Can the ratios between the clearance rates on various components be mathematically described as a function of these controlling factors? How important is the contribution of preferred food items to the development of the copepod population in addition to physical conditions and predation by higher trophic levels? What is the impact of predation by copepods on the population or stock of the preferred prey items?

If selectivity occurs, the relative abundance of the various components is likely to influence the ratio between clearance rates on the various components considerably. Following the optimal foraging theory (Emlen, 1966; review by Pyke *et al.*, 1977), any technique used to selectively capture and ingest a preferred food item is likely to be energetically profitable only down to a certain limit of relative abundance of this item to other potential food items. In copepod feeding studies, the dependence of total ingestion rate, size selectivity pattern and observed incipient limiting concentration on the concentration and size distribution of the particulate matter offered has been demonstrated in both laboratory and field experiments (Paffenhöfer, 1988). Price & Paffenhöfer (1986) observed reduced clearance value for *Eucalanus elongatus* feeding on low and high algal concentrations in agreement with earlier developed optimal foraging models for copepod feeding (Lam & Frost, 1976; Lehman, 1976).

Readily available and most used techniques to quantify particulate matter composition in aquatic systems are analysis of chemical variables such as concentrations of chlorophyll, particulate organic carbon or particulate matter dry weight. Determination of particle volume distributions by electronic counting has to be combined with labour intensive microscopic phytoplankton counts to give information of the contribution of live phytoplankton to total particulate matter (Tackx *et al.*, 1989). Flow cytometry is not yet applicable on samples with a complex particulate matter composition such as those obtained in estuaries. Another reason for the preference for chemical data is that they are relatively easy to incorporate into carbon flow models. Can these chemical variables also be used to understand and model copepod feeding behaviour? Considering the importance of physical aspects such as mechanoreception and prey size, and the complexity of calanoid copepod feeding strategies, it seems unlikely that all factors controlling this behaviour could be functionally summarized in terms of chemical variables.

Flexibility in feeding mode under natural circumstances is well documented for *A. tonsa*. This species has been shown to perform peak tracking (Richman *et al.*, 1977), concentrate living phytoplankton in its ingestion as compared to in situ phytoplankton abundance (Tackx *et al.*, 1989), obtain an important part of its carbon ingestion from microzooplankton (Gifford & Dagg, 1988; Robertson, 1982) and prey upon copepod nauplii (Lonsdale *et al.*, 1977; Tackx & Polk, 1982). Turner & Tester (1989) have shown by microscopic counting that *A. tonsa* feeds unselectively on all phytoplankton species encountered in the plume of the Mississippi river. Their study however does not consider selectivity between phytoplankton and other particulate matter components. Gerber & Marshal (1974) report that gut contents of *A. tonsa* from Narragansett Bay consisted of 34% detritus and 36% material fluorescing as chlorophyll. They also report the presence of micro-organisms and diatom frustules in the gut of *A. tonsa*. White & Roman (1992) measured that the proportion of ingested phytoplankton and microzooplankton by *A. tonsa* in Chesapeake Bay (USA) varies seasonally, but that the sum of both covers the carbon requirements for respiration and egg production.

Experimental data on feeding of *A. bifilosa* and *E. affinis* on natural particulate matter are scarce. Richman *et al.*, (1977) report that *E. affinis* performs peak tracking, but feeds less on large particles than *A. tonsa*. Sellner and Olson (1985) showed that both *E. affinis* and *A. tonsa* feed on dinoflagellate blooms at a low rate. At high dinoflagellate concentrations, *E. affinis* feeds on them at a higher rate than *A. tonsa*. Gulati & Doornekamp (1991) measured the grazing of *E. affinis* on natural ^{14}C-labelled particulate matter from the Volkerak–Zoommeer, a newly-created freshwater lake system in the Rhine Delta (The Netherlands). Their findings show that algal concentrations show a sharp spring decline, concomitant with the abundance peak of *E. affinis*. Grazing activity of *E. affinis* often exceeds phytoplankton primary production, which corroborates the hypothesis of Heinle and Flemer (1975) that detritus and associated microbiota contribute considerably to the diet of *E. affinis*. Boak and Goulder (1985) showed that, in the Humber estuary, the ingestion of both attached and free living bacteria, supplies about 12% of the respiratory requirements of *E. affinis*. Schnack (1982) studied the mouthpart morphology of *E. affinis*, *A. tonsa* and *A. bifilosa* from Kiel Bay. Following Itoh's (1970) edge index, all three species are classified as omnivores. The spacing of the second maxillae in *E. affinis* is narrow, adapted to remove small particles from the water. The second maxilla of *A. bifilosa* has long setae, providing a large mesh size to the screen formed by them.

Most of the studies mentioned above were performed on natural particulate matter of coastal regions and semi-estuarine systems, where detritus concentrations are considerable, but nevertheless much lower than in the brackish water zones of true estuaries.

The present study sets out to verify if for two of these estuarine copepods, *A. tonsa* and *E. affinis*, any selective feeding can be detected under natural estuarine circumstances. The first phase of this study, reported upon in this paper, consisted of two research strategies.

(A) We investigated if any relationship can be detected between the nutritional condition of copepods (as measured by gut fluorescence) and particulate matter composition (as characterized by chemical analysis).

(B) We tried to measure the predation rate of estuarine copepod species on various components of particulate matter using a combination of currently used techniques in zooplankton grazing measurements.

Material and methods

In the Gironde, sampling for particulate matter composition (chemical variables) and copepod in situ gut fluorescence were taken during several 24 to 48 hour cycles at fixed stations J, E, P and K (Fig. 1b). During one of these campaigns, two feeding experiments were carried out at station J (Fig. 1b), following the procedure described below. In the Westerschelde, particulate matter and zooplankton samplings, as well as feeding experiments were carried out during several 24 hours campaigns. All Westerschelde experiments were carried out at 10 ± 1‰ salinity, in the area between station 8 and 10 (Fig. 1a).

Methods used in both estuaries are essentially the same. Any differences are mentioned in the description of procedures given below.

Nutritional condition of the copepods as measured by gut fluorescence in relation to particulate matter composition.

Gut fluorescence: Copepods were caught with a 200 μm mesh size net and a subsample of the animals was filtered onto a 5 × 5 cm net (200 μm) and immediately stored in liquid nitrogen (Gironde) or a deep-freeze (Westerschelde). The further procedure for gut fluorescence measurement is described under feeding experiments.

Particulate matter composition: Water was sampled at surface and near the bottom in the Gironde, and also at mid depth in the Westerschelde. Water was filtered on glassfiber GFC filters (6 replicates per sampling). Volumes filtered (25–250 ml) were adjusted to the prevailing particulate matter concentration. Filters were stored in deep-freeze. Total particulate matter dry weight (SPM) was measured on a microbalance after drying for 24 hours at 60 °C (Gironde). On Westerschelde samples, POC was measured with a Coulomat. For the Gironde, POC was calculated as 15% of SPM (Etcheber, pers. comm.). Extraction of pigments was done by keeping the filters in 10 ml of methanol in a deep-freeze. Chlorophyll *a* (Chl-*a*) and phaeopigment (Phaeo) concentration was quantified by fluorescence after centrifugation of the samples. Fluorescence before and after acidification with 2 drops of lN HCL was measured in a fluorimeter using 420 nm excitation and 670 nm emission wavelength. The fluorimeter was calibrated with a chlorophyll a standard. In the Gironde samples, total chloro- pigment concentration (TP) was calculated as Chl-*a* + 1.51 Phaeo. In the Westerschelde, measurements were only performed after acidification, and the concentration of total chloro- pigments directly calculated from the calibration curve. Chl-*a* concentration on the Westerschelde samples was measured with a Waters HPLC.

Predation rate on various food components.

Copepod feeding in natural conditions was measured simultaneously with different methods to quantify the feeding rate on various particulate matter components. Clearance rate measured on each food component was regarded as a measure of the predation pressure exerted on it under natural conditions. The ratio of clearance rate values on different components is consequently a measure of selectivity of feeding.

Counting experiments: Microscopical phytoplankton cell counts were used to quantify feeding on natural live phytoplankton from incubation experiments. Preliminary tests had shown that measuring copepod feeding on total particulate matter using a Coulter counter is not feasible in the Westerschelde because of high particulate matter concentrations and variability (Tackx *et al.*, 1993). Consequently, a Coulter counter TAII was only used to characterize total particulate matter volume distributions.

A 20 liter water sample was taken at surface (Gironde) or mid depth (Westerschelde). Four control and four grazing bottles were filled with 800 ml of water. Animals used were obtained from the same catch as those used for the gut fluorescence measurements. Four groups of 50 adults of one species were sorted live under binocular microscope and placed in 50 ml of natural water. After addition of the animals to the four grazing bottles, the volume in all eight bottles was adjusted to 1000 ml. Incubation was done on deck in a container. An outboard water pump was used to maintain environmental temperature, and bottles were gently rotated every 30 minutes to keep the particulate matter in suspension. Two samples of 250 ml from the remainder of the original 20 liters of water were fixed with lugol's solution for microscopic counting of phytoplankton (time 0 samples). Three 200 ml samples were taken for Coulter analysis. 100, 200 and 560 μm tubes were used to cover a range of 3–100 μm particle spheric equivalent diameter (S.E.D.). After 14–17 hours, samples were taken from each bottle, and treated in the same way as described for time 0 samples. In the laboratory, phytoplankton samples were decanted to 100 ml, and stained with bengal rose. A 2–4 ml subsample was analyzed for phytoplankton species composition and abundance with a Sedival reversed microscope. Calculation of clearance rates on each phyto-plankton species (Fph) was done following Frost (1972), for those species for which a significant difference in concentration between control and grazing bottles was found at the end of the incubation (Mann Whitney, $p < 0.05$). Abundant phytoplankton species were sized, their volumes calculated following the most appropriate geometric form, and Spheric Equivalent Diameter (S.E.D.) calculated. Natural phytoplankton volume distributions were calculated from these measurements and species abundance data.

^{14}C experiments: Feeding on live phytoplankton was also quantified by adding ^{14}C prelabelled algae to the natural water, and measuring the uptake of ^{14}C

in the copepods. Cultures of *Skeletonema costatum*, *Dunaliella* sp., and *Thalassiosira* sp. grown at + 18 °C and 25‰ salinity were gradually reduced to a salinity of + 13‰ during the week preceding the experiment to avoid osmotic shock when later adding them to 10‰ estuarine water. 25 μCi of $NaH_2{}^{14}CO_3$ was added to 200 ml of culture which was incubated in the light for ± 24 hours. On board, one bottle per available culture was filled with natural water (from the 20 liters sampled for the counting experiment) and live zooplankton added to it at an approximate density of 50 adults/liter. After addition of 1 ml of prelabelled culture to the bottle, two replicate 100 ml samples were taken and filtered onto glasssfiber filters for determination of the radioactivity of the phytoplankton at time 0. Filters were wrapped in Al foil and stored in a deep-freeze. The experimental bottles were incubated during one hour in the deck incubator. At time end, two 100 ml samples were filtered and the zooplankton was collected on a 150 μm net for determination of the radioactivity in the phytoplankton and the zooplankton at time end respectively. All samples were stored in a deep-freeze. In the laboratory, copepods were sorted into species and developmental stage under binocular microscope. 10–20 individuals were put into scintillation vials, and 10 ml of scintillation liquid added. Radioactivity was measured after 24 hours in a Beckman scintillation counter. Clearance rates on the prelabelled algae (Fa) were calculated following Daro & Baars (1986).

Gut fluorescence: Feeding on the totality of live phytoplankton and recently formed detritus was measured by the gut fluorescence method (Mackas & Bohrer, 1976). Copepods were caught with a 200 μm mesh size net and a subsample of the animals was filtered onto a 5 × 5 cm net (200 μm) and immediately stored in liquid nitrogen (Gironde) or a deep-freeze (Westerschelde). Gut passage time (GPT) was measured by putting a part of the remainder of the catch in 20 liters of glassfiber filtered water of the same salinity as the one in which the animals were caught. A subsample was taken every 5 minutes during 30 minutes (Westerschelde) to 1 hour (Gironde) and stored as described above. In the laboratory, adult copepods were sorted into species (Gironde) and sex (Westerschelde) under binocular microscope using dim light, and chloro-pigments extracted in methanol (20–40 animals/4 ml) during 12–24 hours in a deep-freeze. Fluorescence after acidification was measured with a fluorimeter, as described above. Ingestion rates (ITP) were calculated following Mackas & Bohrer (1976). Clearance rate on total chloropigments were calculated as: FTP = ITP/TP *in situ* (ml ind^{-1} h^{-1}).

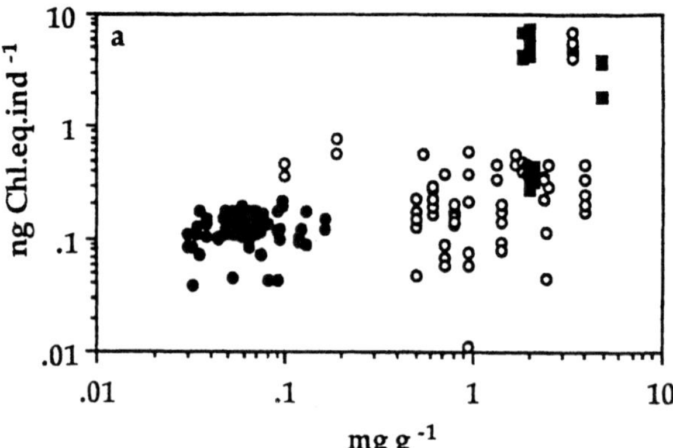

Fig. 2. *E. affinis* GC values as a function Chla/SPM ratio: Gironde March, T= 11–12 °C (black circles); Westerschelde April–May, T= 11–12 °C (white circles) and Westerschelde March, T= 9–10 °C (black squares).

For details on the experimental procedures, and an evaluation of their applicability in estuaries, the reader is referred to Tackx *et al.* (1993).

Results

Nutritional condition of the copepods as measured by gut fluorescence in relation to particulate matter composition.

Figure 2 compares in situ gut contents (GC) of *E. affinis* in the Gironde and the Westerschelde, as a function Chl-*a*/SPM. Gironde observations were made during a 48-hour measurement in March 1991, at station J. Temperature varied between 11 and 12 °C, salinity between 2 and 9‰. For comparison, Westerschelde observations made under the same temperature conditions were selected. These were found during campaigns in April and May of 1990, 1991 and 1992. Observations made in March 1992, when temperature was between 9 and 10 °C are shown separately in Fig. 2. In these observations, Chl-*a*/SPM are a factor 10–100 higher in the Westerschelde than in the Gironde. *E. affinis* GC values at 11–12 °C in the Westerschelde overlap partially with the Gironde data, but about half of the GC values are higher than the maximum values measured in the Gironde. In March (9 °C) GC values in the Westerschelde are considerably higher than the March values of the Gironde.

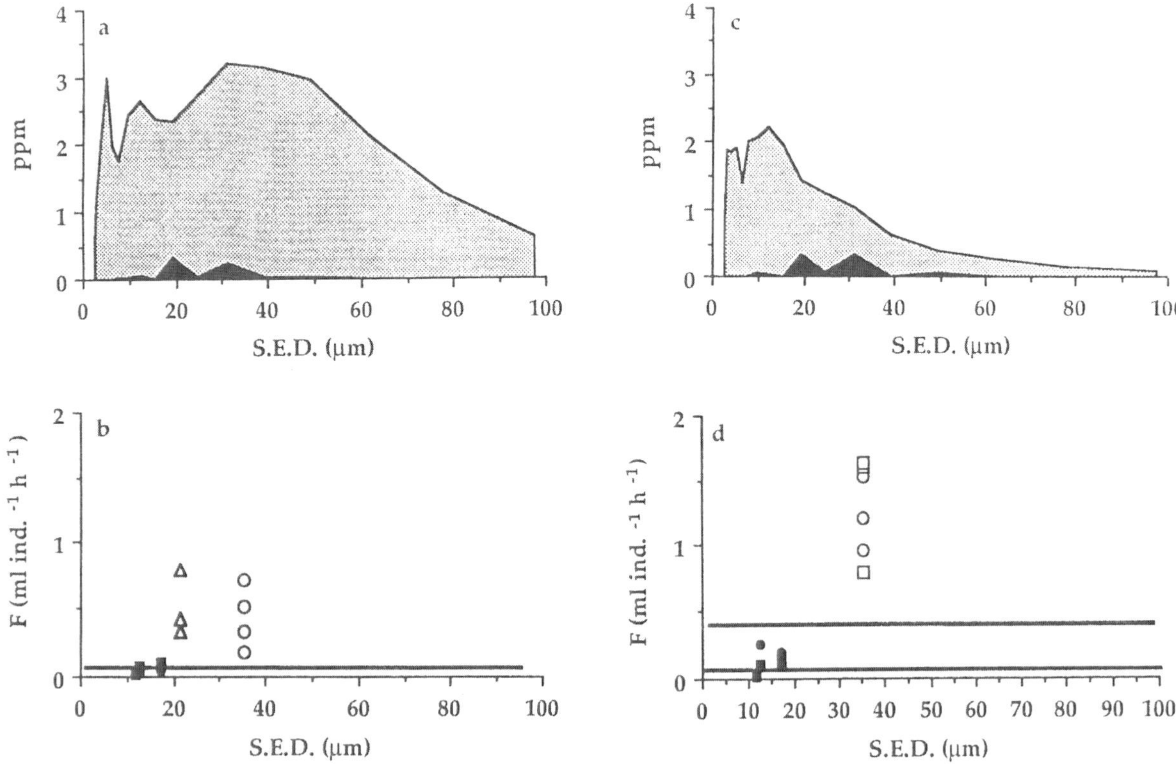

Fig. 3. a,b Westerschelde feeding experiment, example 1. (a) Total particulate matter (shaded) and phytoplankton (black) volume distribution. (b) Fph values for *E. affinis* on *C. commutatus* (blank circles) and on *P. micans* (blank squares). Maximum Fa for *E. affinis* females (black circles) and males (black squares). Maximum FTP measured for *E. affinis* (solid line). (c) and (d). Westerschelde feeding experiment example 2. As (a) and (b), but blank squares represent cleance rates for *A. tonsa* measured on *C. commutatus*. Upper solid line: maximum FTP value for *E. affinis* lower solid line: maximum FTP value for *A. tonsa*.

Predation rate on various food components.

Figure 3a, b–d shows an example of results obtained in the feeding experiments in the Westerschelde with *E. affinis* and *A. tonsa*. Total particle volume distribution and phytoplankton volume distribution are shown in Fig. 3a for the first example. Live phytoplankton occurs between 10 and 40 μm S.E.D., with a peak around 20 and 35 μm S.E.D. This peak consisted mainly of *Prorocentrum micans* (20 μm) and *Coscinodiscus commutatus* (35 μm). Significant differences in cell concentration were found for both these species. Fph values measured for *E. affinis* on *C. commutatus* varied between 0.18 and 0.52 ml ind^{-1} h^{-1}, on *P. micans* between 0.34 and 0.79 ml ind^{-1} h^{-1}. Mean FTP value was 0.05 ($\pm$ 0.02) ml ind^{-1} h^{-1} measured for females. FTP values for males were not measurable in this experiment. Fa values measured on the prelabelled algae varied from 0.02 to 0.10 ml ind^{-1} h^{-1} for females and 0.019-0.088 ml ind^{-1} h^{-1} for males (Fig. 3b.).

In the second example shown in Fig. 3c, d, the two peaks in the phytoplankton distribution were caused by the same phytoplankton species, but only *C. commutatus* was significantly grazed upon by both *A. tonsa* and *E. affinis*. Fph for both copepod species are again considerably higher than FTP and Fa values measured in this experiment.

Figure 4a–d shows the results obtained in the Gironde experiments. In both experiments, *Rhizosolenia delicatula* and *P. micans* caused a peak around 20 μm S.E.D. in phytoplankton volume, *Coscinodiscus* sp. a second important peak around 50 μm S.E.D. In the second experiment, a smaller volume peak of *Ditylum brightwellii* was found around 30 μm S.E.D. Only *R. delicatula* was significantly grazed upon by *E. affinis*. Fph values were again higher than FTP and Fa values. This trend in Fph/FTP and Fph/Fa ratios was found in all experiments in which significant differences in algal cell concentrations were found between control and grazing bottles for one or more species of phytoplankton. The ratio Fph/FTP varied between 0.5

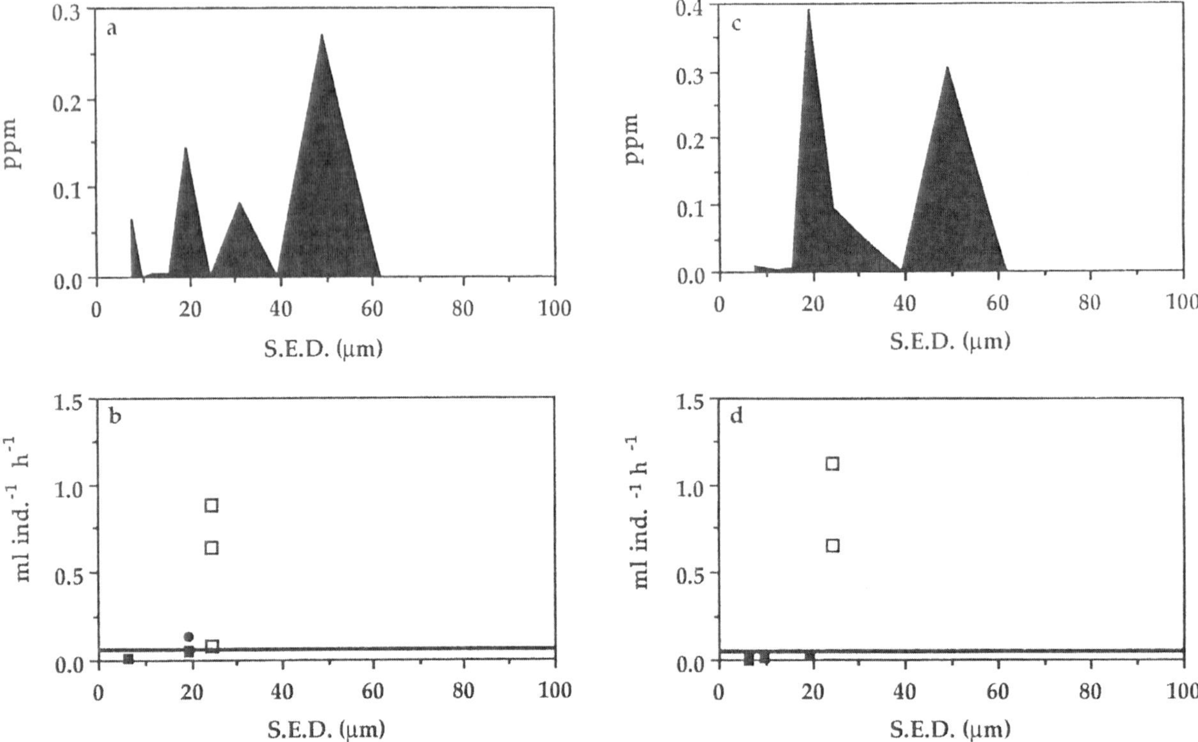

Fig. 4. Gironde experiments with *E affinis*. (a) and (b) Phytoplankton volume distribution (black). Fph on *R. delicatula* (blank squares); Fa values measured for females (black circles) and males (black squares). Maximum FTP value (solid line). (c) and (d): as (a) and (b).

and 33, the majority of the ratios being between 5 and 10 (Fig. 5a). The ratio FA/FTP varied from 0.01 to 2.3, with most values between 0.1 and 0.7 (Fig. 5b). In the Westerschelde experiments, significant differences in cell concentration between control and grazing bottles were repeatedly (4–6 times in 13 experiments) found for 4 species: *C. commutatus, Melosira jurgensii* (in duplets) *Raphoneis amphiceros* (attached to detritus particles or as free cells) and *P. micans*. Other species, for which a significant difference was found only once, were not considered in further data analysis. Fig. 6 gives an overview of the cell concentrations of these 4 species in all the experiments performed. Experiments are ranked following date, although carried out over various years. Experiments 1–7 were carried out in spring, with *E. affinis*; experiments 8–15 in summer, with *A. tonsa*. In experiment 13, feeding of both species was measured. It can be seen that, for *A. tonsa*, significant grazing was detected in most of the cases where the algae occur in peak concentrations. For *E. affinis*, only few of the occurring peaks were significantly grazed upon.

Occasional observations of interest were made in some of the Westerschelde experiments. In two cases where significant feeding on *M. jurgensii* duplets was detected, the concentration of single *M. jurgensii* cells was significantly higher in grazing than in control bottles. In another experiment with *A. tonsa*, microscopic counting revealed a significant difference in concentration between control and grazing bottles for the tintinnid *Tintinopsis beroidea*. Because of the coloring with bengal rose, full and empty lorica could easily be distinguished and counted separately. In this experiment, the concentration of lorica with animals inside was significantly lower in grazing than in control bottles, while that of empty lorica was significantly higher in the former than in the latter (Fig. 7a, b).

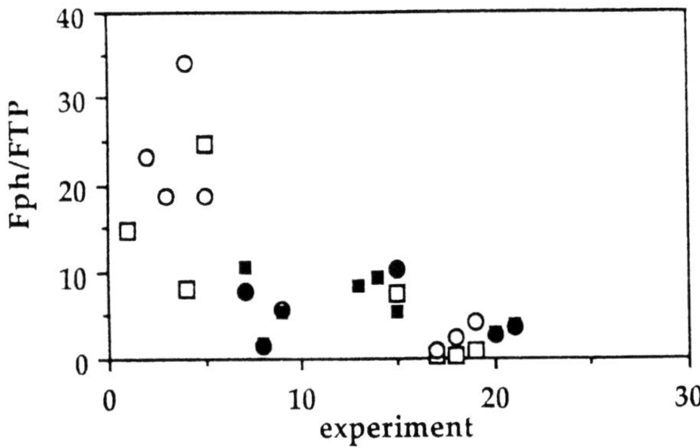

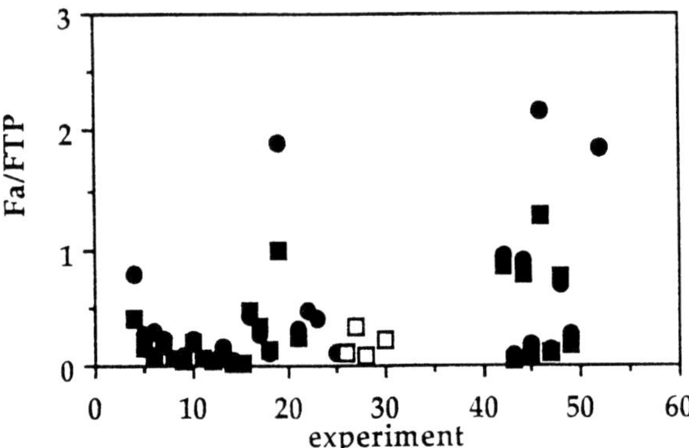

Fig. 5. Ratio Fph/FTP; and ratio Fph/Fa for *A. tonsa* females (blank circles), and males (blank squares) and for *E. affinis* females (black circles) and males (black squares).

Discussion

Nutritional condition of the copepods as measured by gut fluorescence in relation to particulate matter composition.

The comparison of *E. affinis* gut content in the Gironde and the Westerschelde shows that its gut content is influenced by particulate matter composition. Circumstances in which *E. affinis* feeds during spring differ greatly between both estuaries with regard to contribution of live phytoplankton to other particulate material, as characterized by Chl-*a*/SPM ratios. In March, temperature was lower in the Westerschelde than in the Gironde, but relative phytoplankton concentration was higher. March values of *E. affinis* gut content in the Westerschelde were an order of magnitude higher than in the Gironde. In April and May, temperature in the Westerschelde is comparable to March values in the Gironde. The relative contribution of phytoplankton to particulate matter was roughly two orders of magnitude higher in the Westerschelde. During this period, *E. affinis* GC values in both estuaries are partially comparable, but half of the observations are still considerably higher in the Westerschelde than in the Gironde. The overlap in GC values measured in both estuaries shows that the large differences between the Westerschelde and the Gironde March GC values are not due to methodological bias. Although the *E. affinis* population observed in the Gironde was closer to its salinity optimum than the animals studied in the Westerschelde, our results indicated that feeding rate - in terms of chloro pigment ingestion - is generally higher in the Westerschelde than in the Gironde.

In an earlier paper, Irigoien *et al.* (1933) showed that for *A. bifilosa*, in the Gironde, a positive correlation was found between log GC and log Chla/SPM, considering all data collected during several campaigns together. Such a correlation was however not found in the Gironde for *E. affinis*. The present comparison with the Westerschelde data suggests that *E. affinis* GC is influenced by particulate matter composition, but that the Chl-*a*/SPM ratios in the area of the Gironde where *E. affinis* is abundant are too low for this influence to either take place or show. *A. bifilosa* lives more downstream than *E. affinis* in the Gironde, where Chl-*a*/SPM ratios are more comparable to those in the brackish water zone of the Westerschelde. The variation of GC values with particulate matter composition reflects that the copepods depend on particulate matter composition to obtain a certain amount of chloro- pigments in their gut. However, it does not allow any conclusions about selectivity. If copepods would be totally selective on chloropigment containing material, their GC would not vary with Chl-*a*/SPM. If copepods were totally unselective, a significant linear relationship between GC and Chl-*a*/SPM should be found. Such significant linear regressions were not found in the Gironde data analyzed per campaign separately (unpublished data). Westerschelde data per campaign are too scarce to perform this analysis. However, Gironde samplings being performed at fixed stations, many other factors (salinity, light) influencing GC were likely to mask a linear relationship if this existed.

Predation rate on various food components.

The series of experiments conducted with natural Westerschelde and Gironde particulate matter represent different feeding conditions to the copepods with relation to absolute and relative concentration of variously sized and chemically composed potential food items. This aspect was however not explicitly considered in

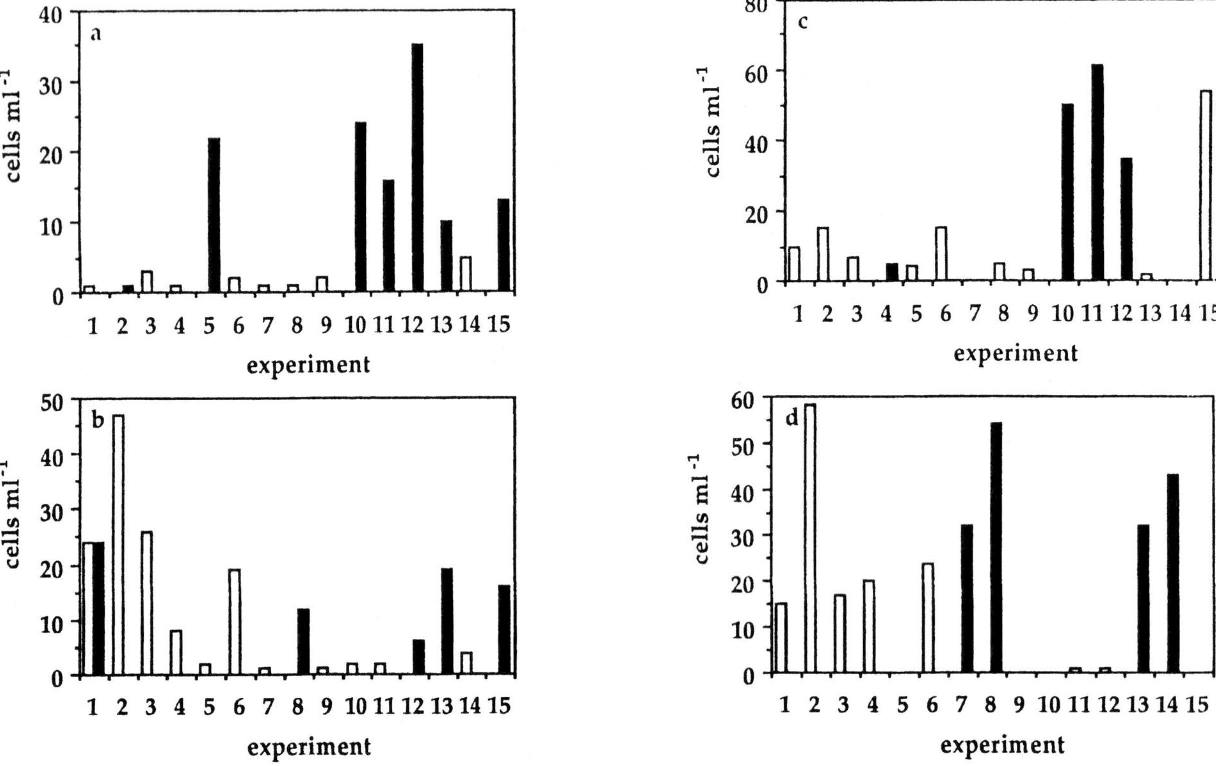

Fig. 6. Cell concentration of (a) *M. sulcata*, (b) *C. commutatus*, (c) *R. amphiceros* and (d) *P. micans* in Westerschelde experiments. Black bars represent experiments in which a significant difference in concentration was found between control and grazing bottles for the phytoplankton species concerned. experiment 1–7: *E. affinis*; 8–15: *A. tonsa*.

the data analysis, because the aim was to quantify predation pressure on various food items measured within each natural situation, not among different situations. Factors affecting the functional response of the copepod's feeding, such as absolute and relative concentration of various food items, size etc. were considered as inherent circumstances to each experiment separately.

In both the Gironde and the Westerschelde feeding experiments Fph values are higher than FTP values. This suggests that the phytoplankton species for which significant concentration differences are repeatedly found are selected above the totality of natural phytoplankton and chloro-pigment containing material. On the other hand, Fa values are comparable, or lower than FTP values, which shows that the prelabelled algae added to the natural water are not selected. A number of methodological factors have to be considered with regard to these results.

(1) Fph values could be overestimated, because of large variability in the obtained cell counts. Variability in Fph values could be reduced by taking time 0 samples from each bottle seperately, shortening incubation time and applying calculations following Marin *et al.* (1986).

On the other hand, agreement with literature data suggest that Fph values in our experiments are realistic or might even be underestimated.

Fph values measured for *E. affinis* and *A. tonsa* values in our experiments, range from 0.5 to 3 ml ind^{-1} h^{-1}. Turner & Tester (1989) report F values between 0 and 2.95 ml ind^{-1} h^{-1} for *A. tonsa* measured by microscopic counting on total natural phytoplankton in the Mississippi. Durbin *et al.* (1990) measured feeding quantified by microscopic counting of *A. tonsa* on *Thalassiosira weisflogii* and in mesocosm studies using natural plankton as food. They report F values between 0.6 and 3 ml ind^{-1} h^{-1} on *Thalassiosira weisflogii* at cell concentrations between 250 and 1990 cells ml^{-1} and between 0.18 and 4.37 ml ind^{-1} h^{-1} in the mesocosm experiments.

Phytoplankton carbon concentrations, calculated following Eppley (1974), varied between 0.02 and 0.04 μg ml^{-1}, with the phytoplankton species being

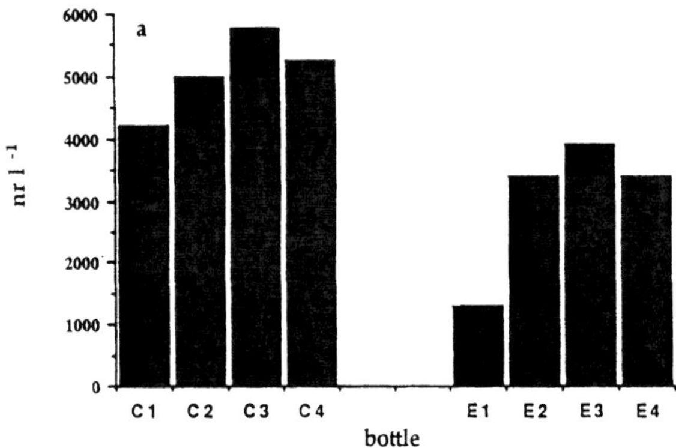

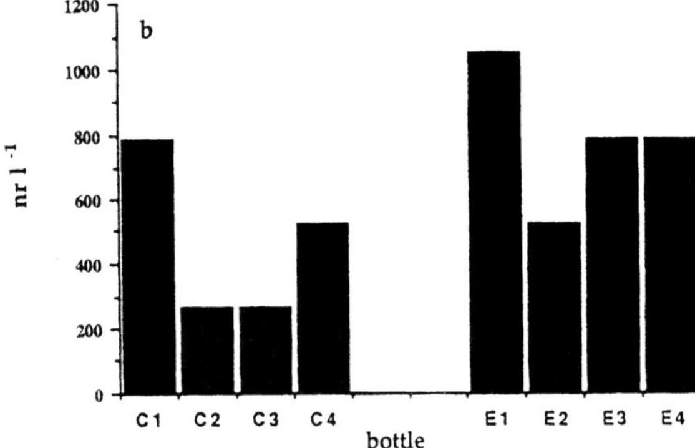

Fig. 7. Westerschelde experiment with *A. tonsa*. Concentration of *T. beroidea* (a) full lorica and (b) empty lorica in control (C1–C4) and grazing bottles (E1–E4) at the end of the incubation.

significantly grazed upon contributing for 10 to 65% of this carbon. Kiørboe *et al.* (1985) report maximum clearance rates of 1.7 ml ind^{-1} h^{-1} for *A. tonsa* feeding on monocultures of a small algae (13 × 8 μm) *Rhodomonas baltica* at concentrations of 0.15 μg ml^{-1}, with clearance rates declining at both lower and higher concentrations. Paffenhöfer and Knowles, (1988) showed that clearance rates of *A. tonsa* feeding on *Thalassiosira weisflogii* decreased rapidly at algal concentration below 0.2 ppm. In our experiments, algal concentrations measured as peaks of distinct species, were between 0.1 and 0.5 ppm at the start of the experiments. Between 32 and 64% of the original algal concentration was removed in the grazing bottles for the phytoplankton species grazed upon. In how far this considerable reduction has limited feeding on these phytoplankton species, is difficult to assess. In any case, this would have resulted in an underestimation of the Fph values measured in the experiments as compared to *in situ* values.

Microscopic cell counting on estuarine samples nevertheless is subjected to a number of shortcomings. On many samples obtained in the Gironde and some ebb samples in the Westerschelde microscopic counting is not feasible because of the high detritus concentrations. When counting can be done, only abundant phytoplankton species, with a S.E.D. > ± 10 μm can be counted in sufficient number to obtain reliable data. This leaves the question whether these species are the only ones which are (sometimes) selected, or the only ones for which we can adequately quantify Fph values, and hence detect selectivity.

(2) Fa values could be underestimated. The majority of Fa values measured on prelabelled algae in this study lies between 0.01 and 0.1 ml ind^{-1} h^{-1}. The maximum value measured (for *E. affinis* females) is 0.27 ml ind^{-1} h^{-1}. Gulati (1991) measures clearance rates between 0.4 and 1 ml ind^{-1} h^{-1} for *E. affinis*, using natural ^{14}C labelled lake plankton < 33 μm as tracer. Daro (1978) measures clearance rates between 0.44 and 1.57 ml ind^{-1} h^{-1} by *A. tonsa* on the natural ^{14}C labelled phytoplankton of a marine lagoon in Ostend (Belgium). The low Fa values could be explained by the fact that the prelabelled phytoplankton species are usually smaller than the natural phytoplankton species which are apparently selected. The tendency for adult copepods to select larger particles is well known (Paffenhöfer, 1988). In addition, practical circumstances did not allow a well controlled adaptation of the algal cultures to salinity and temperature conditions in estuarine water. Damage of the cells by too abrupt physiological changes cannot be excluded. Verification of the cells condition after the experiment was not possible, as the small cells are difficult to observe under microscope when mixed with natural estuarine particulate matter. These experiments clearly need to be repeated using larger sized phytoplankton species, which are well adapted to estuarine circumstances to solve this methodological uncertainty.

(3) FTP values could be underestimated. Discussion is still going on in literature about the amount of ingested chloro pigments that is degraded into non-fluorescent material in the gut of the copepod (Helling & Baars, 1985; Lopez *et al.*, 1988). Losses are reported to be as high as 85% (Mayzaud & Razouls, 1992). Even if FTP values measured in our experiments are corrected for a 85% loss they still remain consistently lower than Fph values. Possible complications with the gut fluorescence method may also arise from the complexity of estuarine particulate matter composition. Measurements on the Gironde particulate matter show that pheaopigment concentration is generally much higher

than that of Chl-*a*. Also in the Westerschelde, concentration of total pigments sometimes reaches values up to 25 μg l^{-1}. It can not be excluded that other components besides chloro-pigments interfere with the fluorescence measurement.

The significant production of single *M. jurgensii* single cells could reflect a selective preferential feeding by *A. tonsa* on larger sized duplets of this species, spilling of single cells during the feeding process. The measured feeding on *T. beroidea* indicates that feeding on microzooplankton by *A. tonsa* takes place in natural Westerschelde conditions. The cooccurrent higher concentration of empty lorica in the grazing bottles as compared to controls suggests a capacity for *A. tonsa* to remove the animals without damaging the lorica. Robertson (1983) reports that, in predation experiments with *A. tonsa* feeding on the tintinnids *Favella panamensis* and *Tintinnopsis tubulosa*, broken lorica were never found in the experimental containers. In his experiments, intact loricae were observed in some copepod feacal pellets, suggesting that the tintinnids are ingested as a whole. In laboratory experiments in which *A. tonsa* was fed the tintinnid *Favella* sp. together with a dinoflagellate *Heterocapsa triquetra*, Stoecker & Sanders (1985) found empty and crushed *Favella* lorica both in the faecal pellets and free in the grazing bottles. In our experiments, silica shells of the *Coscinodiscus commutatus* and of *C. centralis* were sometimes observed in the faecal pellets, but no intact tintinnid lorica.

Conclusions

The discrepancy between Fa and Fph values measured in this study necessitates further methodological checkes before any solid conclusions can be drawn. Besides the above mentioned problems with interpretation of gut fluorescence and counting method data obtained in estuaries, a major drawback to this study was the impossibility to measure feeding on the total particulate matter (FSPM) by Coulter counter, or on detritus separately. Notwithstanding these complications, the present data suggest that in the Westerschelde, *A. tonsa* quite regularly feeds selectively on either live phytoplankton or microzooplankton, when these organisms occur in blooms. For *E. affinis* on the contrary, this selectivity was only occasionally measured. The demonstrated dependence of *E. affinis* GC on particulate matter composition can be interpreted as a reflection of both unselective or selective feeding.

To circumvent the methodological complications encountered with natural estuarine samples, future experiments will involve more controlled setups. In these, the hypothesis concerning the feeding mode of estuarine copepods deduced from the present fieldwork, will be tested using combinations of algal cultures and natural estuarine water.

Acknowledgements

The authors greatly appreciate the practical help of R. Vanthomme and M. Bogaert in shipboard work and laboratory analysis. We also want to express our thanks to the members of the crews of the Luctor and the Côte d'Aquitaine for their enthusiastic collaboration, and to Prof. G. Billen and his colleagues, for putting the fluorimeter at our disposition. Mrs J. McCourt corrected the English. Dr G.A. Paffenhöfer and an anonymous referee gave valuable comments on an earlier version of this paper. This research was part of the JEEP 92-MAST project and funded by the C.E. Communication no. 736 of the NIOO-CEMO.

References

Boak, A. C. & R. Goulder, 1983. Bacterioplankton in the diet of the calanoid copepod *Eurytemora* sp. in the Humber estuary Mar. Biol. 73: 139–149.

Castel, J., 1985. Distribution et évolution du plancton de l'estuaire de la Gironde. Océanis, 6: 535–577.

CNEXO, 1977. Etude ecologique de l'estuaire de la Gironde. Rapport final. 470 pp.

Daro, M. H., 1978. A simplified ^{14}C method for grazing measurements on natural plankton populations. Helg. Wiss. Meeresunters. 31: 241–248.

Daro, M. H. & M. Baars, 1985. ^{14}C method for grazing measurements, some theoretical considerations. Hydrobiol. Bull. 159: 159–170.

Durbin, A. G., Durbin, E. G. & E. Wlodarczy]k, 1990. Diel feeding behaviour in the marine copepod *Acartia tonsa* in relation to food availability. Mar. Ecol. Prog. Ser. 68: 23–45.

Emlen, J. M., 1966. The role of time and energy in food preference, Am. Nat. 100: 611–617.

Frost, B. W., 1972. Effects of size and concentration of food particles on the feeding behaviour of the marine planktonic copepod *Galanus pacificus*. Limnol. Oceanogr. 17: 805–815.

Gerber, R. P. & N. Marshall, 1974. Ingestion of detritus by the lagoon pelagic community at Eniwetoc Atoll. Limnol. Oceanogr. 19: 815–824.

Gifford, D. J. & M. J. Dagg, 1988. Feeding of the estuarine copepod *Acartia tonsa* Dana: carnivory vs. herbivory in natural microplankton assemblages. Bull. Mar. Sc. 43: 458–468.

Gulati, R. & A. Doornekamp, 1991. The spring-time abundance and feeding of *Eurytemora affinis* (Poppe) in Volkerak-Zoommeer, a newly created freshwater lake system in the Rhine delta (The Netherlands). Hydrobiol. Bull. 25: 51–60.

Heinle, D. R. & D. A. Flemer, 1975. Carbon requirements of a population of the estuarine copepod *Eurytemora affinis*. Mar. Biol. 31: 235–247.

Heinle, D. R., R. P. Harris, J. F. Ustach & D. A. Flemer, 1977. Detrius as food for estuarine copepods. Mar. Biol. 40: 341–353.

Helling, G. R. & M. A. Baars, 1986. Changes of the concentrations of chlorophyll and phaeopigments in grazing experiments. Hydrobiol. Bull. 19: 41–48.

Hummel, H, G. Moerland & K. Bakker, 1988. The concomitant existance of a typical coastal and a detritus food chain in the Westerschelde estuary. Hydrobiol. Bull. 22: 35–41.

Itoh, K., 1970. A consideration on feeding habits of planktonic copepods in relation to the structure of their oral parts. Bull. Plankton Soc. Japan, 17: 1–10.

Kiørboe, T., F. Mohlenberg & K. Hamburger, 1985. Bioenergetics of the planktonic copepod *Acartia tonsa*: relation between feeding, egg production and respiration, and composition of specific dynamic action. Mar. Ecol., PS. 26: 85–97.

Lonsdale, D. J., D. R. Heinle & C. Siegfried, 1979. Carnivorous feeding behaviour of the adult calanoid copepod *Acartia tonsa* Dana. J. Exp. Mar. Biol. Ecol. 36: 235–248.

Lopez, M. D. G., M. E. Huntley & P. F. Sykes, 1988. Pigment destruction by *Calanus finmarchicus*: Impact on the estimation of water column fluxes. J. Plankton Res. 10: 715–734.

Mackas, D. & R. Bohrer, 1976. Fluorescence analysis of zooplankton gut contents and an investigation of diel feeding patterns. J. exp. Biol. Ecol. 25: 77–85.

Marin, V., M. E. Huntley & B. Frost, 1986. Measuring feeding rates of pelagic herbiovres: analysis of experimental design and methods. Mar. Biol. 93: 49–58.

Mayzaud, P. & S. Razouls, 1992. Degradation of gut pigment during feeding by a subantarctic copepod: Importance of feeding history and digestive acclimation. Limnol. Ocean. 37: 393–404.

Paffenhöfer, G. A., 1988. Feeding rates and behaviour of zooplankton. Bull. Mar. Sci. 43: 430–445.

Paffenhöfer, G. A., J. R. Strickler & M. Alcaraz, 1982. Suspension feeding by herbivorous calanoid copepods: a cinematographic study. Mar. Biol. 67: 193–139.

Paffenhöfer, G. A & D. Stearn, 1988. Why is *Acartia tonsa* restricted to nearshore environments? Mar. Ecol. P.S., 42: 33–38.

Paffenhöfer, G. A. & K. B. Van Sant, 1985. The feeding responce of a marine planktonic copepod to quantity anf quality of particles. Mar. Ecol. Progr. Ser. 27: 55–65.

Paffenhöfer, G. A. & K. D. Lewis., 1990. Perceptive performance and feeding behaviour of calanoid copepods. J. Plankt. Res. 12: 933–946.

Price, H. J., 1988. Feeding mechanisms in marine and freshwater zooplankton. Bull. Mar. Sci. 43: 327–343.

Price, H. J. & G. A. Paffenhöfer, 1986. Effect of concentration on the feeding of a marine copepod in algal monocultures and mixtures. J. Plankt. Res. 8: 119–128.

Richman, S., D. R. Heinle & R. Huff, 1977. Grazing of adult estuarine calanoid copepods of the Chesapeake Bay. Mar. Biol., 42: 69–84.

Robertson, J. R., 1983. Predation by estuarine zooplankton on tintinnid ciliates. Est. Coast. Shelf Sc. 16: 27–36.

Schnack, S. B., 1982. The structure of the mouth parts of copepods in kiel Bay. Meeresforshung/Rep. Mar. Res. 29: 89–101.

Soetaert, K. & P. Van Rijswijk, 1993. Spatial and temporal patterns of the zooplankton in the Westerschelde estuary. JEEP 92, Report workshop Plymouth Jan. 1992: 77–80.

Stoecker D. K. & N. K. Sanders, 1985. Diffferencial grazing by *Acartia tonsa* on a dinoflagellate and a tintinnid. J. Plankton Res. 7: 85-100.

Tackx M. & P. Polk, 1982. Feeding of *Acartia Dana* (Copepoda, Calanoida) on nauplii of *Canuella perplexa* T. et A. Scott (Copepoda, Harpacticoida) in the sluice dock at Ostend. Hydrobiologia 94: 131–133.

Tackx, M. L. M., C. Bakker, J. W. Francke & M. Vink, 1989. Size and phytoplankton selection by Oosterschelde zooplankton. Neth. J. Sea Res. 23: 35–43.

Tackx, M. L. M., X. Irigoien, M. H. Daro & M. Bogaert, 1993. Zooplankton trophic studies. JEEP 2. Report/manual. 42 pp.

Turner, J. T. & P. A. Tester, 1989. Zooplankton feeding ecology: non-selective grazing by the copepods *Acartia tonsa Dana*, *Centropages velificatus* De Oliveira, and *Eucalanus pileatus* Giesbrecht in the plume of the Mississipi river. J. Exp. Mar. Biol. Ecol. 126: 21–43.

White, J. R. & M. R. Roman, 1992. Egg production by the calanoid copepod *Acartia tonsa* in the mesohaline Chesapeake bay: the importance of food resources and temperature. Mar. Ecol. Prog. Ser. 86: 230–249.

Hydrobiologia **311**: 85–101, 1995.
C. H. R. Heip & P. M. J. Herman (eds), *Major Biological Processes in European Tidal Estuaries.*

Long-term changes in the population of *Eurytemora affinis* (Copepoda, Calanoida) in the Gironde estuary (1978–1992)

Jacques Castel
Laboratoire d'Océanographie Biologique, Université de Bordeaux I, F-33120 Arcachon, France

Key words: Long-term, *Eurytemora affinis*, population structure, Gironde estuary

Abstract

Long-term changes in the population of *Eurytemora affinis* were investigated during 15 years (1978–1992) at a fixed station in the Gironde estuary (South West France). Total numbers, sex-ratio, % ovigerous females, proportion of copepodites and clutch-size were taken into account and their long-term distribution was related to temperature, river flow, salinity, suspended matter and chlorophyll *a* concentrations after the seasonal effect was removed for all variables.

There was a great decrease of the river flow during the period of investigation, due to a general deficit in pluviosity. On the contrary, the general trend for salinity was an increase from 1978 to 1992. The same pattern was observed for chlorophyll, indicating an intrusion of marine phytoplankton. Water temperature increased significantly during the study period, following the general increase in air temperature observed in the area. For the suspended particulate matter (SPM), an increase was observed between 1978 and 1981–1982, then a very sharp decrease occurred from 1984 onwards.

Numbers of *E. affinis* were inversely correlated with temperature and salinity, and positively correlated with the river flow. *E. affinis* was negatively correlated with chlorophyll concentration because of the covariation with salinity. No clear long-term trend was observed for the sex-ratio. A significant correlation was found between females carrying egg-sacs and SPM concentration, probably due to a decrease of the predation pressure in very turbid waters. The percentages of copepodites tended to decrease with time and were inversely correlated with temperature. Clutch-size significantly decreased during the 15-year period. This trend was mainly explained by temperature and salinity.

It is concluded that, in the absence of strong human alteration in the Gironde estuary the long-term distribution of zooplankton can be explained by the natural environmental variability. The dominant factors are the river flow which governs the movement of the populations along the estuary and the temperature which influences the reproductive processes. Any change in climate will have consequences on the pelagic community.

Introduction

Although estuaries are now intensively studied in the perspective of management, long-term series are not so numerous, especially for zooplankton. Continuous zooplankton data for a period of greater than 3–4 years are limited in number. Moreover, very few reports concentrate on long-term changes of the structure of a single population. In most tidal estuaries in the northern hemisphere *Eurytemora affinis* is the dominant autochthonous species. This species has been extensively studied in the field as well as in the laboratory and many aspects of its biology are now understood. Thus any long-term change in *E. affinis* population should be interpretable and should give information on the modification of the environment.

Haertel *et al.* (1969) reported on a 5 years study from the Columbia River estuary (Oregon, USA). Their study was related to the possibility of modification of the water flow and temperature regimes due to the construction of dams, as well as to changes in water quality resulting from human population and industrial growth in the drainage basin. *Eurytemora affinis* was found to be the major zooplankter, reaching highest densities in the oligohaline area. The temporal heterogeneity and year-to-year variation of *Eurytemora* abundance was explained by variations in the river flow. The authors also showed that high temperature may be responsible for the late summer-early autumn depression of the population. Regression analysis indicated a close correlation between phosphate levels and *Eurytemora* abundance during the season of phytoplankton abundance.

In Europe, long-term (9 years) dynamics of mesozooplankton densities was studied at Seili (Northern Baltic) by Vuorinen & Ranta (1987). Rotifers and the Copepods *Acartia* spp. and *Eurytemora affinis hirundoides* were the dominant taxa. The meso-zooplankton responded to an increase in salinity caused by changing hydrography occurring in the Baltic Sea; 1/3 of the taxa increased in numbers, while most of the taxa (2/3) decreased in numbers. *E. affinis* showed a statistically significant negative trend with time as a consequence of the salinity change. However, the authors could not assess the significance of other factors such as eutrophication to the documented abundance changes.

At the same sampling site, and during 9 subsequent years, Viitasalo *et al.* (1990) monitored the abundance of crustacean mesozooplankton in relation to changes in salinity and eutrophication. There was a rise of temperature during the study period. The chlorophyll values were increasing as were the nutrient levels in winter while salinity was decreasing. The most important zooplankton species, *Acartia bifilosa*, which has a neritic origin, increased with time while the other dominant species, *Eurytemora affinis*, showed no obvious trend. These unexpected results could not be attributed to any single environmental factor. The authors suggested that salinity is the most important environmental factor for zooplankton species living mostly below the thermocline, while temperature and eutrophication are more important for species living above the thermocline.

The conclusions from these long-term series show that salinity probably is the major factor affecting *Eurytemora* populations. High temperature may be responsible for summer depression in *Eurytemora* abundance. Eutrophication could have an effect but it is not clear because eutrophication is the result of a mixing of different factors. The Gironde estuary is not submitted to eutrophication. However, as every ecosystem, the Gironde is changing. Most of the long-term dynamics of the hydrology is under the dependence of variation in climate which, during last years, is characterized by a general increase in temperature and a decrease in pluviosity.

This paper reports on the long-term changes of the *Eurytemora affinis* population taking place in the Gironde estuary (South West France) in 1978–1992. The hypothesis is that the population should respond to the climate change observed during the study period: increase in temperature and deficit in pluviosity plus variations of covariate factors (river flow, salinity). Chlorophyll concentration and suspended particulate matter are also taken into account as they directly or indirectly govern the nutritional quality of the particles ingested by the copepods. The response of the population is examined for copepodites and adults of both sexes. The reproductive potential of the population is also estimated (as the proportion of ovigerous females and the number of eggs per egg-sac) as it has been shown that reproduction responds well to any variation in the environment and to food quality (Hart, 1987).

Description of the study area

The Gironde estuary (Lat. 45 °20′N, Long. 0 °45′W) covers an area of 625 km^2 at high water (Fig. 1). The freshwater is brought to the estuary by the rivers Garonne and Dordogne which join 70 km from the inlet. The mean combined discharge of the Garonne and the Dordogne varies between 800 and 1000 m^3 s^{-1}. During average tidal conditions, freshwater flushing time ranges from 20 days (high river discharge) to 86 days (low river flow). Tidal current velocities vary considerably and can reach 2.5 m s^{-1} during spring tide ebb. According to the terminology of Pritchard (1955), the Gironde is a 'Type B' estuary (intrusion of a salt wedge with tides) during high river flow and a 'Type C' (partially mixed) during low river flow.

The mean upstream limit of the saline intrusion (0.5 psu) is located about 75 km from the inlet during low river discharge and 40 km from the inlet during high river flow. The density gradient, i.e. the salinity gradients, result in a residual circulation system. The higher the fluvial discharge in the main channel of the lower estuary, the stronger the upstream residual circulation on the bottom.

One of the main feature of the Gironde is the high turbidity of the water. Suspended particulate matter concentrations may exceed 1 g l^{-1} in a large part of the estuary. The Garonne and Dordogne rivers supply between 1.5 and 3 10^6 t of suspended sediment annu-

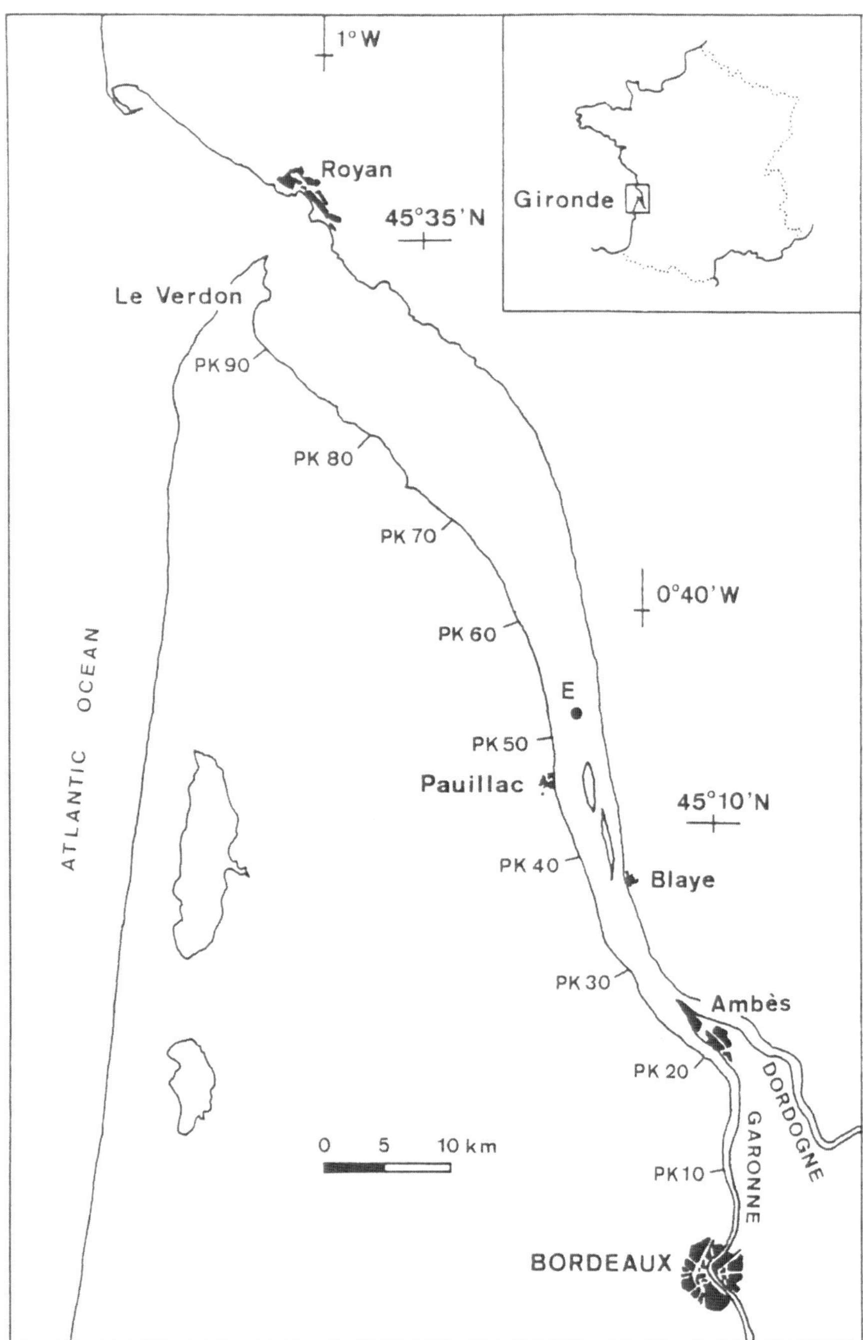

Fig. 1. Map of the Gironde estuary showing the sampling station E (52 km downstream the city of Bordeaux).

ally, the influx being closely correlated to river discharge. A well developed turbidity maximum forms at the upstream limit of the salinity intrusion, in the zone of convergence of bottom residual currents. This maximum migrates to the lower estuary during the winter and spring high river flow, and to the upper estuary during summer and fall low river flow. During low river flow, the mass of turbid water stretches downstream at ebb and upstream at flood, thus covering a zone of more than 70 km in length. Conversely, during high river discharge, the turbidity maximum zone is much more restricted in space, leading to very high suspended matter concentrations (several g l^{-1}). Because of the circulation and transportation patterns, a particle entering the estuary may remain in the maximum turbidity zone for one year or more before being expelled to the sea. A synthesis on the hydrology of the Gironde estuary is given by Jouanneau & Latouche (1981).

Materials and methods

One sampling station (Station E, 52 km from the city of Bordeaux) was selected in the oligo-mesohaline zone of the estuary (Fig. 1). This station is considered as representative of the autochthonous zooplankton community largely dominated by *Eurytemora affinis* (Castel, 1981; Castel & Veiga, 1990; Castel & Feurtet, 1992).

The zooplankton data were collected from May 1978 to November 1992. Samples were taken monthly in winter and fortnightly in summer. Sampling was generally done from February to November during the years 1978–1984. From 1985 onwards, sampling was performed between March and November and duplicate campaigns were made only in July (Table 1). The sampling ranged over 175 months and the number of samples totalled 1556. Zooplankton was collected with a standard WP2 net (200 μm mesh size). One tow was made just below the surface and another near the bottom. Each tow was 1 to 2 min long. Samples were obtained at approx. 2 h intervals during a tidal cycle except in 1992 (half tidal cycle) and were made against the current. The volume of water filtered through the net was monitored with a TSK mechanical flowmeter or with a Hydrobios digital flowmeter. The catch was preserved in 4% formalin.

At the same time and with the same periodicity the following variables were measured: temperature, salinity, suspended matter concentration (dry weight after filtration on GF/C). Chlorophyll *a* in the water was measured by spectrophotometry after concentration on GF/C filters and extraction in 90% aceton.

E. affinis were sorted according to the following categories: copepodites, adult males and females, and females carrying egg sac. Number of eggs per egg-sac were also counted.

For the numerical analyses surface and bottom samples and measurements were averaged. When two sampling campaigns were made during the same month, the values obtained were averaged. Missing values were interpolated by computing the mean between the preceding value and the following one. Two subsequent missing values were not interpolated.

The first step in the analysis was to graph the original data against time. To investigate long-term trends in the zooplankton numbers, the abundance estimates were adjusted by removing the seasonal effect. In order to remove the seasonal variation in plankton abundances and environmental parameters, residuals were calculated by substracting the corresponding monthly average of the fifteen-year period from each observation (Chatfield, 1984). Missing values were considered to be equal to the mean (residual = 0). The residuals computed for river flow, suspended matter and *Eurytemora* abundance were log-transformed. Linear correlation was used to evaluate the covariance of changes in zooplankton abundances and environmental factors; Pearson correlations were calculated between time and zooplankton abundance residuals and environmental residuals.

Results

Hydrography

Water temperature correlated positively with time ($r = 0.457$, $p < 0.01$). This trend is illustrated by the evolution of the maximum temperatures. For instance a maximum value of 22 °C was recorded on 29 August 1978 and a value as high as 26.5 °C was noted on 23 July 1991. However, the temperature data show that the year 1984, winter 1985 and winter-spring 1986 have been colder than the long-term average for the whole period. (Fig. 2). The mean water temperature during the sampling period (February–November) varied between 14.7 °C in 1978 to 17 °C in 1992. The general increase in water temperature was essentially due to variation of climate reflected by an increasing trend of air temperature in the district of Bordeaux (cor-

Table 1. Number of sampling campaigns undertaken in the Gironde estuary during the study period (1978–1992). N° Zp: number of zooplankton samples taken each year. * refers to campaigns during which only environmental variables could be measured. Samples for chlorophyll *a* concentrations were not taken in 1978.

	1978	1979	1980	1981	1982	1983	1984	1985	1986	1987	1988	1989	1990	1991	1992	Sum
J	0	0	0	0	0	0	1	0	0	0	0	0	0	0	0	1
F	1*	1	1	1	1	1*	0	0	0	0	0	0	0	0	0	6
M	0	1	1	1	1	1	1	1	1	0	1	1	1	1	1	13
A	0	1	1	1	1	1	1	1	1	1	1	1	1	1	1	14
M	1	1	1	1	1	1	1	1	1	1	1	1	1	1	1	15
J	2	2	2	2	2	2	1	1	1	1	1	1	1	1	1	21
J	2	2	2	2	2	2	2	2	2	2	2	2	2	2	2	30
A	2	2	2	2	2	2	1	1	1	1	1	0	0	0	0	17
S	0	1	0	0	0	1	1	1	1	1	0	1	1	1	1	10
O	1	0	1	1	1	0	0	0	0	1	1	1	1	1	1	10
N	1	1	0	0	0	1	1	1	1	1	1	1	1	1	1	12
D	0	0	1	1	1	0	0	0	0	0	0	0	0	0	0	3
Sum	10	12	12	12	12	12	10	9	9	9	9	9	9	9	9	152
N° Zp	108	144	126	140	138	126	80	90	90	90	90	90	90	82	72	1556

Table 2. Average monthly values (1978–1992) of environmental variables and parameters of the population of *Eurytemora affinis* in the middle Gironde estuary.

	F	M	A	M	J	J	A	S	O	N
Temperature (°C)	7.77	9.83	12.74	16.50	19.62	22.45	22.81	20.72	16.78	12.02
Salinity (psu)	1.24	2.15	2.33	2.44	2.65	4.65	7.28	8.65	7.28	4.68
River flow ($m^3\ s^{-1}$)	1523	1322	1375	1212	838	485	307	328	610	862
SPM conc. ($mg\ l^{-1}$)	1353	1449	1259	1204	1041	865	737	852	1171	1534
Chlorophyll *a* ($\mu g\ l^{-1}$)	5.49	7.35	7.28	7.78	6.48	6.98	7.11	7.12	7.25	8.15
E. affinis (N° m^{-3})	5232	10621	11519	14328	7789	4500	1728	1093	1540	2649
Males (%)	52.24	54.71	52.96	52.39	50.00	48.08	43.28	44.32	52.85	55.28
Ov. females (%)	23.70	27.73	25.06	24.77	28.23	33.33	33.51	31.68	38.55	35.20
Copepodites (%)	64.30	74.73	75.77	71.87	69.71	56.19	56.66	49.79	50.67	60.54
N° eggs/sac	19.10	12.13	13.45	13.74	9.78	9.07	8.89	8.16	9.02	9.72

Table 3. Correlation coefficients between hydrographic and food variables (the seasonal effect removed from all data sets). Residuals for river flow and SPM concentration are log transformed. The level of significance is indicated by: * $p<0.05$, ** $p<0.01$.

	Salinity	River flow	SPM conc.	Chlorophyll *a*
Temperature (°C)	0.436**	−0.288**	−0.214**	0.148
Salinity (psu)		−0.603**	−0.173*	0.212*
River flow ($m^3\ s^{-1}$)			0.063	−0.138
SPM ($mg\ l^{-1}$)				0.055

Table 4. Correlation coefficients between demographic parameters of *Eurytemora affinis* and hydrographic or food variables (seasonally detrended). Residuals for *E. affinis* abundance, river flow and SPM concentration are log transformed. The level of significance is indicated by: *p<0.05, **p<0.01.

	Temperature	Salinity	River flow	SPM conc.	Chlorophyll *a*
E. affinis (N° m^{-3})	−0.239**	−0.375**	0.375**	0.032	−0.232**
Males (%)	−0.045	−0.071	0.032	0.105	−0.217*
Ov. females (%)	−0.077	0.045	−0.176*	0.300**	−0.184*
Copepodites (%)	−0.319**	−0.126	−0.032	0.001	0.130
N° eggs/sac	−0.366**	−0.428**	0.253**	0.200**	−0.141

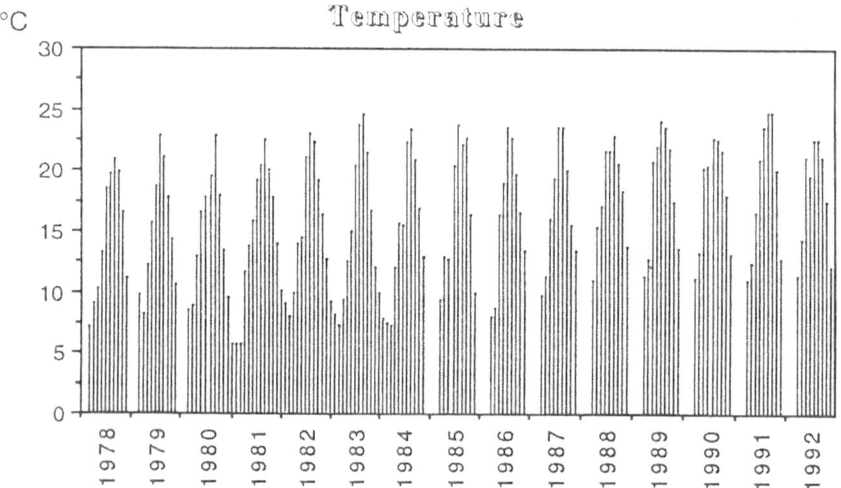

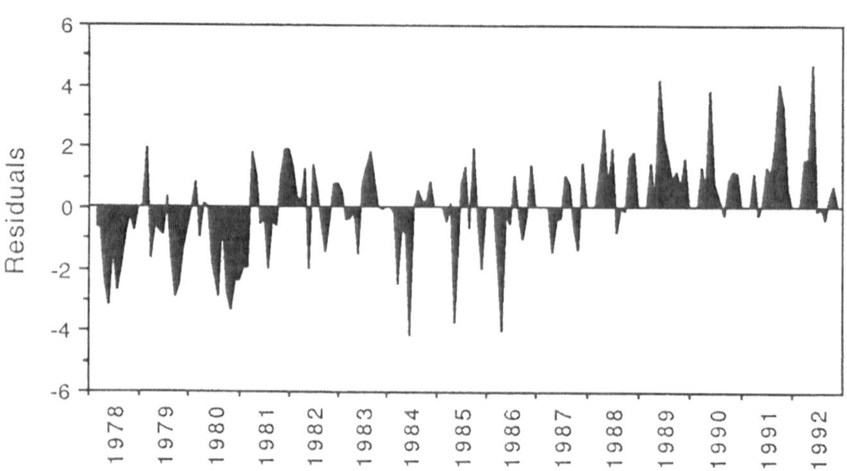

Fig. 2. Averaged surface and bottom water temperatures in the Gironde estuary, station E. The upper panel gives the original data ($n = 147$) and the lower graph displays deviations from the long-term monthly averages.

relation between air and water temperature residuals: $r=0.568$; $p<0.01$).

On average, the river flow is the lowest in August–September and the highest in February (Table 2). There was a general decrease of the river flow during the period of investigation, except in 1988 and late 1992 when a significant increase was observed (Fig. 3). The most striking feature is a protraction of the period of low water discharge. The residuals clearly show the general decreasing trend of the river flow with the exceptions of 1988 and 1992. The correlation with time was −0.417 ($p<0.01$) for the period 1978–1991. The general decrease of the river flow was mainly due to a deficit in pluviosity occurring in 1985–1987 and from 1989 to early 1992 (average monthly deficit: −10 mm and −15 mm respectively). The correlation between river flow and pluviosity (seasonally detrended) was 0.465 ($p<0.01$).

The salinity regime at the sampling station falls in the oligo-mesohaline range. Contrary to the river flow, the general trend for salinity was an increase from 1978 to 1991 (correlation with time, $r=0.541$, $p<0.01$). Exceptions were found in 1988 and late 1992 (Fig. 4) as in the case of river flow. As expected the general trend of the salinity evolution is inverted compared with that of the river flow with a significant correlation (Table 3). Besides these long-term changes there were the normal annual fluctuations in salinity (the highest values in late summer-early autumn and lowest values in winter; Table 2). There was a good positive correlation between salinity and temperature (seasonally adjusted).

For the suspended particulate matter (SPM), the trend was slightly different, although the general tendency was a decrease with time ($r=-0.341$, $p<0.01$). In a first phase a general increase was observed between 1978 and 1981–1982. Values were higher than the mean in 1982–1984, and then from 1984 onwards, a very sharp decrease occurred (Fig. 5). The suspended matter concentration depends on the river flow and resuspension. On a seasonal basis the variations of both river flow and SPM concentrations are similar (Table 2), however, in the long-term, the correlation between SPM and river flow was not significant (Table 3). Conversely, a good correlation ($r=0.543$, $p<0.01$) was found between SPM concentration (in mg l^{-1}) and net velocity (m s^{-1}) computed from the current velocity measurements made during each tidal cycle (data not shown, but see Castel, 1993). This indicated that resuspension processes due to erosion were probably important in determining the level of SPM concentration.

Chlorophyll *a* concentrations are high in spring and in autumn (Table 2). In the middle part of the Gironde estuary autochthonous primary production is quite nil, due to very high turbidity (Irigoien & Castel, submitted). The spring maximum originates from freshwater phytoplankton (mainly as degraded cells) and from resuspension of benthic algae due to high river flow. The autumn maximum, occurring at high salinities, is mainly the result of the intrusion of marine phytoplankton. Thus, chlorophyll has two origins: riverine and marine-polyhaline. During the study period chlorophyll concentrations tended to increase from 1979 onwards (Fig. 6) but the correlation with time was not strong ($r=0.176$; $p<0.05$) due to very high values found in 1980–1981. These high values corresponded to input of freshwater phytoplankton. On the whole, chlorophyll *a* concentrations were positively correlated with salinity (Table 3), indicating that an increase of marine influence was accompanied by an increase of phytoplankton biomass.

Eurytemora populations

Eurytemora affinis is clearly the most dominant species in the oligohaline zone of the Gironde estuary (Castel, 1981), as in most estuaries. It is most abundant in spring, with a maximum in May (Table 2). Lowest densities are observed in September when temperature and salinity are high. During the study period, the *Eurytemora* population showed an oscillatory behaviour, especially when considering the peaks of abundance, with a maximum in 1984 and in 1991 (Fig. 7). However, the trend for the mean abundance (residuals) was not the same as the trend for maximum abundance. During the first part of the study, the trend for Eurytemora was inversely related to that of the suspended matter (compare Figs 7 and 5). However, the correlation was non existent over the whole sampling period (Table 4). During the second phase (1984 onwards), *Eurytemora* abundances followed the general trend observed for the river flow. During this period, the river flow decreased significantly. Furthermore, the current velocity decreased, especially the ebb current (Castel, 1993). It is likely that during this period the *Eurytemora* population was affected by high salinity. Over the whole sampling period, *Eurytemora* was inversely correlated with temperature, salinity and chlorophyll *a* and positively correlated with the river flow (Table 4). The surprising negative correlation between copepod

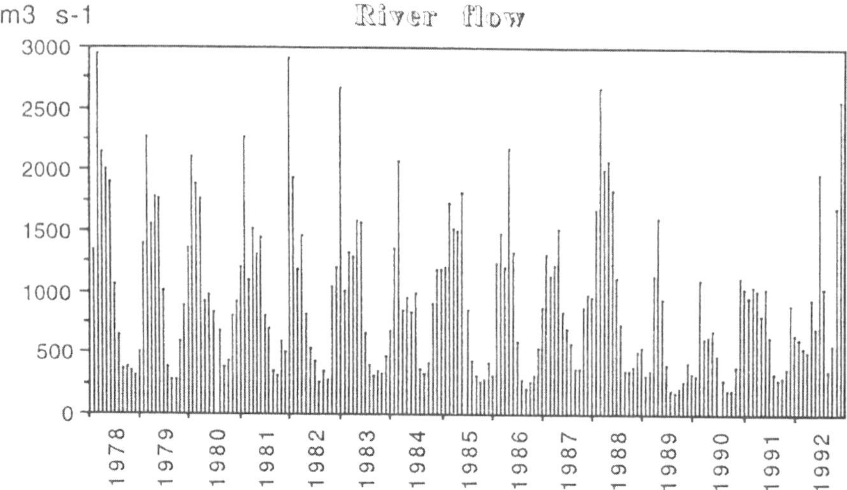

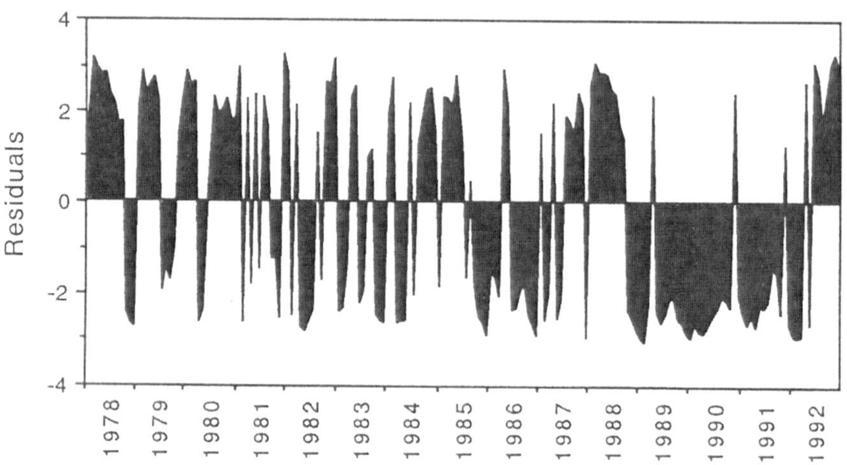

Fig. 3. River flow (Garonne + Dordogne) in the Gironde estuary. The upper panel gives the original data ($n = 180$) and the lower graph displays deviations (log transformed) from the long-term monthly averages.

abundance and chlorophyll *a* was probably due to a covariation between chlorophyll *a* and salinity.

The proportion of males (relative to the adult population) generally exceeds 50% in spring and autumn, the lowest values (<45%) being observed in August–September (Table 2). No clear long-term trend was observed for the sex ratio (correlation with time $r = 0.077$), the residuals being distributed at random around the mean (Fig. 8). Only the chlorophyll *a* concentration did correlate negatively with the percentage of males (Table 4).

On average, the proportion of females carrying eggs is relatively low in spring (~25%), increases in summer (~30%) and shows the highest values (~35%) in autumn (Table 2). On the long-term, high percentages of ovigerous females were recorded between 1979 and 1982, in 1986–1987 and in late 1992 (Fig. 9). There was no correlation between residuals and time ($r = 0.089$). Conversely, a significant positive correlation was found between residuals and SPM concentrations (Table 4). Residuals were also negatively correlated with river flow and chlorophyll *a* but to a lower extent.

The percentages of copepodites are maximum in spring (~70%) i.e. during the period of maximum abundance of the population (Table 2) and show the lowest values (~50%) in September–October when the salinity intrusion is maximum. During the study

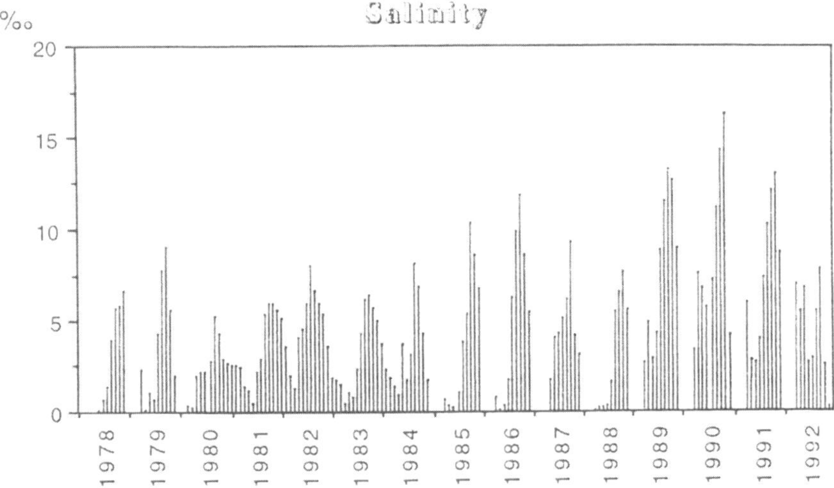

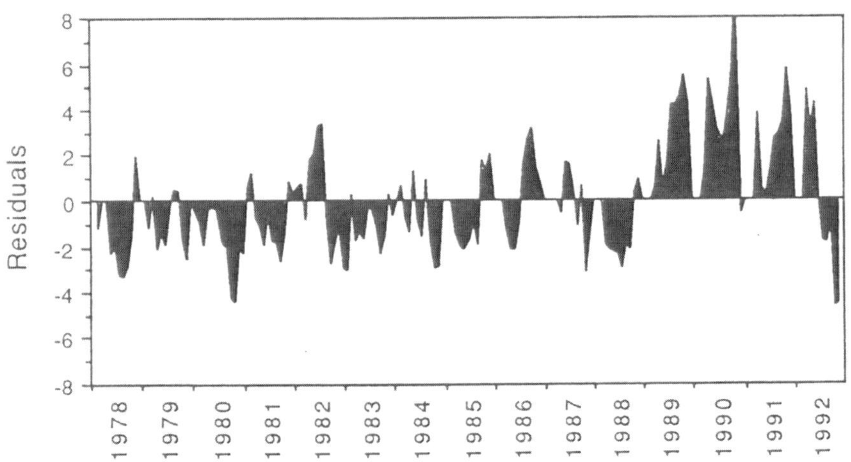

Fig. 4. Averaged surface and bottom salinities in the Gironde estuary, station E. The upper panel gives the original data ($n = 147$) and the lower graph displays deviations from the long-term monthly averages.

period very low values were observed in 1981–1982. Despite the great variations of the residuals observed during the 1978–1992 period (Fig. 10) the percentages of copepodites tended to decrease with time ($r = 0.187$; $p<0.05$). A significant negative correlation was found between residuals and temperature (Table 4).

On a seasonal basis the number of eggs per egg sac is largest just before the period of maximum abundance of *Eurytemora* populations and the lowest values are recorded in summer (Table 2). The clutch-size (Fig. 11) significantly decreased during the 15-year period (correlation with time 0.665; $p<0.01$). Highest numbers of eggs per female were recorded in 1979 (average = 16.9) and the lowest in 1991 (average = 6.8). Residuals were inversely correlated with temperature and salinity, and positively correlated with river flow and SPM concentrations (Table 4).

Discussion

Two main factors can explain the long-term trend of zooplankton population densities in the Gironde estuary during the period 1978–1992: river flow and temperature. There was a clear decrease of the river flow (except in 1988 and late 1992) probably caused by the general evolution of the climatological conditions (deficit in pluviosity). Data for the other regions

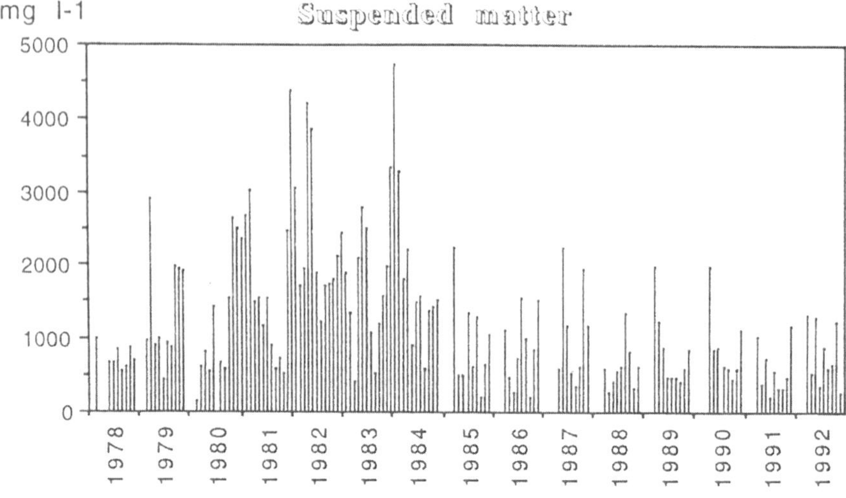

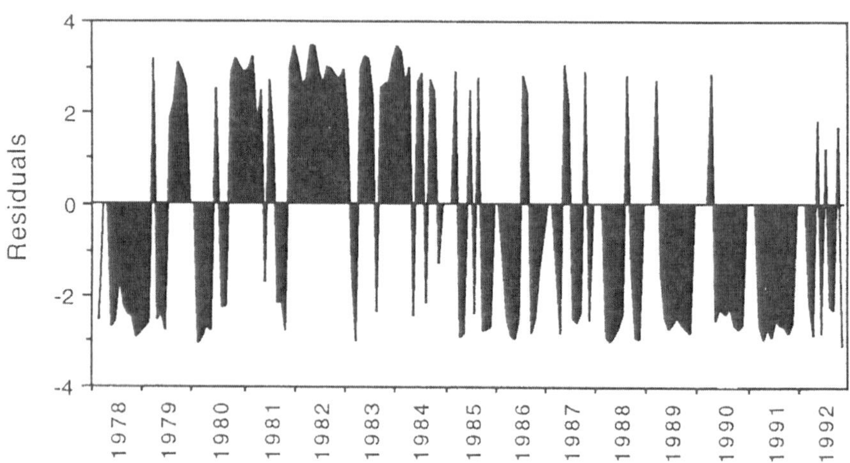

Fig. 5. Averaged surface and bottom suspended matter concentrations in the Gironde estuary, station E. The upper panel gives the original data ($n = 146$) and the lower graph displays deviations (log transformed) from the long-term monthly averages.

drained by the rivers are lacking (especially snowing in the Pyrenees mountains where the Garonne originates) but it is likely that the decrease in rainfall occurred in the whole drainage basin. The impact of the construction of dams on the rivers of the drainage basin cannot be neglected but information is lacking. The second factor is the water temperature which increased during the study period in parallel with air temperature. Generally, freshwater flowing into the estuary is warmer than seawater in summer and colder in winter. Thus it could be supposed that the trend found in water temperature would result from a change of the origin of water masses at the sampling point. Sampling was not carried out during the coldest months (i.e. December and January) when marine waters are warmer than the freshwater ones. On the contrary sampling was made during periods of thermal homogeneity or when marine waters were colder than fluvial waters. Since the general trend observed during the study period was an increase of the marine influence, the results should have shown a tendency to a decrease of water temperature. However, the difference between fluvial and marine water temperature (which never exceeds 2 °C) probably has to be taken into account to explain some anomalies during particular short periods.

To the decreasing trend of river flow are associated variations of ecological factors such as salinity and suspended matter concentration. The mean salini-

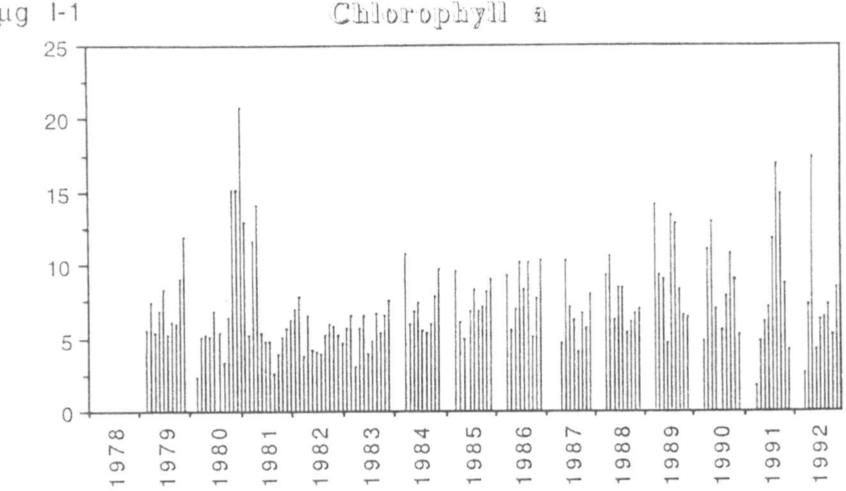

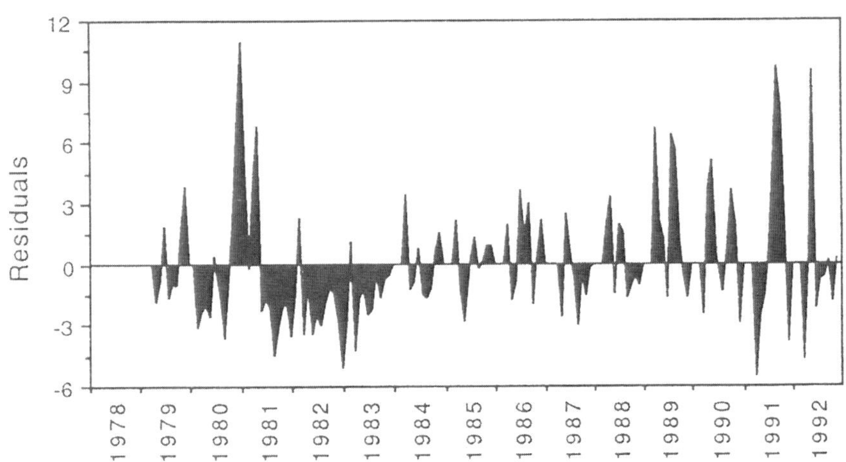

Fig. 6. Averaged surface and bottom chlorophyll *a* concentrations in the Gironde estuary, station E. The upper panel gives the original data ($n = 136$) and the lower graph displays deviations from the long-term monthly averages.

ty normally increased in correlation with the reduction of river flow. As all the silt and clay in suspension in the Gironde originates from the rivers, a consistent correlation was found between river flow and SPM concentration. Conversely, chlorophyll *a* concentration was mainly positively correlated to salinity. *Eurytemora affinis* lives preferentially in the oligohaline zone of the estuary. Its downstream distribution is limited by salinity. The surprising negative correlation between *Eurytemora* abundance, fecundity and chlorophyll *a* concentration is probably due to the strong covariance between chlorophyll concentration and salinity.

Since sampling was carried out at a fixed station, it is likely that the observed long-term trends were due to the movement of the water masses along the estuary. Figure 12 summarizes the typical distribution of SPM and chlorophyll concentrations, and of *Eurytemora* abundance along the salinity gradient. Sampling was made in spring 1992. Mean river discharge (952 m^3 s^{-1}) during the sampling period was very close to the long-term annual average (940 m^3 s^{-1}). In the upstream part of the Gironde, the high chlorophyll concentration is due to phytoplankton of freshwater origin. Downstream the maximum turbidity zone an important autochthonous primary production occurs in the marine-polyhaline zone of the estuary (Irigoien & Castel, submitted). The turbidity maximum is found in the middle part of estuary where almost no primary

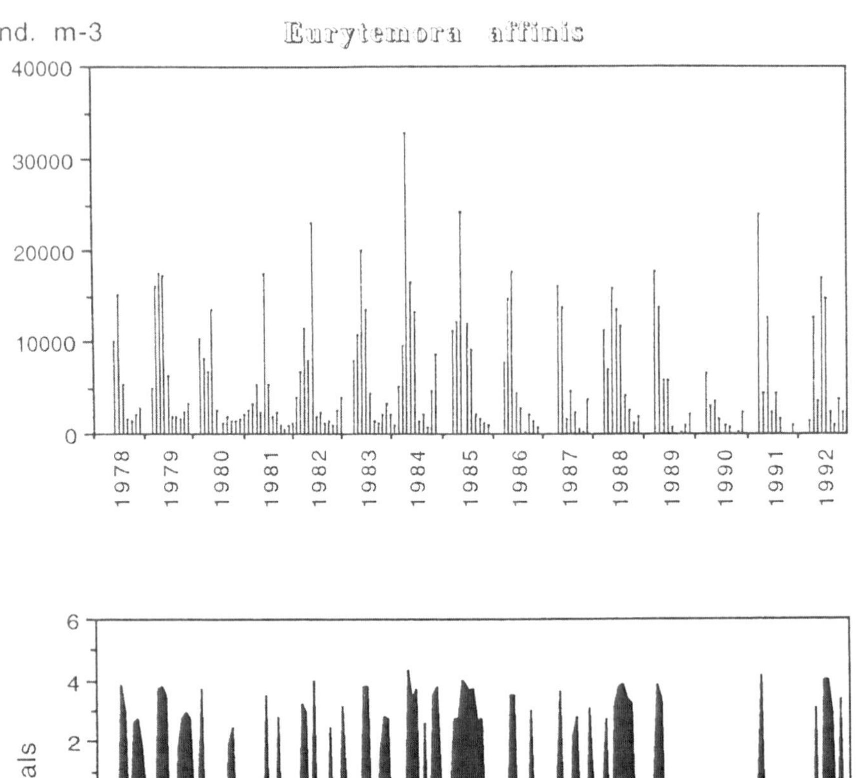

Fig. 7. Averaged surface and bottom densities of the Copepod *Eurytemora affinis* in the Gironde estuary, station E. The upper panel gives the original data ($n = 144$) and the lower graph displays deviations (log transformed) from the long-term monthly averages.

production occurs. Highest abundance of *Eurytemora* is found just upstream the maximum of SPM concentration or in the turbidity cloud if the SPM concentration is not too high (Castel, 1984; Castel & Feurtet, 1989 and Fig. 12). This distribution is the result of transport processes (Castel & Veiga, 1990) leading to a kind of 'amassment' in the zone of high turbidity (Soltanpour-Gargari & Wellershaus, 1984). Thus the long-term trend of the abundance of *E. affinis* at the fixed station can be explained by the relative position of the population and of the turbidity maximum.

This succession: freshwater phytoplankton – *Eurytemora* – turbidity maximum – marine-polyhaline phytoplankton is maintained all year round, but the longitudinal location and the extent of the compartments vary according to the river flow. During average conditions (Fig. 12) SPM concentration and *Eurytemora* abundance are high in the middle estuary, whereas chlorophyll *a* concentration is low.

During the first phase of the study (1978–1981), the river flow was very high. The turbidity maximum was pushed downstream the study area (over a distance of 15 km according to P. Castaing & J. M. Jouanneau, pers. comm.). During such periods of very high river discharge high chlorophyll concentrations from freshwater origin are to be found in the sampling area. This latter situation was met in 1980–81. *E. affinis* population, which was relatively abundant at the very begin-

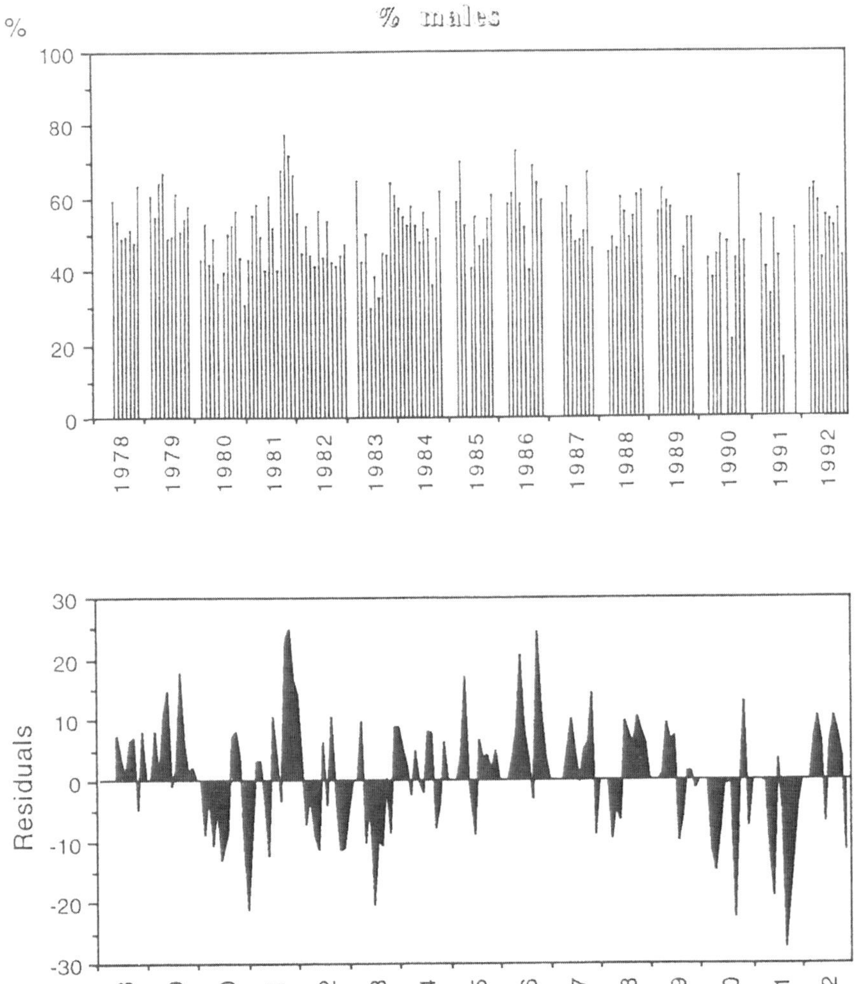

Fig. 8. Eurytemora affinis. Averaged surface and bottom percentages of males (relative to the adult population) in the Gironde estuary, station E. The upper panel gives the original data ($n = 144$) and the lower graph displays deviations from the long-term monthly averages.

ning of the study, was flushed downstream. Then, in a second phase, as a consequence of the decrease of the river flow, the maximum turbidity zone moved upstream and was situated in the sampling area in 1982–1984 (Fig. 5). This upstream migration is corroborated by the fact that, during this period, the net velocity increased and was oriented upstream (Castel, 1993). Following the migration of the turbidity zone, *Eurytemora* became very abundant at the study site (Fig. 7) whereas chlorophyll *a* concentration dropped drastically (Fig. 6). The last phase, with very low SPM concentrations, can be explained by the continuing upstream migration of the maximum turbidity zone in relation to the decrease of the river flow. In fact, during the period 1985–1991, the turbidity zone was often observed near Bordeaux in summer. In parallel to the decrease in SPM concentration, the marine-polyhaline phytoplankton developed at the sampling site. The abundance of *Eurytemora* decreased at the sampling point, the population being pushed upstream together with the turbidity maximum. Samples taken in 1989–1991 showed that the abundance was significantly higher 20 km upstream the station E (unpublished data). Furthermore, the decrease in abundance was probably accentuated by the high salinities occurring at that period. Finally, in 1992, the river flow tended to increase again. This trend was confirmed in 1993 and early 1994. As expected it was observed an increase in

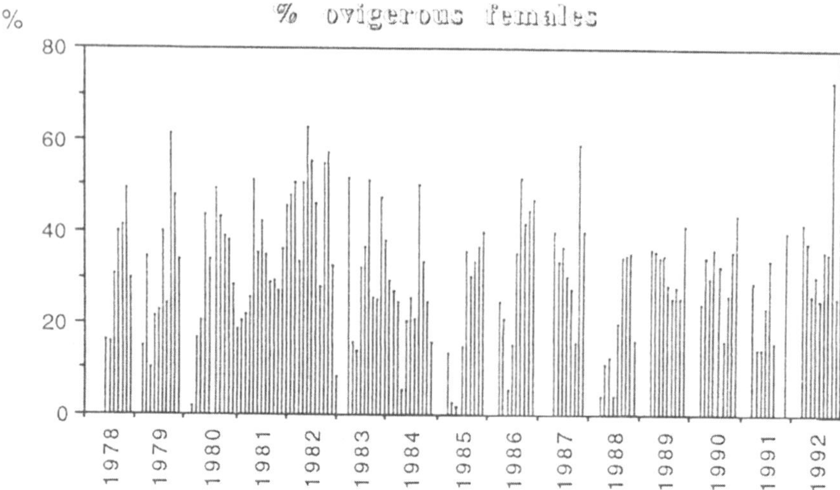

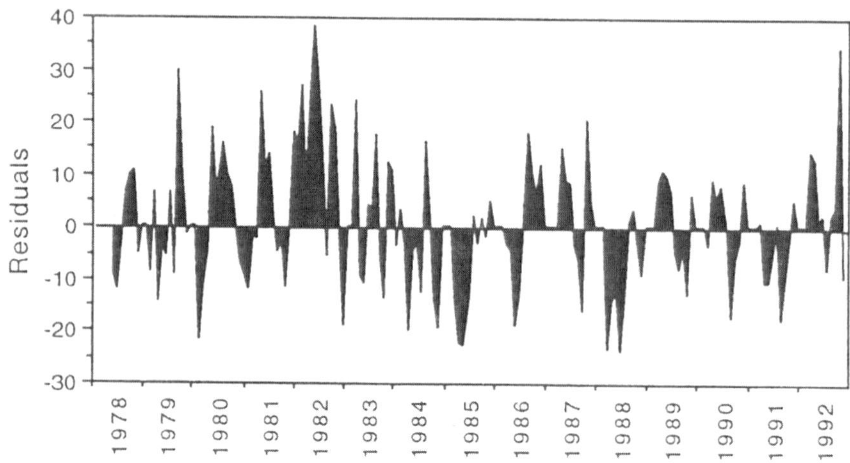

Fig. 9. Eurytemora affinis. Averaged surface and bottom percentages of ovigerous females in the Gironde estuary, station E. The upper panel gives the original data ($n = 144$) and the lower graph displays deviations from the long-term monthly averages.

Eurytemora abundance and SPM concentration whereas chlorophyll *a* concentration felt down (unpublished data for 1993 and 1994).

The general decrease in river flow and the subsequent changes of covariate factors (salinity, chlorophyll *a* and SPM concentration) had no great influence on the structure of *Eurytemora* population at the sampling point. The strongest correlation was found between % ovigerous females and SPM concentrations. According to Ladiges (1935), De Pauw (1973) and Heckman (1986) egg-carrying females tend to concentrate in shallower waters near the banks of estuaries. Observations are lacking for the Gironde, but if this observation is general, it is likely that egg-carrying females are resuspended by tidal currents as it is the case for the sediment, and are found in greater proportion in zones of high turbidity. Furthermore, it has been shown (Sandström, 1980; Vuorinen *et al.*, 1983) that ovigerous females are preferentially eaten by juvenile fish. Presumably the predation efficiency is less when SPM concentration is high, due to a decrease of visibility in turbid water, thus leading to higher proportion of egg-carrying females compared with clear water. This may also explain the correlation between clutch-size and SPM concentration, females bearing large numbers of eggs being probably more visible for predators. Clutch-size was inversely correlated with salinity. The same relationships were found for the total population,

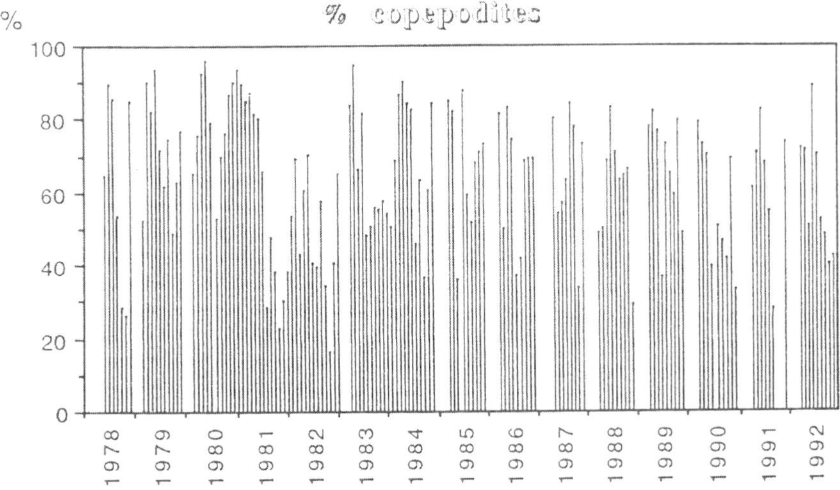

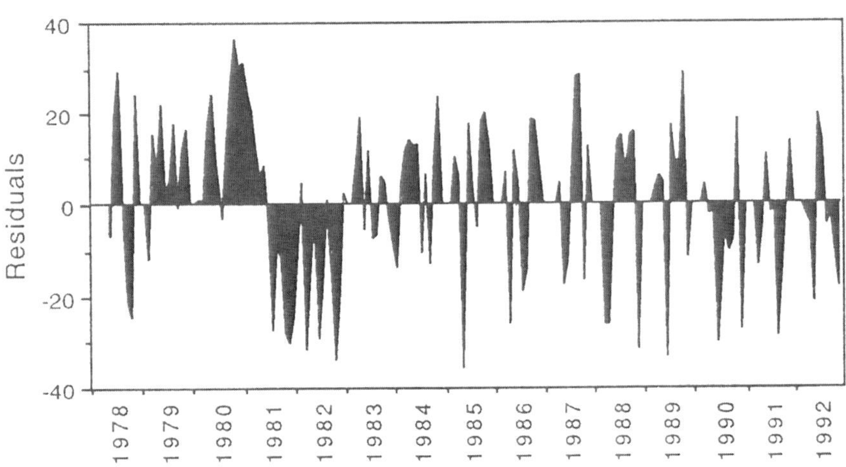

Fig. 10. Eurytemora affinis. Averaged surface and bottom percentages of copepodites in the Gironde estuary, station E. The upper panel gives the original data ($n = 144$) and the lower graph displays deviations from the long-term monthly averages.

indicating that the distribution centre of *Eurytemora* is situated in the upper estuary, in the oligohaline zone. The centre moves along the estuary according to the hydrological conditions.

The general increase in temperature observed in the sampling area clearly affected the population. Significant negative correlations with temperature were found for the clutch-size, the proportion of copepodites and the total numbers, in decreasing order of significance. The decrease in number of eggs as a function of temperature has been observed for *E. affinis* in laboratory conditions (Poli & Castel, 1983) and in the field (Castel & Feurtet, 1992; Hirche, 1992). This is probably due to variations in female body-size (Crawford & Daborn, 1986) which is partly under the dependence of temperature; in the Gironde female body size is clearly bigger in winter than in summer (Castel & Feurtet, 1989). Probably as a consequence of the decreasing trend of the clutch-size, the proportion of copepodites and the total number of *Eurytemora* declined with increasing temperature. Conversely the sex-ratio and the % ovigerous females were not affected.

The influence of long-term variation of river flow and temperature were not unexpected but the amplitude of the consequences on zooplankton populations are far from being negligible. The scale of variation of *Eurytemora* abundance is 5 to 1 (yearly average basis) and of the clutch-size is 2.5 to 1. The sampling was

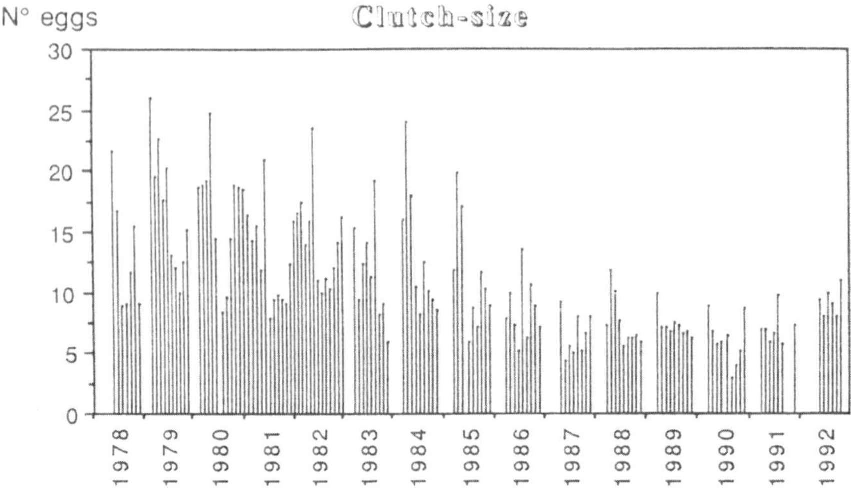

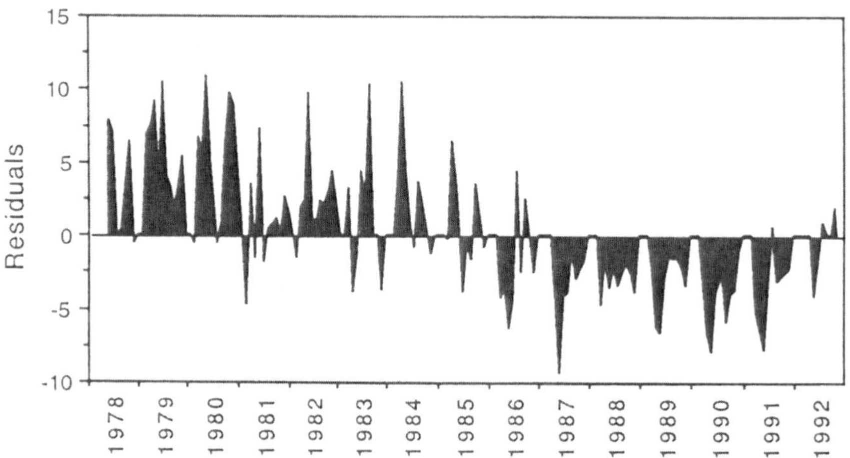

Fig. 11. *Eurytemora affinis*. Mean number of eggs per egg sac in the Gironde estuary, station E. The upper panel gives the original data ($n = 144$) and the lower graph displays deviations from the long-term monthly averages.

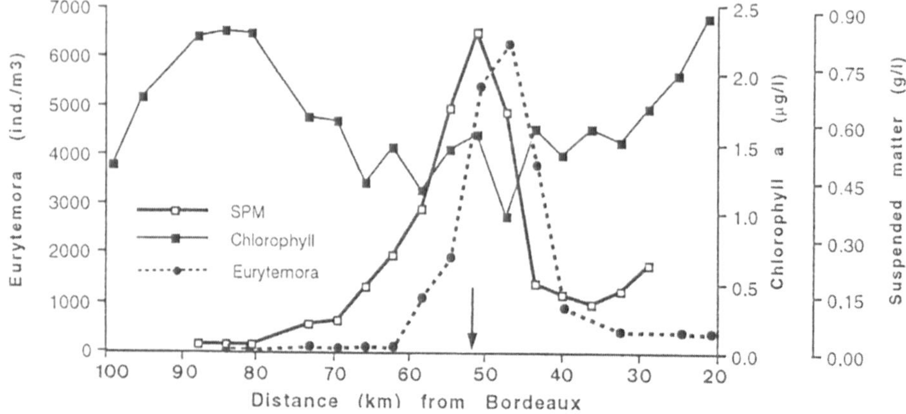

Fig. 12. Spring distribution of chlorophyll *a* and SPM concentrations, and *E. affinis* abundance in the Gironde estuary (April–June 1992; average values, from Irigoien, unpublished data). The arrow indicates the position of the station chosen for the long-term study.

made at a fixed station. The population could have migrated upstream where the ecological conditions are better. Nevertheless the population would have been limited by the freshwater boundary. Since maximum densities of *Eurytemora* in the upper part of the estuary are comparable to that found in the reference station (Castel & Feurtet, 1992; unpublished data) and owing to the morphology of the estuary (constant decrease of the water volume from the mouth to the rivers), it is clear that the total biomass of *Eurytemora* living in the estuary has decreased during the study period. Thus, it can be predicted that any long-term change in climate would have consequences on the functioning of the ecosystem.

Acknowledgments

Thanks are due to A. De Rességuier, J. Séverac and G. Oggian, captains of the R.V. 'Ebalia' (Institut de Géologie du Bassin d'Aquitaine, Université de Bordeaux I). Technical assistance along these 15 years was provided by Mrs A. M. Castel, Drs C. Courties, J. D'Elbée, V. Escaravage, A. Feurtet, X. Irigoien, J. M. Poli, B. Sautour, J. Veiga. Several colleagues are warmly acknowledged for providing unpublished data: A. Féral, Port Autonome de Bordeaux (river flow), P. Castaing & J. M. Jouanneau, University of Bordeaux (Water temperature, SPM concentration), B. Chaurial (chlorophyll determinations). This investigation was supported by the Institut Français de Recherche pour l'Exploitation de la Mer (IFREMER grants 1978–1992), CNRS/INSU(Programme National d'Océanographie Côtière 'Séries à long terme') and CEC MAST programme 'Major Biological Processes in European tidal Estuaries' (MAST grant 0024-C).

References

Castel, J., 1981. Aspects de l'étude écologique du plancton de l'estuaire de la Gironde. Océanis, Paris 6: 535–577.

Castel, 1984. Dynamique du Copépode *Eurytemora hirundoides* dans l'estuaire de la Gironde: influence du bouchon vaseux. J. Rech. Océanogr. 9: 112–114.

Castel, J., 1993. Long-term distribution of zooplankton in the Gironde estuary and its relation with river flow and suspended matter. Cah. Biol. Mar. 34: 145–163.

Castel, J. & A. Feurtet, 1989. Dynamics of the copepod *Eurytemora affinis hirundoides* in the Gironde estuary: origin and fate of its production. Proc. 22nd E.M.B.S. Barcelona, Topics in marine biology, J. D. Ros (ed.), Scient. Mar. 53: 577–584.

Castel, J. & A. Feurtet, 1992. Fecundity and mortality rates of the copepod *Eurytemora affinis* in the Gironde estuary. Proc. 25th E.M.B.S., Ferrara, Marine Eutrophication and population dynamics, G. Colombo, I. Ferrari, V. U. Ceccherelli & R. Rossi (eds), Olsen & Olsen: 143–149.

Castel, J. & J. Veiga, 1990. Distribution and retention of the copepod *Eurytemora affinis hirundoides* (Nordquist, 1888) in a turbid estuary. Mar. Biol. 107: 119–128.

Chatfield, C., 1984. The analysis of time series. An introduction, 3rd edn., Chapman & Hall, London, 286 pp.

Crawford, P. & G. R. Daborn, 1986. Seasonal fluctuations in body size and fecundity in a copepod of turbid estuaries. Estuaries 9: 133–141.

De Pauw, N., 1973. On the distribution of *Eurytemora affinis* (Poppe) (Copepoda) in the Western Scheldt estuary. Verh. int. ver. Limnol. 18: 1462–1472.

Haertel, L., C. Osterberg, H. Curl Jr & K. P. Park, 1969. Nutrient and plankton ecology of the Columbia River estuary. Ecology 50: 962–978.

Hart, R. C., 1987. Observations on calanoid diets, seston, phytoplankton relationships, and inferences on calanoid food limitation in a silt-laden reservoir. Arch. Hydrobiol. 111: 67–82.

Heckman, C., 1986. The anadromous migration of a calanoid copepod, *Eurytemora affinis* (Poppe, 1880) in the Elbe estuary. Crustaceana 50: 176–181.

Hirche, H.-J., 1992. Egg production of *Eurytemora affinis* – Effect of k-Strategy. Estuar. coast. Shelf Sci. 35: 395–407.

Irigoien, X. & J. Castel, 1993. Light limitation and distribution of chlorophyll pigments in a highly turbid estuary: the Gironde (SW France). Estuar. Coast Shelf Sci., submitted.

Jouanneau, J. M. & C. Latouche, 1981. The Gironde estuary. Contributions to sedimentology n ° 10 (H. Füchtbauer, A. P. Lisitzyn, J. D. Millerman & E. Seibold eds), E. Schweizerbart'sche Verlagsbuchhandlung, 115 pp.

Ladiges, W, 1935. Über die Bedeutung der Copepoden als Fischnarung im Unterelbegebiet. Z. Fischerei 33: 1–84.

Poli, J. M. & J. Castel, 1983. Cycle biologique en laboratoire d'un copépode planctonique de l'estuaire de la Gironde: *Eurytemora hirundoides* (Nordquist, 1888). Vie Milieu 33: 79–86.

Pritchard, D. W., 1955. Estuarine circulation patterns. Proc. Am. Soc. civ. Engrs 81: 1–11.

Sandström, O., 1980. Selective feeding by Baltic herring. Hydrobiologia 69: 199–207.

Soltanpour-Gargari, A. & S. Wellershaus, 1984. *Eurytemora affinis* – the estuarine plankton copepod in the Weser. Veröff. Inst. Meeresforsch. Bremerh. 20: 103–117.

Viitasalo, M., I. Vuorinen & E. Ranta, 1990. Changes in crustacean mesozooplankton and some environmental parameters in the Archipelago Sea (Northern Baltic) in 1976–1984. Ophelia 31: 207–217.

Vuorinen I., M. Rajasilta & J. Salo, 1983. Selective predation and habitat shift in a copepod species – support for the predation hypothesis. Oecologia (Berl.) 59: 62–64.

Vuorinen, I. & E. Ranta, 1987. Dynamics of marine mesozooplankton at Seili, Northern Baltic Sea, in 1967–1975. Ophelia 28: 31–48.

Hydrobiologia **311**: 103–114, 1995.
C. H. R. Heip & P. M. J. Herman (eds), Major Biological Processes in European Tidal Estuaries.

Secondary production of the brackish copepod communities and their contribution to the carbon fluxes in the Westerschelde estuary (The Netherlands)

Vincent Escaravage[1,*] & Karline Soetaert[2,**]
[1] *Netherlands Institute of Ecology, Centre for Estuarine and Coastal Ecology, Vierstraat 28, NL-4401 EA Yerseke, The Netherlands*
[2] *Free University of Brussels, Pleinlaan 2, B-1050 Brussels, Belgium*
Present addresses:
**National Institute for Coastal and Marine Management/RIKZ, P.O. Box 8039, NL-4330 EA Middelburg, The Netherlands*
***Netherlands Institute of Ecology, Centre for Estuarine and Coastal Ecology, Vierstraat 28, NL-4401 EA Yerseke, The Netherlands*

Key words: secondary production, copepod, estuary, Westerschelde

Abstract

The zooplankton community of the brackish part of the Westerschelde estuary (November 1989–October 1990) was dominated by two calanoid copepods, *Eurytemora affinis* and *Acartia tonsa*. *Eurytemora* was present during a longer period of the year and was much more important in terms of total abundances and biomasses than *Acartia*.

The secondary production of these species was estimated by means of the growth rate method, using weight-specific growth rates obtained from laboratory cultures (*Eurytemora*) or from the literature (*Acartia*).

Due to the substantially higher growth rates of *Acartia* compared to *Eurytemora*, total yearly productions of both communities were comparable, notwithstanding the large discrepancies in biomass. They amounted to about 5 and 6 g C m^{-2} y^{-1} by *Acartia* and *Eurytemora* respectively.

The food needed to realise this production was estimated to be about 14 and 17 g C m^{-2} y^{-1} by *Acartia* and *Eurytemora* respectively. Provided that the copepods are able to selectively ingest the phytoplankton, *in situ* net production provides sufficient carbon for zooplankton demands for a short period of the year only. As phytoplankton standing stock is very low and net phytoplankton productivity is negative from late fall to early spring, nutritional demands of the copepods have to be fulfilled by other than algal food at least during this period of the year.

Although the copepods in the brackish part can have an important impact on some food items, their contribution to total carbon fluxes in the brackish zone is negligible: each year some 6% of all consumed carbon in the brackish part of the estuary passes through the copepod food web.

Introduction

The Schelde drains large areas of Belgium, the Netherlands and France and is subjected to massive inputs of industrial and domestic sewage. This makes this river one of the most polluted in Europe (Duursma *et al.*, 1988). Due to extensive engineering works, the estuarine part of the Schelde (the Westerschelde) is the only remaining estuary in the delta area of the South-West Netherlands.

River discharge of the Schelde varies moderately on a seasonal basis (50 to 200 m^3 s^{-1}) and typically is an order of magnitude lower than tidal exchange. Thus the seawater is gradually diluted in the estuary and the salinity zones are relatively stable throughout the year. Westerschelde estuarine waters have a rather long residence time estimated as 50 to 70 days (Heip, 1988; Soetaert & Herman, 1995a).

In the upstream part of the estuary a stable turbidity maximum exists. Many of the organic and inorganic pollutants are temporarily retained in this zone and the consequently high bacterial degradation, combined with high nitrification activity there results in badly aerated conditions of the water masses (Billen *et al.*, 1988; Goosen *et al.*, 1992; Soetaert & Herman, 1995c). Chlorophyll concentrations typically are highest in the most upstream part of the estuary, due to import from the river. The unfavourable light climate in the turbid, deep water masses there and the increasing salinity results in a quick decline in phytoplankton standing stock and a sharp switch from freshwater to marine phytoplankton communities more downstream. Thus algal biomass is lowest in the brackish part of the estuary and increases towards the sea (Kromkamp *et al.*, 1995; Soetaert *et al.*, 1994). As a consequence of oxygen deficiency, copepods are absent from the turbidity maximum zone and hence they cannot profit from the high algal stocks there (Soetaert & Van Rijswijk, 1993). However, mesozooplankton biomass is peaking in the (impoverished) brackish zone immediately downstream as soon as the oxygen conditions are improving (Soetaert & Van Rijswijk, 1993; Bakker *et al.*, 1977). These high biomasses in the brackish zone can be ascribed to two species of calanoid copepods: *Eurytemora affinis* (Poppe), a perennial species and *Acartia tonsa* (Dana), only of some importance in late summer-early fall. *E. affinis* reaches maximum biomass in spring (500 mg DW m^{-3}), in summer the population declines and is then replaced by *Acartia tonsa* which gives maximum biomass of 71 mg DW m^{-3} (Soetaert & Van Rijswijk, 1993). Marine species that enter the Westerschelde in spring are declining rapidly in the estuary (Soetaert & Herman, 1994) and do not contribute significantly to the copepod biomass in the brackish part.

Due to the elevated biomasses, brackish copepod species are able to play a significant role in the estuarine trophic food web as they constitute a link to higher exploitable levels (Mauchline, 1970; Burkill & Kendall, 1982). The production of the winter species *Eurytemora affinis* was estimated by Escaravage & Soetaert (1993) but as yet we have no knowledge about the production of *Acartia tonsa*, the summer dominant species in the estuary.

Measurements of zooplankton secondary production can be made by estimating growth and mortality (yield) in cohorts over consecutive sampling intervals (Parslow & Sonntag, 1979). However, as many marine and estuarine zooplankton populations are continuously reproducing, cohorts cannot be identified. The growth-rate methods (Rigler & Downing, 1984) are a good altemative for the calculation of production in continuously reproducing populations (Kimmerer, 1987).

In this paper, weight specific growth rate estimates obtained from original laboratory experiments and from the literature were combined with biomasses measured in the field for the calculation of the copepod production in the brackish part of the Westerschelde. Energy requirements inferred from these production estimates were then compared with the different food stocks available to the copepod population.

Material and methods

Zooplankton and auxiliary environmental data were collected in the Westerschelde from November 1989 to October 1990 on 37 surveys with a mean time interval of ten days between each sampling date. During each survey three stations were sampled in the brackish part of the estuary (Fig. 1). Zooplankton was sampled by means of a pump (capacity of 200 l min^{-1}), 2.5 m above the bottom, 2.5 m below the surface and from mid-depth. From each depth 100 litres of water was poured over a 55 μm mesh. These three samples were then combined and fixed in buffered 4% formaldehyde. The organisms were separated from the high amount of suspended particles by means of density gradient separation (Heip *et al.*, 1985). Developmental stages of the copepods *Eurytemora affinis* and *Acartia tonsa* were enumerated and their biomasses were calculated by length-weight regression. The cephalothorax length of 30 individuals per copepod stage was measured using a digitizing tablet. Length-weight regressions were assessed by weighing 100 pre-measured copepods on a Cahn electronic balance (precision 0.1 μg) after a 24 hours drying process at 60 °C. For the calculations we used the average biomass of the three stations. Conversion of dry weight into carbon content was made assuming that 50% of the dry weight consists of organic carbon (Lenz, 1974).

The temperature dependent weight-specific growth rate of the Westerschelde population of *E. affinis* was ascertained from small-scale culture experiments described in Escaravage & Soetaert (1993). In these experiments, the copepods were fed with natural particulate matter and kept at six temperatures in the range

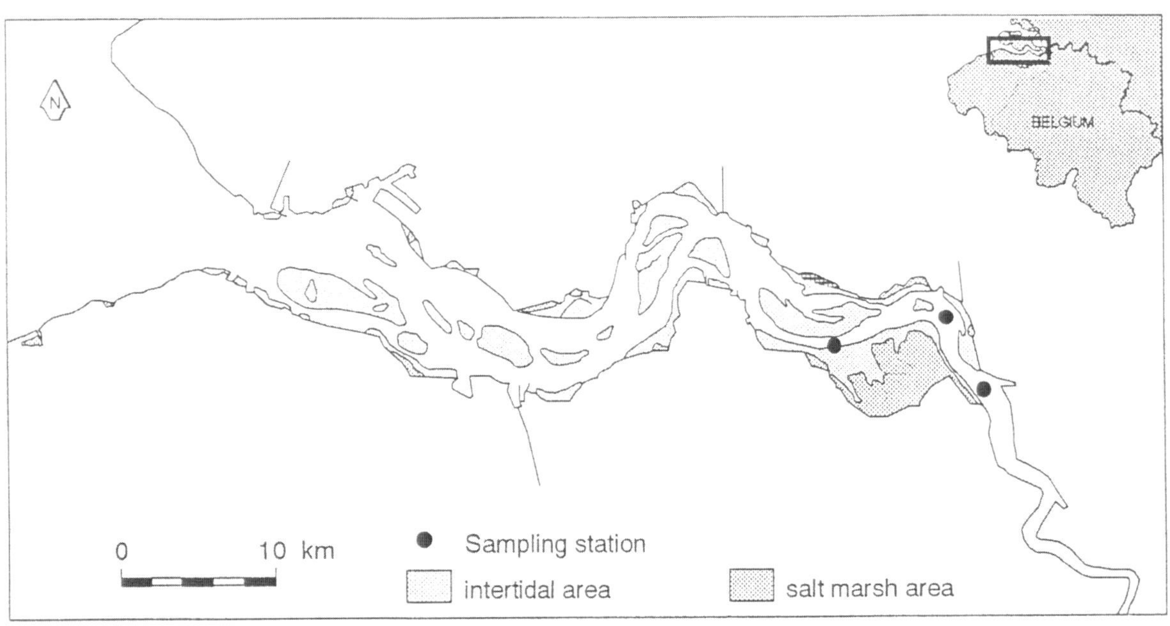

Fig. 1. Sampling positions of the three brackish stations.

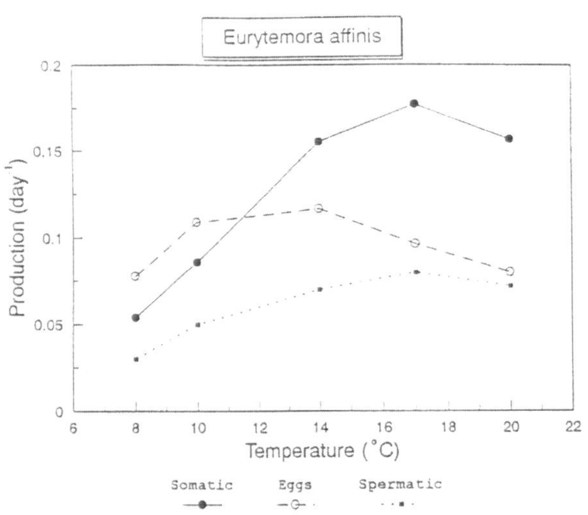

Fig. 2. Juvenile, eggs and spermatophores productions by *Eurytemora affinis* (from Escaravage & Soetaert, 1993).

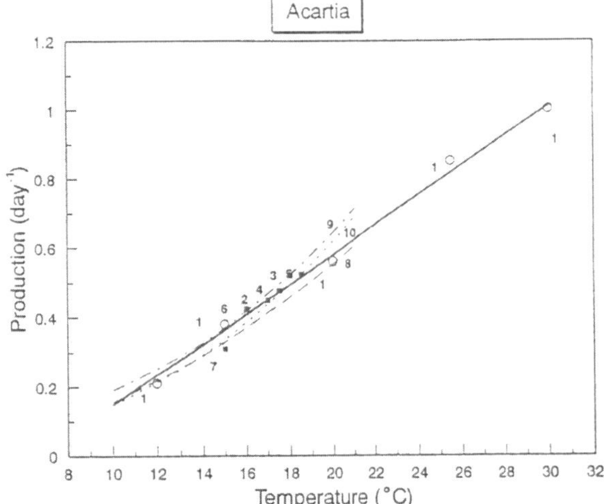

Fig. 3. Growth (G) and Fecundity (F) rates versus temperature for *A. tonsa* (or, when specified, some related species). [1] Miller *et al.* (1977)(G), [2] Bergreen *et al.* (1988) (G), [3] Kiørboe *et al.* (1985 (F), [4] Parrish & Wilson (1978) (F), [1 Stottrup & Jensen (1990) (F), [6] Raymont & Miller (1962) (F), [7] Klein Breteler *et al.* (1982) (G), [8] Uye (1981) (F-*A. steueri*), [9] Uye (1981) (F-*A. clausi*), [10] Sekigushi *et al.* (1980) (F). The solid line indicates the regression used.

of 2 to 20 °C. An attempt was made to use the same culturing procedure for *A. tonsa*. However, we only succeeded to maintain this copepod up till about the third copepodid stage, after which it died. It was then decided to resort to the existing literature on *A. tonsa* development rate, although these were mainly measured on copepods fed artificially in excess.

Based on the weight-specific growth rates (g_i) and the copepod biomasses of stage i (B_j), the integrated production (IP) for the time interval [t_1-t_2] was calculated by the formula of Polishchuk (1990):

$$IP = \sum_i g_i(t) \frac{B_i(t_2) - B_i(t_1)}{Ln[B_i(t_2)/B_i(t_1)]}(t_2 - t_1), \quad (1)$$

where $(t_2\text{-}t_1)$ = the period in between sampling, $B_i(t_j)$ = the biomass of stage i at time t_j and $g_i(t)$ the average growth rate during the sampling period ($g_i(t) = [g_i(t_1)+g_i(t_2)]/2$).

Eurytemora growth characteristics

The weight-specific growth rates realised by the copepods between 8 and 20 °C (Fig. 2) were reported in Escaravage & Soetaert (1993). The juvenile production was measured from hatching to adulthood, the reproductive activity was followed from maturation to death by collecting (counting and weighing) reproductive products (eggs for females, spermatophores for males). The optimal temperature for the egg production was at 14 °C, whereas male spermatophore production and juvenile growth was maximal at 17 °C. The male production rate attained 40 to 90% of the egg production rate.

Acartia sp. growth characteristics

According to Miller *et al.* (1977), several species of the copepod genus *Acartia* (including *A. tonsa*) grow exponentially throughout their life and growth increases with increasing temperature. Thus a common weight-specific growth rate can be used for all developmental stages at a certain temperature.

In Fig. 3 we gathered several growth rate values and several equations describing the temperature dependence of the weight-specific growth rate of *Acartia* (*tonsa* and *clausi*), obtained from the literature. In all these studies the copepods were supplied with an excess of food. There is a large homogeneity in the weight-specific growth rate of *A. tonsa* but also between *tonsa* and *clausi*.

Results

Field data

Average chlorophyll concentrations were in between 1 and 18 μg l^{-1}. They were bimodal, peaking in May and in July (Fig. 4a). The average temperature in the brackish part varied from 6 °C in January to 22 °C in August. Salinity varied from 9 to 19‰ (Fig. 4b).

The populations of the calanoid copepods *Eurytemora affinis* and *Acartia tonsa* were well separated in time (Fig. 5). *Acartia* populations mainly developed between June and October. Their highest densities and biomass were observed immediately after the second chlorophyll peak, when temperature was highest. The rest of the year the copepod community was exclusively dominated by *Eurytemora affinis*. They started to decline about one month before chlorophyll was at its maximum. Highest densities were observed when the temperature was about 15 °C. Not only did *Eurytemora affinis* reach higher abundances than *Acartia tonsa*, the species was present during a significantly larger part of the year. Hence, the yearly integrated abundance of *Eurytemora* equalled three fold the integrated abundance of *Acartia* (Fig. 5a). As individuals of *Eurytemora* were significantly larger than *Acartia*, their dominance in terms of biomass is even more pronounced (Fig. 5b; Table 1).

Temperature dependent growth rates

For *Eurytemora affinis*, the temperature dependence of the weight-specific growth or production rates (g) was best fitted by:

$g = -0.002\ T^2 + 0.06\ T - 0.37$ for the juvenile growth ($r^2 = 0.98$)

$g = (-0.971\ T^2 + 26.629\ T - 67.768)/1000$ for the egg production ($r^2 = 0.97$)

$g = (-0.577\ T^2 + 19.701\ T - 90.346)/1000$ for the spermatophore production ($r^2 = 0.96$)

with g in d^{-1} and T (temperature) in °C.

For *Acartia* tonsa, we used a simple formula, expressing the growth rate as a linear function of temperature (solid line in Fig. 3). Our regression was based on data from Heinle (1969) as reproduced in Miller *et al.* (1977). It takes a central position among the set of estimates presented in Fig. 3.

The obtained regression was:

$g = 0.043\ T - 0.28 \quad (r^2 = 0.99)$

with g in d^{-1}, T in °C.

Typically, the *Acartia* females continue to produce eggs at a rate very similar to the specific growth rate of the juveniles (Landry, 1978; Sekigushi *et al.*, 1980). Hence we used the weight-specific growth rates of juveniles as estimates for the female reproductive rates. As no information exists about the male production in

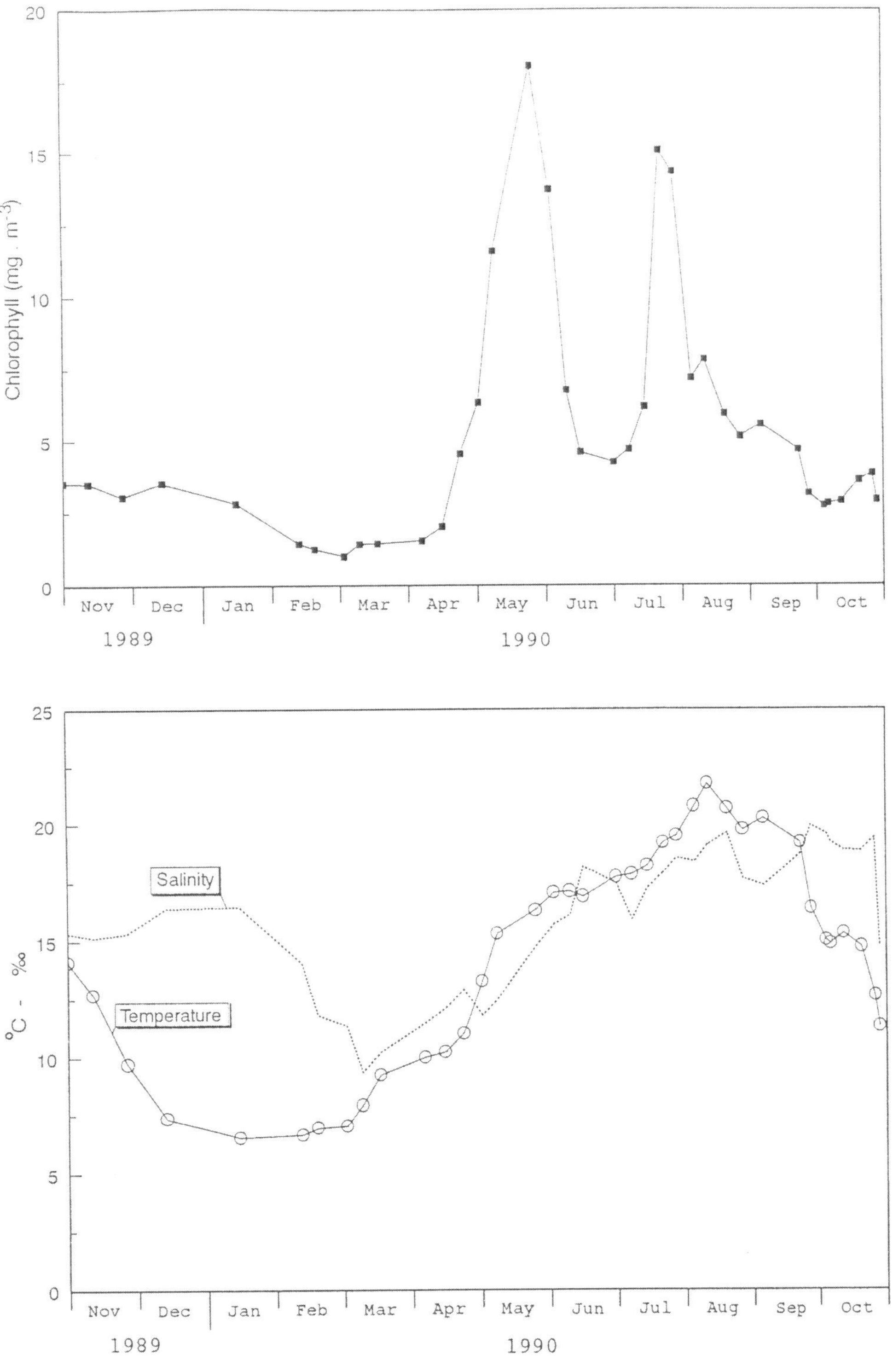

Fig. 4. Chlorophyll (above), salinity and temperature characteristics (below) of the brackish area (mean of three sampling stations).

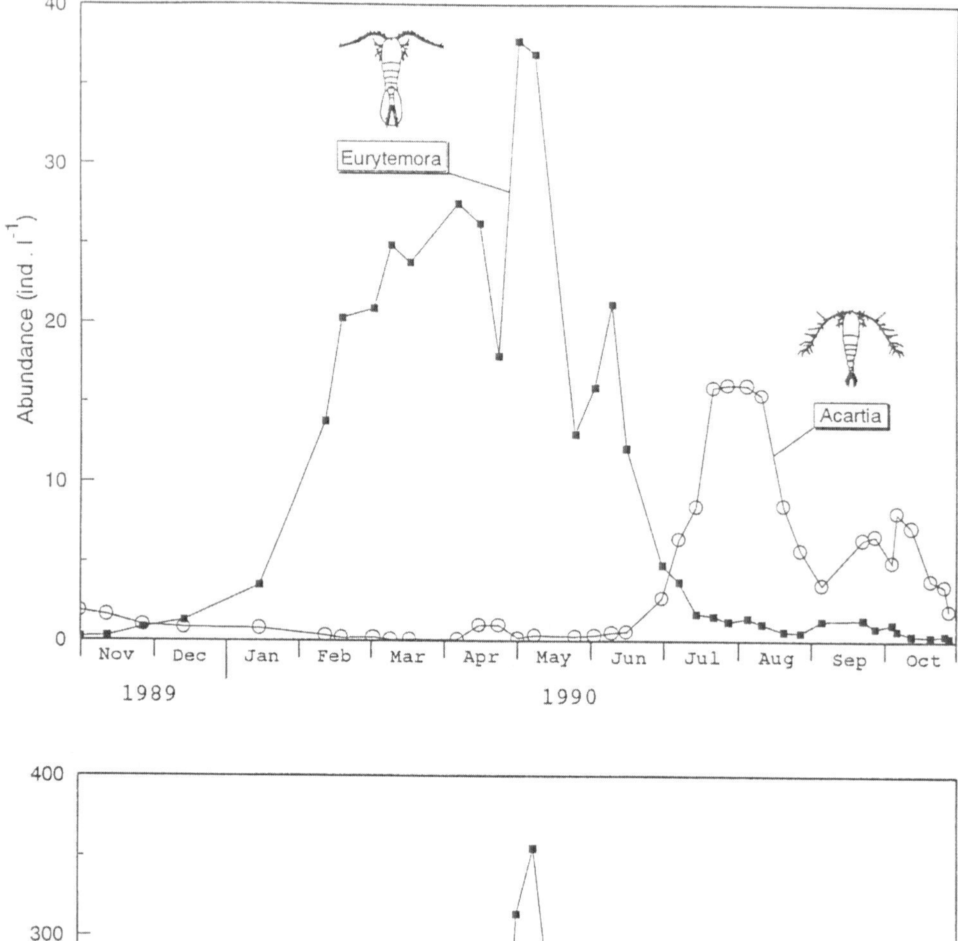

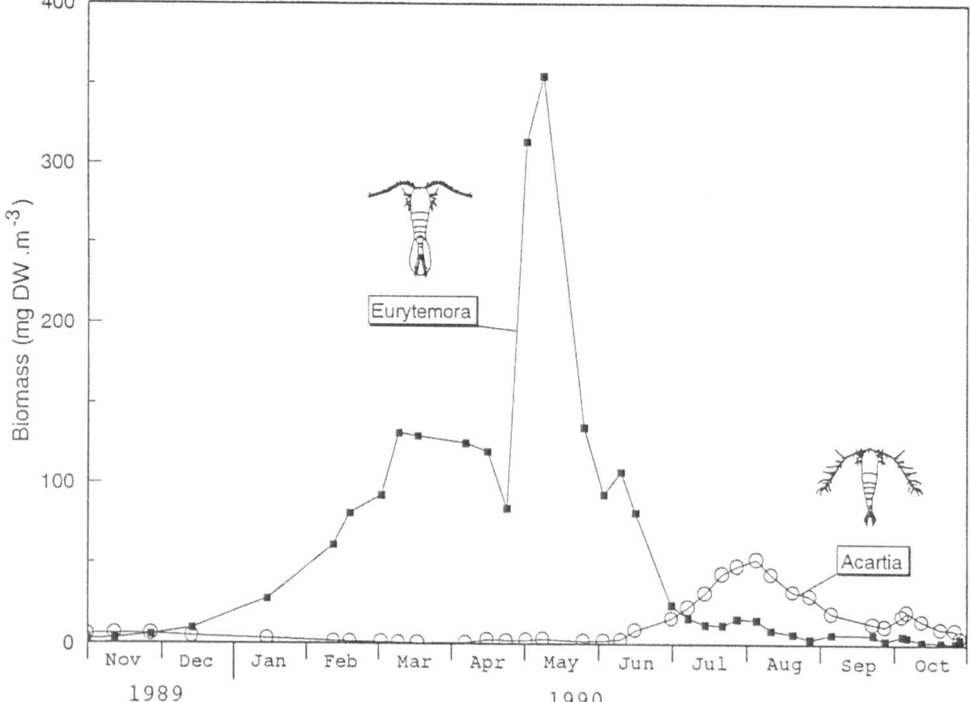

Fig. 5. a: Mean abundance (ind l^{-1}) of the copepods *Eurytemora affinis* and *Acartia tonsa* (excluding nauplii). b: Mean copepod dry weight (mg DW m^{-3}, including nauplii).

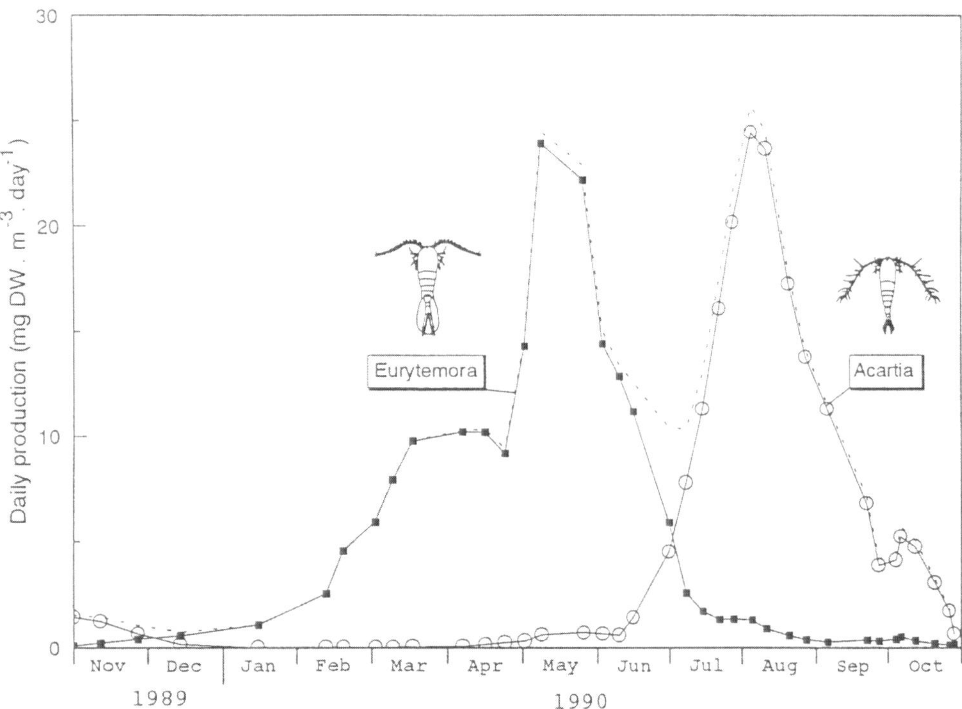

Fig. 6. a: Daily production (mg DW m^{-3} d^{-1}) of the copepods *Eurytemora affinis* and *Acartia tonsa*. The sum is indicated with a dashed line.

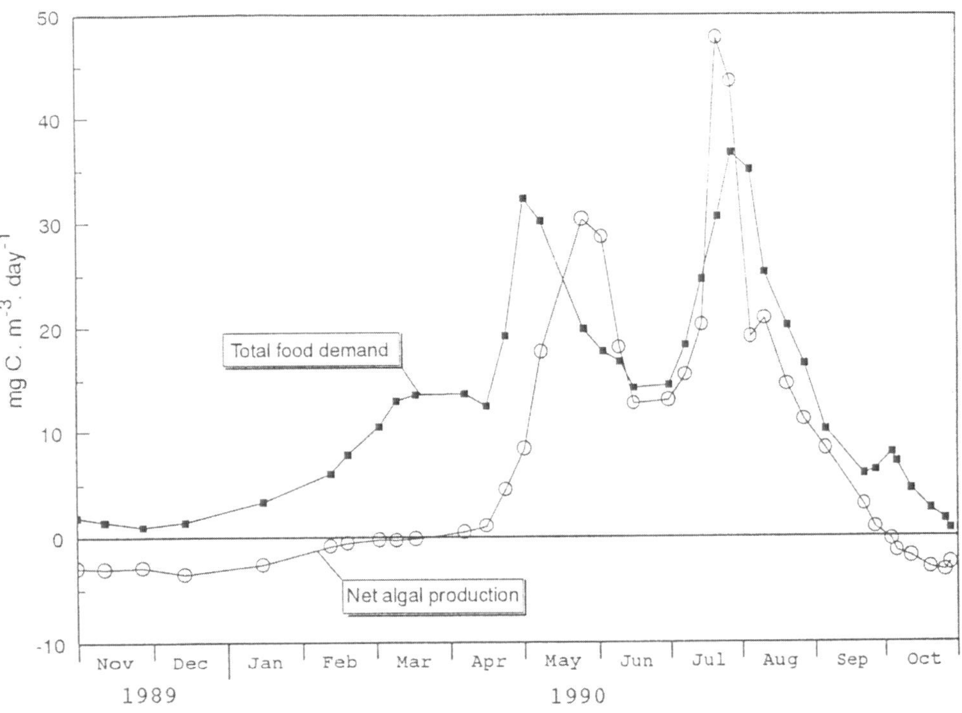

Fig. 7. Total copepod food demand and net *in situ* phytoplankton production (model result).

Table 1. Biomass, production and estimated food demand of the two most important species in the brackish part of the Westerschelde, *Eurytemora affinis* and *Acartia tonsa*. For the conversion of dry weight to carbon, a conversion factor of 0.5 g C $(gDW)^{-1}$ was used.

	Eurytemora affinis	*Acartia tonsa*
Mean biomass (mg DW m^{-3})	57	14
Mean daily production (mg DW m^{-3} d^{-1})	5.0	3.8
Annual production (mg DW m^{-3} y^{-1})	1810	1391
% Production (juveniles/females/males)	66/18/16	63/37/-
Mean daily food demand (mg C m^{-3} d^{-1})	6.5	5.4
Annual food demand (mg C m^{-3} y^{-1})	2380	1986

Acartia, males were not considered in the production estimates.

Estimated in situ copepod production and food demand

Field biomass measurements and temperature dependent growth or production rates were incorporated in Equation 1 to produce for each time interval an estimate of the copepod production. For *Eurytemora*, we distinguished three compartments: juveniles, females and males, each with their own rates of growth. *Acartia* biomass, excluding males, was considered as a whole. The two species had similar annual production values (Table I). Juveniles accounted for about 60% of total production activity in the two populations. The production realised by the males represented 16% of total *Eurytemora* production. As no male production was considered for *Acartia*, our estimate could underestimate the *in situ* production. All in all, annual productions amounted to about 1.4 and 1.8 gram Dry Weight per m^3 by *Acartia* and *Eurytemora* respectively. Assuming an average depth of 7 metres and a conversion of 0.5 gram carbon per gram dry weight, this amounts to a production of about 5 and 6 g C m^{-2} y^{-1} by *Acartia* and *Eurytemora* respectively.

The temporal evolution of daily production is given in Fig. 6. The maximum daily production of *Eurytemora affinis* was slightly lower than of *Acartia* but the species was productive during a larger part of the year. The total copepod production (ignoring the less dominant species) was bimodal and peaked in late spring and late summer.

Based on these production estimates we roughly estimated the total ingestion of the populations.

Barthel (1983) proposed an estimate of the gross growth efficiency (P/I) based on the following formula (Tranter, 1976): $[P/I = U'/100 - R/I]$. For *Eurytemora*, U', the assimilation efficiency was estimated as 89%, the mean ingestion rate (I) was estimated as 1.5 d^{-1}, the mean respiration rate (R) was 0.77 d^{-1} at 15 °C (Barthel, 1983). Thus the mean gross growth efficiency was estimated as 38% and the ingestion of *Eurytemora* was calculated as: $I = P/0.38$.

For *Acartia tonsa*, the relationship between ingestion rate (I, in d^{-1}) and production (P, same units) was estimated in the literature as:

$P = 0.23I - 0.09$	(Kleppel, 1992, where P is the egg production)
$P = 0.36I + 0.10$	(Kiørboe *et al.*, 1985)
$P = 0.44I + 0.081$	(Bergreen *et al.*, 1988)

The 'average' regression then gives $P = 0.35I + 0.03$. For comparison with the equation obtained for *Eurytemora*, we ignored the latter term and hence ingestion of *Acartia* was estimated as $I = P/0.35$.

The average estimated *in situ* food demand (Table 1) was about 2.0 and 2.4 g C m^{-3} y^{-1} or 14 and 17 g C m^{-2} y^{-1} by *Acartia* and *Eurytemora* respectively. The total food demand of the copepods was highest in the period from April to August (Fig. 7); it peaked in late spring and late summer.

Discussion

The *Eurytemora* / *Acartia* species succession is a phenomenon common to the brackish zone of many estuaries. It has for instance been observed in the Gironde (Castel, 1991), Ems-Dollart (Gaedke, 1990) and in the Patuxent River (Heinle, 1969). According to Bradley

(1975), a competition with *A. tonsa* might be an important factor influencing the distribution of *E. affinis* in summer. The feeding mode of *Acartia* on phytoplankton was assumed to give this copepod a competitive advantage in summer (Bakker & De Pauw, 1975). In the Westerschelde, the decline of the *E. affinis* population causes a decrease in total copepod grazing. At the same time the contribution of *Acartia* to total ingestion is increasing (Fig. 6). This 'gap' in total copepod grazing could indicate that apart from competitive displacement between both species, another factor should be responsible for the decline of *Eurytemora*. Several other hypotheses have been put forward to explain disappearance of the species, e.g predation of *Eurytemora nauplii* by *Acartia* (Bakker & De Pauw, 1975), selective predation of mysids or larger organisms preferentially on *Eurytemora* (Heinle & Flemer, 1975; Castel & Veiga, 1990) or seasonal changes in food quality or reproductive status (Hirche, 1992). As yet it is unclear which of these factors is responsible for the decline of *Eurytemora* populations in the Westerschelde estuary. *Acartia tonsa* has a growth efficiency which is almost four times as large as that of *Eurytemora*, and this discrepancy increases with temperature. These strong differences in growth rates could explain the fastness with which the transition between both species takes place. But in the Westerschelde, growth rates of *Acartia tonsa* are higher even at low temperatures, well before the species is seen to increase in relative abundance. Thus it appears that different responses of growth to temperature alone cannot explain the offset of *Acartia* increase and here too other, unknown factors, should be invoked.

According to Huntley & Lopez (1992), a single exponential function can describe the temperature dependence of growth for all marine copepod species. In the Westerschelde, this assumption does not hold: the (estuarine) copepod *Eurytemora affinis* has growth rates which are substantially lower than what is predicted using the general (marine) equation from Huntley & Lopez (1992), whereas those of *Acartia tonsa* are somewhat higher. Similar low growth rates for *Eurytemora affinis* were demonstrated from other studies (Poli & Castel, 1983; Burkill & Kendal, 1982; Heinle & Flemer, 1975; Vuorinen, 1982) and annual production/biomass ratios are remarkably similar between estuaries (Escaravage & Soetaert, 1993). Thus it appears that the lower growth of *Eurytemora*, when compared to marine species is a general phenomenon.

A possible source of error in our estimates lies in the use of laboratory-defined growth rates for the calculation of field production estimates. Both temperature and food availability are known to play a significant role in the copepod production activity (Burkill & Kendall, 1982; Durbin *et al.*, 1983; Klein Breteler & Gonzalez, 1988). According to Miller *et al.* (1977), isochronal and thus exponential growth does not occur by *Acartia* species when excess food is not present. Thus we could have overestimated growth rates of this species. However, for the Westerschelde, it has been shown that individuals of *E. affinis* were growing at very similar rates whether cultured in excess algal food or by using natural food as culture medium (Escaravage & Soetaert, 1993). Like *E. affinis*, *A. tonsa* is an opportunistic grazer, combining microzooplankton, phytoplankton and detritus diets (Roman, 1984; Kleppel, 1992). Moreover, growth rates obtained in cultures of this species were very comparable, even when fed other algae (Fig. 3). Thus it seems reasonable to assume that laboratory-defined growth rates of *A. tonsa* can be used for the estimation of secondary production in the field, as suggested by Huntley & Lopez (1992). The fact that we were unable to culture *Acartia* in our small-scale experiments whereas *Eurytemora* grew well, was merely due to the absence of stirring in the culture vessels. Individuals of *Eurytemora* were able to feed from particles on the bottom whereas individuals of *Acartia* were not.

The diet of calanoid copepods most often consists of phytoplankton, but they can obtain food from any known stock of organic matter (Poulet, 1983; review in Kleppel, 1993). Thus in some attempts to define a food budget of a filter-feeding zooplankton community, it has been necessary to assume that part of the zooplankton diet is obtained from detrital material in the water column. This is particularly so during the winter months when there is very little primary productivity and a low algal standing stock. Based on production measurements of the dominant estuarine copepod *Eurytemora affinis* and primary production measured from light and chlorophyll data, Heinle & Flemer (1975) estimated that algal production in the Patuxent river estuary (USA) was too small to satisfy the carbon requirements of this copepod. They postulated that the population of *Eurytemora affinis* thrives on the abundant detritus in the area. This hypothesis was adopted by Hummel *et al.* (1988) for the Westerschelde estuary. They suggested the existence of two food chains, one based on detritus in the brackish part of the estuary, the other one based on primary production in the marine part of the estuary. Since that paper much scientific effort has been directed towards the Westerschelde estuary and the magnitude of the different food stocks and primary production is now better known.

The total amount of particulate organic carbon in the brackish part of the Westerschelde was highest from October till April (about 4 g C m^{-3} in 1989–1990), it was lowest from May to September (about 1 g C m^{-3} in 1990). If we exclude coprophagy, then in the worst possible case (combining the highest food demand of 0.04 g C m^{-3} d^{-1} with the lowest POC values) copepods would need 25 days to ingest all available POC. This is longer than the residence time of the water mass (and of POC) in this part of the estuary (Soetaert & Herman, 1995a). Hence, although not all particulate organic carbon may be available as food for the copepods, it seems likely that the high stocks of POC could meet at least part of the copepods nutritional demand.

By means of a dynamic simulation model, Soetaert *et al.* (1994) and Soetaert & Herman (1995b) were able to estimate net phytoplankton growth in the Westerschelde estuary. The brackish part of the estuary was characterised by low standing stocks of phytoplankton; it was the site of a rapid transition of freshwater-based towards marine-based phytoplankton communities (Soetaert *et al.*, 1994; Kromkamp *et al.*, 1995). Moreover, net primary production was low here and amounted only to about 15 to 30 g C m^{-2} y^{-1}. This is about the food demand of the two copepod species together (estimated as 31 g C m^{-2} y^{-1}). To see in how far the *Eurytemora* and *Acartia* communities can depend on an algal diet, we compared net phytoplankton production values with total copepod food demand during the year. Algal net production was estimated by multiplying net production/chlorophyll ratios derived from the model (Soetaert *et al.*, 1994), with observed chlorophyll values. The results are given in Fig. 7. Provided that copepods can graze all primary production, *in situ* produced phytoplankton would be sufficient for growth during a limited period of the year only, when algal biomass is at its highest (May–July). In fall, winter and spring, *in situ* algal production cannot meet nutritional demands of the copepods. In reality, only part of the phytoplankton is consumable by the copepods, implying that even less is available.

Finally we can calculate the impact of the brackish water community on the carbon budget of the Westerschelde estuary, using the simulation model of the Westerschelde (Soetaert & Herman, 1995 a, b, c). Each year some 100.10^3 ton of carbon is net imported in the entire estuary by means of the river or through waste discharges (Wollast, 1976; Soetaert & Herman, 1995b), while some 20.10^3 ton of carbon are primary produced. The estuary is only a small net exporter of carbon (some 6.10^3 ton of carbon per year, Soetaert & Herman, 1995b) but much of the refractory carbon leaves the estuary at the seaside, while more reactive organics (e.g. marine copepods, Soetaert & Herman, 1994) are imported from the sea (Soetaert & Herman, 1995c). In the brackish part of the estuary, some 65.10^3 ton of carbon is imported, mainly from the upstream part; some 4.10^3 ton of carbon is primary produced (mainly benthic algae). Total carbon consumption in this part is estimated as 30.10^3 ton of carbon per year, mainly by pelagic and benthic mineralisation processes. Considering a total volume of the brackish zone of 394.10^6 m^3 (model compartments 5–8), the amount of carbon passing through the brackish mesozooplankton can be estimated as $1.7.10^3$ ton C per year, or some 6% of all consumed carbon, some 2% of all imported or *in situ* produced carbon.

Acknowledgements.

Culture experiments were done by the first author as part of a project paid by the EEC. Field processing was done by the second author as part of a NFWO grant at the Free University of Brussels. Thanks are due to P. Van Rijswijk, who helped with sampling and processing. C. Bakker is thanked for critically reading the manuscript. This is article no. 737 from the NIOO-CEMO.

References

Bakker, C. & N. De Pauw, 1975. Comparison of plankton assemblages of identical salinity ranges in estuarine tidal and stagnant environments. II. Zooplankton. Neth. Journ. Sea Res, 9(2): 145–165.

Bakker, C., W. J. Phaff, M.v Ewijk-Rosier & N. De Pauw, 1977. Copepod biomass in an estuarine and a stagnant brackish environment of the S.W. Netherlands. Hydrobiologia, 52(1): 3–13.

Barthel, K. G., 1983. Food uptake and growth efficiecy of *Eurytemora affinis* (Copepoda: Calanoida). Mar. Biol., 74: 269–274.

Bergreen, U., B. Hansen & T. Kiørboe, 1988. Food size spectra, ingestion and growth of the copepod *Acartia tonsa* during development: implications for determination of copepod production. Mar. Biol., 99: 341–352.

Billen, G., C. Lancelot, E. De Becker & P. Servais, 1988. Modelling microbial processes (phyto- and bacterioplankton) in the Schelde Estuary. Hydrobiol. Bull. 22: 43–55.

Bradley, B. P., 1975. The anomalous influence of salinity on temperature tolerances of summer and winter populations of the copepod *Eurytemora affinis*. Biol. Bull., 148: 26–34.

Burkill, P. H. and T. F. Kendall, 1982. Production of the copepod *Eurytemora affinis* in the Bristol Channel. Mar. Ecol. Prog. Ser., 7: 21–31.

Castel, J., 1991. Comparative field study of the ecological structure of major european tidal estuaries: the Gironde estuary. Major Biological Processes in European Tidal Estuary. JEEP 92, Plymouth Jan. 29-Feb. 1 1992: 55–67.

Castel, J. & J. Veiga, 1990. Distribution and retention of the copepod *Eurytemora affinis* hirundoides in a turbid estuary. Mar. Biol., 107: 119–128.

Durbin, E. G., A. G. Durbin, T. J. Smayda & P. G. Verity, 1983. Food limitation of production by adult *Acartia tonsa* in Narragansett Bay, Rhode Island. Limnol. Oceanogr. 28 (6): 1199–1213.

Duursma, E. K., A. G. A. Mercks & J. Nieuwenhuize, 1988. Exchange processes in estuaries such as the Westerschelde, an overview. Hydrobiol. Bull. 22(1): 7–20.

Escaravage, V. & K. Soetaert, 1993. Estimating secondary production for the brackish Werterschelde copepod population *Eurytemora affinis* (Poppe) combining experimental data and field observations. Cah. Biol. Mar., 34: 201–214.

Gaedke, U., 1990. Population dynamics of the calanoid copepods *Eurytemora affinis* and *Acartia tonsa* in the Ems-Dollart-Estuary: a numerical simulation. Arch. Hydrobiol., 118(2): 185–226.

Goosen, N., P. van Rijswijk, J. Peene & J. Kromkamp, 1992. Annual patterns of bacterial production in the Scheldt estuary (SW-Netherlands). Major Biological Processes in European Tidal Estuary. JEEP 92 report, Plymouth Jan. 29–Feb. 1 1992: 109–113.

Heinle, D. R., 1969. Temperature and zooplankton. Chesapeake Sci., 10: 186–209.

Heinle, D. R. & D. A. Flemer, 1975. Carbon requirements of a population of the estuarine copepod *Eurytemora affinis*. Mar. Biol. 31: 235–247.

Heip, C., M. Vincx & G. Vranken, 1985. The ecology of marine nematodes. Oceanogr. Mar. Biol. Ann. Rev., 23: 399–489.

Heip, C., 1988. Biota and abiotic environment in the Westerschelde estuary. Hydrobiol. Bull. 22: 31–34.

Hirche, H. J., 1992. Egg production of *Eurytemora affinis* - effect of k-strategy. Est. coast. Shelf Sci. 35: 395–407.

Hummel, H., G. Moerland & C. Bakker, 1988. The concomitant existence of a typical coastal and a detritus food chain in the Westerschelde estuary. Hydrobiol. Bull. 22: 35–41.

Huntley, M. E. & M. D. G. Lopez, 1992. Temeperature dependent production of marine copepods: a global synthesis. Am. Nat., 140(2): 201–242.

Kiørboe, T., F. Mohlenberg & F. Hamburger, 1985. Bioenergetics of the planktonic copepod *Acartia tonsa*: relation between feeding, egg production and respiration, and composition of specific dynamic action. Mar. Ecol Prog. Ser. 26: 85–97.

Kimmerer, W. J., 1987. The theory of secondary production calculations for continuously reproducing populations. Limnol. Oceanogr., 32(1): 1–13.

Klein Breteler, W. C. M., H. G. Fransz & S. R. Gonzalez, 1982. Growth and development of four calanoid copepod species under experimental and natural conditions. Neth. J. Sea. Res., 16: 195–207.

Klein Breteler, W. C. M. & S. R. Gonzalez, 1988. Influence of temperature and food concentration on body size, weight and lipid content of two calanoid copepod species. Hydrobiologia, 167/168 (Dev. Hydrobiol. 47): 201–210.

Kleppel, G. S., 1992. Environmental regulation of feeding and egg production by *Acartia tonsa* of Southern California. Mar. Biol., 112: 57–65.

Kleppel, G. S., 1993. On the diets of calanoid copepods. Mar. Ecol. Prog. Ser. 99: 183–195.

Kromkamp, J., J. Peene, P. van Rijswijk, A. Sandee & N. Goosen (1995). Nutrients, light and primary production by phytoplankton and microphytobenthos in the eutrophic, turbid Westerschelde estuary (The Netherlands). Hydrobiologia 311 (Dev. Hydrobiol. 110): 9–19.

Landry, M. R., 1978. Population dynamics and production of a planktonic marine copepod, *Acartia clausi* in a small temperate lagoon on San Juan Island, Washington. Int. Rev. ges. Hydrobiol., 63(1): 77–119.

Lenz, J., 1974. On the amount and size distribution of suspended matter in Kiel Bight. Ber. dt. wiss. Kommn. Meeresforsch., 23: 209–225.

Mauchline, J., 1970. The biology of *Schistomysis ornata* (Crustacea: Misidacea). J. mar. biol. Ass. U.K. 50: 169–175.

Miller, C. B, D. R. Heinle & J. K. Johnson, 1977. Growth rules in the marine copepod genus *Acartia*. Limnol. Oceanogr. 22(5): 326–335.

Parrish, K. K. & D. F. Wilson, 1978. Fecundity studies on *Acartia tonsa* (Copepoda: Calanoida) in standardized culture. Mar. Biol. 46: 65–81.

Poli, J. M. & J. Castel, 1983. Cycle biologique en laboratoire d'un copepode planctonique de l'estuare de la Gironde: *Eurytemora hirundoides* (Nordquist, 1888). Vie Milieu 33 (2): 79–86.

Polishchuk, L. V., 1990. A comment on the 'theory of secondary production calculations for continuously reproducing populations' (Kimmerer). Limnol. Oceanogr. 32(6): 1645–1651.

Poulet, S. A., 1983. Factors controlling utilization of non-algal diets by particle-grazing copepods. A review. Oceanol. Acta, 6(3): 221–234.

Parslow, J. & N. C. Sonntag, 1979. Technique of systems identification applied to estimating copepod population parameters. J. Plankt. Res. 1(2): 137–151.

Raymont, J. E. G. & R. S. Miller, 1962. Production of zooplankton with fertilization in an enclosed body of sea water. Int. Rev. Gesam. Hydrobiol., 47: 169–209.

Rigler, H. & J. A. Downing, 1984. The calculation of secondary productivity. In: Downing J. A. & F. H. Rigler (eds), A manual on methods for the assessment of secondary productivity in fresh waters: 19–58.

Roman, M. R., 1984. Utilization of detritus by the copepod *Acartia tonsa*. Limnol. Oceanogr., 29(25): 949–959.

Sekigushi, H., I. A. McLaren & C. J. Corkett, 1980. Relationship between growth rate and egg production in the copepod *Acartia clausi hudsonica*. Mar. Biol. 58: 133–138.

Soetaert, K. & P. van Rijswijk, 1993. Spatial and temporal patterns in Westerschelde zooplankton. Mar. Ecol. Prog. Ser., 97: 47–59.

Soetaert, K. & P. M. J. Herman, 1994. One foot in the grave: zooplankton drift into the Westerschelde estuary (The Netherlands). Mar. Ecol. Prog. Ser. 105: 19–29.

Soetaert, K. & P. M. J. Herman, 1995a. Estimating estuarine residence times in the Westerschelde (The Netherlands) using a box model with fixed dispersion coefficients. Hydrobiologia 311 (Dev. Hydrobiol. 110): 215–224.

Soetaert, K. & P. M. J. Herman, 1995b. Carbon flows in the Westerschelde estuary (The Netherlands) evaluated by means of an ecosystem model (MOSES). Hydrobiologia 311 (Dev. Hydrobiol. 110): 247–266.

Soetaert, K. & P. M. J. Herman, 1995c. Nitrogen dynamics in the Westerschelde estuary (SW Netherlands) estimated by means of the ecosystem model MOSES. Hydrobiologia 311 (Dev. Hydrobiol. 110): 225–246.

Soetaert, K., P. M. J. Herman & J. Kromkamp, 1994. Living in the twilight: estimating net phytoplankton growth in the Westerschelde estuary (The Netherlands) by means of an ecosystem model (MOSES). J. plankt. Res. 16: 1277–1301.

Stottrup, J. G. & J. Jensen, 1990. Influence of algal diet on feeding and egg production of the calanoid copepod *Acartia tonsa* Dana. J. Exp. Mar. Biol. Ecol. 141: 87–105.

Tranter, D. J., 1976. Herbivore production. In: D. H. Cushings and J. J. Walsh (eds), The ecology of the seas. Chapter 9: 186–224.

Uye, S., 1981. Fecundity studies of neritic copepods *Acartia clausi* Giesbrecht and *A. steueri* Smirnov: a simple empirical model of daily production. J. exp. mar. Biol. Ecol. 50: 255–271.

Vuorinen, I., 1982. The effect of temperature on the rate of development of *Eurytemora hirundoides* (Nordquist) in laboratory culture. Ann. Zool. Fenn. 19: 129–134.

Wollast, R., 1976. Transport et accumulation de polluants dans l'estuaire de l'Escaut. In J. C. Nihoul & R. Wollast (eds). l'Estuaire de l'Escaut. Projet Mer Rapport final. Bruxelles. Service du Premier Ministre 10: 191–201.

Hydrobiologia **311**: 115–125, 1995.
C. H. R. Heip & P. M. J. Herman (eds), *Major Biological Processes in European Tidal Estuaries.*

Feeding rates and productivity of the copepod *Acartia bifilosa* in a highly turbid estuary; the Gironde (SW France)

Xabier Irigoien* & Jacques Castel
Laboratoire d'Océanographie Biologique, Université de Bordeaux I, F-33120 Arcachon, France
* *Present address: Instituto de ciencias del Mar, P. Joan de Borbó s/n, 08039 Barcelona, Spain*

Key words: Acartia bifilosa, Acartia tonsa, suspended particulate matter, feeding, productivity

Abstract

Acartia spp. are the dominant copepod species in the Gironde estuary, seaward of the turbidity maximum area. *Acartia bifilosa* develop a large population in spring and early summer whereas *Acartia tonsa* appear in late summer. High values and high variability of chlorophyll *a*/suspended particulate matter ratio are found seaward of the turbidity maximum area. Feeding rates of *A. bifilosa* were measured by fluorometry. Phytoplankton ingestion was found to be highly variable, between 8 to 80% of copepod carbon body weight. Except for adult females, copepods were heavier in summer than in winter. P/B ratios, estimated by the instantaneous growth rate method, varied from 0.03 d^{-1} to 0.14 d^{-1}. The gut contents and P/B ratios of *Acartia bifilosa* were related to chl *a*/SPM ratio. From those data, and a few obtained for *A. tonsa*, it is concluded that only in summer months phytoplankton ingestion is enough to maintain secondary production.

Introduction

Suspended particulate matter (SPM) concentrations in the Gironde estuary reach very high values (>1 g l^{-1}). Such high concentrations influence most of the biological processes. Primary production in the turbidity maximum area is very low (CNEXO, 1977) and consequently zooplankton may be food limited (Castel & Feurtet, 1989). However a stock of 66 × 10^3 t of particulate organic carbon remains in the estuary and possibly provides the copepods with an alternative food source (Heinle *et al.*, 1977), but this would be of only very low nutritive value (Lin, 1988).

The mesozooplankton of the Gironde is dominated by the copepod *Eurytemora affinis* (Poppe) in the oligo-mesohaline zone and by *Acartia bifilosa* Giesbrecht in the meso-polyhaline area. In summer, large populations of *Acartia tonsa* (Dana) also develop. *E. affinis* and *A. tonsa* have been extensively studied in several estuarine systems as well as under laboratory conditions. Conversely, works about *A. bifilosa* are scarce (Ciszewski & Witek, 1977; Viitasalo, 1992a and b).

The capacity of *Acartia* species (particularly *A. tonsa*) to feed on a large spectrum of food sources other than phytoplankton is well documented. Detritus (Roman, 1984), microzooplankton (Gifford & Dagg, 1988; Stoecker & Egloff, 1987; Robertson 1983), and copepod nauplii (Lonsdale *et al.*, 1979; Tackx & Polk, 1982) have been considered as possible food items for *Acartia* sp. It has been shown however, that *Acartia* sp. are able to select phytoplankton cells (Paffenhöfer *et al.*, 1982; Price *et al.*, 1983; Price 1988; Tackx *et al.*, 1989). Consequently the food quantity and quality have been frequently suggested as factors limiting production and could account for differences between P/B ratios in laboratory cultures and in the natural environment (Landry, 1978; Durbin *et al.*, 1983; Ambler, 1985; Beckman & Peterson, 1986; Kimmerer & McKinnon, 1987; Durbin & Durbin, 1992; Christou & Verriopoulos 1993).

Whatever the feeding behaviour of *Acartia* (selective or nonselective) one may expect a limitation of the productivity of the population in highly turbid waters. If nonselective feeding is the most important behaviour, the low nutritional quality of particles will affect copepod productivity. If, on the contrary, selective feeding is predominant, productivity will still be limited by the low quantity of phytoplanktonic cells (CNEXO, 1977; Tackx *et al.*, 1994; Sabbe pers. comm.).

In this work we investigated the influence of the high SPM concentration on the phytoplankton distribution, phytoplankton ingestion and on secondary production of *Acartia bifilosa* (some data about *Acartia tonsa* are also provided), in order to understand the impact of turbidity on the biology of the species.

Material and methods

Field campaigns combined sampling at fixed stations and along transects covering the salinity gradient. At fixed stations measurements and sample collections were made at 1 m above the bottom and 0.5 m below the water surface. Along transects all measurements and sample collection were made at 0.5 m below the surface. The average water depth in the Gironde is around 3.5 m.

Two stations were chosen (Fig. 1):

- Station J in the oligohaline area which is strongly influenced by high SPM concentrations.
- Station A in the polyhaline area, is less influenced by the turbidity maximum.

Station J was investigated as a comparison point for environmental variables in May and October 1990, March and July 1991. Station A was sampled in May 1990, April and July 1991. In each case samples for environmental variables were collected every two hour for a period ranging from 14 to 48 hours. The gut fluorescence of copepods was measured only at station A, except in October 1990 for *A. tonsa*.

In 1992 only station A was investigated each month from March to December (except August). Samples were taken for environmental variables and copepod abundance at 2-hourly intervals during a half tidal cycle. Productivity experiments were performed in April, May, June, July, September and December 1992.

Transects were performed from fluvial waters (Ambès in Fig. 1) to the marine water (Le Verdon) with samples taken every two nautical miles. All the transects started at low tide and were done during the flood, against the current, in order to have measurements in the different water masses. Sampling was conducted in August and October 1991, and in April, May, June, July, September and December 1992. During the transects, environmental variables were measured, and during the summer months of '92, copepod abundance was measured.

Temperature and salinity were measured with a Kent Eil 5005 probe. Pigments in the water were measured by fluorometry on 90% acetone extracts (Neveux, 1983). Chl *a* was transformed in C using a ratio C:Chl a of 50:1 (Dagg & Grill, 1980). Suspended particulate matter concentration (SPM) was estimated as dry weight (60 °C, 24 h) after filtration on Whatman GF/C. In May 1990, particle size measurements were performed from samples taken at stations J and A. Water was fixed with formaldehyde (final concentration 3%) and 800 particles per sample were measured under a microscope (magnification: 40×10) provided with a camera lucida and a digitalizing tablet connected to an Apple computer (KOALA software, developed by J. C. Duchêne, Arago laboratory, Banyuls, France).

Copepod (copepodid 5 and adult) gut fluorescence (Mackas & Bohrer, 1976) was measured during the 1990 and 1991 campaigns. Ingestion (I) was calculated by multiplying the Gut content (G) by the gut passage time (GPT) (Mackas & Bohrer, 1976). GPT for *A. bifilosa* was determined in the laboratory for different temperatures. Copepods were adapted to the experimental temperature 36 h before the experiment. Adults were isolated in a 15 ml vial and 0.5 ml of a *Isochrysis galbana* culture were added. Produced fecal pellets were counted every two hours for a period of 24 h, and GPT was calculated as $\mathrm{GPT} = X/t\ n$, where GPT is the gut passage time in min^{-1}, X is the number of fecal pellets, t is the time interval between counting and n is the number of fecal pellets in the copepod (Peterson *et al.*, 1990). Five replicates were performed per experiment.

Copepod abundances were estimated during 1992. Along transects samples were collected with a bucket (100 l) and filtered through a 63 μm net. At station A, four samples were performed during a half tidal phase with a 200 μm net.

The instantaneous growth rate for *Acartia* sp. was determined following the method of Kimmerer & McKinnon (1987). Zooplankton were sampled using a WP2 net (200 μm mesh size). Copepods were separated into size classes using mesh sizes of 280–200 μm and 250–200 μm. In each size class two subsamples were taken, one was immediately fixed (To), and the other one was incubated for 3 days with a concentration of approximately 20 copepods per litre. Each incubation was performed in 5 l closed bottles filled with nat-

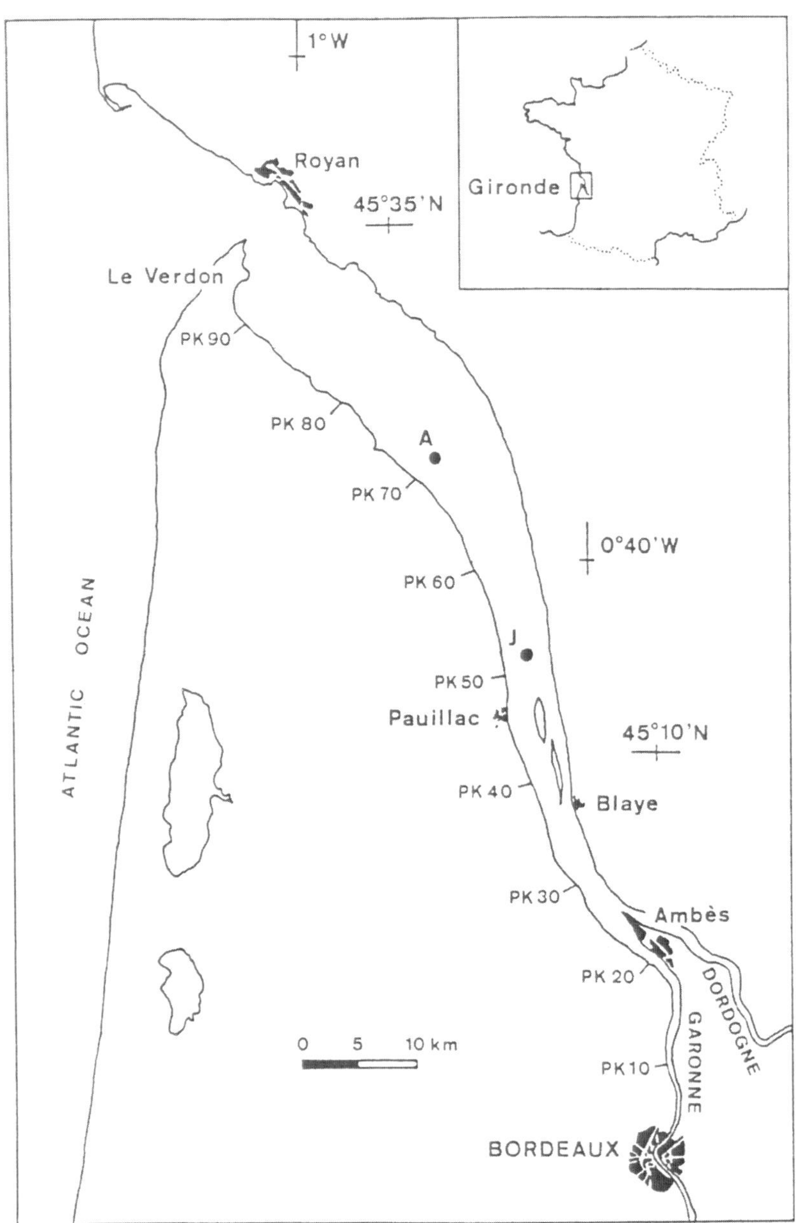

Fig. 1. Map of the Gironde estuary showing the sampling stations (A and J). Pk refers to the distance, in Km, from the city of Bordeaux.

ural water sieved through 63 μm, at the environmental temperature. At the end of the incubation (T1), to avoid weight losses, samples were fixed with glutaraldehyde, (Kimmerer & McKinnon, 1986). The prosome length of the copepods from the two subsamples (To and T1) was measured under a dissecting microscope (nearest 0.02 mm by optical micrometer). Three to five replicates were performed per experiment, and differences between To and T1 were tested by the Kolmogorov & Smirnov test. To transform sizes into biomass, a length–weight relationship was established for winter months (copepods from March 92) and another one for summer months (copepods from July 92). Twenty to fifty copepods per size class (fixed with glutaraldehyde), were measured, (prosome), rinsed in distilled water and dried (24 h at 60 °C) before weighing, (Mettler ME 22 microbalance, ±0.1 μg). The growth rate was calculated as the weight increase per unit

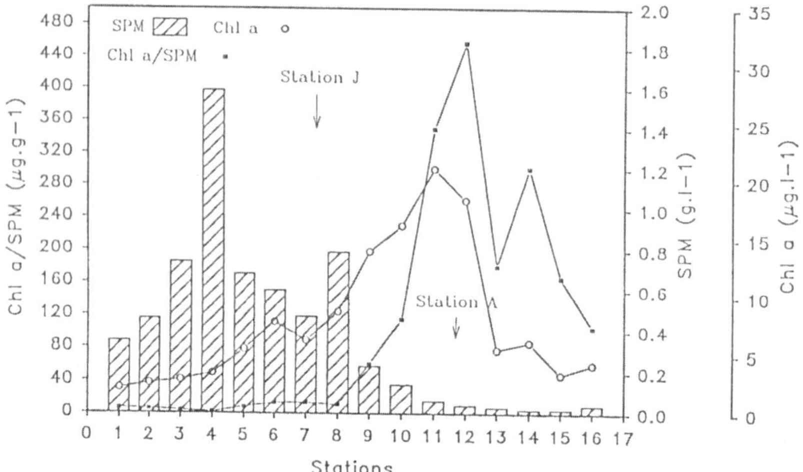

Fig. 2. Chl *a* (mg l^{-1}), SPM (g l^{-1}) and Chl *a*/SPM ratio (mg g^{-1}) along the Gironde estuary during a phytoplanktonic bloom period. August 1991. Sampling every 2 nautical miles. Station 1 (Ambès) upstream, station 16 (Le Verdon), seaward (see Fig. 1). Redrawn from Irigoien *et al.* (1993).

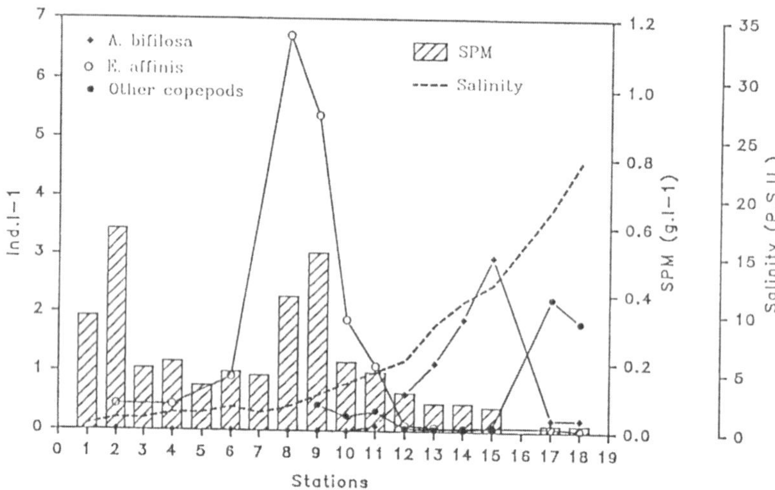

Fig. 3. Copepod abundance (ind l^{-1}), salinity (p.s.u.) and SPM (g l^{-1}) along the Gironde estuary on June 1992. Station 1 (Ambès) upstream, station 18 (I,e Verdon), seaward (see Fig. 1).

time (Kimmerer & McKinnon, 1987). This instantaneous growth rate was employed to determine the P/B ratio.

In June, July and September the Chl *a* concentration was measured in the replicates and in the control bottle of the growth experiments. Filtration rate in the incubation bottle was calculated following Newell (1973) as $F = [(\ln Ct) - (\ln Ci)] \times (V/t\ n)$, where F is the filtration rate in ml cop^{-1} h^{-1}; Ct is the concentration of Chl *a* at the end of the experiment in the control bottle; Ci is the concentration of Chl *a* at the end of the experiment in the incubation bottle; V is the volume (ml) of the bottles; n is the number of animals and t is the time. Ingestion was calculated multiplying Chl *a* concentration by the filtration rate.

Results

Pigment distribution

In the turbidity maximum area (station J) Chl *a* and pheopigments concentration were strongly linked to SPM ($r > 0.9$ for every sampling campaigns at station J). Generally, both pigment and SPM concentrations showed maximum at low tide and minimum

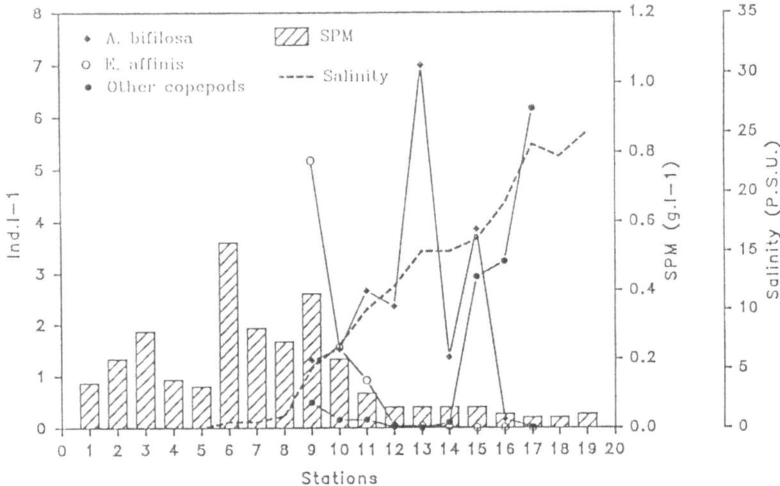

Fig. 4. Copepod abundance (ind l^{-1}) salinity (p.s.u.) and SPM (g l^{-1}) along the Gironde estuary on July 1992. Station 1 (Ambès) upstream, station 19 (Le Verdon), seaward (see Fig. 1).

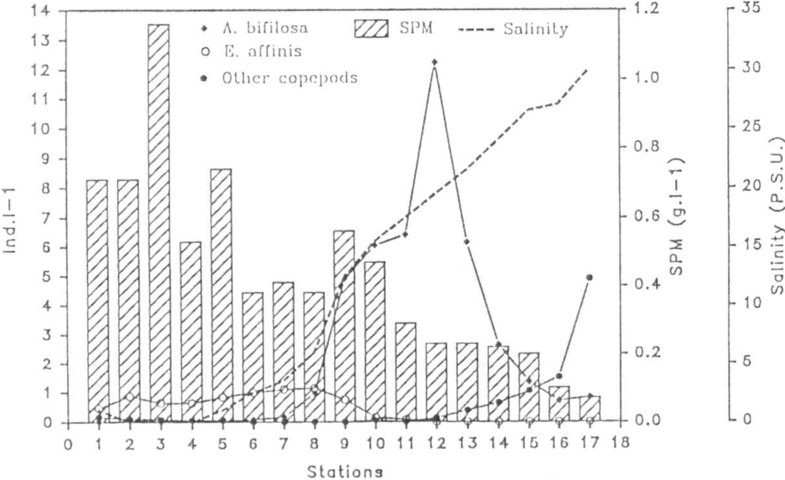

Fig. 5. Copepod abundance (ind l^{-1}) salinity (p.s.u.) and SPM (g l^{-1}) along the Gironde estuary on September 1992. Station 1 (Ambès) upstream, station 17 (Le Verdon), seaward (see Fig. 1).

at high tide (see Irigoien *et al.* (1993) for more details).

At station A the relationship between Chl pigments and SPM was lower (r from -0.36 in July 91 to 0.95 in April 92 with an average value of 0.6), and disappeared altogether during a phytoplanktonic bloom occurring in July 1991. This bloom did not appear in the oligohaline area, which is more influenced by the turbidity maximum (Fig. 2). Pheopigments were always correlated to SPM ($r>0.9$ for every sampling campaigns, A + J together).

Conversely, the Chl *a* percentage of total pigment [Chl *a*/(Chl *a* + 1.51 pheopigments)] showed an inverse relationship with SPM (log Chl *a* percentage = -0.39 log SPM + 1.45, $r=-0.55$, $p<0.01$, $n=543$, data from all campaigns, A + J together).

The Chl *a*/SPM ratio was used as an index of phytoplankton availability along the transect. The general pattern of this ratio yielded very low values in the turbidity maximum and peak values seaward of the turbidity maximum (Fig. 2). The importance of this peak varied according to the seasons, with values ranging from 10 $\mu g\ g^{-1}$ in December 1992 to 450 $\mu g\ g^{-1}$ in August 1991.

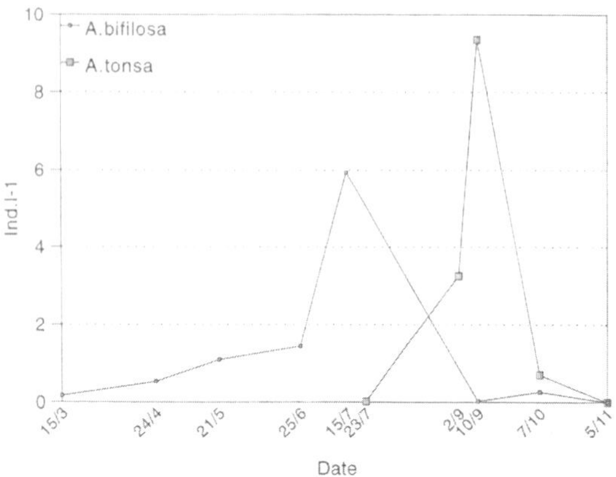

Fig. 6. *Acartia bifilosa* and *Acartia tonsa* (cop l^{-1}) at station A during 1992. Each point is the average of four samples taken with a 200 μm net during a tidal phase.

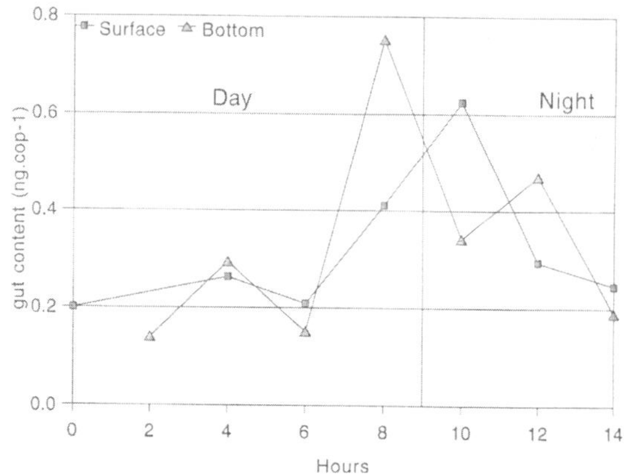

Fig. 9. *A. bifilosa.* Variation of gut contents (ng Chl *a* equivalents. cop^{-1}) during a 14 h cycle. July 1991, station A (see Fig. 1).

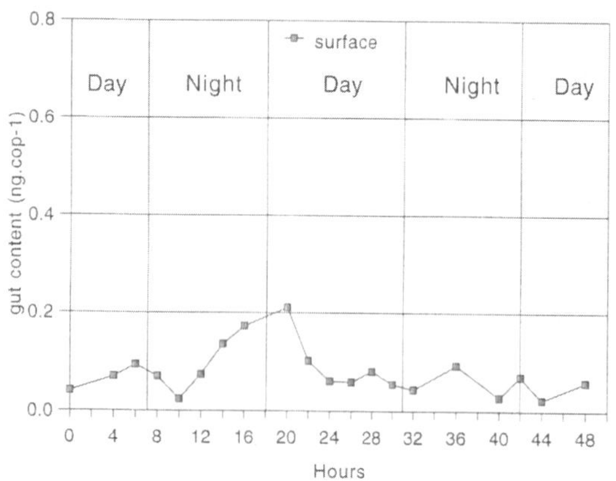

Fig. 7. *A. bifilosa.* Variation of gut contents (ng Chl *a* equivalents cop^{-1}) during a 48 h cycle. April 1991, station A (see Fig. 1).

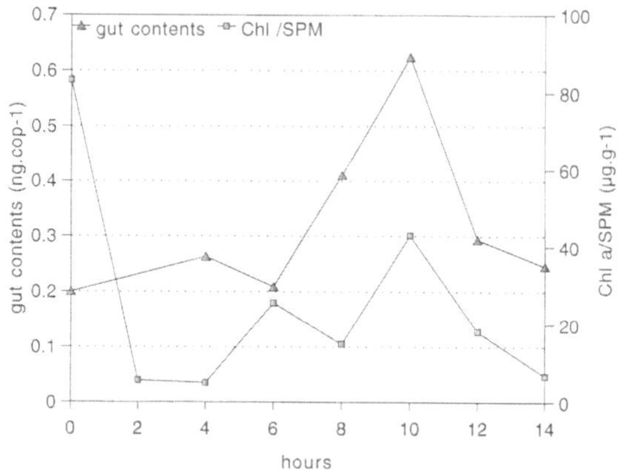

Fig. 10. *A. bifilosa.* Example of variation of gut contents (ng Chl *a* equivalents. cop^{-1}) and Chl *a*/SPM ratio ($\mu g\ g^{-1}$) during a 14 h cycle. July 1991, surface, station A (see Fig. 1).

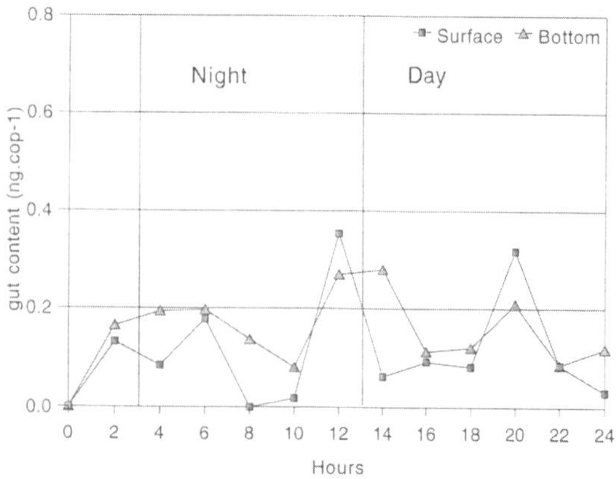

Fig. 8. *A. bifilosa.* Variation of gut contents (ng. Chl *a* equivalents cop^{-1}) during a 24 h cycle. May 1990, station A (see Fig. 1).

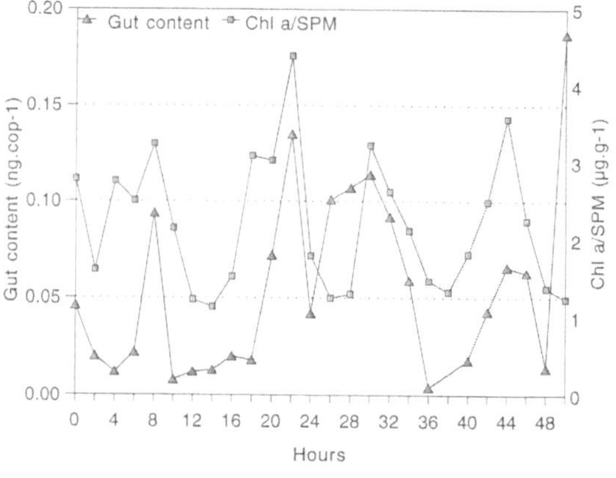

Fig. 11. *A. tonsa.* Example of variation of gut contents (ng Chl *a* equivalents. cop^{-1}) and Chl *a*/SPM ratio ($\mu g\ g^{-1}$) during a 48 h cycle. October 1990, station J (see Fig. 1).

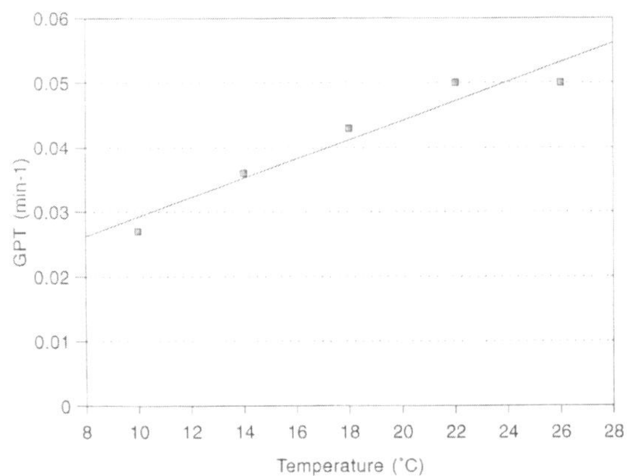

Fig. 12. A. bifilosa. Gut passage time (min^{-1}) against temperature (°C) in laboratory conditions. Fitted curve is $y = 0.0059\ x^{0.68}$, $r = 0.98$, $p < 0.01$, $n = 5$.

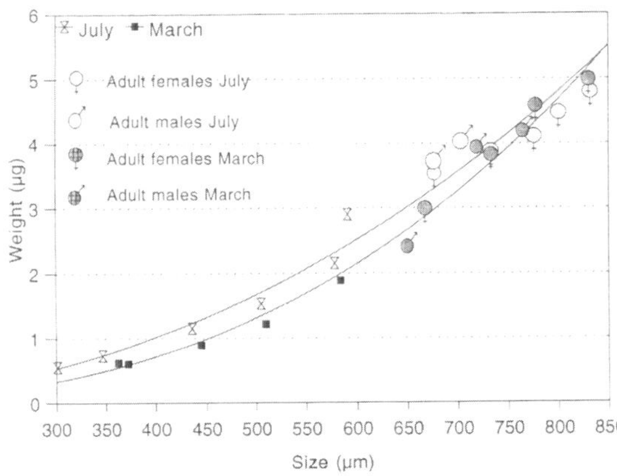

Fig. 13. A. bifilosa. Length (µm)–Weight (µg) relationships in the Gironde estuary in March 92 (filled symbols) and July 1992 (empty symbols).

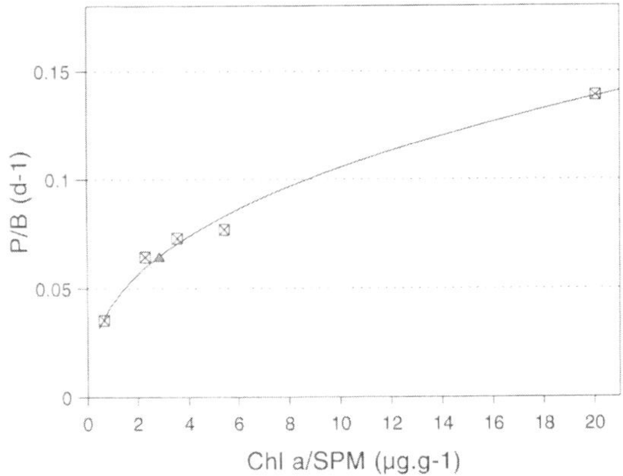

Fig. 14. A. bifilosa. P/B ratio in relation to Chl *a*/SPM ratio. Data from 1992 in point A. The triangle shows the P/B ratio obtained for *A. tonsa* in September 1992 (this point has not be taken into account to estimate fitted curve).

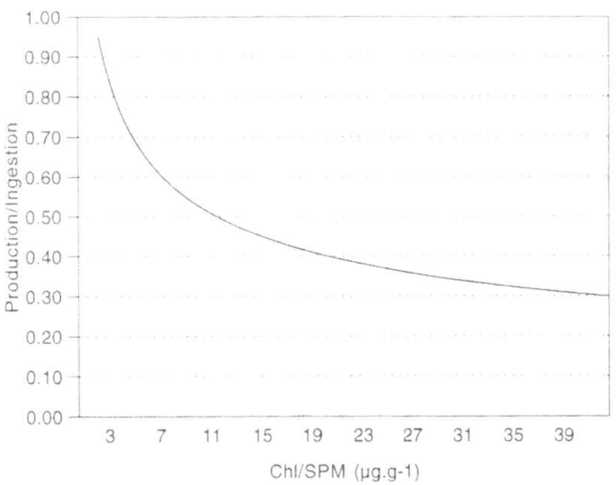

Fig. 15. A. bifilosa. Relation between Production/Ingestion ratio and Chl *a*/SPM ratio. Curve fitted from equation gut fluorescence – Chl *a*/SPM ratio and P/B – Chl *a*/SPM equation, see the text.

Particles size distribution

Suspended particulate matter was mainly composed of small particles. Average diameter size varied from 4 to 11 µm. The distribution of size frequency showed a strong skewness: More than 50% of particles were smaller than 6 µm, maximum frequencies being situated around 1.5 to 2.5 µm. At station J average size (µm) was related to SPM ($g\ l^{-1}$): $y = 6.41 + 0.47\ x$, $r = 0.73$, $n = 13$, $p < 0.05$. At station A no relationship between average size and SPM was found ($r = -0.17$), however maximum sizes were related to the Chl *a*/SPM ratio ($y = 37.58 + 0.5\ x$, $r = 0.55$, $n = 13$, $p = 0.05$); this relationship between Chl *a*/SPM ratio and maximum sizes did not appear at station J ($r = -0.13$).

The average particle size during the tidal cycle was practically the same at station J (6.84 µm) as at station A (6.82 µm). The mean of maximum sizes was not significantly greater at station A, (41.66 µm), than at station J (40.76 µm), (One way ANOVA).

Copepod distribution

Three separate assemblages of species were found in the Gironde (Figs 3, 4, 5). *Eurytemora affinis* appears to be strongly related to the turbidity maximum, showing maximum abundances in early spring (Castel & Feurtet, 1989). *Acartia spp* are found seaward of the maximum turbidity zone with a maximum abundance in summer (Castel, 1993). At the mouth of the estuary a neritic zooplankton community was found. The dominant species were *Euterpina acutifrons* and *Temora longicornis*.

In the estuary there was a succession of two *Acartia* species, with *A. bifilosa* dominating in early summer and *A. tonsa* in late summer (Fig. 6). Maximum abundances of *A. bifilosa* were found in July, with more than 6×10^3 ind m^{-3}, and maximum abundances of *A. tonsa* were found in September with a peak of 9.5×10^3 ind m^{-3}. *A. bifilosa* appeared in the 5 to 22 g l^{-1} salinity range, with maximums at 12.5 g l^{-1} in June and at 15 g l^{-1} in July. In September *A. tonsa* appeared in the 2.5 to 30 g l^{-1} salinity range, with some isolated individuals at 0 g l^{-1} and a maximum abundance situated at 19.5 g l^{-1}.

Copepod feeding

No significant differences in gut content were found between day and night samples and neither between surface and bottom samples, but some peaks of gut fluorescence have been found during sunrise or sunset (Figs 7, 8, 9). There was a large difference in the gut fluorescence values between months, ranging from an average value of 0.078 ng cop^{-1} in April (Fig. 7) to an average value of 0.327 ng cop^{-1} in July (Fig. 9). A good correlation was found between the gut contents of *A. bifilosa* and the Chl *a*/SPM ratio, ($\log y = 0.78 \log x - 1.65$, $r = 0.62$, $p < 0.01$, $n = 138$, using data from all the campaigns) (Fig. 10). There was also an inverse relationship between *A. bifilosa* gut contents and SPM concentration, ($\log y = -0.52 \log x - 1.33$, $r = -0.50$, $p < 0.01$, $n = 138$). *A. tonsa*'s gut fluorescence, measured only in October 1990, was also related to Chl *a*/SPM ratio (Fig. 11). Direct observations under the dissecting microscope showed that *Acartia*'s guts in highly turbid samples were practically empty. Accordingly there was no relationship between *A. bifilosa* gut contents and Chl *a* concentration ($r = 0.017$). Under laboratory conditions *A. bifilosa*'s gut passage time increased with temperature, from 0.027 min^{-1} at 10 °C to 0.05 min^{-1} at 22 °C. GPT was the same at 26 °C as at 22 ° (Fig. 12). Daily ingestion was estimated applying the GPT corresponding to the temperature measured in the field.

The daily ingestion of *A. bifilosa* varied from 163 ng C cop^{-1} d^{-1} in April 1991 to 1455 ng C cop^{-1} d^{-1} in July 1991, which represented approximately from 8 to 80% of copepod body weight. The daily ingestion of *A. tonsa* was about 181 ng C cop^{-1} d^{-1} in October 1990. A thorough description of those results has been published by Irigoien *et al.* (1993).

During incubation experiments in June and September 92 there were no significant differences between Chl *a* concentration at To and T1 in the bottles, but in July 1992 significant differences between the control and incubation bottles were found (One-way Anova). Filtration rates in July varied from 0.18 to 0.27 ml cop^{-1} h^{-1} and estimated daily ingestion varied from 265 to 463 ng C cop^{-1} d^{-1}, i.e. about 38 to 57% of copepods carbon body weight.

Copepod productivity

The length (L, in μm)–weight (W, in μg) relationship was $W = 6.88\ 10^{-11}\ L^{2.71}$ ($r = 0.99$, $p < 0.01$, $n = 12$) in March 92 and $W = 1.52 \times 10^{-9}\ L^{2.24}$ ($r = 0.99$, $p < 0.01$, $n = 13$) (Fig. 13). Except for adult females, copepods were heavier in July than in March (analyse of Covariance after log–log transformation of data, $p < 0.05$). Condition factor calculated as $CF = W \times 10^{-1}\ L^3$ was 0.95 in March and 1.41 in July for copepodids sized around 570 μm and 0.87 in March and 0.84 in July for adult females sized around 830 μm.

Estimated P/B ratio for *A. bifilosa* varied from 0.03 d^{-1} in December 92 to 0.14 d^{-1} in July 92. *Acartia*'s P/B ratios were better correlated to Chl *a*/SPM ratio ($y = 0.043x^{0.39}$ $r^2 = 0.99$, $p < 0.01$, $n = 5$) (Fig. 14), than to temperature ($y = 0.002x^{1.33}33$, $r^2 = 0.7$, $p > 0.05$, $n = 5$). Mean P/B ratio for *A. bifilosa* from April to December was 0.08 d^{-1} giving an annual P/B ratio of 28 yr^{-1}.

We employed obtained equations for gut fluorescence and productivity against Chl *a*/SPM ratio to estimate the ratio between production and ingestion as a function of Chl *a*/SPM ratio. In this way ingestion was calculated applying an average gut passage time of 0.041 min^{-1} to gut contents obtained from gut fluorescence. Production was calculated applying the P/B obtained from P/B, Chl *a*/SPM equation to an average carbon biomass of 2 μg C ind^{-1}. The obtained curve shows that a reasonable gross-growth efficiency (Produced carbon/Ingested carbon), about 0.35 and below, is only reached with a Chl *a*/SPM ratio higher than 28 $\mu g\ g^{-1}$ (Fig. 15). When low values of Chl *a*/SPM ratio are present an important underestimate of measured ingested carbon appears.

Discussion

There is no data in the literature about *A. bifilosa* ingestion. Conversely, *A. tonsa* has been extensively studied. Since there is a strong similarity in our results between the gut fluorescence and the ingestion of *A. bifilosa* and *A. tonsa* we assume that our data concerning *A. bifilosa* can be compared to those found in the literature about *A. tonsa*.

A. bifilosa phytoplankton ingestion was generally, except during the phytoplankton bloom, much lower than obtained for *A. tonsa* in the laboratory by Roman (1977), 312 to 5060 ng C cop^{-1} d^{-1}. Gut fluorescences were also lower (except during the bloom) than those measured by Stearns *et al.* (1987) in the Newport estuary (0.498 to 1.46 ng cop^{-1}). This can be explained by the generally low concentration of Chl *a* in the Gironde estuary (Irigoien *et al.*, 1993). The absence of night-day cycles in the feeding activity, frequently noticed in the literature (Mackas & Bohrer, 1976; Simard *et al.*, 1985; Nicolajsen *et al.*, 1983; Stearns, 1986), can be explained by the low penetration of the light in the water column due to the high turbidity. Another explanation could be that our sampling schedule was not adapted to detect night-day activity cycles in an estuary such as the Gironde. Even in the clearest waters ($\pm$100 mg l^{-1}), changes in light penetration, at an anchor station are not only due to night-day cycles but also due to changes in the SPM concentration during tidal cycles.

P/B ratios for *A. bifilosa* (0.03 to 0.14 d^{-1}) are lower than those obtained in the laboratory with food in excess (see review by Escaravage & Soetaert, 1995). On the other hand they are in the same order of magnitude as those calculated by Ciszewski & Witek (1977) for *Acartia bifilosa* in the Gdansk bay, (0.035 to 0.12 d^{-1}), and slightly lower than those obtained by Castel & Feurtet (1989) for *Eurytemora affinis* in the Gironde estuary (0.03 d^{-1} to 0.2 d^{-1}). Although the approach to estimate productivity in those works was different – instantaneous growth rates in the present study and in Ciszewski & Witek's (1977) work, or mortality rates (Heinle, 1966) in Castel & Feurtet (1989) – the values are not strongly different. The similarity of productivities obtained by Escaravage & Soetaert (1993) for *E. affinis* in the Westerschelde by the growth rate method (33 yr^{-1}), and productivities obtained by Castel & Feurtet (1989) for the same species in the Gironde by the mortality rate method (32.8 yr^{-1}), allows us to conclude that the results provided by these two different approaches are comparable.

The presence of the Chl *a*/SPM peak seaward of the turbidity maximum can be explained by the mixing depth/euphotic depth approach (Z_m/Z_{eu}), (Sverdrup, 1953). With a mean depth of 3.5 m in the Gironde, turbidity has to be lower than 200 mg l^{-1} to yield a Z_m/Z_{eu} ratio lower than 5, allowing primary production, (Cole & Cloern, 1984; Grobelaar, 1985). Except for some values characteristic of neritic waters at the mouth of the estuary, all the Chl *a*/SPM values greater than 30 μg g^{-1} appeared with Z_m/Z_{eu} ratios lower than 5, consequently the Chl *a*/SPM ratio can probably be related to primary productivity (Irigoien & Castel, submitted). In the Gironde areas with turbidities lower than 200 mg l^{-1} are scarce, only in the summer months (July, August) does the area seaward of the turbidity maximum show low turbidities for long periods.

The SPM in the Gironde is mainly formed by small particles (mean diameter: 4–11 μm, maximum frequency: 1.5–2.5 μm). This distribution is similar to those obtained by Weber *et al.* (1991) in the Gironde employing a laser difractometer. Seaward of the maximum turbidity zone (MTZ) the Chl *a*/SPM ratio was generally higher than in the most turbid areas. The distribution of particle concentration, size and quality can perhaps explain the distribution and secondary production of *Acartia* spp. in the Gironde. It is known that *A. tonsa* cannot feed efficiently on small particles because it does not create low-amplitude mandible II motions (Price *et al.*, 1983) and that it actively selects large and good quality particles (Paffenhöfer, 1988). From Itoh's (1970) edge index calculation, Castel (1981) suggested that *A. bifilosa*, is an omnivorous species, that feeds on larger particles than *E. affinis*. The energetic cost to find large particles, phytoplankton or others (e.g. microzooplankton or nauplii), 'diluted' in a large amount of small mineral particles in the turbidity maximum, may be excessive. Seaward of the turbidity maximum large particles such as microzooplankton and phytoplankton will be less 'diluted' and more readily available to *Acartia* sp, especially in the summer months, when primary production is possible. This may explain the difference in the weight of copepods between July and March 1992 as shown by the condition factors. The similarity in the condition factor for adult females in July and March would perhaps be explained by the increased energetic cost of egg production in July.

Long term time series analysis (Ibanez *et al.*, 1993) shows that *Acartia* sp. abundances at station J are negatively correlated to SPM concentration. Since SPM and salinity are also negatively correlated it is not possible to distinguish between their action. In the literature, *Acartia bifilosa* is present in a large range of salinities, 2 to 15 g l^{-1} (Cannicci, 1962), 10 to 25 g l^{-1} (Castel & Courties, 1977), or around 30 g l^{-1} (Villate *et al.*, 1993), but it is possible that the energetic cost to find large particles in the MTZ and to control osmoregulation (Lance, 1965) is too great and results in a lack of colonisation of the upper Gironde estuary by *Acartia* sp.

Kiørboe *et al.* (1985) found that gross-growth efficiency (the quotient carbon production/carbon food ingested) for *A. tonsa* was around 0.35. Similar values have been found for other *Acartia* species (Saiz *et al.*, 1992) and other calanoids (Peterson *et al.*, 1990). If one considers the phytoplanktonic carbon ingestion alone, the obtained production/ingestion (P/I) curve shows that only with Chl *a*/SPM ratios higher than 28 $\mu g\ g^{-1}$ gross-growth efficiency will be lower than 0.35. This trend was confirmed by the production and ingestion results of July 1992 which gives a gross-growth efficiency of 0.3 with a Chl *a*/SPM ratio around 20 $\mu g\ g^{-1}$. Only in the summer months, when primary production is possible, was ingested phytoplanktonic carbon enough to cover *Acartia*'s energy requirements. Throughout the rest of the year other sources of carbon must be considered, since at Chl *a*/SPM values lower than 28 $\mu g\ g^{-1}$ phytoplanktonic carbon ingestion alone is not sufficient to explain productivity. These results are quite similar to those of White & Roman (1992), who found that *A. tonsa* feeds mainly on microzooplankton during the winter and on phytoplankton in the summer.

Phytoplankton seems to be very important in the development of *A. bifilosa*; maximum population densities are related to the seasons, (summer), and area, (seaward of the turbidity maximum), where primary production is possible; and productivity is related to phytoplankton availability, (Chl *a*/SPM ratio).

Acknowledgments

This research was supported by a CEC MAST programme ('Major Biological Processes in European tidal estuaries' contract 0024.C). X. Irigoien had a doctoral scholarship from the Basque country government. Thanks are due to the captain and crew of the R.V. 'Côte d Aquitaine' and to B. Sautour, I. Auby, and L. Pereira Santos for their help in the field. We are indebted to N. J. Stevenson for revision of the English language.

References

Ambler, J. W., 1985. Seasonal factors affecting egg production and viability of eggs of *Acartia tonsa* Dana from east lagoon, Galveston Texas. Estuar. coast. Shelf Sci. 20: 743–760.

Beckman, R. B. & W. T. Peterson, 1986. Egg production by *Acartia tonsa* in Long Island sound. J. Plankton Res. 8: 917–925.

Cannicci, G., 1962. Instabilità delle associazioni planctoniche in alcuni stagni salmastri della penisola italiana e della Sardegna in rapporto alla caratteristiche dell' habitat. Pubbl. Staz. Zool. Napoli 32: 349–367.

Castel, J., 1981. Aspects de l'étude écologique du plancton de l'estuaire de la Gironde. Oceanis 6: 535–577.

Castel, J. & C. Courties, 1977. Le zooplancton. In Etude écologique de l'estuaire de la Gironde, EDF-CNEXO (eds): 220–295.

Castel, J., 1993. Long-term distribution of zooplankton in the Gironde estuary and its relation with environmental factors. Cah. Biol. mar. 34: 145–163.

Castel, J. & A. Feurtet, 1989. Dynamics of the copepod *Eurytemora affinis hirundoides* in the gironde estuary: Origin and fate of its production. Topics in Marine Biology. Ros, J. D. (ed.), Scient. Mar. 53: 577–584.

Ciszewski, P. & Z. Witek, 1977. Production of older stages of copepods *Acartia bifilosa* Giesb. and *Pseudocalanus elongatus* Boeck in Gdansk bay. Pol. Arch. Hydrobiol. 24: 449–459.

Christou, E. D. & G. C. Verriopoulos, 1993. Analysis of the biological cycle of *Acartia clausi* (Copepoda) in a meso-oligotrophic coastal area of the eastern Mediterranean Sea using time-series analysis. Mar. Biol. 115: 643–651.

CNEXO., 1977. Etude écologique de l'estuaire de la Gironde Rapport final, 470 pp.

Cole, B. E. & J. E. Cloern, 1984. Significance of biomass and light availability to phytoplankton productivity in San Francisco Bay. Mar. Ecol. Prog. Ser. 17: 15–24.

Dagg, M. J. & D. W. Grill, 1980. Natural feeding rates of *Centropages typicus* females in the New York bight. Limnol. Oceanogr. 25: 583–596.

Durbin, E. G. & A. G. Durbin, 1992. Effects of temperature and food abundance on grazing and short-term weight change in the marine copepod *Acartia hudsonica*. Limnol. Oceanogr. 37: 361–378.

Durbin, E. G., A. G. Durbin, T. Smayda & P. G. Verity, 1983. Food limitation of production by *Acartia tonsa* in Narraganset bay, Rhode Island. Limnol. Oceanogr. 28: 1199–1213.

Escaravage, V. & K. Soetaert, 1993. Estimating secondary production for the brackish Westerschelde copepod population *Eurytemora affinis* (Poppe) combining experimental data and field observations. Cah. Biol. mar. 34: 201–214.

Escaravage, V. & K. Soetaert, 1995. Secondary production of the brackish copepod communities and their contribution to the carbon fluxes in the Westerschelde estuary (The Netherlands). Hydrobiologia 311 (Dev. Hydrobiol. 110): 103–114.

Gifford, D. J & M. J. Dagg, 1988. Feeding of the estuarine copepod *Acartia tonsa* Dana: Carnivory *vs* herbivory in natural microplankton assemblages. Bull. mar. Sci. 43: 458–468.

Grobbelar, J. U., 1985. Phytoplankton productivity in turbid waters. J. Plankton Res. 7: 653–663.

Heinle, D. R., 1966. Production of a calanoid copepod *Acartia tonsa*, in the Patuxent river estuary. Chesapeake Sci. 7: 59–74.

Heinle, D. R., R. P. Harris, J. F. Ustach & D. A. Flemer, 1977. Detritus as food for estuarine copepods. Mar. Biol. 40: 341–353.

Ibanez, F., J. M. Fromentin & J. Castel, 1993. Application de la méthode des sommes cumulées à l'analyse des séries chronologiques océanographiques. C. r. Acad. Sci., Paris, Life Sci. 316: 745–748.

Irigoien, X. & J. Castel, 1995. Light attenuation and distribution of Chlorophyll pigments distribution in a highly turbid estuary, the Gironde (SW France). Estuar. (submitted). Estuar. Coast. Shelf Sci.

Irigoien, X., J. Castel & B. Sautour, 1993. *In situ* grazing activity of planktonic copepods in the Gironde estuary. Cah. Biol. mar. 34: 225–237.

Itoh, K., 1970. A consideration on feeding habits of planktonic copepods in relation to the structure of their oral parts. Bull. Plankton Soc. Japan 17: 1–10.

Kimmerer, W. J. & A. D. McKinnon, 1986. Glutaraldehyde fixation to maintain biomass of preserved plankton. J. Plankton Res. 8: 1003–1008.

Kimmerer, W. J. & A. D. McKinnon, 1987. Growth, mortality and secondary production of the copepod *Acartia tranteri* in Westernport bay, Australia. Limnol. Oceanogr. 32: 14–28.

Kiørboe, T., F. Mohlenberg & K. Hamburger, 1985. Bioenergetics of the planktonic copepod *Acartia tonsa*: relation between feeding, egg production and respiration, and composition of specific dynamic action. Mar. Ecol. Prog. Ser. 26: 85–97.

Lance, J., 1965. Respiration and osmotic behaviour of the copepod *Acartia tonsa* in diluted sea water. Comp. Biochem. Physiol. 14: 155–165.

Landry, M. R., 1978. Population dynamics and production of a planktonic marine copepod, *Acartia clausi*, in a small temperate lagoon on San Juan island, Washington. Int. Revue ges. Hydrobiol. 63: 77–119.

Lin, R. G., 1988. Etude du potentiel de dégradation de la matière organique particulaire au passage eau douce-eau salée: cas de l estuaire de la Gironde. Thèse Doct., Univ. Bordeaux I 209 pp.

Londsdale, D. J., D. R. Heinle & C. Siegfried, 1979. Carnivorous feeding behaviour of the adult calanoid copepod *Acartia tonsa*. J. exp. mar. Biol. Ecol. 36: 235–248.

Mackas, D. & R. Bohrer, 1976. Fluorescence analysis of zooplankton gut contents and a investigation of diel feeding patterns. J. exp. mar. Biol. Ecol. 25: 77–85.

Neveux, J., 1983. Dosage de la chlorophyll *a* et des phéopigments par fluorimétrie. Manuel des analyses chimiques en milieu marin. CNEXO ed. 11: 193–203.

Newell, R. C., 1973. Biology of intertidal animals. Marine Ecological Surveys. Faversham, Kent, 781 pp.

Nicolajsen, H., F. Mohlenberg & T. Kiorboe, 1983. Algal grazing by the planktonic copepods *Centropages hamatus* and *Pseudocalanus* sp. Diurnal and seasonal variation during the spring phytoplankton bloom in the Oresund. Ophelia 22: 15–31.

Paffenhöfer, G. A., 1988. Feeding rates and behaviour of zooplankton. Bull. mar. Sci. 43: 430–445.

Paffenhöfer, G. A., J. R. Strickler & M. Alcaraz, 1982. Suspension feeding by herbivorous calanoid copepods: a cinematographic study. Mar. Biol. 67: 193–199.

Peterson, W., S. Painting & R. Barlow, 1990. Feeding rates of *Calanoides carinatus*: A comparison of five methods including evaluation of the gut fluorescence method. Mar. Ecol. Prog. Ser. 63: 85–92.

Price, H. J., 1988. Feeding mechanisms in the marine and freshwater zooplankton. Bull. mar. Sci. 43: 327–343.

Price, H. J., G. A. Paffenhöfer & J. R. Strickler, 1983. Modes of cell capture in calanoid copepods. Limnol. Oceanogr. 28: 116–123.

Robertson, J. R., 1983. Predation by estuarine zooplankton on tintinnid ciliates. Estuar. coast. Shelf Sci. 16: 27–36.

Roman, M. R., 1977. Feeding of the copepod *Acartia tonsa* on the diatom *Nitzchia closterium* and brown algae (*Fucus vesiculosus*) detritus. Mar. Biol. 42: 149–155.

Roman, M. R., 1984. Feeding of detritus by the copepod *Acartia tonsa*. Limnol. Oceanogr. 29: 949–959.

Saiz, E., M. Alcaraz & G. A. Paffenhofer, 1992. Effects of small-scale turbulence on feeding rate and gross-growth efficiency of three *Acartia* species (Copepoda: Calanoida). J. Plankton Res. 14: 1085–1097.

Simard, Y., G. Lacroix & L. Legendre, 1985. In situ twilight grazing rhythm during diel vertical migrations of a scattering layer of *Calanus finmarchicus*. Limnol. Oceanogr. 30: 598–606.

Stearns, D. E., 1986. Copepod grazing behavior in simulated natural light and its relation to nocturnal feeding. Mar. Ecol. Prog. Ser. 30: 65–76.

Stearns, D. E., W. Litaker & G. Rosenberg, 1987. Impacts of zooplankton grazing and excretion on short-interval fluctuations in chlorophyll *a* and nitrogen concentration on a well mixed estuary. Estuar. coast. Shelf Sci. 24: 305–325.

Stoecker, D. K. & D. A. Egloff, 1987. Predation by *Acartia tonsa* Dana on planktonic ciliates and rotifers. J. exp. mar. Biol. Ecol. 110: 53–68.

Sverdrup, H. U., 1953. On conditions for the vernal blooming of phytoplankton. J. Cons. perm. int. Explor. Mer 18: 287–295.

Tackx, M., X. Irigoien, N. Daro, J. Castel, L. Zhu, X. Zhang & J. Nijs, 1995. Copepod feeding in the Westerschelde and the Gironde. Hydrobiologia 311 (Dev. Hydrobiol. 110): 71–83.

Tackx, M. L. M., C. Bakker, J. W. Francke & M. Vink, 1989. Size and phytoplankton selection by Oosterschelde zooplankton. Neth. J. Sea Res. 23: 35–43.

Tackx, M. & P. Polk, 1982. Feeding of *Acartia tonsa* Dana (Copepoda, Calanoida) on nauplii of *Canuella perplexa* T. et A. Scott (Copepoda, Harpacticoida) in the sluice dock at Ostend. Hydrobiologia 94: 131–133.

Viitasalo, M., 1992a. Calanoid resting eggs in the Baltic sea: implications for the populations dynamics of *Acartia bifilosa* (Copepoda). Mar. Biol. 114: 397–405.

Viitasalo, M., 1992b. Mesozooplankton of the gulf of Finland and northern Baltic proper – a review of monitoring data. Ophelia 35: 147–168.

Villate, F., A. Ruiz & J. Franco, 1993. Summer zonation and development of zooplankton populations within a shallow mesotidal system: the estuary of Mundaka. Cah. Biol. Mar. 34: 131–143.

Weber, O., J. M. Jouanneau, P. Ruch & M. Mirmand, 1991. Grain – size relationship between suspended matter originating in the Gironde estuary and shelf mud – patch deposits. Mar. Geol. 96: 159–165.

White, J. R. & M. R. Roman, 1992. Egg production of the calanoid copepod *Acartia tonsa* in the mesohaline Chesapeake bay: the importance of food resources and temperature. Mar. Ecol. Prog. Ser. 86: 239–249.

Hydrobiologia **311**: 127–137, 1995.
C. H. R. Heip & P. M. J. Herman (eds), *Major Biological Processes in European Tidal Estuaries.*

Production rates of *Eurytemora affinis* in the Elbe estuary, comparison of field and enclosure production estimates

Andrea Peitsch
Institut für Hydrobiologie und Fischereiwissenschaft, D-22765 Hamburg, Zeiseweg 9, Germany

Key words: zooplankton, estuary, production rates, food limitation

Abstract

Production rates of the calanoid copepod *Eurytemora affinis* were estimated from field studies in the Elbe estuary and from an enclosure experiment. As one basic parameter of production rates, the body length, was compared between both investigations. Most of the copepodid stages in the enclosure experiment reached a significant greater length than the copepodids in the estuary. The differences in length between copepods from the field and the experiment could mainly be explained by a four times higher chlorophyll-*a* level in the enclosure experiment. The better food supply also results in a higher individual growth rate for all instars in the enclosure experiment. Therefore the population of *Eurytemora affinis* in the Elbe estuary was regarded as food limited during certain times of the year, especially in late spring and summer.

Maximum daily production rate in the enclosure experiment (40 μg dw l^{-1} d^{-1}) was four times higher than in the estuary (12 μg dw l^{-1} d^{-1}). The mean daily P:B ratio in the enclosure was 0.301 d^{-1} compared to 0.11 d^{-1} in the estuary.

Introduction

In many european estuaries *Eurytemora affinis* is a typical inhabitant of the low salinity stretches (Castel & Feurtet, 1986; De Pauw, 1973; Soltanpour-Gargari & Wellershaus, 1987). As well it is the dominant zooplankton species of the oligohaline region in the Elbe estuary. It accounts for 90 to 99% of the crustacean plankton throughout the year (Peitsch, 1992). Because of that dominant role in the zooplankton community the population dynamics and production rates of *Eurytemora affinis* are very important for the understanding of biological processes in the brackish water region.

Basic variables determining production rates are the abundance, the mortality rates, the developmental times and the individual weight. The weight of the instars of the copepods is in most cases derived from length measurements. For the estimation of biomass of *Eurytemora affinis* the body length is a reliable parameter (Kankaala & Johansson, 1986). Length and growth rate are controlled by two factors, temperature and food level. High content of suspended matter in the turbidity zones of estuaries restrict the light penetration and therefore the primary production. In different estuaries the maximum abundance of *Eurytemora affinis* was found frequently in the zone of maximum turbidity (Castel & Feurtet, 1989; Soltanpour-Gagari & Wellershaus, 1984) although the low primary production, even in spring time, is not sufficient to cover the carbon demand of *Eurytemora affinis* (Heinle & Flemer, 1975). Bacteria, detritus and ciliates are proposed to be additional food components in the diet of *Eurytemora affinis* (Heinle *et al.*, 1977).

In this study the influence of temperature and food level (chlorophyll-*a*) on length, growth rate and production rates was regarded particularly. A comparison of production rates under field conditions in the Elbe estuary and under controlled conditions during an enclosure experiment was made.

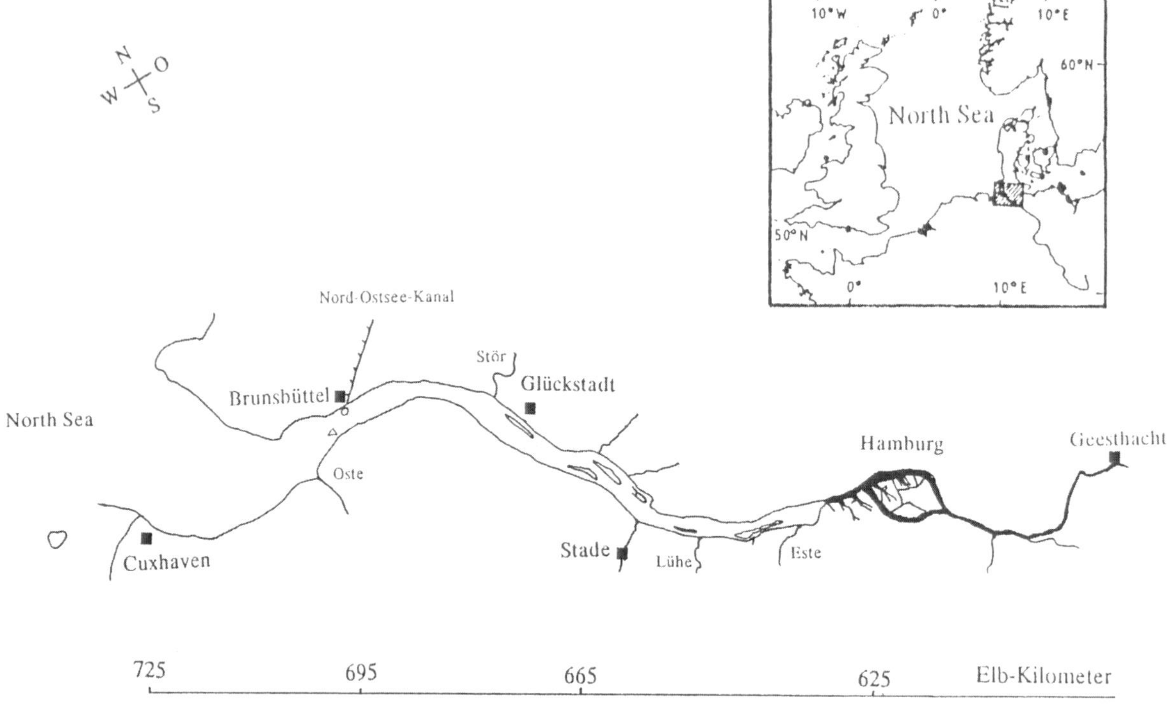

Fig. 1. Map of the Elbe estuary (o - Enclosure experiment, △ - field study).

Study site

The Elbe estuary is a coastal plain estuary, which opened funnel-shaped to the North Sea (Fig 1). The river Elbe has a total length of 1143 km. Since 1960 a weir at km 586 (city of Geesthacht) forms an upstream boundary to the tide. Today the total length of the estuary is 140 km. The estuary is to a large extent channelized since the beginning of the century. Estuarine conditions are characterized by a steep salinity gradient, a zone of high turbidity and strong current velocities. At mean river discharge (700 m^3 s^{-1}) salinity gradient reaches from the mouth of the estuary to the city of Glückstadt (km 670) while the maximum turbidity zone is located between km 665 and 700.

Current velocities could reach maximum values of 2.3 m s^{-1}. During tidal flow normally maxima of 1 to 2 m s^{-1} are measured (Siefert, 1970). River discharge is one important factor, which shifts the salinity gradient and the zone of high turbidity seasonally.

Methods

Sampling strategy

In the estuary six samples from low to high tide were taken twice a week from April to September 1989 at an anchor station. Sampling was carried out from a platform being located in the oligohaline region (km 695, Fig. 1). Using the flow, which passes the cross section during the sampling interval (flood phase), a weighted average of the abundance (of all six samples) was calculated for each sampling day (Peitsch, 1993).

In 1988 the enclosure experiment was carried out from August 16 to September 8 at the entrance of the North-Ostsee- Kanal (km 695, Fig. 1). Figure 2 shows the design of the enclosure (Brockmann *et al.*, 1974). The bags have a volume of nearly 3.3 m^3. The water was stirred and a transparent cover allowed light penetration to the water surface.

A original body of estuarine water with natural populations of phyto- and zooplankton was enclosed. The development of the *Eurytemora affinis* population was followed over a period of four weeks. Duplicate samples were taken at 0.00 and 12.00 every day during

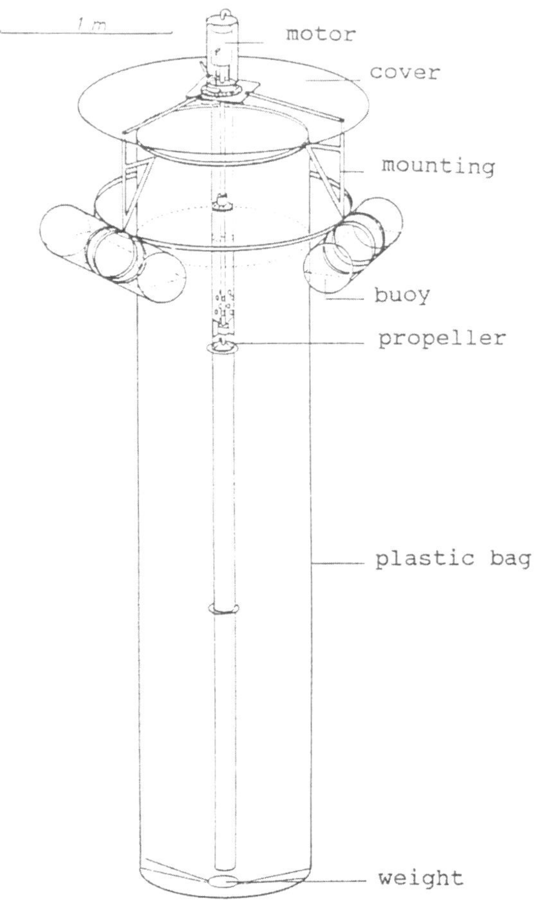

Fig. 2. Design of enclosure bag (Brockmann *et al.*, 1974).

the first three weeks. During the last week zooplankton was sampled only once a day (12.00).

Sampling procedure

Zooplankton samples were collected using an electric driven pump with a capacity of 40 l min^{-1}. In the estuary integrated samples were taken from the bottom of the river to the water surface. Samples from the enclosure experiment were taken at a depth of 1 m. 10 l of water thus obtained were poured through a 55 μm mesh and preserved in 4% buffered formaldehyde. Abundance of the different instars of *Eurytemora affinis* was recorded as numbers per liter.

Length of all or a maximum of 30 individuals of each instar were measured using a digitizing tablet connected to a computer. Total length of naupliar stages and prosoma length of copepodid stages and adults were considered.

Estimation of parameters

Abundance (Ni)

For the use of the cohort analysis it was necessary to get only one value of abundance for each sampling day, so a weighted average was calculated taking the different flow during sampling into account. A description of the calculation is given in Peitsch (1993).

Than the cohort analysis of Rigler & Cooley (1974) was applicable to this set of data. This method was choosen, because it was developed to derive population data of multivoltine species, especially copepods, with overlapping cohorts from field samples. Therefore the center of gravity or a mean time (Mn) of a pulse at regular time intervals was calculated:

$$Mn = \frac{(\text{days} * \text{animals})}{\text{animals}}$$

Mn = mean pulse

Stage duration time was calculated by the cohort analysis. Mortality rates and growth rates were calculated for the instars belonging to the same cohort, assuming that there was no change of these parameters within a cohort.

Mortality rate (m_i)

Daily mortality rates were estimated from the change in abundance of successive instars within a cohort. Following the formula of Landry (1978) abundance is somewhat between the instars initially entering the stage and molting to the next instar, calculated by the area under the curve divided by stage duration time.

$$\left(\frac{A_{i+1} * D_i}{A_i * D_{i+1}}\right) = e - m_i \times DT_i$$

A_i = Area under the curve of abundance of instar i (in copepod-days)

A_{i+1} = area under the curve of abundance of instar $i+1$ (in copepod-days)

D_i = stage duration time of instar i

D_{i+1} = stage duration time of instar $i+1$

m_i = mortality rate of instar i

DT_i = time interval between the midpoints of the durations of successive stages

Mortality rates of adults could not be received from cohort analysis, because life time of adults is very variable. Mortality rates of adults were iteratively calculated by a formula of Christiansen (1988):

$$m_{Ad} = \frac{1}{t_2 - t_1} * \ln\frac{N_{Ad1}}{N_{Ad2}} + \frac{N_{C5} * t_1 * (e^{(t2-t1)*(m_{Ad}-m_{c5})} - 1}{D_{c5} * (m_{Ad} - m_{c5})}$$

m_{Ad} = instantenous mortality rate of adults
t_2-t_1 = time interval between successive sampling dates
A_{Ad1} = abundance of adults at time 1
N_{Ad2} = abundance of adults at time 2
N_{c5} = abundance of copepodid V
m_{c5} = mortality rate of copepodid V
D_{c5} = instar duration time of copepodid V

Growth rate (g_i)

The daily weight specific growth rate of an instar was estimated following a formula given by Winberg & Edmondson (1971):

$$g_i = \frac{1}{ti} * \ln\left(\frac{W_i}{W_{i-1}}\right)$$

g_i = growth rate of instar i
W_i = mean weight of instar i
W_{i-1} = mean weight of instar i-1
ti = time interval between the midpoints of the durations of successive stages

Biomass

The individual weight (W) of an instar was calculated from length (L)-weight regressions for naupliar and copepodid stages estimated by Christiansen (1988) for a population of *Eurytemora affinis* in the Schlei fjord, northern Germany:

Naupliar stages: $W(\mu g) \equiv 6.3\, L^{2.06}$ (μm)
Copepodid stages and adults: $W(\mu g) \equiv 12.9\, L^{2.92}$ (μm)

Biomass was calculated using the individual weight, mean mortality and mean growth rates of naupliar stages respectively copepodid stages (Landry, 1978). The daily changes of abundance were derived by linear interpolation.

$$B_i = \int_0^{1\text{day}} N_i * W_i\, e^{(g_i - m_i)t} d_t = \frac{N_i * W_i(e^{(g_i - m_i)} - 1)}{(g_i - m_i)}$$

B_i = biomass of instar i
N_i = numbers of instar i
W_i = mean individual weight of instar i
m_i = mean daily mortality rate of naupliar respectively copepodid stages or adults
g_i = mean daily growth rate of naupliar respectively copepodid stages

Estimation of production

For calculation of daily production rates a formula of Landry (1978) was used.

$$P = \sum_{NII}^{Ad} g_i * B_i$$

P = daily production rates
g_i = mean growth rate of instar i
B_i = biomass of instar i

Production of adult females was set equal to egg production while the production of males is assumed half as high as for females. Egg mortality was assumed to be equal to mortality of adults.

Results

Environmental conditions

In the Elbe estuary temperature increased from 8 °C in April 1989 to 16 °C at the beginning of June. In summer temperature varied between 18 and 20 °C. In the estuary chlorophyll-*a* level was very low, maximum values of 10 μg l^{-1} were estimated in the beginning of June and only 5 to 6 μg l^{-1} in summer (Wolfstein, pers. comm.). Unfortunatly there were no measurements in spring time, but probably there would have been higher levels.

During the enclosure experiment temperature was only slightly reduced; however, a strong increase in chlorophyll-*a* concentration occurred. Temperature decreased from 18.5 °C at August 16 to about 17 °C at the end of the experiment. The chlorophyll-*a* concentration increased from 5–6 μg l^{-1} at the beginning of the experiment during the first week to 10 μg l^{-1} and to a four times higher level in the end.

Abundance

Results of the distribution and seasonal changes of abundance and the separation of cohorts in the estuary are already published in Peitsch, 1993. From April to September three cohorts were separated. First cohort started at April 28 with an increase in the numbers of eggs and ended at July 4 with the adults. Second cohort with lower abundance started at the end of May and lasted until the beginning of August (adults). The third cohort which could be separated showed the lowest abundance developing during the summer from July 5 to August 22.

Figure 3 shows the abundance of all instars of *Eurytemora affinis* during the enclosure experiment. Starting with the assembly of different instars at the beginning of the experiment a cohort (shaded in Fig. 3) developed from naupliar stage III to the adults. At the beginning of the experiment abundance of eggs, N I and N II could be neglected, at first N III showed reasonable abundance to let the cohort start. It developed until the adults at the end of the experiment.

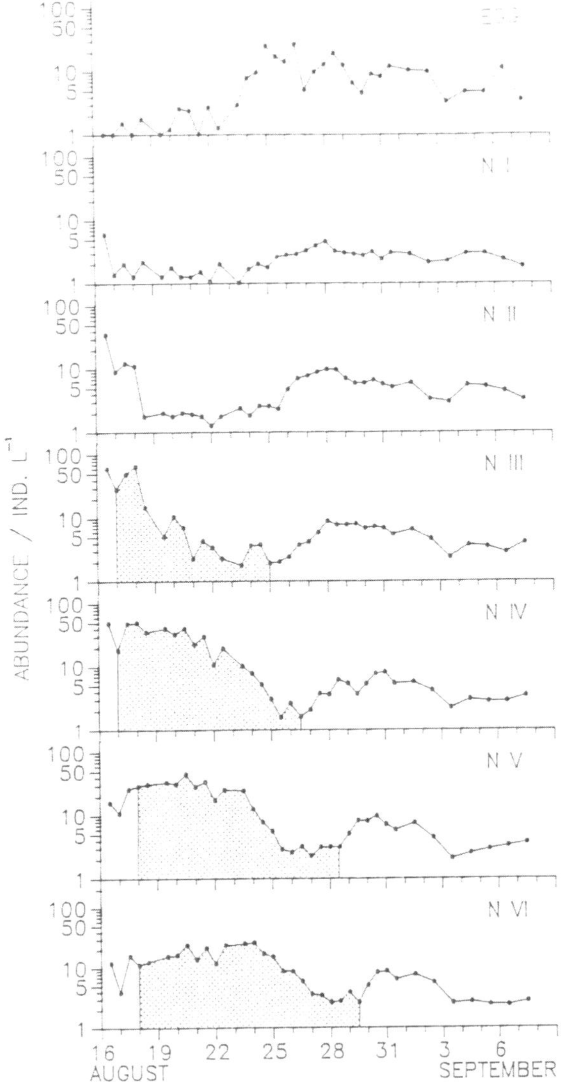

Fig. 3. Abundance of the instars of *Eurytemora affinis* and separated cohort during the enclosure experiment

Length measurements

In the estuary mean length of adults was reduced from 0.9 mm in April and May to 0.8 mm in June and 0.7 mm during the summer (Fig. 4). There was a decrease from the beginning to the mid of June. When the adults passed the first cohort they had achieved a length of about 0.8 mm. During summer (cohort 2 and 3) adults show a shorter length and only little variations.

During the enclosure experiment length of all instars increased. Figure 5 shows the mean length of copepodid V. The length increased from about 0.6 mm during the first week to nearly 0.8 mm at the end of the experiment.

The length of adults showed the same trend as for copepodid V (Fig. 6), an increase during the course of the experiment. Regression analysis demonstrates significant correlation between mean length and temperature, as well as between length and chlorophyll-*a* level. For copepodid V 78.3% of the length variations could be explained by changes in temperature, for adults only 48.3%. The change of chlorophyll-*a* level is responsible for 48.9% (C V) respectively 67.1% (adults) of the length variations.

Table 1 shows the comparison of the mean length of the different instars in cohort 1 and 3 in the field and in the enclosure experiment. Cohort 1 in the field developed at a mean temperature of 16 °C and cohort 3 at 18.6 °C, the mean temperature for the cohort in the enclosure experiment was 18.5 °C. Although the mean temperature for cohort 1 in the field was lower as in the enclosure, the length of copepodid III to V were significant shorter in the estuary as in the enclosure experiment. No significant differences between the adults of cohort 1 and enclosure were detected. The lengths of copepodids III to adults of cohort 3, which had grown at the same temperature as animals from the enclosure experiment, were significant shorter.

The development of the different cohorts could also be expressed as daily individual growth rates

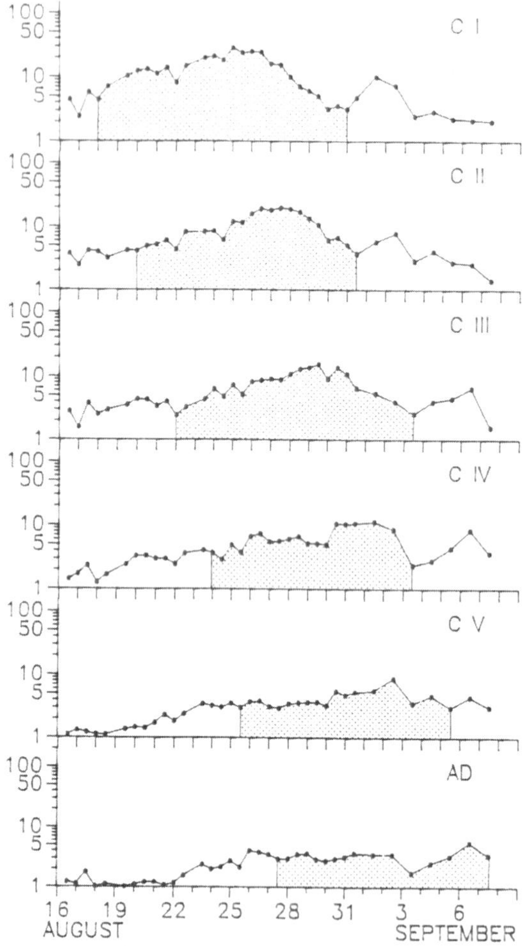

Fig. 3B.

Table 1. Mean length (mm) of different instars of *Eurytemora affinis* in cohort 1 and 3 in the field and from the enclosure experiment

Instars	Cohort 1 (mm)	Cohort 3 (mm)	Enclosure (mm)
CI	0.282*	0.278	0.29
CII	0.367	0.359	0.375
CIII	0.471*	0.46*	0.494
CIVf	0.593*	0.572*	0.611
CIVm	0.563*	0.553*	0.587
CVf	0.708*	0.683*	0.739
CVm	0.654*	0.636*	0.682
Fem.	0.803	0.792*	0.803
Male	0.749	0.724*	0.755

* Significantly shorter than copepods from enclosure

of the instars. The mean growth rates of naupliar stages in cohort 1 and 3 ($gn = 0.151$ d^{-1} and $gn = 0.178$ d^{-1}) and copepodids ($gc = 0.186$ d^{-1} and $gc = 0.206$ d^{-1}) respectively were obviously lower than in the enclosure experiment ($gn = 0.217$ d^{-1} and $gc = 0.356$ d^{-1}).

Mortality rates

High mean daily mortality rates in the Elbe estuary were found for naupliar stages (cohort 1 = 0.259 d^{-1} and cohort 3 = 0.248 d^{-1}), while in contrast mortality rate of copepodids was low. In both cohorts (1 and 3) a mean daily mortality rate of 0.05 was estimated.

In the enclosure experiment mortality of the early instars was lower (0.115 d^{-1}) which was balanced by a higher mortality rate of copepodids (0.177 d^{-1}). Mean mortality rates of all stages together show no differences between the field study and the enclosure experiment.

Because of negative mortality rates between N VI and C I in all cohorts and N II and N III in cohort 3, mean mortality rates for naupliar stages and copepodid stages have been used for the estimation of production.

In the estuary mortality rates of adults were very high in cohort 1 and 2 (0.77 d^{-1}) and only half as high in cohort 3 (0.37 d^{-1}).

Daily production rates and biomass

In the Elbe estuary the maximum of biomass, reaching more than 300 μg dw l^{-1}, was found at the beginning of June. During summer biomass fell to a lower level and remained at about 50 μg dw l^{-1} (Fig. 7).

The daily production rates exceeded 10 μg dw l^{-1} d^{-1} four times, in May and June when a high biomass was produced and in the end of June and beginning of July, when the biomass was low, around 100 μg dw l^{-1}.

During the period of investigation the mean production rate in the estuary was 5.4 μg dw l^{-1} d^{-1}. Mean daily P:B ratio during the sampling period was 0.11 d^{-1}.

During the enclosure experiment biomass increased to over 100 μg dw l^{-1}. This was twice as high as in summer (July and August) in the estuary itself (Fig. 8). Production rate increased to over 40 μg dw l^{-1} d^{-1}, a four times higher level than observed in the field. The mean P:B ratio in the enclosure experiment reached a value of 0.301 d^{-1}.

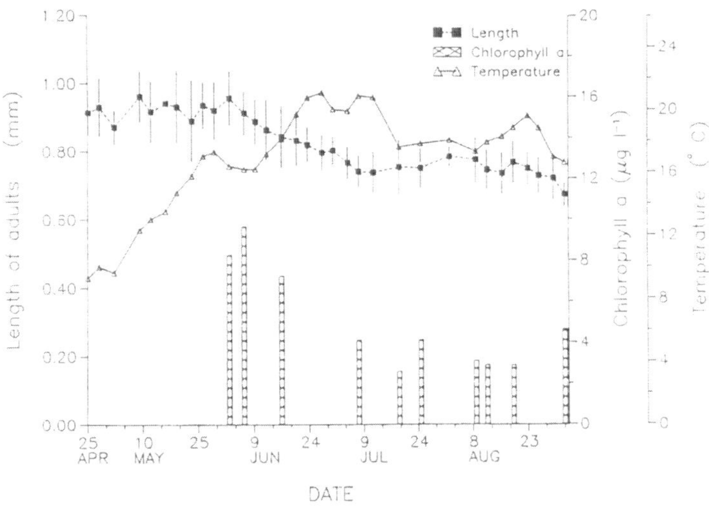

Fig. 4. Mean length of adults, temperature and chlorophyll- *a* level in the Elbe estuary 1989.

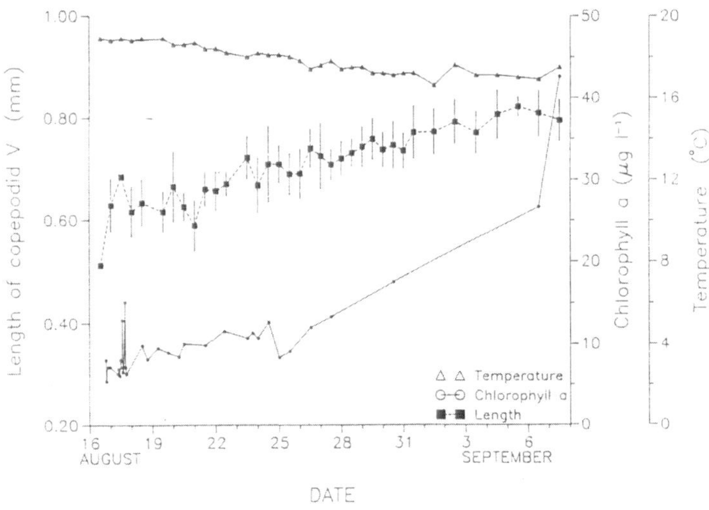

Fig. 5. Mean length of copepodid V, temperature and chlorophyll-*a* level in the enclosure experiment 1988.

Discussion

Representative sampling of zooplankton in estuaries is an important requirement for the estimation of production rates and population parameters in the field. Johnson (1981) chose a 'population center method' for *Acartia tonsa* at Yaquina Bay, considering only the maximum abundance of each instar in longitudinal section at a sampling day. Castel & Feurtet (1989) sampled during half a tidal cycle and then took a mean of abundance on a sampling day. In the Elbe estuary as well half a tidal cycle was sampled but the abundance was calculated as a weighted average, giving a greater importance to the samples at high current velocities (Peitsch, 1993).

Enclosure experiments offer a good opportunity to study a population under nearly natural conditions and to observe the development of a distinct population over a period of time. Duplicate samples from the experiment had a logarithmic variation coefficient of 13–23%, with the exception of the eggs (45.9%). The variation coefficient was low compared the range of 23–53% for diverse water bodies (UNESCO, 1968). Although the bag was stirred turbidity was not as high as in the maximum turbidity zone of the estuary.

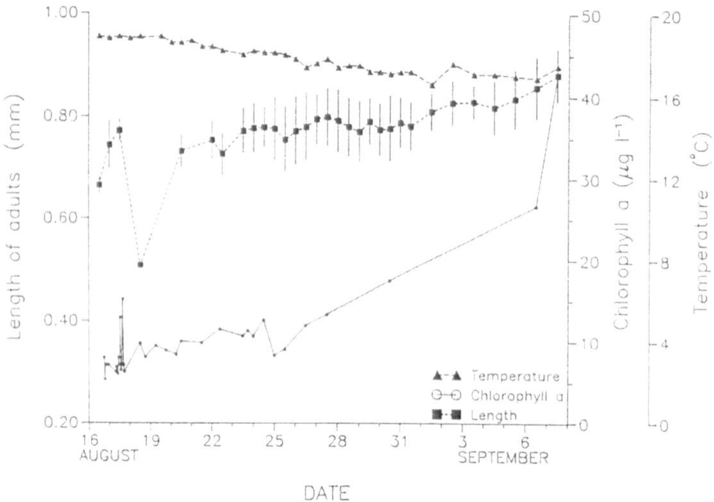

Fig. 6. Mean length of adults, temperature and chlorophyll-*a* level in the enclosure experiment 1988.

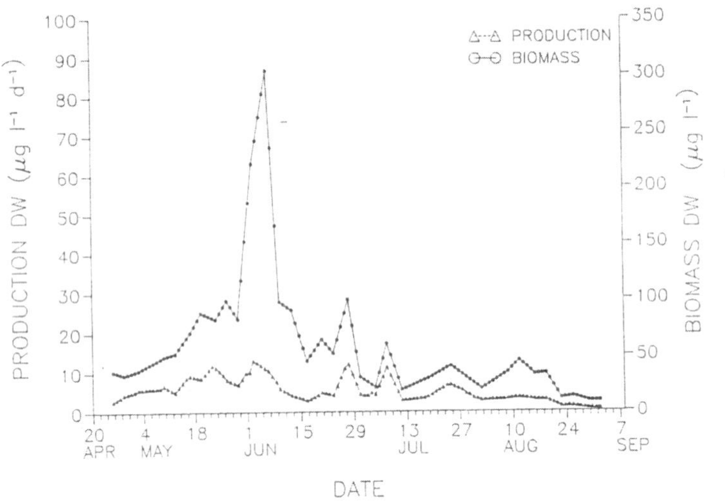

Fig. 7. Daily production rates and biomass of *Eurytemora affinis* in the Elbe estuary from April to September 1989.

Copepods often have a long life cycle of two to four weeks or even longer (depending on species and temperature) and a short reproduction time of the adult females, because of short life times of the adults in the field. Therefore in most cases cohorts of copepods could be recognized. Different methods have been developed to calculate cohort population parameters and production rates in the field (Gehrs & Robertson, 1975; Edmondson, 1968). The cohort analysis of Rigler & Cooley (1974) makes the assumptions that development time and mortality rates are constant within an instar in one cohort. This was criticized by Hairston & Twombly (1985) because rapid change of temperature or abundance of predators will lead to some kind of errors. But nevertheless the authors regarded this method as better than that of Comita (1972) and Gehrs & Robertson (1975). Also, Saunders & Lewis (1987) showed that errors will be low if there are no abrupt changes in the survival rate. Nevertheless the cohort analysis takes some subjective criterions it is robust over a broad range of conditions.

In contrast to production models of Heinle *et al.* (1977) and Allan *et al.* (1976) the method of Landry (1978) takes mortality and growth rates into account. The assumption that the production of males is half as

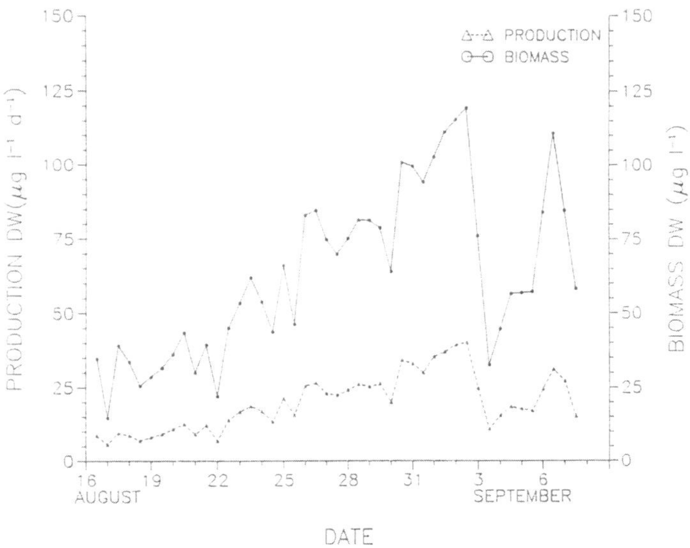

Fig. 8. Daily production rates and biomass of *Eurytemora affinis* in the enclosure experiment during summer 1988.

high as for females will be in the range of 40 to 90% really measured by Escaravage & Soetaert (1993).

Unfortunately it was not possible to run the experiment and the field investigations during the same year. But the environmental conditions (chlorophyll-*a* and temperature) which influence the production and biomass are known and therefore a comparison is reasonable. From the differences between field results and data from the enclosure experiment some conclusions can be drawn.

It is known that length of copepods and temperature show a negative relationship. The copepods of cohort 1 in the field had grown at the lowest temperature and therefore the greatest length could be expected for this group. In contrast, most of the instars in the enclosure experiment had a significant greater length although they developed at a higher mean temperature.

Statistically temperature explains 78% of the observed length variations of copepodid V and 48% of adults. But the difference in temperature at the beginning and at the end of the experiment was only 1.5 °C. So we have to look at the other important factor controlling growth of copepods, the food level. The length of naupliar stages are not regarded particularly, because they are not or only slightly affected in their length by a low food level. Klein-Breteler & Gonzales (1982) could show that at different food concentrations the length development was only influenced for copepodid II to adults. Length variation of another copepod, *Temora longicornis*, could also be explained to 83% by the change of food (Evans, 1981). In the enclosure experiment 49 to 67% of the length variations are statistically explained by the change of chlorophyll-*a* concentration.

The maximum biomass of *Eurytemora affinis* in the Elbe estuary in spring 1989 was comparable to results of the Gironde estuary with 300 μg dw l^{-1} (Castel & Feurtet, 1989), to Rhode River with 218 μg dw l^{-1} (Allan *et al.*, 1976) and the Westerscheldt estuary with 370 μg l^{-1} (Escaravage & Soetaert, 1993).

The P:B ratios estimated for *Eurytemora affinis* in different estuaries are very similar, in the Gironde 0– 0.2 d^{-1} (Castel & Feurtet, 1989), Barther Bodden, Balic Sea 0.2 d^{-1} (Arndt, 1989), Bristol Channel 0.09– 0.13 d^{-1} (Burkill & Kendall, 1982)) Schlei fjord 0.08– 0.13 d^{-1} (Christiansen, 1988) and Elbe estuary 0.11 d^{-1}.

In the enclosure experiment it could be shown that *Eurytemora affinis* is able to have a higher productivity of 0.3 d^{-1} at a high food supply.

Taking the two factors temperature and chlorophyll-*a* level into account the higher individual growth rate of the enclosure population of *Eurytemora affinis* in comparison to the field leads to the conclusion that the field population in the Elbe estuary is food limited during certain times of the year. A limited food supply of this species is also reported from other estuaries (Castel & Feurtet, 1989; Boyd, 1985; Burkill & Kendall, 1982; Heinle *et al.*, 1977).

In the turbidity zone a lot of carbon like detritus and suspended matter is available for the copepods as food. Besides the quantity of food the quality plays an important role for egg production and growth; except carbon, nitrogen, amino acids and proteins are necessary components of the food. Heinle *et al.* (1977) demand that metabolites for egg production must be derived from phytoplankton. Food limitation also leads to other effects on the population. From investigations of Poli & Castel (1983) and Bakker & Van Rijswijk (1987) for *Temora longicornis* it is known that a low food concentration, prolonges developmental times. Additionally egg production depends on food quality and quantity (Heinle, 1969; Ambler, 1985). This was estimated for the egg production of the *Eurytemora affinis* population of the Elbe estuary by Posewang-Konstantin (pers. comm.).

The results of this study show that the copepodid stages and adults of *Eurytemora affinis* in the Elbe estuary had shorter length and lower growth rates compared to animals in the enclosure experiment, caused by food limitation.

Acknowledgments

The work was supported by the Sonderforschungsbereich 327 Tide-Elbe, Deutsche Forschungsgemeinschaft. I thank Vincent Escaravage for his valuable comments on the manuscript.

References

Allan, J. D., T. G. Kinsey & M. C. James, 1976. Abundances and production of copepods in the Rhode River subestuary of Chesapeake Bay. Chesapeake Sci. 17: 86–92.

Ambler, J. W., 1985. Seasonal factors affecting egg production and viability of eggs of *Acartia tonsa* DANA from East Lagoon, Galveston, Texas. Estuar. coast. Shelf Sci. 20: 743–760.

Arndt, H., 1989. Zooplankton production and its consumption by planktivores in a baltic inlet. Proc. of 21. EMBS: 205–214.

Bakker, C. & P. van Rijswijk, 1987. Development time and growth rate of the marine calanoid copepod *Temora longicornis* as related to food conditions in the Osterscheldt estuary (Southern North Sea). Neth. J. Sea Res. 21: 125–141.

Boyd, C. M., 1985. Is secondary production in the Gulf of Maine limited by the availability of food?. Arch. Hydrobiol. Beih. Limnol. 21: 57–65.

Burkill, P. H. & T. F. Kendall, 1982. Production of the copepod *Eurytemora affinis* in the Bristol Channel. Mar. Ecol. Prog. Ser. 7: 21–31.

Brockmann, U. H., K. Eberlein, H. D. Junge, H. Trageser & K. J. Trahms, 1974. Einfache Folientanks zur Planktonuntersuchung *in situ*. Mar. Biol. 24: 163–166.

Castel, J. & A. Feurtet, 1986. Influence des matières en suspension sur la biologie d'un copepode estuarien: *Eurytemora hirundoides* (Nordquist, 1988). Coll. Nat. CNRS 'Biologie des Populations': 391–396.

Castel, J. & A Feurtet, 1989. Dynamics of the copepod *Eurytemora affinis hirundoides* in the Gironde estuary: origin and fate of its production. Topics in marine Biologie; Ros, J. D. (ed.). Scient. Mar. 53: 577–584.

Christiansen, B., 1988. Vergleichende Untersuchungen zur Populationsdynamik von *Eurytemora affinis* POPPE und *Acartia tonsa* DANA, Copepoda in der Schlei. Dissertation Universität Hamburg, Fachbereich Biologie.

Comita, G. W., 1972. The seasonal zooplankton cycles, production and transformation of energy in Severson Lake, Minnesota. Arch. Hydrobiol. 70: 14–66.

De Pauw, N., 1973. On the distribution of *Eurytemora affinis* (POPPE) (Copepoda) in the Western Scheldt Estuary. Verh. int. Ver. Limnol. 18: 1462–1472.

Edmondson, W. T., 1968. A graphical model for evaluating the use of egg ratio for measuring birth and death rates. Oecologia 1: 1–37.

Evans, F., 1981. An investigation into the relationship of sea temperature and food supply to the size of the planktonic copepod *Temora longicornis* MÜLLER in the North Sea. Estuar. coast. Shelf Sci. 13: 145–158.

Escaravage, V. & K. Soetaert, 1993. Estimating secondary production for the brackish Westerschelde copepod population *Eurytemora affinis* (Poppe) combining experimental data and field observations. Cah. Biol. Mar. 34: 201–214.

Gehrs, C. W. & Robertson, 1975. Use of life tables in analyzing the dynamics of copepod populations. Ecology 56: 665-672.

Hairston, N. G. & S. Twombly, 1985. Obtaining life table data from cohort analysis: a critique of current methods. Limnol. Oceanogr. 30: 886–893.

Heinle, D. R., 1969. Culture of calanoid copepods in synthetic seawater. J. Fish. Res. Bd Can. 26: 150–153.

Heinle, D. R. & D. A. Flemer, 1975. Carbon requirements of a population of the estuarine copepod *Eurytemora affinis*. Mar. Biol. 31: 235–247.

Heinle, D. R., R. P. Harris, J. F. Ustach & D. A. Flemer, 1977. Detritus as food for estuarine copepods. Mar. Biol. 40: 341–353.

Johnson, J. K., 1981. Population dynamics and cohort persistence of *Acartia californiensis* (Copepoda: Calanoida) in Yaquina Bay. Oregon. Ph.D. Thesis. Oregon State University, Corvallis, Orc.

Kankaala, P. & S. Johansson, 1986. The influence of individual variation on the length-biomass regressions in three crustacean zooplankton species. J. Plankton Res. 8: 1027–1038.

Klein-Breteler, W. C. M. & S. R. Gonzales, 1982. Influence of cultivation and food concentration on body length of calanoid copepods. Mar. Biol. 71: 151–161.

Landry, M. R., 1978. Population dynamics and production of a planktonic marine copepod *Acartia clausi* in a small temperate lagoon on San Juan Island, Washington. Int. Revue Ges. Hydrobiol. 63: 77–79.

Peitsch, A., 1992. Untersuchungen zur Populationsdynamik und Produktion von *Eurytemora affinis* (Calanoida; Copepoda) im Brackwasserbereich des Elbe-Ästuars. Dissertation Universität Hamburg, Institut für Hydrobiologie und Fischereiwissenschaft.

Peitsch, A., 1993. Difficulties in estimating mortality rates of *Eurytemora affinis* in the brackish water region of the Elbe estuary. Cah. Biol. Mar. 34: 215–224.

Poli, J. M. & J. Castel, 1983. Cycle biologique en laboratoire d'un Copépode planctonique de l'estuaire de la Gironde: *Eurytemora hirundoides* (Nordquist, 1988). Vie et Milieu 33: 79–86.

Rigler, F. H. & J. M. Cooley, 1974. The use of field data to derive population statistics of multi-voltine copepods. Limnol. Oceanogr. 19: 636–655.

Saunders, J. F. I. & W. M. Jr. Lewis, 1987. A perspective on the use of cohort analysis to obtain demographic data for copepods. Limnol. Oceanogr. 32: 511–513.

Siefert, W., 1970. Die Salzgehaltsverhältnisse im Elbe-Mündungsgebiet. Hamburger Küstenforschung 15 (April).

Soltanpour-Gargari, A. & Wellershaus, 1984. *Eurytemora affinis* — the estuarine plankton copepod in the Weser. Veröff. Inst. Meeresforsch. Bremerh. 20: 103–117.

Soltanpour-Gargari, A. & S. Wellershaus, 1987. Very low salinity stretches in estuaries — the main habitat of *Eurytemora affinis*, a planktonic copepod. Contrib. of Alfred Wegener Institute for Polar and Marine Research, 37.

Unesco (1968). Zooplankton sampling (Monographs on oceanographic methodology 2).o.O.: United Nations Eductional, Scientific and Cultural Organisation.

Winberg, G. & W. T. Edmondson (ed.), 1971. A manual on methods for the assessment of Secondary Productivity in Fresh Water. IPB Handbook 17, Oxford and Edinburgh, Blackwell Scientific Publications.

Hydrobiologia **311**: 139–151, 1995.
C. H. R. Heip & P. M. J. Herman (eds), Major Biological Processes in European Tidal Estuaries.

Comparative spring distribution of zooplankton in three macrotidal European estuaries

Benoît Sautour & Jacques Castel
Laboratoire d'Océanographie Biologique, Université de Bordeaux I, F-33120 Arcachon, France

Key words: estuaries, zooplankton, copepods, distribution, biomass

Abstract

The zooplankton of three european estuaries (Ems, Gironde and Westerschelde) was investigated during spring 1992 by means of samples taken along the salinity gradient. The three estuaries are comparable in terms of total area, flushing time and salinity gradient but differ by their level of eutrophication (highest in the Westerschelde), suspended matter concentration (highest in the Gironde) and potential phytoplankton production (highest in the Ems). Copepods and meroplankton dominated the zooplankton in the three estuaries. The dominant copepod species were *Eurytemora affinis* and *Acartia bifilosa.* The distribution of *E. affinis* along the salinity gradient differed between the estuaries. Peaks of abundance were observed at 0 PSU in the Gironde, 6 PSU in the Ems and 9 PSU in the Westerschelde. The downstream shift of the population in the Westerschelde was likely due to anoxic conditions occurring in the oligohaline zone. In the Gironde the downstream distribution of *E. affinis* was limited by the very high suspended matter concentration found in the maximum turbidity zone. Whatever the estuary, the parameters of the population of *E. affinis* and maximum abundance values were similar. However, the influence of the better quality of the available food was suggested in the Ems where individual dry weights and egg production were higher than in the two other estuaries. The influence of a good quality of food in the Ems was confirmed by the development of a large population of *Acartia bifilosa* (as abundant as *E. affinis*) and highest values of adult individual weights.

The meroplankton (essentially Polychaete and cirripede larvae) was much more developed in the Ems than in the Westerschelde and Gironde. This was likely due to the large extent of mudflats and hard substrates in the Ems favouring adult settlement and hence the number of larvae locally produced.

Introduction

In estuaries, the zooplankton community is largely dominated by Calanoid Copepods. Copepod production may be very high (up to 56.2 mg DW m^{-3} d^{-1}, Heinle, 1969) and phytoplankton primary production has been shown to be sometimes too low to support growth of herbivorous copepods (Heinle & Flemer, 1975). In such a case they have to complete their diet with detritus (Heinle & Flemer, 1975; Heinle *et al.*, 1977; Boak & Goulder, 1983).

The distribution of zooplankon taxa in estuaries is classically known to depend on physical constraints: e.g. temperature (Castel *et al.*, 1983), salinity (Gunter, 1961; Jeffries, 1962; Bakker & De Pauw, 1975), flushing and mixing (Soltanpour Gargari & Wellershaus, 1985; Castel & Veiga, 1990). However, the dominant copepod species (*Eurytemora affinis* and *Acartia bifilosa* in the northern hemisphere) are euryhaline and eurythermic (3–30 °C and 2–35 PSU, Castel, 1981) and their distribution could also be affected by biotic factors such as food availability, competition and predation (Bradley, 1975; Burkill & Kendall, 1982; Baretta & Malschaert, 1988).

Meroplanktonic larvae also contribute significantly to the estuarine zooplankton. They often show a marked seasonality related to the breeding activity of the adults living in the benthos (Williams & Collins, 1986). The abundance of the meroplankton moreover depends on the surface of intertidal areas and hard substrates which are colonized by benthic organisms.

The aim of the present study was to compare the distribution of zooplankton in three tidal estuaries: the Ems, Westerschelde and Gironde, showing a clear salinity gradient but differing by their primary production and by the load of suspended particulate matter.

The Ems estuary has relatively low turbidity (100 mg l^{-1}), allowing considerable values of primary production (100 gC m^{-2} y^{-1} de Jonge, 1993). The Westerschelde estuary is less turbid (<100 mg l^{-1}) but supports high amounts of organic pollution which induce an important decrease in dissolved oxygen in the oligohaline zone near Antwerp (Heip, 1989). The average primary production is 50 gC m^{-2} y^{-1} (Soetaert & Herman, 1995). The Gironde estuary is a highly turbid estuary (>500 mg l^{-1}) with a reduced primary production (10 gC m^{-2} y^{-1}, Etcheber, 1986; Lin, 1988).

The spatial and temporal patterns of the zooplankton in the three estuaries are well-known (Baretta & Malschaert, 1988; Soetaert & van Rijkswijk, 1993; Castel, 1981). As expected, the distribution of species is conditioned by the salinity gradient but the abundances of the species are difficult to compare since the authors have used different sampling techniques. In these studies, the importance of chlorophyll content of the water and suspended matter concentration in explaining zooplankton community structure has been assessed but the conclusions are mainly valid for within-site observations.

In the present study samples were taken quasi-simultaneously in the three estuaries during the spring period, i.e. the period of maximum abundance, using the same plankton net. The distribution pattern of the species is related to the salinity gradient and some hypotheses are proposed to explain the general role of feeding conditions on the abundance and biomass of the dominant species.

Methods

Study areas (Fig. 1, Table 1)

The Ems-Dollard estuary (north of the Netherlands) covers an area of about 500 km^2. Freshwater enters the estuary by different sources of which the most important is the river Ems with a mean annual discharge of 125 m^3 s^{-1}. A second freshwater source is the Westerwoldsche Aa river. The water discharge is roughly 10% of that of the river Ems (De Jonge, 1988). The maximal current velocity is 1.5 m s^{-1}. This estuary is well mixed except in the Dollard where a stratification can occasionally be detected. The zone of maximum turbidity (100–400 mg l^{-1}) is observed in the oligohaline region (De Jonge, 1988). Intertidal flats constitute about 36% of the total area of the estuary; in the Dollard they comprise 85% of the surface. The mean water depth in the middle part of the estuary is about 3.5 m; it is much lower (1.2 m) in the Dollard.

The Westerschelde estuary (S. W. of the Netherlands) can be divided into 3 parts: a marine zone (70 km long from Vlissingen to Walsoorden), a central zone (50 km long, from Walsoorden to the river Rupel) and a fluvial zone where the tidal influence is still observed (Peters & Sterling, 1976). The total surface area is about 600 km^2. In the first zone, mixing of the water is intense, while it is only partial in the central zone. The water depth in the navigation channel is 15–20 m. A zone of maximum turbidity (100 mg l^{-1}) is observed close to Antwerpen. The freshwater input due to the river Schelde is about 105 m^3 s^{-1} (Heip, 1989). The maximum current velocity in the brackish zone is 1.65 m s^{-1} (Bakker & De Pauw, 1975). The tidal flats and the salt marshes occupy a surface of 110 km^2 (Oenema *et al.*, 1988).

The Gironde estuary, (S.W. of France) covers an area of 625 km^2 at high water. Freshwater inflow to the estuary is brought by the rivers Dordogne and the Garonne. The mean yearly freshwater discharge is 900 m^3 s^{-1} (Castel, 1993). Maximum current velocity can reach 2 m s^{-1}. The mean water depth is 5–19 m. The waters are partially mixed during low river flow and slightly stratified during high water discharge (Jouanneau & Latouche, 1981). The extent of the intertidal flats are very low (50 km^2) except at the mouth of the estuary. One of the main characteristics of the Gironde is the high turbidity of the water: particulate concentrations of 1 g l^{-1} at the surface and 10 g l^{-1} near the bottom are common values in the oligohaline zone (Jouanneau & Latouche, 1981).

Sampling

As it is difficult to describe zooplankton composition and distribution in terms of geographical location because of the continuous movement of the estuarine water masses, we chose to use salinity as a descriptor of spatial distribution pattern because of its conservative properties (Baretta & Malschaert, 1988). Thus, samples were taken throughout the estuaries along the salinity gradient from 0 to 30 PSU (interval: 3 PSU).

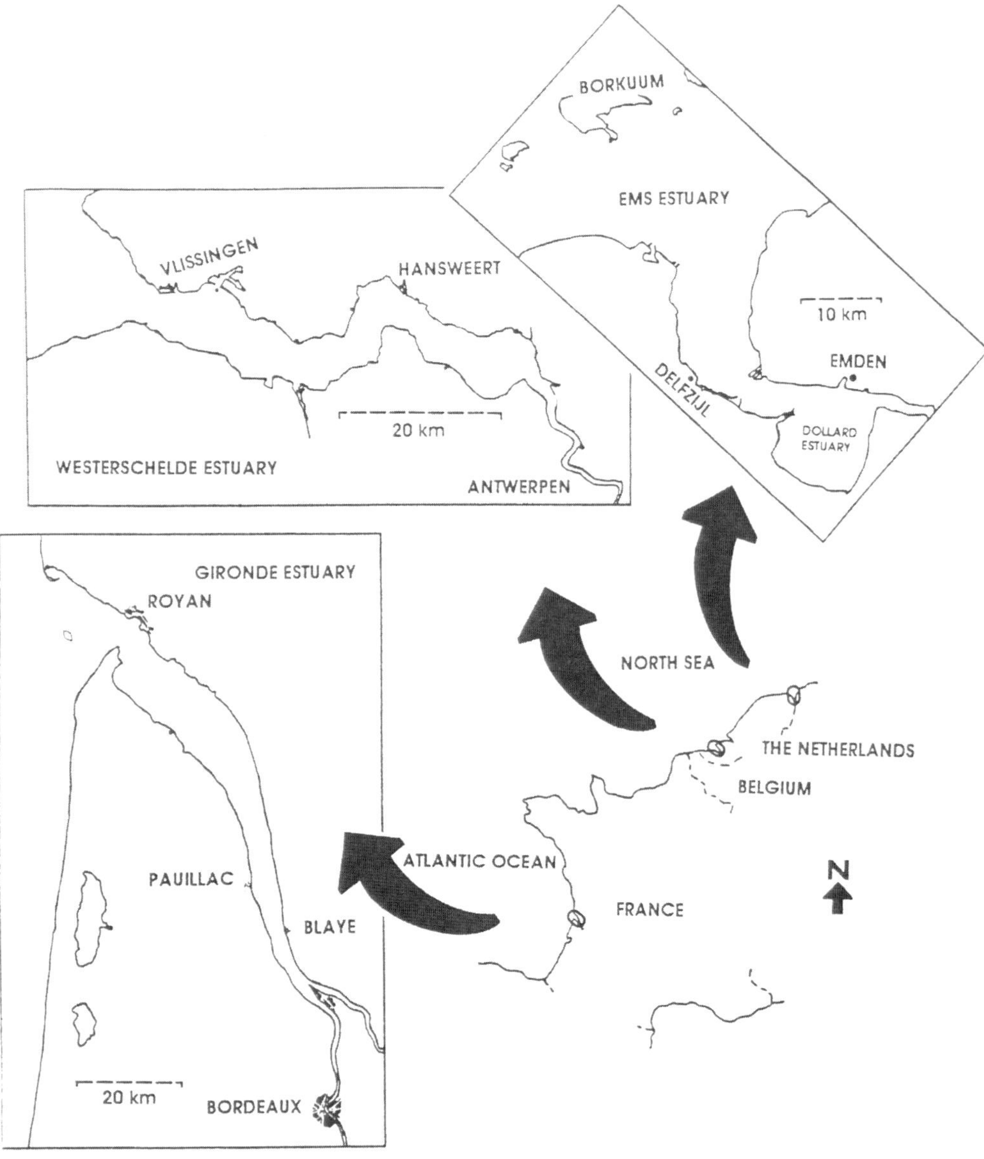

Fig. 1. Geographical location of the sampling areas.

In the Gironde, monthly samples were obtained from March to June 1992. Due to local constraints, sampling was made in March, April and June in the Westerschelde (30 PSU were never sampled), and only in March and May in the Ems. In the Ems and Westerschelde, oblique zooplankton hauls were carried out whereas in the Gironde, surface and bottom samples were taken.

Zooplankton was collected with a standard WP2 net (200 μm mesh size; Fraser, 1966). This mesh size was selected to avoid rapid clogging of the net in these highly turbid zones. Thus, this study only concerns the mesozooplankton. The volume filtered ranged from 4 to 20 m^3 per haul (Hydrodata digital flowmeter, model 438110). Samples were rinsed with filtered estuarine water, concentrated and immediately stored in buffered 4% formalin.

Data analysis

In the laboratory, the different taxa were sorted under a binocular microscope and the developmental stages of the different copepod species were identified and counted. For *Acartia* spp: adults were identified at the species level but the developmental stages were identified at the genus level only. Abundances obtained

Table 1. Average values of the environmental factors suspected to act on the distribution and abundance of estuarine zooplankton. SPM and POC values are given for the maximum turbidity zone. All data from the literature, except temperature. (1) De Jonge (1988,1993), (2) Laane *et al.* (1987), (3) Burdloff (1993), (4) Oenema *et al.* (1988), (5) Heip (1989); (6) Soetaert & Van Rijswijk (1993), (7) Soetaert & Herman (**this volume**), (8) Jouanneau & Latouche (1981), (9) Etcheber (1986), (10) Irigoien & Castel (submitted), (11) Lin (1988).

	Ems	Westerschelde	Gironde
Tidal flat area (km^2)	240(1)	110(4)	50(8)
Tidal flat (% area of the estuary)	51	18	8
Water residence time (days)	38 (1)	75(5)	20–70(8)
SPM conc. ($mg\,l^{-1}$)	100–400(1)	60–400(3, 6)	500–1000(9, 10)
POC (%)	3.00–4.00(2, 3)	3.50(3, 7)	1.45(3, 9)
Chlorophyll *a* ($\mu g\,l^{-1}$)	2–10(1)	4–7(6)	1–8(10)
Production ($gC\,m^{-3}\,y^{-1}$)	48(1)	9(7)	<5(11)
Sampling temperature (°C)	6.5–11.8	8.2–19.0	11.0–18.6

from surface and bottom samples in the Gironde were averaged in order to make comparisons with the integrated samples taken in the Ems and in the Westerschelde. Monthly abundances were averaged per salinity zones.

For *Eurytemora affinis* the number of eggs per egg-sac were counted from pools of 30 ovigerous females taken at random in samples representing the maximum peak abundance of the population. Copepodid instars (copepodids CII to CVI) of the two dominating taxa, *Eurytemora affinis* and *Acartia* spp, were counted and weighed. Weight measurements were made on material preserved for at least one month since a storage period <30 d causes variable weight changes (Kulhman *et al.*, 1982). Samples corresponding to the peak abundance of both taxa were chosen for this analysis (*E. affinis*: in April, at 6 and 9 PSU respectively in the Gironde and in the Westerschelde and at 6 PSU in the Ems in May; *Acartia* spp in May, at 9 and 15 PSU respectively in the Gironde and in the Ems and at 18 PSU in the Westerschelde in April). These samples were supposed to be representative of populations having optimal developmental conditions. Copepods were rinsed with distilled water and dried at 60 °C for 24 hours. Three aliquots of 20 individuals of each instar were weighed on a Mettler ME22 microbalance (sensitivity: 0.1 μg). Body carbon weights were obtained by multiplying individual dry weights by a factor 0.4 (Parsons *et al.*, 1977). Total biomass of each instar was calculated by multiplying its individual carbon weight by its abundance. Results are given in carbon weight per cubic meter.

Comparisons between sex ratios, individual dry weights and total biomass were made either with t-test or with non parametric tests (*Wilcoxon-Mann-Whitney test* or *Kruskal-Wallis test*, when the data were not normally distributed or when the homogeneity of variances was not verified).

Results

Gross taxonomic composition and distribution

The zooplankton of the 3 estuaries was divided into the following 3 groups: copepods, meroplankton and other organisms (Figs 2 and 3).

Copepods dominated the zooplankton in the 3 estuaries, particularly in the oligohaline zone (0.5 to 5 PSU, Venice system of classification of saline waters) and mesohaline (5 to 18 PSU) zones. At the limit of the limnetic zone in the Ems and in the Westerschelde freshwater copepods (*Cyclops* spp and *Eudiaptomus gracilis*) were found. Furthermore freshwater cladocerans and the rotifer *Brachionus plicatilis* were observed in the Westerschelde. In the Gironde, the oligohaline zone was only characterized by *E. affinis*, while the species was quite absent in the Westerschelde at lowest salinities.

E. affinis developed in the mesohaline zone of the Ems and Westerschelde, when its abundances decreased drastically in the Gironde. In the Ems, *A. bifilosa* and meroplankton (polychaetes) were also abundant in this zone. Some neritic taxa appeared spo-

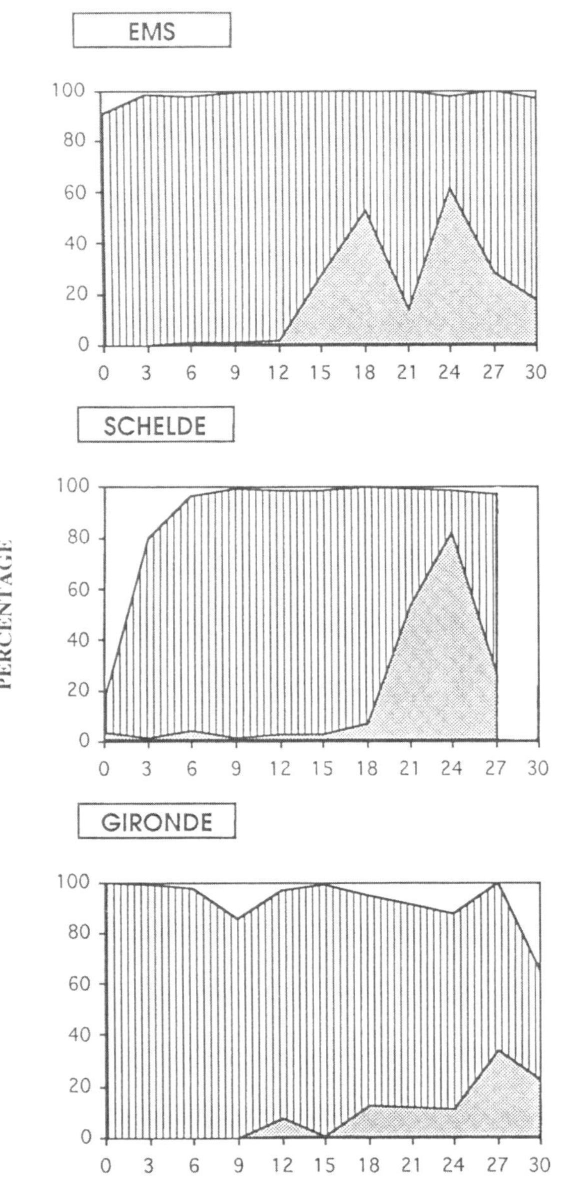

Fig. 2. Relative abundance of copepod (hatched) and meroplankton (grey) communities as a function of salinity in the 3 estuaries during spring 1992.

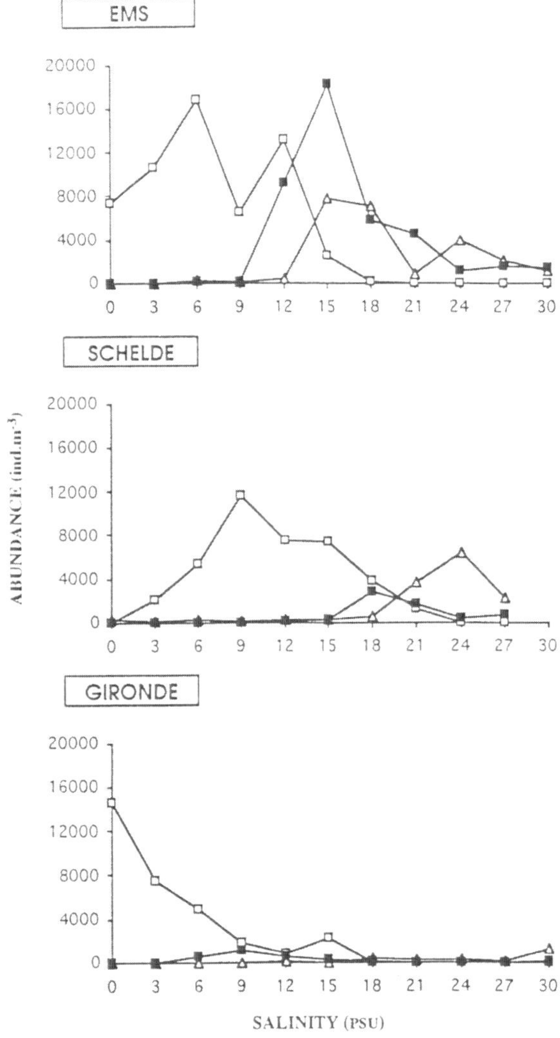

Fig. 3. Eurytemora affinis (-□), *Acartia* spp (-■-) and meroplankton (-△-) abundances as a function of salinity in the 3 estuaries during spring 1992.

radically in the mesohaline zone of the three estuaries: cnidaria, ctenophora (over 12 PSU in Westerschelde and 15 PSU in the 2 other estuaries) and the cladoceran *Podon* sp (over 12 PSU in Ems and Gironde).

The polyhaline zone (18 to 30 PSU) was further characterized by neritic copepods and meroplankton. Copepod abundance was variable (maximum values: 3800 ind m^{-3} in the Ems, 2300 ind. m^{-3} in the Gironde and 5100 ind m^{-3} in the Westerschelde) and their penetration into the estuaries differed from one estuary to the other. These allochtonous species could be divided into 2 groups. The first one was composed of euryhaline copepods (*Centropages hamatus*, *Temora longicornis*, *Paracalanus parvus*, *Pseudocalanus elongatus*, *Euterpina acutifrons*) and the second one was composed of more oceanic copepods (*Oithona helgolandica*, *Calanus helgolandicus*, *Oncaea* spp, *Corycaeus* spp). This second group, plus *E. acutifrons*, did not penetrate in the Westerschelde during spring. The dominating species of the neritic community in the 3 estuaries were always *C. hamatus* and *T. longicor-*

nis. The meroplankton was dominated by polychaete and cirripede larvae and its importance varied between estuaries (Figs 2, 3). The cirripedes were characteristic from 24 to 30 PSU in the Ems and from 21 to 27 PSU in the Westerschelde. At the highest salinities, the Gironde estuary was also characterized by the cladocerans (*Podon* sp) and the Ems also by Gastropod and bivalve larvae.

On average, the most important zooplankton assemblage was constituted by copepods: essentially *Eurytemora affinis* and *Acartia* dominated (*Acartia bifilosa* dominated below 18 PSU and was accompanied by *A. discaudata* and *A. clausi* at higher salinities). Peak abundances of *E. affinis* were in the same order of magnitude in the 3 estuaries: 16800 ind. m^{-3} in the Ems, 14500 ind m^{-3} in the Gironde and 11500 ind m^{-3} in the Westerschelde. Contrarily, the maximum abundance of *A. bifilosa* was much higher in the Ems (18200 ind m^{-3}) than in the 2 other estuaries (1100 ind m^{-3} in the Gironde and 2800 ind m^{-3} in the Westerschelde).

Eurytemora affinis

The peaks of abundance were not observed at the same salinities in the 3 estuaries: 0 PSU in the Gironde, 6 PSU in the Ems and 9 PSU in the Westerschelde. In addition, this species occurred in a wider range of salinity in the Ems and in the Westerschelde than in the Gironde (Fig. 4). The abundance peaks (Fig. 5) were mainly due to the young instars (copepodids CII and CIII), particularly in the Ems. Downstream and upstream the abundance peaks, aged instars (copepodids CIV to CVI) dominated. In the 3 estuaries, the average sex ratio (males/females) calculated for copepodids V was lower than the one determined for adults ($p = 0.05$ *t*-test, in the Ems and Westerschelde, n.s. in the Gironde). The global values (CV + CVI) calculated for the Ems (0.79, SD = 0.60) and the Westerschelde (1.06, SD = 0.65) were not significantly different (*t*-test). The sex ratio determined in the Gironde (1.52, SD = 0.44) was significantly higher ($p = 0.05$) than that determined in the Ems. The highest abundance of ovigerous females were observed downstream the peak of the population in the Gironde and in the Westerschelde; in the Ems their distribution was wider. The highest values of fecundity were found in the Ems estuary (37 eggs $female^{-1}$ SD = 13) and lowest in the Gironde estuary (14 eggs $female^{-1}$ SD = 8); values obtained in the Westerschelde were intermediate (27 eggs $female^{-1}$ SD = 5).

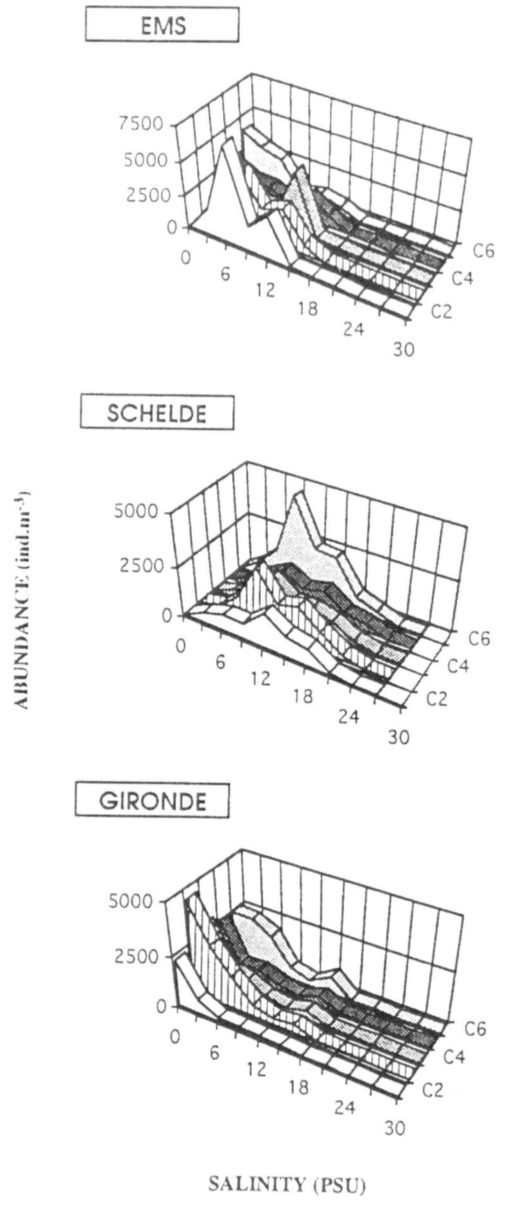

Fig. 4. Copepodid (II to VI) abundances of *Eurytemora affinis* as a function of salinity in the 3 estuaries during spring 1992.

Acartia spp

As for *Eurytemora affinis*, abundance peaks of *A. bifilosa* were not situated at the same salinity in the 3 estuaries: 9 PSU in the Gironde, 15 PSU in the Ems and 18 PSU in the Westerschelde. Although *Acartia* copepodids were not determined at the species level, it is likely that most individuals found in the mesohaline zone belong to *A. bifilosa*. Adults (CVI) dominated in the 3 estuaries (Figs 6 and 7) and were relatively

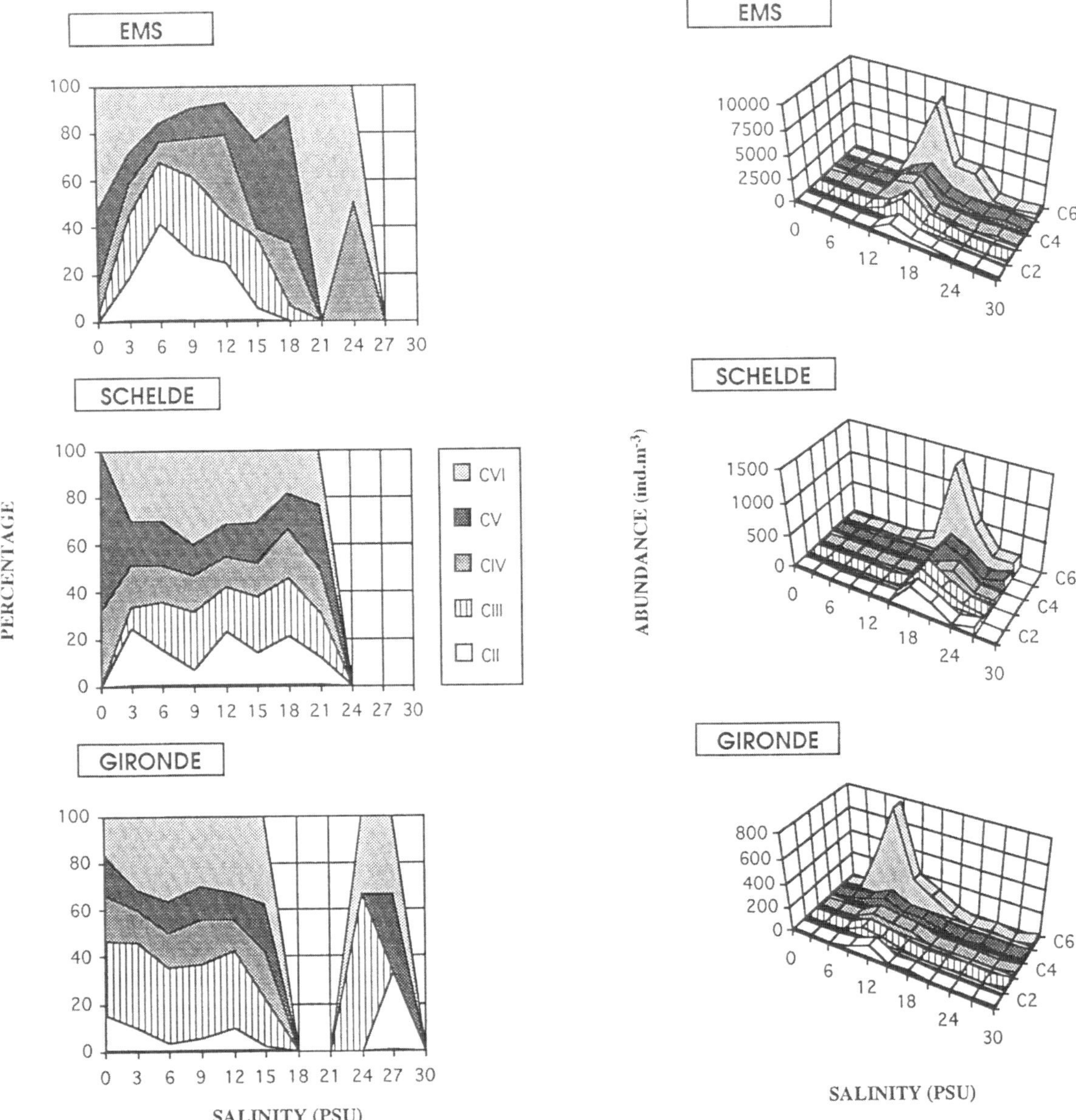

Fig. 5. Relative abundances of copepodids (II to VI) of *Eurytemora affinis* as a function of salinity in the 3 estuaries during spring 1992.

Fig. 6. Copepodid (II to VI) abundances of *Acartia* spp as a function of salinity in the 3 estuaries during spring 1992.

more abundant just upstream the abundance peak of the population. In the bulk of the adult population, males dominated in the Ems and females dominated in the Gironde. The situation was intermediate in the Westerschelde. The global sex ratio (CV + CVI) was not significantly different from one estuary to the other (0.97, SD = 0.33 in the Ems; 0.83, SD = 0.32 in the Gironde; 0.74, SD = 0.59 in the Westerschelde) nor between copepodids V and VI.

The small peaks of copepodids II observed at salinities above 21 PSU were probably produced by the 2 neritic species of this genus: *A. discaudata* and *A. clausi*, since adults were recorded in the estuaries at these salinities.

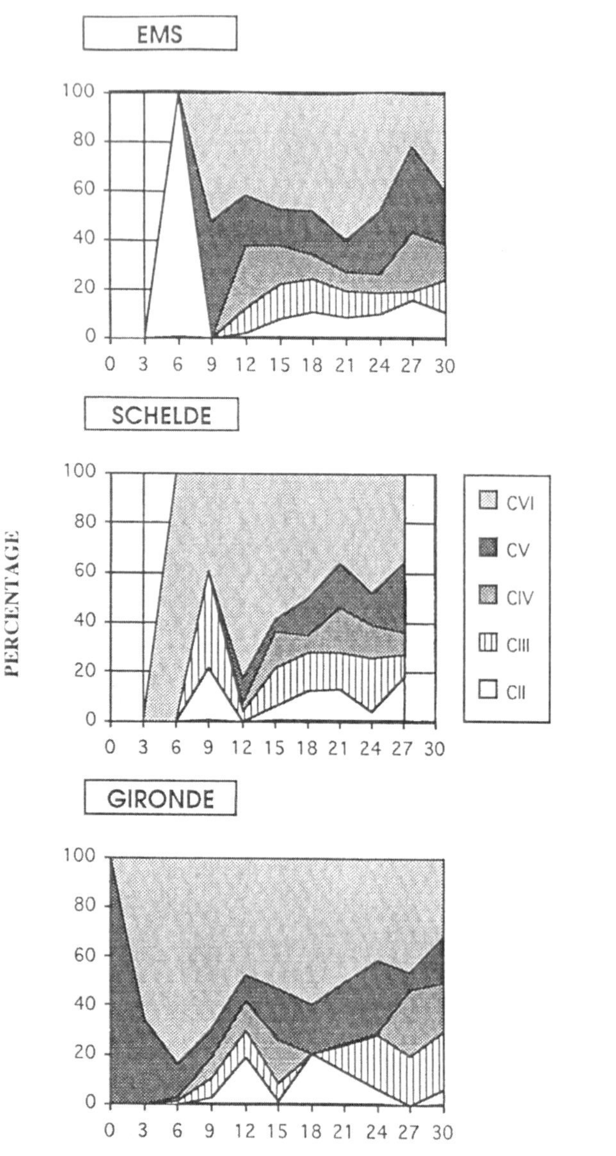

Fig. 7. Relative abundances of copepodids (II to VI) of *Acartia* spp as a function of salinity in the 3 estuaries during spring 1992.

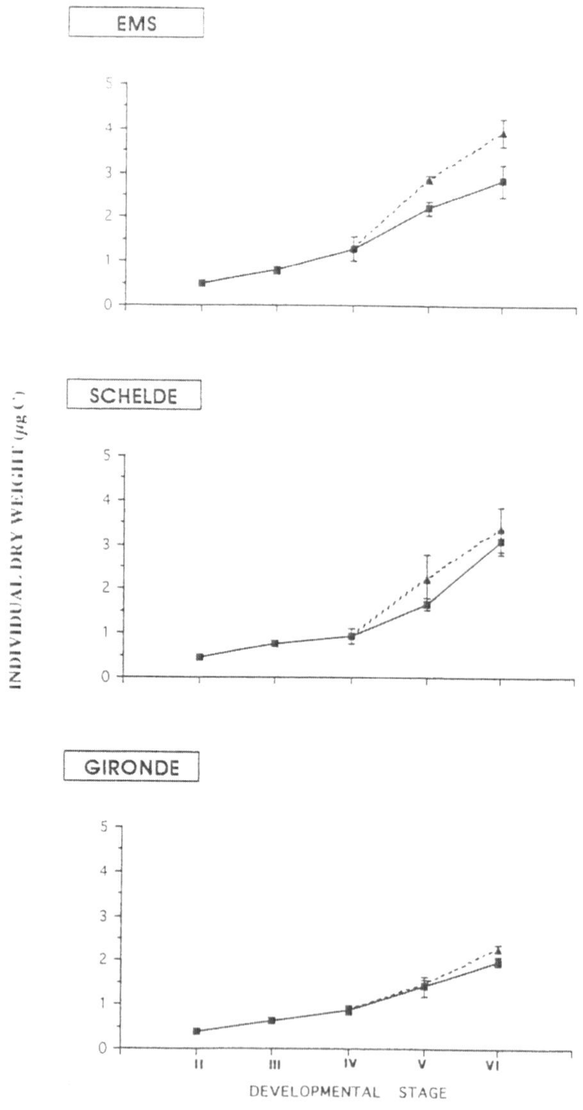

Fig. 8. Individual carbon weight (±standard deviation) of the different stages of *Eurytemora affinis* in the 3 estuaries during spring 1992. (Dotted line = females).

Individual dry weight of the 2 dominant copepod species

For each instar, individual weights of *E. affinis* were lower in the Gironde than in the Ems (Fig. 8, Wilcoxon-Mann-Whitney test, $p = 0.025$). Individual weights in the Westerschelde were intermediate but close to the Ems' (significant difference was only observed between Ems and Westerschelde for CV and between Gironde and Westerschelde for CIII, CV females and CVI, $p = 0.025$). Whatever the estuary, individual weights of females were higher than those of males ($p = 0.025$: stages CV and CVI in the Ems and CVI in the Gironde); however, the difference was not significant in the Westerschelde.

For the young stages (CII to CV) of *A. bifilosa* no difference in weight was found between the Ems and the Gironde, these weights were slightly lower than in the Westerschelde (Wilcoxon-Mann-Whitney test, $p = 0.025$, Fig. 9). Females were heavier in the Ems than in the Westerschelde and males and females were heavier in the Westerschelde than in the Gironde

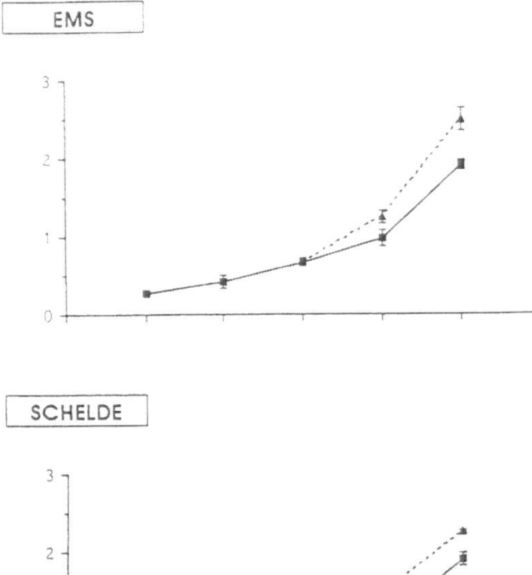

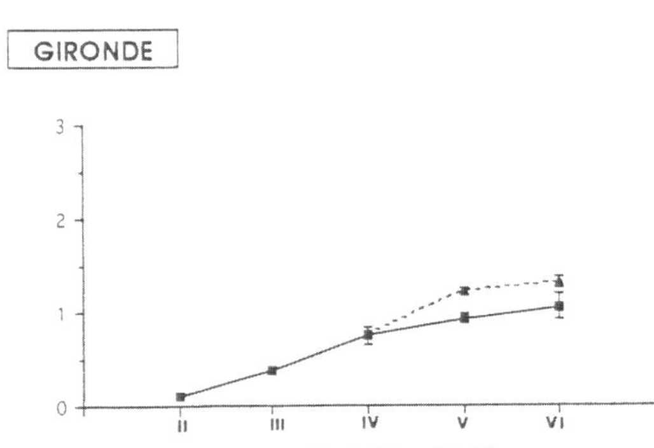

Fig. 9. Individual carbon weight (±standard deviation) of the different stages of *Acartia* spp in the 3 estuaries during spring 1992. (Dotted line = females).

($p = 0.025$). In all cases females had higher weight than males ($p = 0.025$).

Copepod biomass in the oligohaline and mesohaline zones

Copepods being strongly dominant (Fig. 2) in the oligohaline and mesohaline zones, copepod biomass (= *E. affinis* + *A. bifilosa*) was a good approximation of the zooplanktonic biomass. Copepod biomass was significantly different between the 3 estuaries ($p = 0.025$, Kruskal-Wallis test). Highest copepod biomass was found in the Ems (Fig. 10) where the contribution of the dominant species *E. affinis* and *A. bifilosa* was quite similar: 20.54 mgC m^{-3} and 24.53 mgC m^{-3} respectively. In the Westerschelde, biomass was lower than in the Ems (17.68 mgC m^{-3} for *E. affinis* and 4.05 mgC m^{-3} for *A. bifilosa*). The situation was different in the Gironde where the biomass of *A. bifilosa* was very low (1.21 mgC m^{-3}) and the biomass of *E. affinis* was only important in the oligohaline zone (maximum at 0 PSU: 12.84 mgC m^{-3}). In this estuary, the copepod biomass in the mesohaline zone was lower than 7 mgC m^{-3}).

Discussion

The present study was aimed to compare the composition and distribution of the zooplankton in 3 european tidal estuaries showing similarities in terms of total area, flushing time and salinity gradient but differing by their level of eutrophication, SPM concentration and potential phytoplankton production (Table 1). The Ems is not very turbid and is well aerated which allows for the highest primary production. The Westerschelde is subject to anoxia (due to anthropic activities) in its oligohaline zone. The Gironde is a highly turbid estuary with very low primary production, especially in the maximum turbidity zone.

Sampling was concentrated during the spring period which corresponds to the maximum abundance of the dominant species *Eurytemora affinis*: April in the Ems (Baretta & Malschaert, 1988), March–May in the Westerschelde (Soetaert & Van Rijswijk, 1993), April–May in the Gironde (Castel, 1993). The second dominant species, *Acartia bifilosa*, shows a less constant temporal pattern, being most abundant in March–July in the Ems (Baretta & Malschaert, 1988), in late winter in the Westerschelde (Soetaert & Van Rijswijk, 1993), in July–August in the Gironde (Castel, 1993). The reasons of these different seasonal patterns are unknown but this can explain why *A. bifilosa* was much more abundant in the Ems than in the two other estuaries. The third taxon considered in this study, the meroplankton, is always most abundant in spring (Baretta & Malschaert, 1988; Soetaert & Van Rijswijk, 1993; Castel, 1981).

The population of the copepod *Eurytemora affinis* was well developed in the three estuaries. It dominated the zooplankton in the oligohaline zone in the Ems and the Gironde thus confirming previous findings (Baretta & Malschaert, 1988; Castel & Veiga, 1990). In

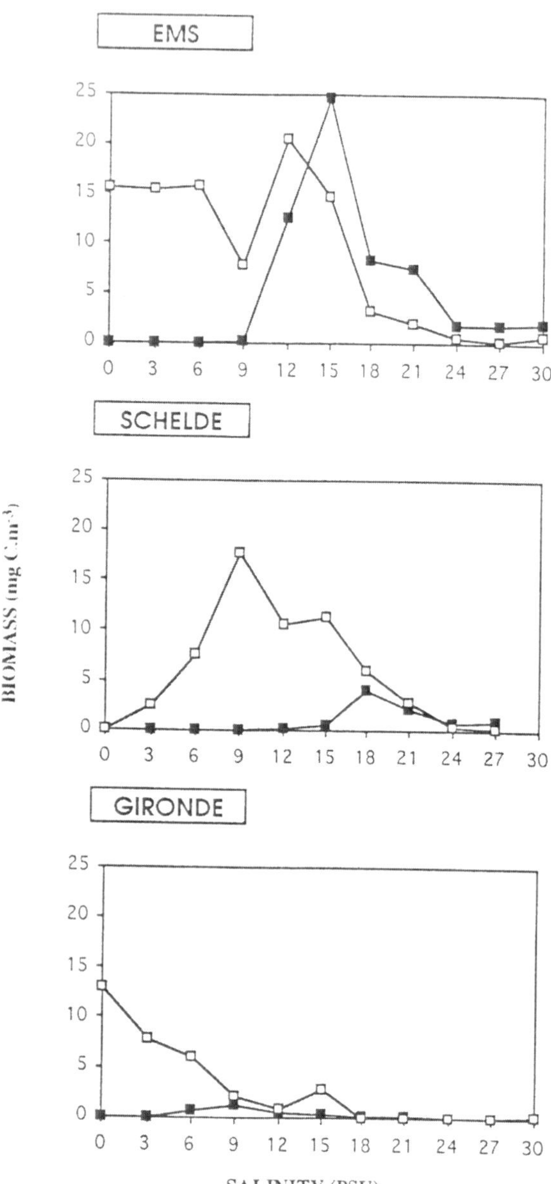

Fig. 10. Biomass (mgC m^{-3}) of *Eurytemora affinis* (□) and *Acartia* spp (-■-) in the 3 estuaries during spring 1992.

the Westerschelde *E. affinis* was most prominent more downstream, at higher salinities. The same observation was made by Soetaert & Van Rijswijk (1993). The downstream shift of the population is likely due to anoxic conditions occurring in the low salinity zone of the Westerschelde. Rotifers and the copepods *Cyclops* spp and *Eudiaptomus gracilis* were the only zooplankters resisting the harsh conditions in the oligohaline zone in the Westerschelde but they were found in low numbers.

E. affinis occurred in a very limited zone (0–15 PSU) in the Gironde. The peak of the population was always found just upstream the turbidity maximum which occurred at 3–6 PSU (data not shown but see Castel, (1995)). It has been shown by Castel & Veiga (1990) that the copepods behave as passive particles and are trapped in the turbidity zone. Castel & Feurtet (1992) also found high mortality rates during periods of high turbidity, confirming the results obtained by Sellner & Bundy (1987) showing that SPM concentrations over 350 mg l^{-1} affect the survival of *E. affinis*. These observations suggest that, in the Gironde, *E. affinis* is limited downstream by the maximum turbidity zone. The broader distribution of *E. affinis* was observed in the Ems estuary, where it appeared from 0 to 27 PSU. The abundance peak of *E. affinis* in the Ems was observed near its salinity optimum (12 PSU in laboratory conditions; Heinle, 1969). However, as shown for the 2 other estuaries, this optimum is not clear in the field. Remarkably enough the abundance peak of *E. affinis* was nearly the same in the three estuaries whatever its location along the salinity gradient. Thus, salinity is unlikely the only variable influencing the distribution of *E. affinis* (Bradley, 1975; Castel & Veiga, 1990). Organic loading has certainly to be taken into account.

Whatever the estuary, common features were observed for the parameters of the populations of *E. affinis*. As usual for copepods, young instars dominated. Adults were more tolerant to lower and higher salinities than copepodids. This is confirmed by studies made on the variations of tolerance to salinity (Von Vaupel-Klein & Weber, 1975). Ovigerous females were generally found downstream the abundance peaks of the populations (this was not clear in the Ems). In addition, the sex ratio calculated for copepodids CV was lower than the sex ratio of the adults indicating a higher male mortality. This could reflect differential mortalities in stage CV during the final molt (as observed for Acartia by Lee & Mc Alice, 1979), or a shorter life span of the males at maturity (Corkett & Mc Laren, 1978). A preferential predation on males (CVI) is unlikely (Sandström, 1980; Vuorinen *et al.*, 1983).

The similar maximum abundance observed in the three estuaries suggests that the nutritive requirements of *E. affinis* are met in the field, even when primary production is low. In this latter case *E. affinis* has to feed on detritus (Heinle & Flemer, 1975: Burkill & Kendall, 1982) which is generally considered of lower nutritive value.

Differences were observed between the individual dry weight measured in the three estuaries. The lightest individuals of *E. affinis* were found in the Gironde, where primary production is generally low. Heaviest individual weights were determined in the Ems where highest values of primary production are to be found. The variations in body weight are unlikely correlated to temperature (temperature during sampling for biomasses: 12.0 °C, 9.0 °C and 14.9 °C for the Ems, Westerschelde and Gironde respectively). Neither suspended matter concentration nor chlorophyll *a* were measured during the present study, but previous observations made in the Gironde indicate a strong influence of the nutritional quality of the particles on the individual weight of *E. affinis*. Feurtet & Castel (1988) showed that adult body weight drops significantly in the maximum turbidity zone where the ratio chlorohyll *a*/SPM, taken as an index of food availability, is the lowest. Furthermore, the condition factor (weight/length^3) is inversely related to SPM concentration (Castel & Feurtet, 1987) both on a spatial and a temporal scale.

The influence of the quality of the available food was also suggested for the production of eggs (highest clutch-size in the Ems, lowest in the Gironde). Presence of phytoplankton is known to favour eggs production (Heinle *et al.*, 1977, Chervin, 1978). Comparing samples taken in spring 1992 in the Elbe, Shannon, Ems and Gironde, and using principal component analysis, Burdloff (1993) showed a good correlation between number of eggs and proteins + carbohydrates + lipids/SPM ratio. Using a stepwise multiple regression analysis, Castel & Feurtet (1992) showed that the ratio chlorophyll *a*/SPM was the most effective variable explaining clutch-size variations in the Gironde estuary. Number of eggs per sac was also correlated (negatively) with temperature. Together, the two variables explained 62% of the variance. In the present study clutch-size was not correlated with temperature (mean temperature during sampling for egg counts: 12.8 °C, 8.7 °C and 14.2 °C for the Ems, Westerschelde and Gironde, respectively) suggesting a strong influence of the nutritional conditions.

The trend found for the number of eggs per egg sac is supported by the trend observed for the sex-ratio. Highest values of the sex-ratio (males/females) was found in the Ems and the lowest in the Gironde. Females *Eurytemora* require repeated fertilization for continued reproduction (Heinle, 1970; Katona, 1975). Thus, for copepods, a predominance of males over females is considered to promote fertilization of the females (Green *et al.*, 1993).

The copepod *Acartia bifilosa* was the second dominant species. It was as abundant as *E. affinis* in the Ems whereas it was collected in much lower number in the Westerschelde and Gironde. The bulk of the population was situated in the mesohaline zone in the Ems and Gironde and at the limit of the polyhaline zone in the Westerschelde. *A. bifilosa* is strongly euryhaline. In the Ems, *A. bifilosa* is a typical winter-spring species occurring at all salinities between 0 and 35 PSU during this period. In summer its occurrence is restricted to salinities between 15 and 30 PSU (Baretta & Malschaert, 1988). In the Westerschelde, maximum abundance of *A. bifilosa* are recorded in late winter between 18 and 30 PSU (Soetaert & Van Rijswijk, 1993). In the Gironde, *A. bifilosa* is generally found between 5 and 22 PSU with a maximum peak around 15 PSU in July (Castel, 1981; Irigoien & Castel, (1995)). Thus the present observations on *A. bifilosa* distribution along the salinity gradient are in accordance with the previous studies. The difference in the maximum abundance reached in the three estuaries could be attributed to a specific influence of temperature as well as differences in nutritional conditions.

The relative abundances of adults *A. bifilosa* were high upstream the density peaks of the populations; Lance (1964) has shown that they are more resistant to high salinity than young instars. The distribution of males was restricted to narrow zones (males are less resistant to dilution than females, Lance, 1964). Males of *A. bifilosa* were highly dominant in the Ems, less in the Westerschelde and females dominated in the Gironde. As for *E. affinis*, high densities of males are probably an adaptive characteristic involving a better fertilization of females. This is consistent with our observations showing the highest population abundance in the Ems and the lowest in the Gironde.

The lowest individual weights were found in the Gironde. Weights of copepodids CII to CIV were significantly higher in the Westerschelde than in the Ems. The weight of copepodids CV were not significantly different and adults were heavier in the Ems than in the Westerschelde. This indicated a faster growth of *A. bifilosa* in the Ems, leading to higher adult biomass. This reinforces the hypothesis that *A. bifilosa* was under the best conditions for developing in the Ems and that nutritional conditions are likely to influence its growth. In the Gironde estuary, Irigoien & Castel

(1995) showed that the productivity (P/B ratio) is positively correlated with the chlorophyll *a*/SPM ratio.

The meroplanktonic community was much more developed in the Ems than in the Westerschelde and the Gironde. In the Gironde, the very low abundance of meroplankton was likely due to the reduced intertidal zones. In the Ems, on the contrary, a very abundant autochthonous meroplankton community (polychaetes and cirripedes) was observed in the mesohaline and polyhaline zones. In accordance with the previous observations of Baretta & Malschaert (1988) we found Polychaet larvae being dominant in the salinities between 15 and 21 PSU and the cirripede larvae being most abundant at 24–30 PSU. This was likely due to the large extent of mud flats in the Dollard estuary, which provide a habitat for polychaetes and to the presence of hard substrates downstream providing settlement areas for cirripedes.

In conclusion, our observations suggest that salinity is not the only factor determining the distribution of species in estuaries. This was especially clear for *E. affinis* which showed abundance peak of the same value at different salinities. It is likely that other factors such as organic loading and eutrophication processes (Westerschelde) or high concentration of suspended particulate matter (Gironde) partly determine the species distribution. However, the difference in individual weights, observed for *E. affinis* and *A. bifilosa*, and of clutch-size for *E. affinis*, suggest that the nutritional conditions were not the same in the three estuaries. This hypothesis is supported by the fact that zooplankton developed best in the Ems where highest primary production is to be found.

Acknowledgments

This investigation was supported by the CEC Programme 'Major Biological Processes in European tidal Estuaries' (MAST grant 0024 – C). Thanks are due to the field workers in the Ems and Westerschelde estuaries: V. Escaravage, H. Francke, H. L. Kleef and C. Van Sprundel. X. Irigoien helped us with sampling in the Gironde. Technical assistance in the laboratory was provided by A.-M. Castel. We are indebted to M. Tackx for constructive comments on an earlier version of the manuscript.

References

Bakker, C. & N. De Pauw, 1975. Comparison of plankton assemblages of identical salinity ranges in estuarine tidal and stagnant environments. II. Zooplankton. Neth. J. Sea Res. 9: 145–165.

Baretta, J. W. & J. F. P. Malschaert, 1988. Distribution and abundance of the zooplankton of the Ems estuary (North Sea). Neth. J. Sea Res. 22: 69–81.

Boak, A. C. & R. Goulder, 1983. Bacterioplankton in the diet of the calanoid copepod *Eurytemora affinis* in the Humber estuary. Mar. Biol. 73: 139–149.

Bradley, B. P., 1975. The anomalous influence of salinity on temperature tolerances of summer and winter populations of the copepod *Eurytemora affinis*. Biol. Bull. mar. biol. Lab., Woods Hole 148: 26–34.

Burdloff, D., 1993. Potentiel nutritif du bouchon vaseux: impact sur les copépodes. D.E.A., Univ. Bordeaux I, 28 pp.

Burkill, P. H. & T. F. Kendall, 1982. Production of the copepod *Eurytemora affinis* in the Bristol Channel. Mar. Ecol. Prog. Ser. 7: 21–31.

Castel, J., 1981. Aspects de l'étude écologique du zooplancton de l'estuaire de la Gironde. Oceanis 6: 535–577.

Castel, 1993. Long-term distribution of zooplankton in the Gironde estuary and its relation with river flow and suspended matter. Cah. Biol. Mar. 34: 145–163.

Castel, J., 1995. Long-term changes in the population of *Eurytemora affinis* (Copepoda, Calanoida) in the Gironde estuary (1978–1992). Hydrobiologia 311 (Dev. Hydrobiol. 110): 85–101.

Castel, J., C. Courties & J. M. Poli, 1983. Dynamique du copépode *Eurytemora hirundoides* dans l'estuaire de la Gironde: effet de la température. Oceanol. Acta N °SP: 57–61.

Castel, J. & A. Feurtet, 1987. Influence des matières en suspension sur la biologie d'un copépode estuarien: *Eurytemora hirundoides* (Nordquist, 1888). Coll. Nat. CNRS 'Biologie des population', Univ. Claude Bernard, Lyon: 391–396.

Castel, J. & A. Feurtet, 1992. Fecundity and mortality rates of the copepod *Eurytemora affinis* in the Gironde estuary. Proc. 25th Europ. Mar. Biol. Symp., Ferrara, Olsen & Olsen: 143–149.

Castel, J. & J. Veiga, 1990. Distribution and retention of the copepod *Eurytemora affinis hirundoides* in a turbid estuary. Mar. Biol. 107: 119–128.

Chervin, M. B., 1978. Assimilation of particulate organic carbon by estuarine and coastal copepods. Mar. Biol. 49: 265–275.

Corkett, C. J. & I. A. Mc Laren, 1978. The biology of *Pseudocalanus*. Adv. Mar. Biol. 15: 1–231.

De Jonge, V. N., 1998. The abiotic environment. In Baretta, J. W. & P. Ruardij (eds), Tidal Flats Estuaries. Simulation and analysis of the Ems estuary. Ecological studies 71, Springer, Berlin, 353 pp.

De Jonge, V. N., 1993. Physical processes and dynamics of microphytobenthos in the Ems estuary (The Netherlands). PhD Thesis, Univ Groningen, 176 pp.

Etcheber, H., 1986. Biogéochimie de la matière organique en milieu estuarien: comportement, bilan, propriétés. Cas de la Gironde. Mém. Inst. Géol. Bassin d'Aquitaine, 19, 379 pp.

Feurtet, A. & J. Castel, 1988. Biologie du copépode *Eurytemora affinis hirundoides* dans la Gironde: données morphométriques. In Aspects récents de la biologie des crustacés. Actes colloq. 8, IFREMER: 223–227.

Fraser, J. H., 1966. Zooplankton sampling. Nature 211: 915–916.

Green, E. P., R. P. Harris & A. Duncan, 1993. The seasonal abundance of *Calanus helgolandicus* and *Pseudocalanus elongatus* off Plymouth. J. mar. biol. Ass. U.K. 73: 109–122.

Gunter, G., 1961. Some relations of estuarine organisms to salinity. Limnol. Oceanogr. 6: 182–190.

Heinle, D. R., 1969. Culture of calanoid copepods in synthetic seawater. J. Fish. Res. Bd Can. 26: 150–153.

Heinle, D. R., 1970. Population dynamics of exploited cultures of calanoid copepods. Helgoländer wiss. Meeresunters. 20: 360–372.

Heinle, D. R. & D. A. Flemer, 1975. Carbon requirements of a population of the estuarine copepod *Eurytemora affinis*. Mar. Biol. 40: 341–353.

Heinle, D. R., R. P. Harris, J. F. Ustach & D. A. Flemer, 1977. Detritus as food for estuarine copepods. Mar. Biol. 40: 341–353.

Heip, C., 1989. The ecology of the estuaries of Rhine, Meuse and Scheldt in the Netherlands. In J. D. Ros (ed.), Topics in marine biology. Scient. Mar. 53: 457–463.

Irigoien, X. & J. Castel, 1995. Feeding rates and productivity of the copepod *Acartia bifilosa* in a highly turbid estuary: the Gironde (SW France). Hydrobiologia 311 (Dev. Hydrobiol. 110): 115–125.

Irigoien, X. & J. Castel, 1994. Light limitation and distribution of chlorophyll pigments in a highly turbid estuary: the Gironde (SW France). Estuar. coast. Shelf Sci. (submitted).

Jeffries, H. P., 1962. Salinity – space distribution of the estuarine copepod genus *Eurytemora*. Int. Revue ges. Hydrobiol. 47: 291–300.

Jouanneau, J. M. & C. Latouche, 1981. The Gironde estuary. In H. Füchtbauer, A. P. Lisitzyn, J. D. Millerman & E. Seibold (eds), Contributions to sedimentology n ° 10, E. Schweizerbart'sche Verlagsbuchhandlung, 115 pp.

Katona, S. K., 1975. Copulation in the copepod *Eurytemora affinis* (Poppe, 1880). Crustaceana 28: 89–95.

Kulhmann, D., O. Fukuhara & H. Rosenthal, 1982. Shrinkage and weight loss of marine fish food organisms preserved in formalin. Bull. Nansei Reg. Fish. Res. Lab. 14: 13–18.

Laane, R. W. P. M., H. Etcheber & J. C. Relexans, 1987. Particulate organic matter in estuaries and its ecological implication for macrobenthos. Mitt. Geol.-Paläont. Inst. Univ. Hamburg, SCOPE/UNEP Sonder. 64: 71–91.

Lance, J., 1964. The salinity tolerances of some estuarine planktonic crustaceans. Biol. Bull. mar. biol. Lab., Woods Hole 127: 108–118.

Lee, W. Y. & B. J. Mc Alice, 1979. Seasonal succession and breeding of three species of *Acartia* (Copepoda: Calanoida) in a Maine estuary. Estuaries 2: 228–235.

Lin, R. G., 1988. Etude du potentiel de dégradation de la matière organique particulaire au passage eau douce-eau salée: cas de l'estuaire de la Gironde. Doct. Thesis, Univ. Bordeaux I, 209 pp.

Oenema, O, R. Steneker & J. Reynders, 1988. The soil environment of the tidal area in the Westerschelde. Hydrobiol. Bull. 22: 21–30.

Parsons, T. R., M. Takahashi & B. Hargrave, 1977. Biological oceanographic processes. Pergamon Press, Oxford, 332 pp.

Peters, J. J. & A. Sterling, 1976. Hydrodynamique et transport des sédiments de l'estuaire de l'Escaut. In L'estuaire de l'Escaut. Project Mer. Nihoul & Wollast (ed.) 10: 1–65.

Sandström, O., 1980. Selective feeding by Baltic herring. Hydrobiologia 69: 199–207.

Sellner, K. H. & M. H. Bundy, 1987. Preliminary results of experiments to determine the effects of suspended sediments on the estuarine copepod *Eurytemora affinis*. Cont. Shelf Res. 7: 62–64.

Soetaert, K. & P. M. J. Herman, 1995. Carbon flows in the Westerschelde estuary (The Netherlands) evaluated by means of an ecosystem model (MOSES). Hydrobiologia 311 (Dev. Hydrobiol. 110): 247–266.

Soetaert, K. & P. van Rijswijk, 1993. Spatial and temporal patterns of the zooplankton in the Westerschelde estuary. Mar. Ecol. Prog. Ser. 97: 47–59.

Soltanpour-Gargari, A. & S. Wellershaus, 1985. *Eurytemora affinis* – one year study of abundance and environmental factors. Veröff. Inst. Meeresforsch. Bremerh. 20: 183–198.

Von Vaupel-Klein, J. C. & R. E. Weber, 1975. Distribution of *Eurytemora affinis* (Copepoda: Calanoida) in relation to salinity: field and laboratory observations. Neth. J. Sea Res. 9: 297–310.

Vuorinen, I., M. Rajasilta & J. Salo, 1983. Selective predation and habitat shift in a copepod species – support for the predation hypothesis. Oecologia 59: 62–64.

Williams, R. & N. R. Collins, 1986. Seasonal composition of meroplankton and holoplankton in the Bristol channel. Mar. Biol. 92: 93–101.

Hydrobiologia **311**: 153–174, 1995.
C. H. R. Heip & P. M. J. Herman (eds), *Major Biological Processes in European Tidal Estuaries.*

Comparative study of the hyperbenthos of three European estuaries

Jan Mees[1], Nancy Fockedey[1] & Olivier Hamerlynck[1,2]
[1]*Marine Biology Section, Zoology Institute, University of Ghent, K. L. Ledeganckstraat 35, B-9000 Gent, Belgium*
[2]*Centre for Estuarine and Coastal Ecology, Netherlands Institute for Ecology, Vierstraat 28, NL-4401 EA Yerseke, The Netherlands*

Key words: hyperbenthos, estuary, Mysidacea, multivariate analysis, Westerschelde, Gironde, Eems

Abstract

The hyperbenthic fauna of the subtidal channels of the Eems (N. Netherlands), Westerschelde (S.W. Netherlands), and Gironde (S.W. France) estuaries was sampled within a 15-day period in summer 1991. In each estuary, quantitative samples were taken at regularly spaced stations covering the entire salinity gradient from marine conditions at the mouth to nearly freshwater conditions upstream. The diversity of the samples and the distribution of the species along the main estuarine gradients were assessed. Hyperbenthic communities were identified using different multivariate statistical techniques. The species composition and the density and biomass of the dominant species of each community were compared among communities.

Spatial patterns in density, biomass and diversity of the hyperbenthos were similar in the three estuaries: diversity was highest in the marine zone where density and biomass were lowest. Diversity decreased upstream and was lowest in the brackish part where density and biomass reached maximal values. In Eems and Gironde there was a slight increase in diversity towards the freshwater zone. Within each estuary two (Westerschelde) or three (Eems and Gironde) communities could be distinguished and their position along the unidirectional salinity-turbidity-temperature gradient was similar: a marine community in the high salinity zone, a brackish water community in the middle reaches and a third community (absent in the Westerschelde) in the stations with the lowest salinities. Qualitative and quantitative differences in the corresponding hyperbenthic communities among estuaries were evident. Some species were restricted to one or two of the estuaries studied, while others, especially the abundant species in the brackish part, were common to all three. Still, these differences were marginal compared to the overriding similarity of the hyperbenthos in the three estuaries. The distribution of single species in the estuaries varied to some extent but the among estuary differences in density and biomass in comparable salinity zones rarely exceeded an order of magnitude.

In the Westerschelde, the low salinity hyperbenthic community was completely absent. Upstream of the 10 g l^{-1} isohaline the dissolved oxygen concentration dropped to a critical threshold value for hyperbenthic life. The populations of a number of species, which in Gironde and Eems reached highest density and biomass in this zone, seem to have (almost) disappeared from the Westerschelde (e.g. *Gammarus zaddachi* and *Palaemon longirostris*). Other brackish water species did not occur in their 'normal' salinity range and their populations have shifted to higher, atypical salinity zones (e.g. *Neomysis integer, Mesopodopsis slabberi, Pomatoschistus microps* and *Gammarus salinus*.

Introduction

Estuaries are located at the interface between sea and land. As ecosystems they perform several vital functions, e.g. as nursery areas for juvenile fish and shrimp, migration routes for anadromous and catadromous fish, habitats for estuarine residents and spawners, etc.

(Ketchum, 1983). They are highly productive systems around which many human activities are concentrated (shipping, cities, industry). Correlated with this is a high anthropogenic stress (e.g. dredging, eutrophication, pollution,...) which may have important negative effects on the biota and thus the ecological structure of the system. Though interest in the functioning of estuaries has sharply increased in the last decades, thorough baseline studies on several of the food web compartments are still lacking, even for the relatively well studied northwestern European estuaries. Historical data are scanty and virtually no long time series are available on the different functional compartments of estuarine ecosystems (but see Castel 1993 and this volume). For an understanding of pollution impact only extensive sampling campaigns permit comparisons of estuaries subjected to high pollution loads with relatively pristine estuaries. The influence of zoogeographical differences (i.e. latitudinal effects) can be accommodated by choosing estuaries situated both north and south of the estuary under consideration. To date few synoptic studies have been conducted using the same methodology in different estuaries. This is especially true for the hyperbenthos since sampling methodology for this compartment is far from standardised and recognition of the importance of the hyperbenthos is relatively recent. Research on the hyperbenthos has only started in the last few decades (the term was defined by Beyer in 1958) and very few studies have been conducted in European estuaries. For purposes of comparison, scanty records of accidentally caught hyperbenthic animals in zooplankton and macrobenthos surveys are virtually the only source of information. Hyperbenthic animals (mainly mysids, but also amphipods, juvenile shrimp, ...) successfully exploit a diversity of food resources and are an important link in the detritus based food chains. Their size is intermediate between zooplankton and fish and nearly all estuarine fish species are found to feed to some extent on *Neomysis integer* and *Crangon crangon* (e.g. Hartman, 1940, review in Mauchline 1980). Any threats to the estuarine system which affect this fauna will consequently endanger its nursery function for commercially important crustaceans and fish.

For this study three major European estuaries were sampled quasi-synoptically along the longitudinal salinity gradient ranging from marine waters near the mouth to nearly fresh water upstream: the Eems (north Netherlands), the Westerschelde (southwest Netherlands) and the Gironde (southwest France). All samples were taken with a single gear and processed by the same research team. Sampling was concentrated within a short time interval (15 days) to minimise seasonal effects on hyperbenthic community structure. Indeed, seasonal patterns can dominate hyperbenthic community structure due to the presence of temporary hyperbenthic species (Hamerlynck & Mees, 1991). The hyperbenthos of the Westerschelde estuary, which is characterised by a high degree of industrialisation and urbanisation making it one of the most polluted rivers of Europe, has been intensively studied in recent years (Mees & Hamerlynck, 1992, Cattrijsse *et al.*, 1993, Mees *et al.* 1993a, Mees *et al.* 1993b). The hyperbenthos of the Gironde has been studied by Sorbe (1981). No information on the hyperbenthos of the Eems estuary was available to date.

Materials and methods

Study area (Fig. 1)

The Eems-Dollard estuary is situated in the northeast of the Netherlands on the border with Germany. The system is about 33 km long from Eemshaven to Pogum. The surface area of the estuary (excluding the part extending to the Wadden Sea islands downstream Eemshaven) is approximately 255 km^2, including a fresh water tidal area in the Eems of about 37 km^2 (de Jonge, 1988). The tidal influence is artificially stopped upstream of Leer (Germany). In the marine part two major gullies are separated by sandbanks; further upstream (past the mouth of the Dollard) only one channel remains. The major source of freshwater inflow is the river Eems (catchment area of about 12 650 km^2), which has a variable discharge ranging from 25 to 390 m^3 s^{-1}. The Westerwoldsche Aa has no well defined watershed and discharges roughly 10 % (5.1 to 31 m^3 s^{-1}) of the discharge of the river Eems in the southeast corner of the Dollard. Variable (and still smaller) amounts of fresh water enter the estuary from some channels near Delfzijl. The tidal excursion is approximately 15 km. There is no stratification and water turnover is 18 to 36 days. Suspended matter concentrations in the maximum turbidity zone rarely exceed 0.4 g l^{-1} (Baretta & Ruardij, 1988). Dissolved oxygen concentration in the estuary proper rarely drops below 70% of the saturation value, even in the maximum turbidity zone.

The Westerschelde estuary is the lower part of the river Schelde. The estuarine zone of the tidal system extends from the North Sea (Vlissingen) to Antwer-

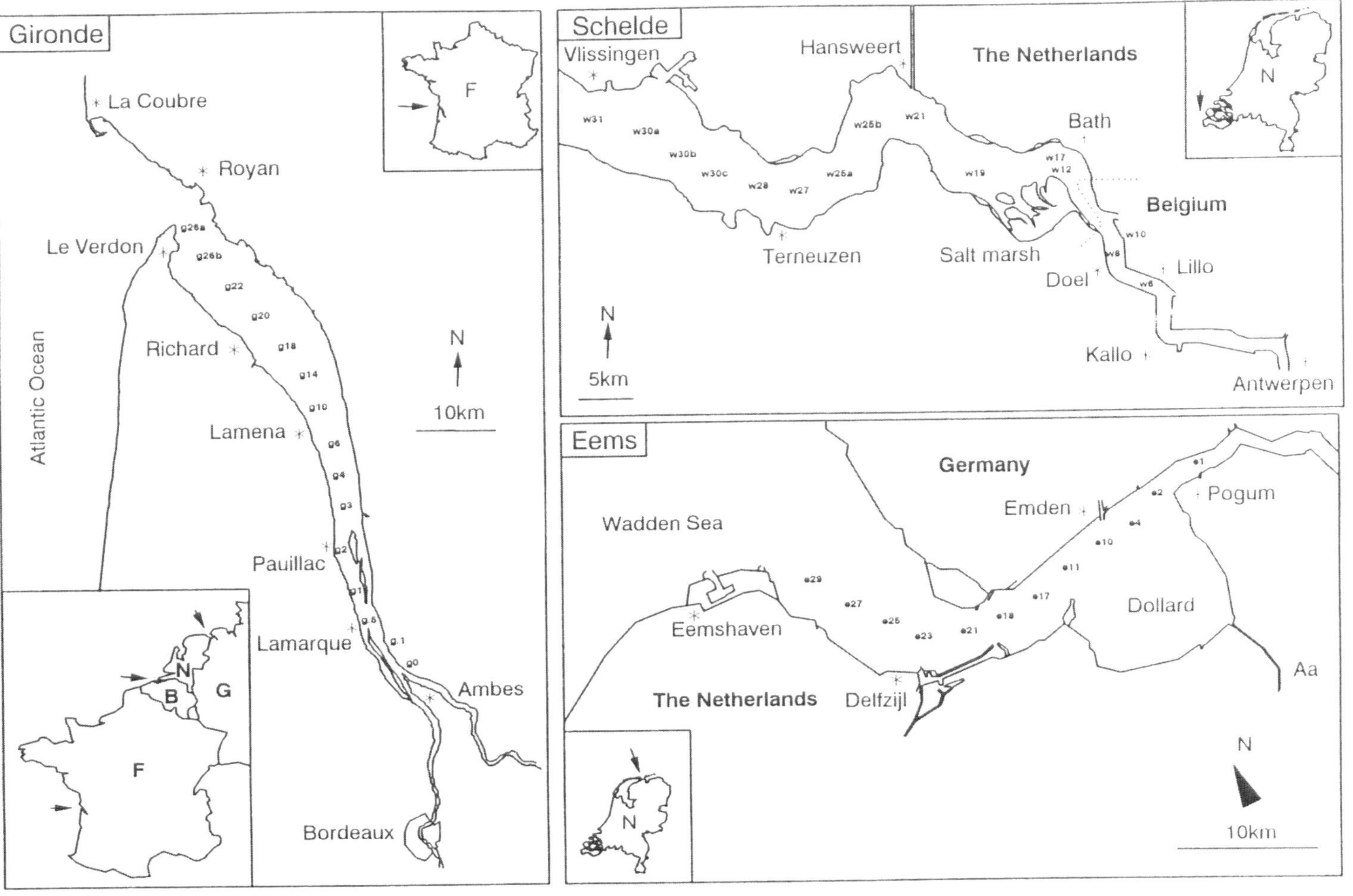

Fig. 1. Study area and sampling sites: the names of the samples are composed of a letter indicating the estuary and a number representing the salinity zone in which it was taken.

pen, 80 km inland. The estuary is rain fed, with a catchment area of some 20000 km^2. Its surface is approximately 300 km^2. The seaward part is a well mixed region characterised by a complex system of channels. There are two major gullies in the marine part and only one main channel in the weakly stratified region more upstream. Tidal influence extends to Gent (160 km from mouth) were it is artificially stopped. The residence time in the brackish part is rather high: about 60 days or 120 tidal cycles in summer (Soetaert & Herman, submitted). Consequently fresh water (average inflow 100 m^3 s^{-1}; range 30 to 500 m^3 s^{-1}) dilution is gradual and downstream transport is relatively slow. Shifts in salinity zone distribution occur in accordance with seasonal variations in the freshwater inflow. The physical, chemical and biological characteristics are discussed in Heip (1989), Herman *et al.* (1991) and Van Eck *et al.* (1991). The estuary is subject to a large anthropogenic stress, e.g. dredging (Belmans, 1988), and carries high pollution loads, both in anorganic and organic contaminants (Duursma *et al.* 1988). Dissolved oxygen concentration decreases sharply upstream the Dutch–Belgian border and the riverine part of the system is anoxic throughout most of the year (Herman *et al.*, 1991). Suspended matter concentrations are never higher than 0.05 g l^{-1} suggesting there is no real maximum turbidity zone in this estuary.

The Gironde estuary on the atlantic coast of France is the estuarine part of the rivers Garonne and Dordogne, which together have a catchment area of about 71 000 km^2 (Jouanneau & Latouche, 1981). The estuary is 70 km long from the inlet near Le Verdon to Bec d'Ambès where both rivers meet. The upstream part is characterised by the presence of numerous islands and sandbanks separating a network of channels. The downstream part consists of two main channels separated by shallower areas and sandbanks. The surface area at flood tide is 625 km^2. In summer tidal influence extends 160 km upstream Pointe de Grave. The water is well mixed: especially in summer there is virtually no stratification. Seasonal variations in salinity are related to freshwater discharge. River flow of the Garonne and Dordogne varies between 200 m^3 s^{-1} in summer to 1500 m^3 s^{-1} in winter (800–1000 m^3 s^{-1} on average). The residence time of a water particle is on average 20 tidal cycles in winter and 140 tidal cycles in summer. Dissolved oxygen concentrations in summer are never lower than 70% of the saturation value. One of the main features of the Gironde is the high turbidity of the water: suspended matter concentrations in the maximum turbidity zone generally exceed 1 g l^{-1} and values of 5 g l^{-1} and higher are regularly recorded (mainly silt and clay particles from freshwater origin).

Sampling

The location of the sampling stations in Eems, Westerschelde and Gironde is shown in Fig. 1. In the Eems 12 evenly spaced (3 km) stations were sampled in salinity zones ranging from 28.6 g l^{-1} near Eemshaven to 1.0 g l^{-1} near Pogum. In the Westerschelde 15 samples were taken from a salinity of 31.0 g l^{-1} near Vlissingen down to a salinity of 6.3 g l^{-1} near Lillo. The stations were selected according to the sampling grid used in Mees *et al.* (1993b). Since no animals were caught in the last station, no further attempts were made to sample more upstream. In the Gironde 15 stations (evenly spaced at 5 km) were selected covering salinity zones ranging from 26.1 g l^{-1} near Le Verdon to truly freshwater (0.0 g l^{-1}) near Bec d'Ambès.

Both Gironde and Eems were sampled in 2 consecutive days (5–6 August and 14–15 August, respectively). In the Westerschelde stations w31 upto w17 were sampled on the 12th of August; stations w12 upto w6 one week later on the 20th of the same month.

The samples were collected with a sledge (Hamerlynck & Mees, 1991) which consists of a heavy metal frame with two mounted monofilament nets. The nets are 4 m long and 1 m wide with a mesh size of 2 × 2 mm in the first 3 m and 1 × 1 mm in the last 1 m. The sledge glides over the bottom and samples the water column from 20 to 100 cm above the sediment. On each occasion it was trawled over a distance of 1000 m (radar readings from fixed points) at an average ship speed of 4.5 knots relative to the bottom. All samples were taken during daytime when hyperbenthic animals are known to be concentrated near the bottom. The contents of both nets were pooled for the present study. Thus the recorded densities are numbers of individuals (N) per 1000 m^2; the maximal volume of water filtered through the nets is 800 m^3. Where possible the 10 m isobath was followed. Actual sampling depths varied between 10.5 m and 7.5 m in the Eems, between 6.1 m and 15.8 m in the Westerschelde, and between 6 m and 15 m in the Gironde. The samples were rinsed over a 1 mm sieve and immediately preserved in a buffered formaldehyde solution, 7% final concentration.

At the end of each trawl Secchi disc depth was recorded and salinity, dissolved oxygen concentration,

Table 1. Common species excluded from the community analyses on the basis of size, with total number caught in each estuary

	Eems	Westerschelde	Gironde
Adult Caridea			
Crangon crangon	6268	474	850
Palaemonetes varians	4	1	
Palaemon longirostris	3		887
Adult Brachyura			
Carcinus maenas	39	37	
Liocarcinus holsatus	77	87	5
Liocarcinus pusillus			16
Portumnus latipes	6	1	
Macropodia species		1	
Rhithropanopeus harrisii	4		
Pinnotheres pisum			1
Adult Pisces			
Anguilla anguilla			9
Clupea harengus	19	3	
Sprattus sprattus	2	2	1
Osmerus eperlanus	335		3
Trisopterus luscus	1		
Gasterosteus aculeatus		4	3
Liparis liparis	3		
Gymnocephalus cernuus	9		
Zoarces viviparus	2		
Pomatoschistus microps	453	3	335
Pomatoschistus minutus	141	442	189
Pomatoschistus lozanoi	15	92	1
Limanda limanda	1		
Pleuronectes platessa	4		
Solea solea	24		1

pH, conductivity and temperature were measured near the bottom.

Laboratory procedures

After sorting, all animals present in the samples were identified, if possible to species level, and counted. Different developmental stages of some crustacean groups were considered as different functional species (zoeae, postlarvae and adults for caridean shrimp; zoeae, megalopae and adults for anomuran and brachyuran crabs). For gobies of the genus *Pomatoschistus* only *P. microps* could always be identified to species level. Small individuals (less than 25 mm standard length) of *P. minutus* and *P. lozanoi* were pooled as *Pomatoschistus* species. Other identification problems concerned postlarval clupeoids (probably a mixture of *Clupea harengus* and *Sprattus sprattus*) and amphipods of the genus Bathyporeia (pooled as Clupeidae species and *Bathyporeia* species, respectively). Possibly the counts of zoeae and megalopae of *Liocarcinus holsatus* also include larvae of other crabs of the same genus (e.g. *L. pusillus* in the Gironde). Several rare larval stages of brachyuran crabs could not be identified at all. Single records of a caprellid (*Caprella*) and an isopod (*Cymothoa*) in the Gironde could only be identified to genus level, though the former probably is *C. aequilibra* (Sorbe, 1978). For animals with more or less continuous growth, a maximum of 60 individuals per species and per sample (30 from each net) were measured to the nearest 0.1 mm using a binocular microscope and drawing mirror. Except for crabs (carapace width) standard lengths (from the tip of the rostrum to the last abdominal segment) were used. Biomass was then derived from the length-frequency distributions and length-ashfree dry weight (AFDW) regressions obtained from Westerschelde and Voordelta populations (Mees unpublished, Mees *et al.* 1994). Densities of species growing in discrete stages were converted to biomass with average AFDW values.

Statistical analysis

Diversity of each sample was calculated as Hill's diversity numbers of the order 0, 1, 2 and ∞ (Hill, 1973), with

N_0 = the number of species,
N_1 = e^H with $H = -\Sigma p_i \ln(p_i)$ (p_i is the relative abundance of the i dominant species),
N_2 = Σp_i^2 and
N_∞ = $1/p_1^{-1}$ (the reciprocal of the relative abundance of the most abundant species).

Diversity calculations were not considered meaningful if less than 10 animals were caught (station g26).

All multivariate analyses were performed on both density and biomass matrices. First, in order to assess differences between estuaries, the datamatrices combining the samples of the 3 estuaries were analysed. Then, to refine the identification of communities within each estuary, the analyses were repeated on smaller data matrices comprising only the samples of a single estuary.

Table 2. List of species and abbreviations used in the community analyses. Middle column: first letter of the estuary(ies) in which they occurred.

Species	E	W	G	Abbreviation
Sagitta elegans	E	W	G	Sagi eleg
Gastrosaccus spinifer	E	W	G	Gast spin
Schistomysis spiritus	E	W	G	Schi spir
Schistomysis kervillei	E	W	G	Schi kerv
Mesopodopsis slabberi	E	W	G	Meso slab
Neomysis integer	E	W	G	Neom inte
Praunus flexuosus	E	W		Prau flex
Eurydice pulchra		W	G	Eury pulc
Idotea linearis	E	W		Idot line
Synidotea laevidorsalis			G	Syni Spec
Sphaeroma rugicauda		W		Spha rugi
Sphaeroma serratum			G	Spha serr
Cymothoa species			G	Cymo Spec
Daphnia magna	E	W		Daph magn
Caprella linearis	W			Capr line
Caprella species			G	Capr Spec
Pariambus typicus	W			Pari typi
Gammarus crinicornis	E	W	G	Gamm crin
Gammarus salinus	E	W	G	Gamm zali
Gammarus zaddachi	E		G	Gamm zadd
Gammarus duebeni	E			Gamm dueb
Gammarus locusta	E			Gamm locu
Melita palmata			G	Meli palm
Atylus swammerdami	E	W	G	Atyl swam
Pleusymtes glaber		W	G	Pleu glab
Corophium volutator	E	W	G	Coro volu
Corophium acherusicum		W		Coro ache
Corophium lacustre		W		Coro lacu
Bathyporeia species	E	W		Bath Spec
Jassa falcata		W		Jass falc
Hyperia galba		W		Hype galb
Crangon crangon postlarva	E	W	G	Cran Post
Crangon crangon zoea	E	W	G	Cran zoea
Palaemonetes varians postlarva	E	W		Pala varP
Palaemonetes varians zoea	E			Pala varZ

Table 2. (Continued).

Species	E	W	G	Abbreviation
Palaemon longirostris postlarva	E		G	Pala lonP
Palaemon longirostris zoea	E		G	Pala lonZ
Pagurus species megalopa			G	Pagu Mega
Porcellana species zoea		W	G	Porc Zoea
Carcinus maenas megalopa	E	W	G	Carc Mega
Carcinus maenas zoea	E	W		Carc Zoea
Liocarcinus holsatus small adults	E	W	G	Lioc hols
Liocarcinus holsatus megalopa		W		Lioc Mega
Liocarcinus holsatus zoea	E	W	G	Lioc Zoel
Liocarcinus species zoea type 2			G	Lioc Zoe2
Liocarcinus species zoea type 3			G	Lioc Zoe3
Liocarcinus species zoea type 4	E			Lioc Zoe4
Macropodia species megalopa		W	G	Macr Mega
Eriocheir sinensis megalopa	E		G	Erio Mega
Unidentified zoea Westerschelde		W		Wtyl Zoea
Unidentified zoea Gironde type 1			G	Gtyl Zoea
Unidentified zoea Gironde type 2			G	Gty2 Zoea
Nymphon rubrum		W		Nymp rubr
Anguilla anguilla glass eels			G	Angu angu
Clupeidae species postlarva	E	W	G	Clup Spec
Syngnathus rostellatus	E	W	G	Syng rost
Pomatoschistus microps postlarva	E	W	G	Poma micr
Pomatoschistus species postlarva	E	W		Poma Spec

Density and biomass data were subjected to fourth root transformation prior to analysis. Three multivariate techniques, each yielding specific information, were applied to the data (Field *et al.*, 1982). The sampling sites were classified into clusters according to species composition using the classification technique TWINSPAN (Hill, 1979). This is a hybrid (the first step involves a reciprocal averaging ordination) divisive clustering technique which also gives indicator species and preferential species for each division. Pseudospecies cutlevels (7 in each case) were chosen to equalise the number of observations within each cutlevel, except for the lowest cutlevel which contained all the zero observations and the two highest cutlevels which contained approximately half as many observations as the other levels (in this way some extra weight was given to the most abundant species). The minimum group size for division was set to 5 and the analysis was stopped at the fifth division. An agglomerative clustering method (group average sorting or GAS of Bray-Curtis dissimilarities) was also applied to the data. The output (dendrograms) of these analyses were compared with the TWINSPAN results and the degree of similarity between clusters, and (within clusters) between samples could be assessed. The relationship between species, stations and environmental variables was investigated by means of a Canonical

Correspondence Analysis or CCA (Jongman *et al.*, 1987; Ter Braak, 1988), a technique performing regression and ordination of the data concurrently. Preliminary analyses showed that pH did not correlate well with any axis and that conductivity was strongly and positively correlated with salinity. Thus both parameters were not used in further analyses. Secchi disc depth was transformed reciprocally and thus becomes a light extinction measure, correlated with turbidity of the water. Whereas the first two techniques emphasised discontinuities in the data, the CCA emphasised continuities along the estuarine gradients. Plotting of the TWINSPAN/GAS clusters on the CCA ordination planes aided in evaluating the divisions imposed.

Results

Accidentally caught individuals one or several orders of magnitude larger than an 'average' hyperbenthic animal can seriously distort analyses with biomass data. Adult individuals of epibenthic shrimp and crab species and adult demersal or pelagic fish species, although often very abundant in the samples (Table 1), were excluded from the community analyses. Though these animals apparently make use of the hyperbenthal as a habitat they are inefficiently sampled with the sledge and are normally studied using beam trawls. Only small adults of *Liocarcinus holsatus* (carapaxlength <10 mm) and postlarval gobies (S.L.<25 mm), clupeoids (not yet displaying adult pigmentation nor habitus, S.L.<25 mm), pipefish (S.L.<60 mm) and glass eels were considered to be representative residents of the hyperbenthal. Other species eliminated from the data matrices are: Porifera species (epibenthic freshwater sponges in Eems and Gironde), Hydrozoa species (epibenthic, in every sample of Eems and Westerschelde, rarely in the Gironde), *Aurelia aurita* (planktonic, high densities in Eems and Westerschelde), Cyanea species (planktonic, high densities in Gironde), Anthozoa species (epibenthic, rare), *Pleurobrachia pileus* (planktonic, high densities in Eems, Westerschelde and Gironde), Nematoda species (benthic, mainly among peat in the brackish Eems samples), *Lanice conchilega* aulophore larvae (planktonic, 9 and 3 individuals in Westerschelde and Gironde, respectively), *Nereis* species (benthic, rarely caught in all three estuaries), and a variety of rarely and accidentely caught species: *Macoma ballhica*, *Cerastoderma edule*, *Mytilus edulis*, *Hydrobia ulvae*, *Sepiola* species, Bryozoa species, *Asterias rubens* and Cirripedia species. Also excluded were regularly encountered groups originating from land, air or fresh water: Aranea species, Diptera species, Lepidoptera species and Coleoptera species (adults and larvae).

From a total of 101 recorded species, 58 were thus retained after data reduction (Table 2). Eighteen were recorded from all three estuaries and most of these were very abundant. Four species were only encountered in the Eems, eleven only in the Gironde and ten only in the Westerschelde. Most of these species were rare and have previously also been recorded from the other estuaries in other studies. Exceptions are *Synidotea laevidorsalis* in the Gironde and *Gammarus locusta* in the Eems, which were quite common constituents of the hyperbenthos and have never been recorded from one of the other estuaries. Four species occurred both in Gironde and Eems but were absent from the Westerschelde. Four were only absent from the Eems and seven from the Gironde.

Environmental gradients

The environmental variables measured at each station are presented in Table 3. The most pronounced gradient in the three estuaries was salinity (see materials and methods). The three estuaries displayed the characteristic summer temperature gradient with lowest values near the mouth gradually increasing upstream. Geographical differences between the estuaries are obvious, with temperature increasing with decreasing latitude from Eems over Westerschelde to Gironde. The temperature difference between the mouth of the estuary and the 8 g l^{-1} isohaline in the Westerschelde (maximal difference of 3.0 °C) is high in comparison to that in Gironde and Eems (difference of 3.4 and 1.3 °C over a longer gradient). This may reflect the one week gap in the sampling scheme. An alternative explanation may be thermal pollution by the nuclear power plant of Doel. Secchi disc visibility decreased with increasing distance from the mouth. The marine reaches of the Gironde were characterised by very high light penetration. Upstream of the maximum turbidity zone in the Gironde there was a slight increase in light penetration. The maximum turbidity zone was not reached in Westerschelde (supposedly situated around Antwerpen) and in the Eems its upstream border was not reached. Dissolved oxygen concentrations of the water ranged from oversaturation in the marine part to about 80% of the saturation value in the 17 to 12 salinity zone of the Westerschelde. Then a rapid decline in the oxygen content was observed in the three inner-

Table 3. Environmental variables measured at the end of each trawl.

Eems	e29	e27	e25	e23	e21	e18	e17	e11	e10	e4	e2	e1
Salinity (g l^{-1})	28.63	26.54	25.34	22.87	21.00	18.47	16.65	11.44	10.03	4.45	1.96	0.94
Secchi depth (cm)	120	50	45	20	20	15	20	10	10	5	5	5
Temperature (°C)	19.8	19.6	19.4	19.2	19.6	19.7	19.8	20.1	20.1	20.3	20.4	20.5

Westerschelde	w31	w30a	w30b	w30c	w28	w27	w25a	w25b	w21	w19	w17	w12	w10	w8	w6
Salinity (g l^{-1})	31.00	30.30	29.50	29.50	28.00	26.60	25.40	25.00	21.10	19.30	17.00	11.90	10.30	8.10	6.30
Secchi depth (cm)	125	120	125	125	100	100	140	140	80	90	100	40	50	60	–
Temperature (°C)	20.0	20.0	19.8	20.1	20.2	20.2	20.1	20.1	20.3	20.5	20.6	22.3	22.8	22.7	–

Gironde	g26a	g26b	g24	g20	g18	g14	g10	g6	g4	g3	g2	g1	g.5	g.1	g0
Salinity (g l^{-1})	26.10	26.00	24.00	20.00	18.00	14.00	10.00	6.00	4.00	3.40	2.00	1.40	0.50	0.10	0.00
Secchi depth (cm)	440	440	260	90	40	40	30	10	5	5	3	5	3	5	10
Temperature (°C)	21.0	21.2	21.6	22.4	22.6	22.8	23.5	24.0	24.2	24.1	24.4	24.4	24.4	24.4	24.4

most stations: 49% in w10, over 38% in w8 down to 22% in w6. In the other two estuaries dissolved oxygen concentration never dropped below 70% saturation.

General trends in density, biomass and diversity

In Westerschelde and Gironde hyperbenthic density (Fig. 2) and biomass (not figured) were lowest in the most seaward stations. They increased upstream, decreasing again towards the most riverine stations.

In comparison to the other estuaries, the Eems was characterised by a rather uniform density over a wide salinity range: density was low in the outermost station (<5 individuals or 5 mg AFDW per m^2) but, in contrast to the other estuaries, was already high at the 27 g l^{-1} isohaline. Densities remained at about the same level (between 10 and 20 ind m^2, 10 to 47 mg AFDW per m^2) upto 17 g l^{-1}. A drop in density (again less than 5 ind m^2) was observed at the 11 g l^{-1} isohaline (mouth of the Dollard). Density and biomass then increased to a maximum of 26 individuals or 66 mg AFDW per m^2 at the 2 g l^{-1} isohaline and decreased again in the last station. Some of the density peaks reflect the appearance and disappearance of dominant species (Fig. 3): the peak around 21 g l^{-1} was mainly due to high densities of *Schistomysis kervillei*, the peak around 2 g l^{-1} reflects the abundance maximum of *N. integer*, which was of overriding importance throughout most of the estuary.

In the Westerschelde two peaks were evident (Fig. 2): a first in the 19 g l^{-1} salinity zone, a second around 10 g l^{-1}. These were an order of magnitude higher than maximal densities observed in the other estuaries (250 and 105 individuals per m^2; 555 and 208 mg AFDW per m^2). The two peaks probably do not represent two distinct zones of higher density but are an artefact of the discontinuous sampling scheme. Both peaks correspond to the maximum abundance of *N. integer* and, depending on the geographical location of the oxygen depletion zone, the population maximum can be found in different salinity zones on different sampling days. In the Westerschelde *N. integer* are always concentrated near the limit of viable oxygen concentrations (about 40% of the saturation value) regardless of salinity (Mees *et al.*, 1993a; Mees *et al.*, 1993b). Density became very low at 8 g l^{-1} and in the 6 g l^{-1} sample no hyperbenthic animals were found. In Gironde and Eems the abundance maximum of *N. integer* was correlated with the tidally shifting salinity zone around 2–4 g l^{-1}. In the marine stations (w31 to w21) of the Westerschelde density and biomass were below 3.5 ind per m2 and 3.5 mg AFDW per m^2 respectively.

Densities in the Gironde were only substantial upstream of the 20 g l^{-1} isohaline. Very few animals were caught in the most seaward Gironde samples g26a, g26b, g24 and g20 (7, 7, 69 and 29 individuals, respectively). Three peaks were evident: the first peak

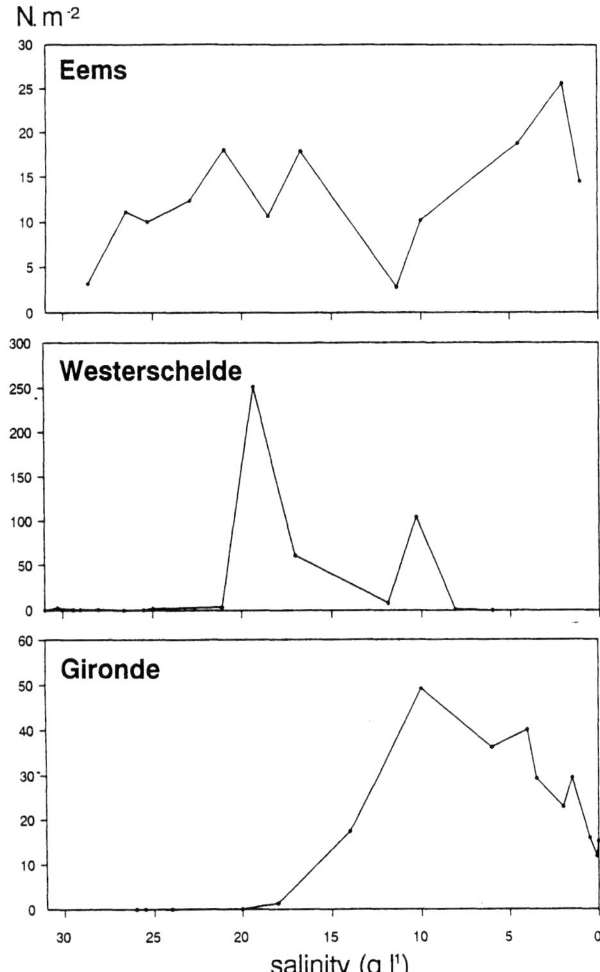

Fig. 2. Total hyperbenthic density along the salinity gradients in the three estuaries.

(50 individuals or 63 mg AFDW per m^2 at 10 g l^{-1}) coincided with the maximal abundance of *Mesopodopsis slabberi* and submaximal abundances of *N. integer* (Fig. 3), the second peak (41 individuals or 113 mg AFDW per m^2 at 4 g l^{-1}) corresponded with maximal densities of *Pomatoschistus microps* and high numbers *N. integer*, and the third peak (32 ind. m^2 or 99 mg AFDW at 1 g l^{-1}) corresponded to the abundance maximum of *N. integer* (Fig. 3). Since *M. slabberi* is a very slender species, the first peak became lower than the other two in terms of biomass.

All diversity measures (Fig. 4) were highest in the marine part of the estuaries. In the high salinity zone many species were present, with no species clearly dominating the fauna. This is also evident from Fig. 3 where the 'others' section accounted for a higher percentage of the community than in the brackish part. Diversity decreased towards the brackish reaches where fewer species occurred in higher numbers (cf. L_∞ approaches a value of 1 in the stations with maximal abundance of *N. integer*). In Eems and Gironde diversity increased slightly towards the fresh water.

The species

The hyperbenthos was dominated by crustaceans, especially mysids. Other important groups were gammaridean amphipods, isopods (in the Gironde), caridean shrimp, larval stages of brachyuran crabs and postlarval fish. Chaetognaths, daphnids, pycnogonids, caprellid and hyperiid amphipods and larval stages of anomurans were observed occasionally. The distribution of the 11 most abundant species along the salinity gradients of the three estuaries is depicted in Fig. 5. Below, the distribution patterns are described based on the density data only (biomass data of the individual species are available on request).

Mysidacea

Gastrosaccus spinifer occurred in the three estuaries from the marine reaches upto the 10 g l^{-1} isohaline. Maximal densities amounted to 3400 ind per 1000 m^2 in the Westerschelde (at 19 g l^{-1}) and 650 ind in Gironde and Eems (at 18 and 29 g l^{-1}, respectively). One adult female from the Eems was infested with the parasitic isopod *Prodajus ostendensis*. *Mesopodopsis slabberi* was abundant throughout the Westerschelde from the mouth upto the 10 g l^{-1} isohaline where it reached a maximal density of 60 individuals per m^2. The same pattern was observed in the Gironde (maximal densities of 39 ind m^2 at 10 g l^{-1}), but here the species was still present in the most upstream stations (upto 0 g l^{-1}). In the Eems *M. slabberi* also occurred over the entire transect but it only reached important densities in a narrower salinity band (between 29 and 10 g l^{-1}) where a maximum of 20 ind m^2 was recorded. In the Gironde only one individual of *Schistomysis kervillei* was caught in the most downstream station. This essentially coastal species was present in the Westerschelde from the mouth upto 21 g l^{-1} (maximum of 250 ind 1000 m^2 at 30 g l^{-1}). In the Eems the species moved much further up the estuary (upto 11 g l^{-1}) and reached maximal abundance at lower salinities (7000 ind 1000 m^2 at the 21 g l^{-1} isohaline). *S. spiritus* is also a typical coastal species which was present in the three estuaries from the mouth to 10, 21 and 14 g l^{-1} in Eems (maximal abundance of 800

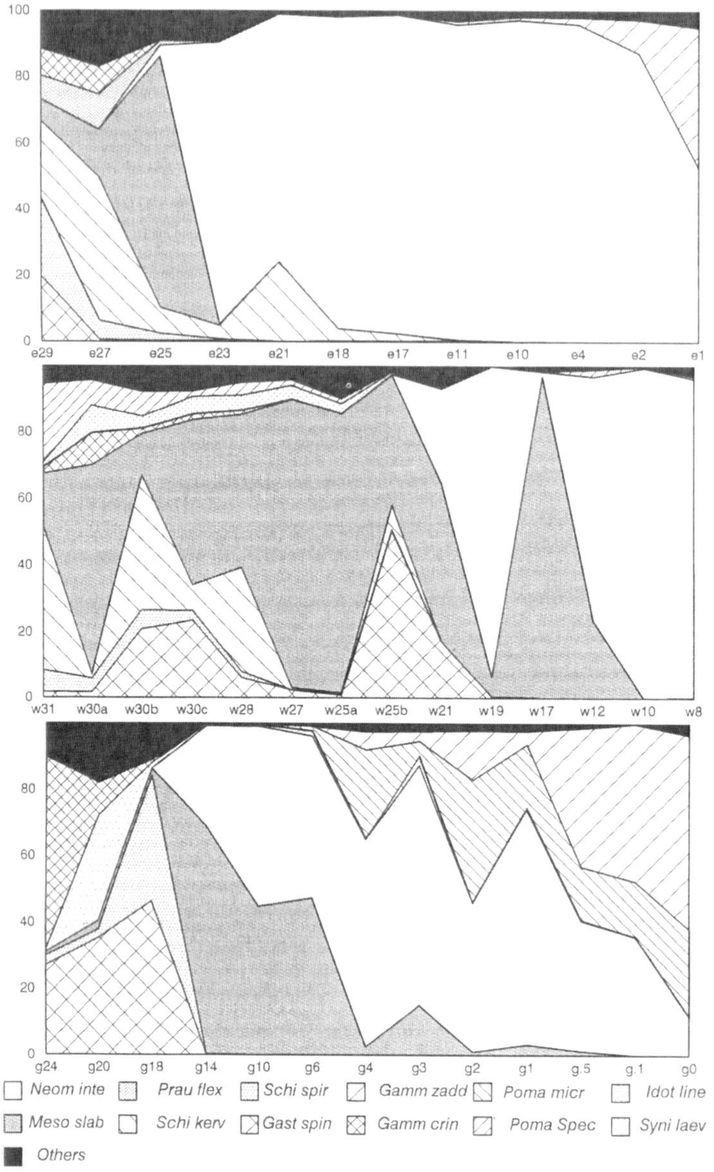

Fig. 3. Relative species composition of the hyperbenthic community along the salinity gradient in Eems (top), Westerschelde (middle) and Gironde (bottom).

individuals per 1000 m2), Westerschelde (55 ind) and Gironde (800 ind), respectively. The estuarine endemic *Neomysis integer* was, especially in biomass terms, the most important constituent of the hyperbenthos in the three estuaries. It was restricted to the brackish reaches, never being caught in fully marine conditions. The zone of maximal abundance in Eems and Gironde was situated in the vicinity of the maximum turbidity zone (as exemplified for the Gironde in Fig. 6). In the Eems the species was present from 25 g l^{-1} upto 1 g l^{-1} (maximum of 18 ind m^2 at 4 g l^{-1}). In the Gironde it colonised the salinity zone from 18 to 0 g l^{-1} (maximum of 26 ind m^2 at 4 g l^{-1}). In the Westerschelde *N. integer* was only present from 21 to 8 g l^{-1} with much higher maximal densities of 193 and 103 ind m^2 at the 19 and 10 g l^{-1} isohaline depending on the sampling date. *Praunus flexuosus* was only recorded in Eems and Westerschelde. In the former estuary densities amounted to 284 ind m^2 at 27 g l^{-1} (range 29 to 18 g l^{-1}), in the Westerschelde density was always

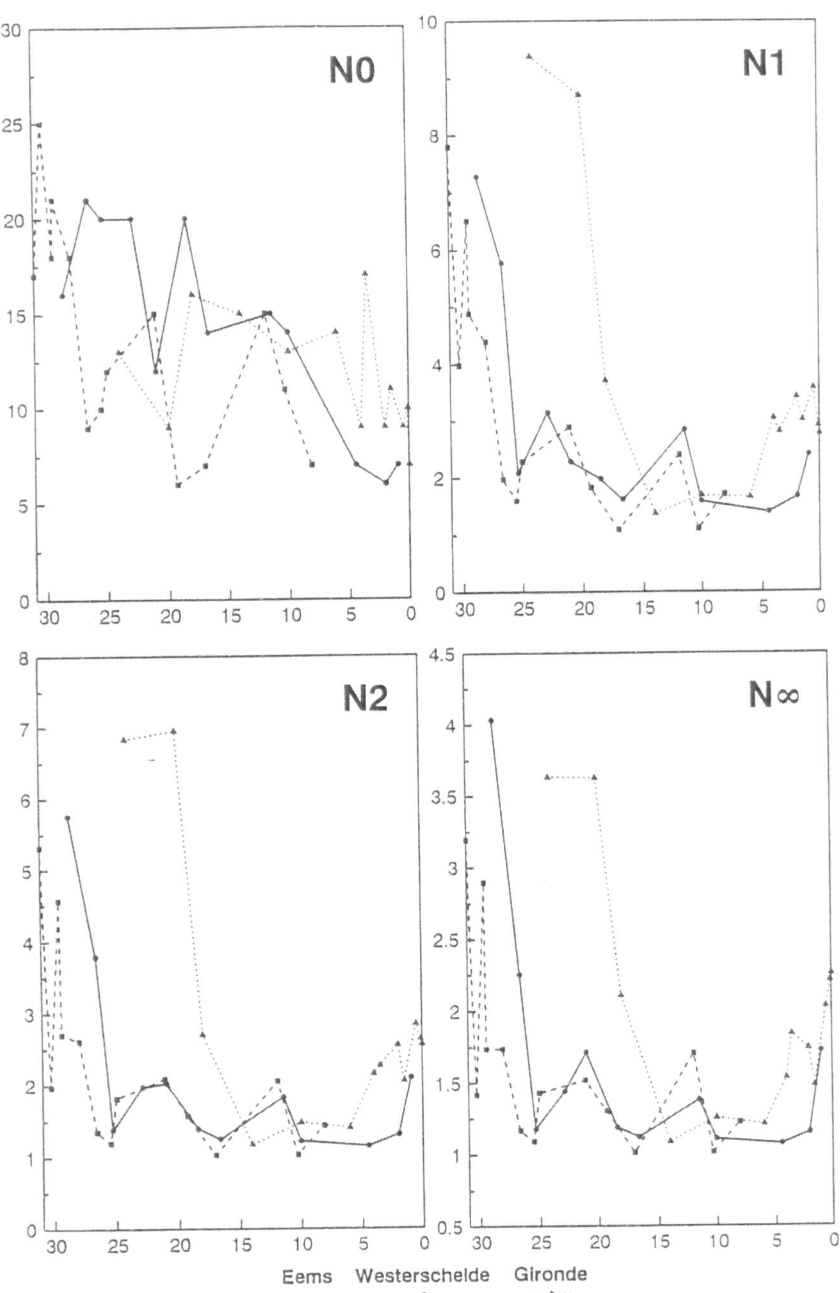

Fig. 4. Diversity numbers of Hill plotted against salinity.

low (maximum of 33 ind 1000 m^2) in the salinity zone from 25 to 17 g l^{-1}.

Amphipoda Gammaridea

Gammarus crinicornis is a marine species which does not penetrate the estuarine system very far. It occurred in all three estuaries with maximal densities of 646 (at 27 g l^{-1}), 103 (30 g l^{-1}) and 8 (24 g l^{-1}) individuals per 1000 m^2 in Eems, Westerschelde and Gironde respectively. *Gammarus salinus* is a brackish water species which in all three estuaries replaced *G. crinicornis* upstream. The species occurred in the Eems between 25 and 4 g l^{-1} (maximum of 52 ind 1000 m^2 at 10 g l^{-1}) and in the Gironde between 14 and 0.5 g

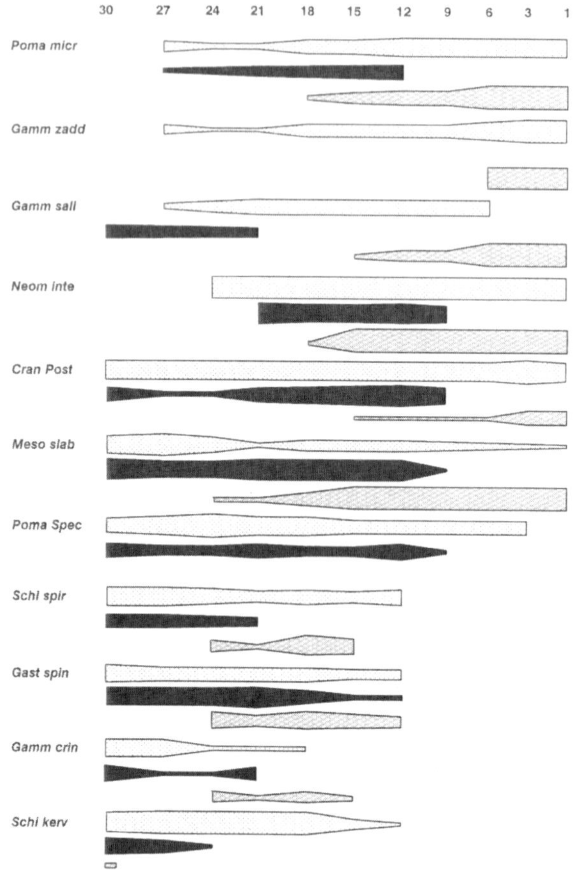

Fig. 5. Density distribution of the 11 most abundant hyperbenthic species along the salinity gradient of Eems (top), Westerschelde (middle, in black), and Gironde (bottom). Species are ordered following the two-way table of the TWINSPAN. Width of the bar gives indication of abundance reached in the salinity zone (top line).

l^{-1} (maximum of 1858 ind 1000 m^2 at 4 g l^{-1}). During this survey, *G. salinus* was only caught in low numbers (a total of 28 ind) in the Westerschelde although the species is known to be the most common amphipod in the hyperbenthos of the brackish part of this estuary (Cattrijsse *et al.*, 1993; Mees *et al.*, 1993b). *Gammarus zaddachi* replaced *G. salinus* in the oligohaline reaches of Eems and Gironde. Densities were very high in the salinity zone between 2 and 0 g l^{-1}: 5 and 7 ind m^2, respectively. The species was not caught in the Westerschelde. Two other *Gammarus* species were only caught in the Eems: *G. duebeni* (only a few individuals in the 27 g l^{-1} station) and *G. locusta* (between 29 and 23 g l^{-1}, maximum of 160 ind 1000 m^2 at 27 g l^{-1}). *Atylus swammerdami* is a marine species. It was present in the Westerschelde from 31 to 27 g l^{-1}. In Gironde and Eems the species penetrated further into the estuary: upto 20 and 11 g l^{-1}, respectively. Densities were never higher than 30 ind 1000 m^2. Individuals of *Melita palmata* (in marine waters) and *Pleusymtes glaber* (in brackish stations) were only caught in the Gironde, although they are also known to occur in the Westerschelde (Mees *et al.*, 1993b). *Corophium volutator* is a euryhaline, tube building species regularly encountered in the hyperbenthos of the three estuaries (density was never higher than 50 ind 1000 m^2). In the Westerschelde three other species were found in very low numbers: *C. acherusicum*, *C. lacustre*, and *Jassa falcata*. *Bathyporeia* species were only found in Eems and Westerschelde.

Isopoda

Eurydice pulchra was not observed in the Eems. The species was found in the Westerschelde in salinities ranging from 25 to 10 g l^{-1} with a maximal density of 26 ind 1000 m^2 at 12 g l^{-1}. In the Gironde only 4 individuals were caught between 18 and 3 g l^{-1}. *Idotea linearis* was only found in the marine parts of the Westerschelde (maximum of 85 ind 1000 m^2 at 30 g l^{-1}) and Eems (maximum of 13 ind 1000 m^2 at 29 g l^{-1}). *Synidotea laevidorsalis* was only present in the Gironde and was the only isopod which reached considerable densities and biomass (maximum of 288 ind 1000 m^2 at 3 g l^{-1}) in the brackish part of this estuary. Details on its distribution have been published elsewhere (Mees & Fockedey, 1993). *Sphaeroma rugicauda*, *Sphaeroma serratum*, and *Cymothoa* species were rare constituents of the hyperbenthos.

Caridean shrimp

Crangon crangon was very abundant in the three estuaries. The developmental stages were found to be segregated along the salinity gradient: zoeae did not penetrate the estuary as far as postlarvae. Maximal densities for the zoeae were 12 ind 1000 m^2 at 29 g l^{-1} in the Eems, 24 ind 1000 m^2 at 30 g l^{-1} in the Westerschelde, and 7 ind 1000 m^2 at 26 g l^{-1} in the Gironde. Postlarvae (and adults) were present over the entire sampled salinity range but were most abundant in the brackish part: maximal densities of 976 ind 1000 m^2 (1 g l^{-1}) in the Eems, 1148 ind 1000 m^2 (10 g l^{-1}) in the Westerschelde, and 39 ind 1000 m^2 (1 g l^{-1}) in the Gironde. *Palaemonetes varians* was only recorded in the Eems (zoeae and postlarvae between 23 and 10 g l^{-1}) and the Westerschelde (only postlarvae between 30 and 10 g l^{-1}). *Palaemon longirostris* is a typical species for the oligohaline reaches of estuaries. It was

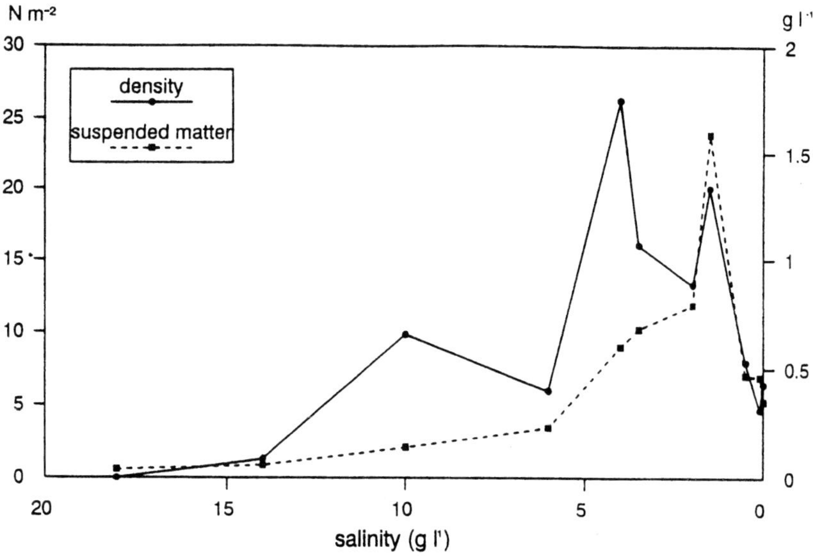

Fig. 6. Density of *Neomysis integer* (right axis) and concentration of suspended matter (left axis) plotted against the salinity gradient of the Gironde.

absent from the Westerschelde. Densities were low in the Eems (maximum of 9 zoeae, 23 postlarvae and 1 adult per 1000 m^2). In the Gironde it was the most abundant caridean: maximum of 44 zoeae (3 g l^{-1}), 324 postlarvae (2 g l^{-1}) and 224 adults (0.1 g l^{-1}) per 1000 m^2.

Larval Brachyura

Larval stages of the shore crab *Carcinus maenas* were rare in the Gironde (only 3 megalopae). In Eems and Westerschelde both zoeae (maximal densities of 15 and 34 ind 1000 m^2) and megalopae (maximal densities of 861 and 17 ind 1000 m^2) were present upto the 10 g l^{-1} isohaline. *Liocarcinus holsatus* zoeae and megalopae were only common in the Westerschelde (298 zoeae and 31 megalopae per 1000 m^2 at 17 and 21 g l^{-1}, respectively). Both stages were present but rare in Eems and Gironde. *Eriocheir sinensis* megalopae were only caught in Eems and Gironde, all other crab larvae were rare and mostly restricted to a single estuary.

Postlarval fish

Pomatoschistus microps was very abundant in the oligohaline reaches of Eems (maximal density of 435 ind 1000 m^2 at 1 g l^{-1}; range 27 to 1 g l^{-1}) and Gironde (maximal density of 6321 ind 1000 m^2 at 4 g l^{-1}; range 18 to 0 g l^{-1}). Densities in the Westerschelde were much lower (maximum of 5 ind 1000 m^2) and the species was only present downstream the 10 g l^{-1} isohaline. Postlarval *Pomatoschistus* species were absent from the Gironde. They were caught along the entire sampled transect of the Eems (maximum of 1260 ind 1000 m^2 at 2 g l^{-1}) and the Westerschelde (maximum of 1260 ind 1000 m^2 at 10 g l^{-1}). *Syngnathus rostellatus* was common in the marine reaches of the Eems (upto 17 g l^{-1}) and in the brackish reaches of Westerschelde (upto 12 g l^{-1}) and Gironde (upto 0.1 g l^{-1}). Larval clupeoids were always caught in low numbers (never more than 85 ind 1000 m^2) between the mouth of the estuary and 2 g l^{-1}, 19 g l^{-1} and 3 g l^{-1} in Eems, Westerschelde and Gironde, respectively. Glass eels *Anguilla anguilla* were only encountered (a total of 7 individuals) in the three lowest salinity samples of the Gironde.

Other taxa

Highest densities of the marine chaetognath *Sagitta elegans* were recorded around the 10 g l^{-1} isohaline in the Eems (300 ind 1000 m^2 as compared to less than 15 per 1000 m^2 in Westerschelde and Gironde). *Daphnia magna* is the only cladoceran large enough to be caught with a 1 mm mesh. The species was found in the brackish stations of Westerschelde and Eems, always in low numbers. Caprellid amphipods and pycnogonids are typical 'aufwuchs' species, rarely encountered in the estuarine hyperbenthos. Hyperiid amphipods (*Hyperia galba*) are commensals on coelenterates and were only encountered in the marine part of the Westerschelde, often in association with their host *Aurelia aurita*. Larvae of anomuran decapods (genera *Pagu-*

rus and *Porcellana*) were recorded occasionally in the marine waters of all three estuaries.

Identification and characterisation of communities

Multivariate analysis with density and biomass data permitted the identification of 8 geographically separated communities. The general pattern in the data is illustrated with the TWINSPAN result using the biomass data (Fig. 7). The first division groups the high salinity samples of the three estuaries on the positive side and all low salinity samples on the negative side. Indicator species for the latter group are the mysid *Neomysis integer* at the 7th and the common goby *Pomatoschistus microps* at the 4th cutlevel; the mysid *Schistomysis kervillei* is indicator for the former group. In the second and third divisions geographical groupings per estuary become prominent in both main clusters: in each group one or two Gironde, Eems and Westerschelde communities can be identified. Stations w8 and e21 are indicated by the program as misclassified and borderline negative respectively. The marine cluster (left side) is divided in three groups, one for each estuary: Gironde West (GW: g26b to g18), Westerschelde West (WW: w31 to w21 and w17) and Eems West (EW: e29 to e23). In the brackish cluster (right side) the Gironde samples are first split from the Eems and Westerschelde samples. Indicators for the Gironde group are postlarval *Palaemon longirostris*, *Syngnathus rostellatus* (second cutlevel) and *Synidotea laevidorsalis*. Within this group the lowest salinity samples (Gironde East, GE: g4 to g0) are further split from the rest (Gironde Mid, GM: g14 to g6) on the basis of *Gammarus zaddachi*. Indicators for the Eems-Westerschelde group are postlarval *Crangon crangon* (fourth cutlevel) and *Pomatoschistus* species (second cutlevel). The third division divides this cluster in an Eems group and a Westerschelde group (Westerschelde Mid, WM with indicator *Eurydice pulchra*). The Eems samples are further divided in an Eems Mid group (e21 to e11) with indicator species *Schistomysis kervillei* and an Eems East group (e10 to e1 plus w8). Further divisions in these 8 clusters are not considered to be ecologically meaningful: the groupings they yield are not consistently found with the other multivariate analyses and their indicator species have limited significance. The TWINSPAN with the density data (not figured) yielded nearly the same picture. Differences only apply to transitional stations showing indecisive behaviour in all analyses and indicated by the program either as misclassified samples or borderline positives or negatives (e.g. w8 rather clustering with the w10 to w19 stations than with the Eems samples, and e21 clustering with the e23 to e29 cluster rather than with the e11 to e18 cluster). The TWINSPAN divisions were confirmed by group average sorting clusteranalyses (not figured): the two main low and high salinity clusters could always be distinguished with high dissimilarity. Groupings of samples with high similarities yielded the same 8 clusters (again with some stations shifting between geographically adjacent clusters). G26b and g26a were found to be highly dissimilar from the rest of the samples, reflecting the fact that few individuals were caught in these stations. The same applies to w8 which, though it invariably clustered with the brackish main group, was always found in an isolated position with lower similarity to the other stations. In the CCA only the first (eigenvalue 0.44) and second (eigenvalue 0.23) axes are important and the ordination plane they form suffices to visualise the structure in the data (Fig. 8). The axes of higher order (eigenvalues lower than 0.10) do not yield additional information and are not discussed. The first axis correlates strongly with the main estuarine gradients: salinity, Secchi disc depth and, to a lesser extent, temperature. The largest vector, which, per definition, explains most of the variance, nearly parallels the first axis and represents the salinity gradient. It is strongly and negatively correlated with the 1/Secchi vector. Projection of the temperature vector on the first axis reflects its correlation with the estuarine temperature gradient. The second axis also has an important temperature component, now being a covariable of latitude. The first axis (with its correlated salinity-turbidity-temperature gradient) thus reflects within-estuary variation, whereas the second axis (with its latitudinal temperature component) rather reflects among-estuary differences. All marine samples are located on the positive side of the first axis. The three western clusters can be found segregated along the second axis, with the EW cluster in the upper right quadrant, the GW cluster in the lower right quadrant and the WW cluster situated upon the first axis in between EW and GW. The mid estuary clusters are located close to the second axis with WM near the centre of the diagram, again in between EM (top) and GM (bottom). The clusters grouping the eastern Eems and Gironde samples are situated in the upper and lower left quadrants respectively. Interestingly, corresponding salinity zones in the three estuaries have the same position along the first axis. The gap along the second axis between the EE and GE clusters suggests the position where we might expect the WE cluster. The position of the species in the ordination plane reveals groups of hyperbenthic species characteristic for the sample clusters identified (Fig. 8 bottom). The analyses with the data of the single estuaries emphasise the continuous, gradual aspect of community replacement. Only the ordination planes formed by the first two canonical axes of the CCA with the density data (Fig. 9) are presented. First axis eigenvalues are 0.38, 0.41 and 0.58 for Eems, Westerschelde, and Gironde, respectively. The three variables are always represented by long vectors lying close to the first axis, temperature and 1/Secchi pointing towards the brackish samples, salinity towards the marine samples (not figured). The horse-shoe effect could readily be removed by detrending after which all stations were located in the same order close to the first axis (results not figured), implying these are truly single-axis ordinations. There are no real gaps between any two stations but within each estuary an eastern, middle and western group (sometimes forming tight clusters) can be delineated confirming the ecological significance of the classifications.

Combining the three techniques, in each estuary two (Westerschelde) or three (Eems and Gironde) communities could be distinguished. The species composition of these communities can differ, but their position along the gradient is similar: a marine or western community in the high salinity zone (GW, EW and WW), a brackish water community in the middle reaches (GM, EM and WM) and a third community in the stations with the lowest salinities (eastern communities GE and EE). As shown by CCA (and also by the progressive agglomerative clustering of the stations along the salinity gradient, not figured) the classification of the gradients into distinct clusters is somewhat artificial: some stations show indecisive behaviour, clustering with one community in one analysis and with another, neighbouring community in the next. Still, despite the fact that in each estuary one or two stations represent a transitional situation where two neighbouring communities meet, the communities are distinct and for practical purposes an objective division can be made.

In summary, it was decided to consider the following 8 communities (see also Fig. 8): three in the Eems: a marine, western community EW from 29 to 23 g l^{-1}, a brackish community in the middle part of the estuary EM from 21 to 10 g l^{-1}, and a oligohaline, eastern community EE from 4 to 1 g l^{-1}; two in the Westerschelde: WW from 31 to 21 g l^{-1} (+w17) and WM from 19 to 8 g l^{-1}; and three in the Gironde: GW from 26 to 18 g l^{-1}, GM from 14 to 6 g l^{-1}, and GE from 4 to 0 g l^{-1}.

The biotic characteristics (species composition, density, biomass) of the 2 or 3 communities within each estuary are – per definition – distinctly different (Fig. 10). Within each estuary the marine community is characterised by many species occurring in low densities. The middle community is characterised by few species reaching very high densities and biomass. The eastern communities of Eems and Gironde have still higher biomass (though somewhat lower density in the Gironde) and again more species contribute. Differences between estuaries especially concern GW which is very poor and WM which is very rich in comparison to the corresponding communities in the other estuaries. The eastern and middle communities are characterised by the same dominant species in each estuary.

Discussion

Though no data on net efficiency of the sledge are available, the sampling gear deployed seems to be suitable for quantitative sampling of the hyperbenthos. Densities reported for the Gironde from previous studies are either lower than, or comparable to, densities reported in this study. Mees & Sorbe (in preparation) using a passive fishing technique with a rectangular plankton net (0.5 mm mesh), estimated average annual density for *Neomysis integer* at 6 ind m^3 in the zone of maximal abundance with maximal densities of 10 to 15 ind m^3 in spring and autumn and summer densities of about 3 ind m^3. This is lower than maximal density reported in this study: 33 ind m^3 at 4 g l^{-1}. Sorbe (1981) reported maximal densities of Gammarus zaddachi at 550 ind 100 m^3 in summer, which is about the same density as found in this study. The same author estimated maximal density of *G. salinus* at 20–30 ind 100 m^3 water which is about 10 times lower than maximal densities reported in this study. The sampling strategy seems to have a sufficiently narrow grid for studying the replacement of hyperbenthic communities along the estuarine gradients (e.g. the continuous aspect of the CCA sample score biplots in Fig. 9).

The three estuaries are remarkably similar qua species composition (especially in the brackish reaches) and general trends in diversity, density and biomass. The following within-estuary patterns were consistently found: diversity was highest in the marine zone, where density and biomass were lowest. Diversity then

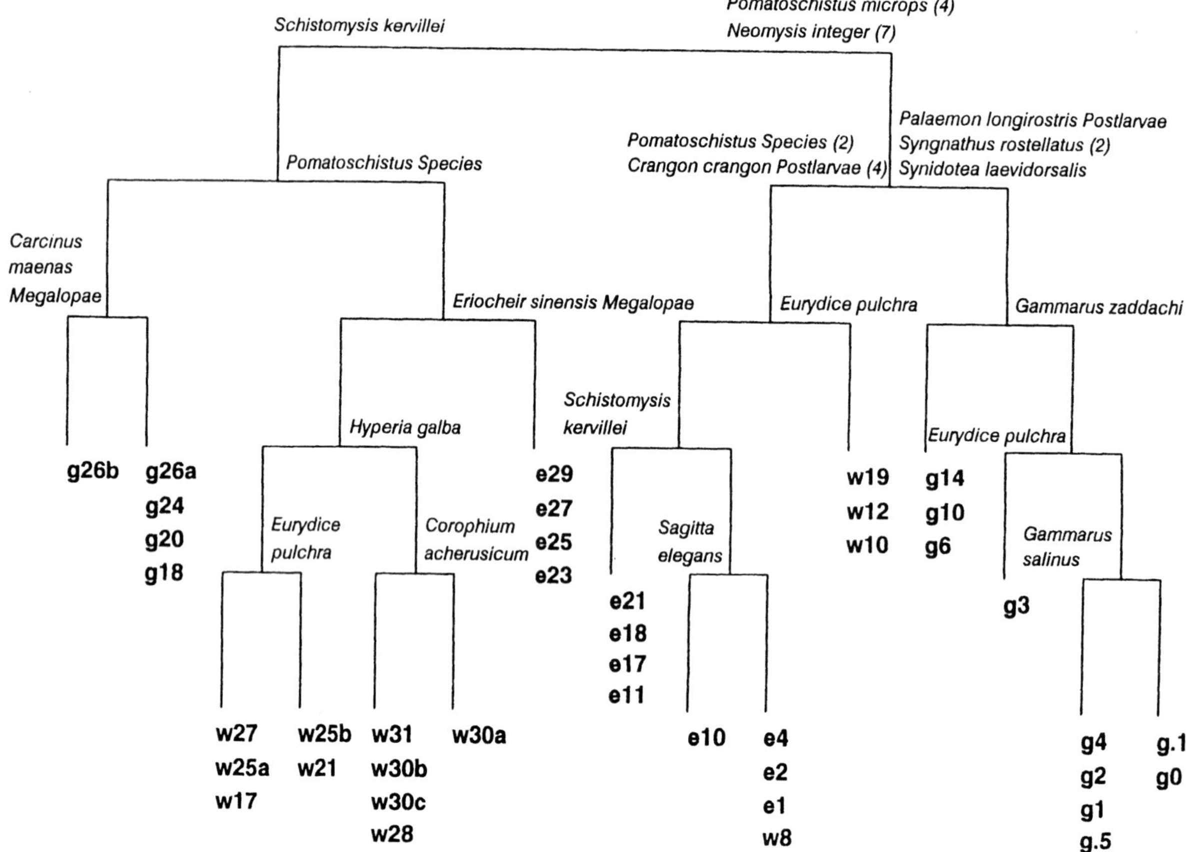

Fig. 7. TWINSPAN with the fourth root transformed biomass data of the three estuaries. Indicator species (and pseudospecies cutlevel) are given for each division.

decreased in an upstream direction and became minimal in the brackish part, where density and biomass reached maximal values. In Eems and Gironde diversity then increased slightly towards the freshwater zone. The poorness of the most seaward Gironde samples is exceptional. The capture of very few animals in these stations coincided with very high Secchi disc values. Possibly net avoidance contributed to the low densities. Other explanations may include the elimination of hyperbenthic populations by visual predators or lower food availability in this area. The overall higher densities reported for the marine part of the Eems as compared to Westerschelde and Gironde are probably related to the adjacent Wadden Sea system. The Wadden Sea itself is highly productive and has estuarine characteristics. No comparable system borders the mouths of the other estuaries, which open onto a typical coastal area. The low density found in the station located in front of the mouth of the shallow Dollard area of the Eems can not be explained.

Each estuary contained distinct communities along the unidirectional salinity-turbidity-temperature gradient. The position of these communities was similar in each estuary: a marine community in the high salinity zone, a brackish water community in the middle reaches and a third community (absent from the Westerschelde) in the stations with the lowest salinities. The similarity of the brackish water faunas among estuaries is high, higher than the within-estuary similarity of brackish and marine fauna. Brackish water species are few and the dominant residents occurred in all three estuaries at about the same density. In his extensive literature review Wolff (1973) concluded that the macrobenthic faunas of the brackish estuaries of N.W. Europe (including the Eems, the Delta area and the Arcachon Bay) were very similar to one another. This similarity can be extended to the hyperbenthic brackish water fauna and can be traced back to a common ancestral area on the western coast of France during the last Pleistocene glaciation, some 18 000–20 000 years ago. As for the macrobenthos, the low

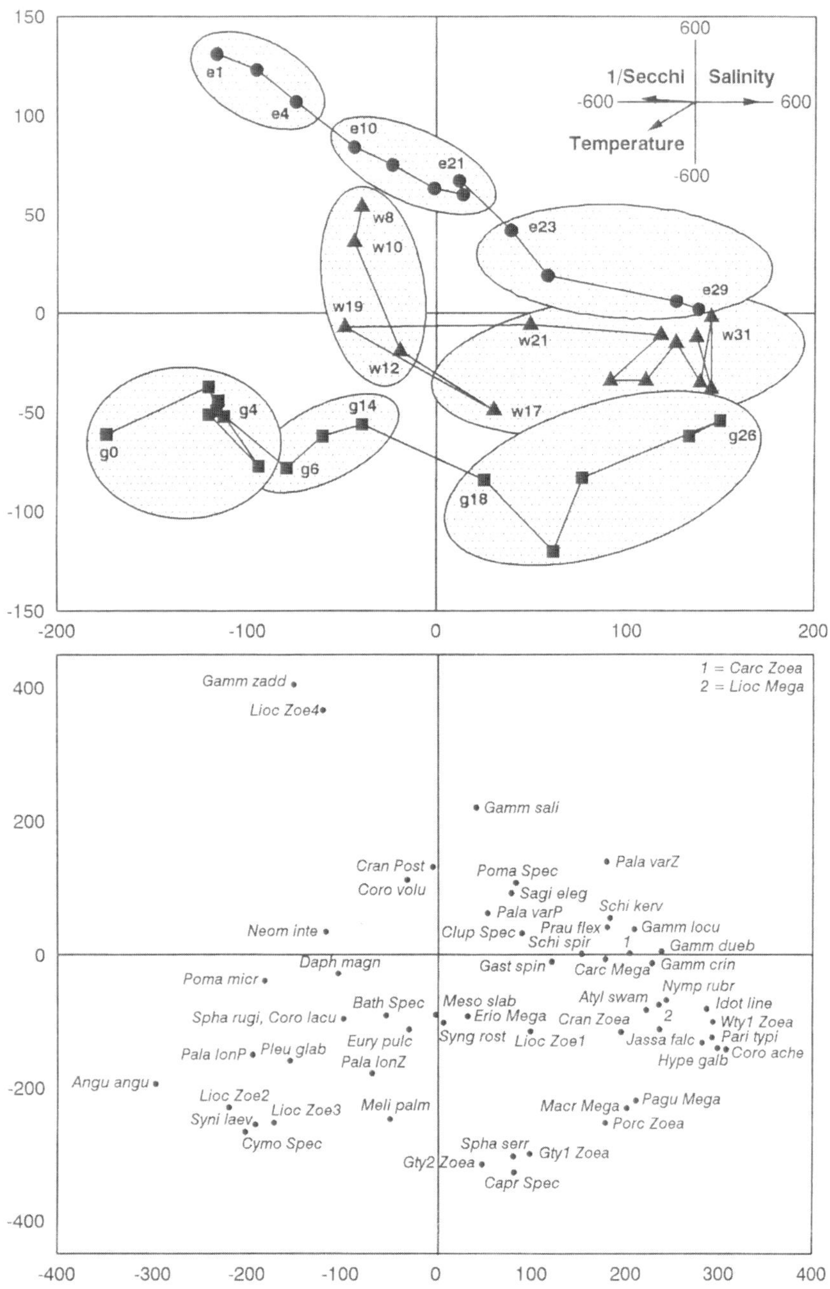

Fig. 8. Species scores (bottom), sample scores (top), and environmental biplot in the ordination plane formed by the first (horizontal) and second (vertical) canonical axes of the CCA with the fourth root transformed biomass data of the three estuaries. Samples are connected following salinity; communities as identified by TWINSPAN and GAS are circled.

number of hyperbenthic brackish water species can be explained by the low predictability of the environment and the geological history of northwestern Europe, and by the physiological stress which confronts the animals colonizing them (McLusky, 1981).

The classical pattern in species succession within *Gammarus*, already described for many European tidal estuaries (Lincoln, 1979; Sorbe, 1978, 1979; Meurs & Zauke, 1988), was confirmed in the present study. *G. crinicornis* is a marine species which penetrates estuaries, *G. salinus* is a brackish water species and

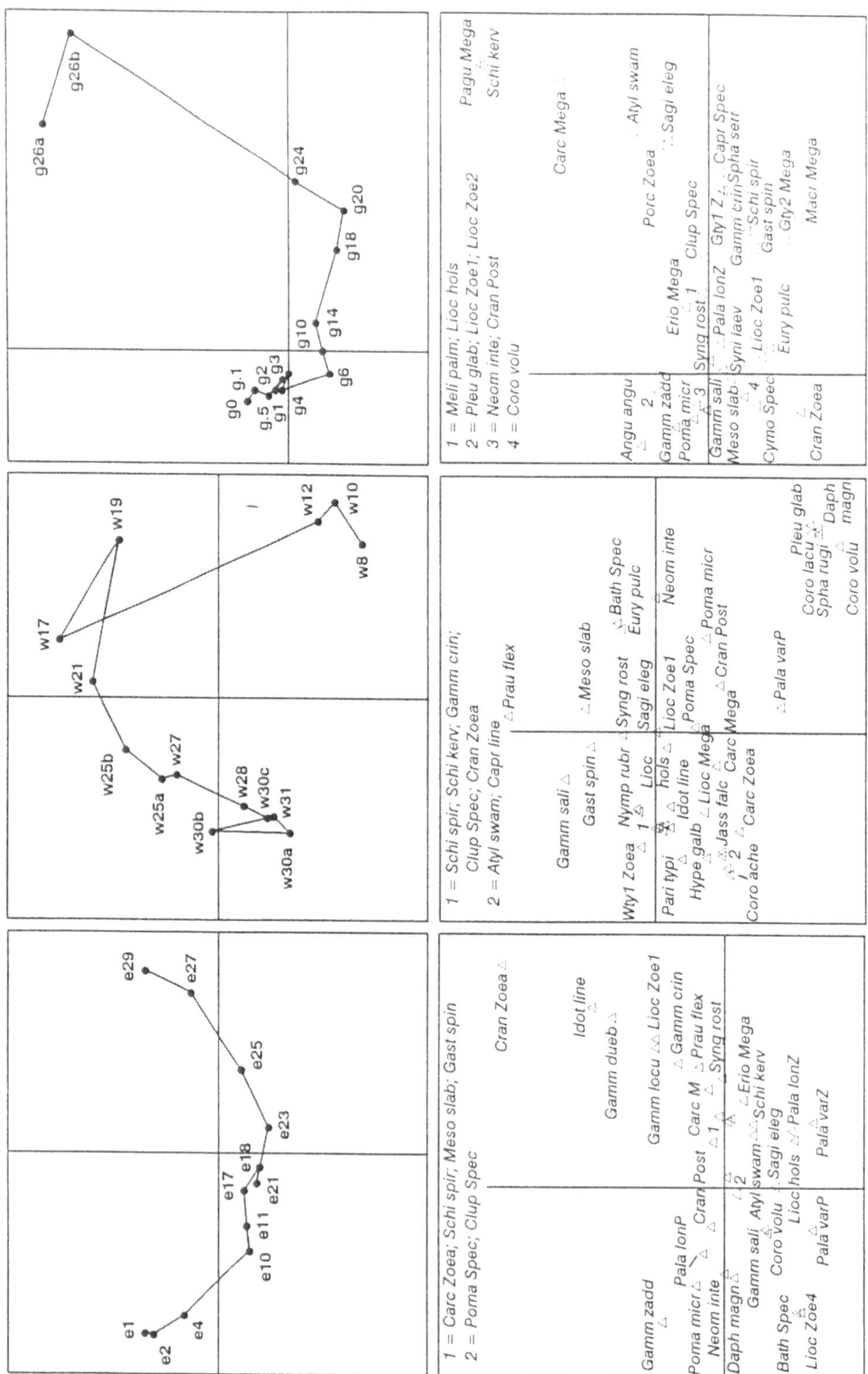

Fig. 9. Species scores (bottom) and sample scores (top) in the ordination plane formed by the first (horizontal) and second (vertical) canonical axes of the CCA with the fourth root transformed biomass data of the three estuaries separately. Samples are connected following salinity.

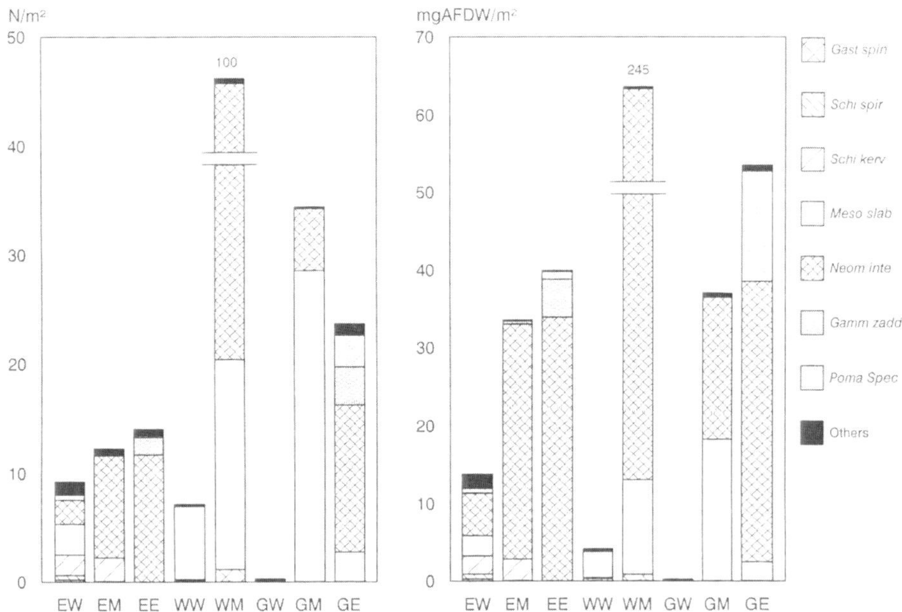

Fig. 10. Faunal composition of the 8 communities as identified by the multivariate analyses: biomass. First letter: *E*ems, *W*esterschelde, *G*ironde; second letter: *W*est, *M*iddle, *E*ast.

G. zaddachi prefers the areas of very low salinity. In contrast to the findings of Meurs & Zauke (1988) an overlap was recorded in the distribution of *G. salinus* and *G. zaddachi* in the Eems and the Gironde.

Still, qualitative and quantitative differences in the corresponding hyperbenthic communities of the three estuaries were obvious. Some species were restricted to one or two of the estuaries studied, while others, especially in the brackish part, were common to all three. The distribution of these common species along the salinity gradient – and the density and biomass they reach in the corresponding salinity zones – varied to some extent between estuaries (for examples see Fig. 5).

The most prominent difference between the estuaries was caused by the oxygen deficiency in the Westerschelde: there was no hyperbenthos present upstream of the 8 g l^{-1} isohaline in the Westerschelde, whereas the density and biomass maximum was situated around the 2 to 4 g l^{-1} isohaline in Eems and Gironde. The low salinity hyperbenthic community was completely absent in the Westerschelde. Upstream of the 10 g l^{-1} isohaline the dissolved oxygen concentration dropped below 40% of the saturation value, which seems to be a critical threshold for hyperbenthic life. The oxygen depletion zone was probably located around even higher salinities on the first sampling day when the *Neomysis* peak was situated around 19 g l^{-1}. Averaged over the year, the abundance maximum of this species in the Westerschelde is located around the 15 g l^{-1} isohaline (Mees *et al.*, 1994). Whereas the distribution of hyperbenthic species in Gironde and Eems seemed to be primarily determined by salinity, the location of the Westerschelde populations was governed by dissolved oxygen concentration. The populations of a number of species, which in Gironde and Eems reached highest density and biomass in the oligohaline zone, seem to have almost disappeared from the Westerschelde (e.g. *Gammarus zaddachi* and *Palaemon longirostris*). Other species did not occur in their 'normal' salinity range and their populations have shifted to higher, atypical salinity zones (e.g. *Neomysis integer*, *Mesopodopsis slabberi*, *Pomatoschistus microps*, *Gammarus salinus*). In the case of *P. microps* and *G. salinus* this is probably accompanied by a considerable decrease in numbers. Most 'marine' species did not penetrate as far into the Westerschelde as they did in Gironde and Eems and had a more limited upstream distribution (e.g. *Schistomysis spiritus*, *S. kervillei*, *Pomatoschistus* species).

Absence of a species from the samples does not necessarily mean it is not present in the estuary. This is certainly true for the rare species and for animals not efficiently caught with the sledge e.g. *Pleusymtes glaber* and *Melita palmata* (Mees *et al.*, 1993b), and *Rhithropanopeus harrisii* (Wolff & Sandee, 1971; Van

Damme *et al.*, 1992) were previously recorded from the Westerschelde and *Sphaeroma rugicauda* is known to be present in the Eems (Michaelis *et al.*, 1992). Still, especially for the abundant species of Eems and Gironde (e.g. Table 1), the low number of individuals (if any) caught in the Westerschelde can give an indication of significantly reduced population sizes in the impoverished Westerschelde. *Gammarus zaddachi* was not recorded from the Westerschelde in this study although occasionally an individual is caught in the tidal marsh of Saeftinghe (Cattrijsse *et al.*, 1993). Den Hartog (1964) discussed the Gammaridea of the Westerschelde and reported *G. zaddachi* from several locations in the Delta area. The species disappeared from the Rhine before 1958 and from all other estuarine waters in the area in 1960, probably due to pollution or an epidemic. The estuarine populations had not recovered by 1963. *Palaemon longirostris* was formerly common in the oligohaline parts of the estuaries of the Delta area (Holthuis, 1950). During intensive sampling in the early seventies only 1 individual was found in the Westerschelde near Bath (Heerebout, 1974) and the disappearance was attributed to pollution. *P. longirostris* is very common in European estuaries of the geographical area covered (e.g. Heerebout, 1974; Marchand, 1981; Sorbe, 1983). It occurs along the entire estuarine gradient (0 to 28 g l^{-1}). It is reasonable to conclude that these species should normally also occur in the Westerschelde and that their absence from the system has to do with pollution.

Neomysis integer occurred in much higher densities in the Westerschelde (242 and 129 ind m^2 at 19 and 10 g l^{-1}) than it did in the other estuaries (30 and 33 ind m^2, for Eems and Gironde, at 4 and 2 g l^{-1} respectively). This suggests a spatial compression of the population between critically low oxygen concentrations on the freshwater side and perhaps species better adapted to higher salinities on the seaward side. The population was concentrated in the immediate vicinity of the edge of the oxygen depletion zone, probably in order to be as close as possible to the large amounts of high quality food (organic detritus with associated bacteria) entering the oxygenated part of the system. The high biomass in this narrow zone indicates *N. integer* is capable of successfully exploiting this detrital material, either directly or through the second trophic level (e.g. the copepod *Eurytemora affinis*). Like most brackish water species *N. integer* is euryhaline and eurythermic. There are no indications that the species' growth or reproduction are hampered by the higher salinities at which it occurs in the Westerschelde (Mees *et al.*, 1994; Mees & Sorbe, in preparation).

Other differences among the estuaries are related to latitude. Some species probably reached their northern or southern distribution limits in one of the systems studied (e.g. the absence of *Idotea linearis* from the Gironde; the occurrence of *Sphaeroma serratum* and the several crab larvae only found in the Gironde). *Schistomysis kervillei* has its southern distribution limit in the north of Spain and, though quite common in the coastal waters adjacent to the Gironde (San Vincente & Sorbe, 1990), the species probably avoids the high summer temperatures in the estuary. Also, the smaller salinity range and lower densities of *Mesopodopsis slabberi* in the Eems may be linked to the lower temperature: possibly its summer migration into the brackish reaches (Mees *et al.*, 1993a) occurs later in the year. The absence of juveniles of coastal *Pomatoschistus* species from the Gironde can not be explained. Both *P. lozanoi* and *P. minutus* are distributed south to the Portuguese coasts (Miller, 1986). *P. lozanoi* seems to avoid estuaries in southern Europe (Hamerlynck, 1990), but *P. minutus* is very common even in the Tagus (Moreira *et al.*, 1991).

Differences among the estuaries in the reproductive state of populations may also be related to temperature. Populations of the same species are liable to have been sampled in a different life cycle phase at each latitude, which may explain the presence/absence and relative densities of certain developmental stages – i.e. functional species or temporary hyperbenthic species (Hamerlynck & Mees, 1991) – in the estuaries (e.g. the lower density of *Crangon crangon* zoeae and *Palaemon longirostris* zoeae in the Eems, the presence of *Pagurus*, *Porcellana* and *Macropodia* megalopae in the Gironde). Other differences may relate to the physical characteristics of the estuaries. Sorbe (1983) found *Palaemonetes varians* to be common in less dynamic areas close to the Gironde, but thinks it doesn't tolerate the high current velocities in the estuary proper. The Eems samples were characterised by large amounts of peat, which may explain the higher net efficiency for larger fish and epibenthic crustaceans (Table 1). Other differences result from recent introductions of brackish water species. Whereas the crabs *Eriocheir sinensis* and *Rhithropanopeus harrisii* were both successful in colonising the three estuaries, the isopod *Synidotea laevidorsalis* is only present in the Gironde (Mees & Fockedey, 1993).

Acknowledgments

This research was supported by the European Community, contract no. MAST-0024-C (JEEP 92 project). We thank Jacques Castel, Xabier Irigoien, Karel Essink, Zwanette Jager, Kris Hostens and the crews of the Côte d'Aquitaine, the R. V. Luctor and the M/S Dr ir Joh. van Veen for their hospitality and help during the sampling campaigns. This is contribution nr. 738 of the Centre for Estuarine and Coastal ecology.

References

Baretta, J. & P. Ruardij, 1988. Tidal flat estuaries. Simulation and analysis of the Ems estuary. Ecological Studies 71, Springer Verlag, Berlin: 353 pp.

Belmans, H., 1988. Verdiepings- en onderhoudsbaggerwerken in wester- en zeeschelde. Water 43: 184–194.

Beyer, F., 1958. A new, bottom-living Trachymedusa from the Oslofjord. Nytt Mag. Zool. 6: 121–143.

Castel, J., 1993. Long-term distribution of zooplankton in the Gironde estuary and its relation with river flow and suspended matter. Cah. Biol. mar. 34: 145–163.

Cattrijsse, A., J. Mees & O. Hamerlynck, 1993. The hyperbenthic Amphipoda and Isopoda of the Voordelta and the Westerschelde estuary. Cah. Biol. mar. 34: 187–200.

de Jonge, V. N., 1988. The abiotic environment. In Tidal flat estuaries. Simulation and analysis of the Eems estuary. Baretta, J. & P. Ruardij (eds) Springer-Verslag. Berlin Heidelberg New York London Paris Tokyo: 14–27.

den Hartog, C., 1964. The Amphipods of the Deltaic region of the rivers Rhine, Meise and Scheldt in relation to the hydrography of the area. Part III: The Gammaridae. Neth. J. Sea Res. 2: 407–457.

Duursma, E. K., A. G. A. Merks & J. Nieuwenhuize, 1988. Exchange processes in estuaries such as the Westerschelde, an overview. Hydrobiol. Bull. 22: 7–20.

Field, J. G., K. R. Clarke & R. M. Warwick, 1982. A practical strategy for analysing multispecies distribution patterns. Mar. Ecol. Prog. Ser. 8: 37–52.

Hamerlynck, O., 1990. The identification of *Pomatoschistus minutus* and *Pomatoschistus lozanoi* (Pisces, Gobiidae). J. Fish Biol. 37: 723–728.

Hamerlynck, O. & J. Mees, 1991. Temporal and spatial structure in the hyperbenthic community of a shallow coastal area and its relation to environmental variables. Oceanologica Acta 11: 205–212.

Hartley, P. H. T., 1940. The Saltash tuck-net fishery and the ecology of some estuarine fishes. J. mar. biol. Ass. U.K., 24: 1–68.

Heerebout, G. R., 1974. Distribution and ecology of the Decapoda Natantia of the estuarine region of the rivers Rhine, Meuse and Scheldt. Neth. J. Sea Res. 8: 73–93.

Heip, C., 1988. Biota and abiotic environment in the Westerschelde estuary. Hydrobiol. Bull. 22: 31–34.

Heip, C., 1989. The ecology of the estuaries of Rhine, Meuse and Scheldt in the Netherlands. Topics in marine biology. Ros (ed.). Scient. Mar. 53: 457–463.

Herman, P. M. J., H. Hummel, M. Bokhorst & G. A. Merks, 1991. The Westerschelde: interaction between eutrophication and chemical pollution? In M. Elliot & J.-P. Ducrotoy (eds). Estuaries and Coasts: Spatial and temporal intercomparisons. Olsen & Olsen: 359–364.

Hill, M. O., 1973. Diversity and eveness: a unifying notation and its consequences. Ecology 54: 427–432.

Hill, M. O., 1979. TWINSPAN. A Fortran program for arranging multivariate data in an ordered two-way table by classification of individuals and attributes. Cornell University, Ithaca, New York, 60 pp.

Jongman, R. H. G., C. J. F. Ter Braak & O. F. R. Van Tongeren, 1987. Data analysis in community and landscape ecology. Pudoc, Wageningen.

Jouanneau, J. M. & C. Latouche, 1981. The Gironde estuary. Contributions to sedimentology, 10. H. Füchtbauer, A. P. Lisitzyn, J. D. Millerman & E. Seibold (eds). E. Schweizerbartsche Verlagsbuchhandlung, 115 pp.

Ketchum, B. H., 1983. Estuarine characteristics. In Ecosystems of the World 26. Ketchum, B. H. (ed.). Estuaries and enclosed seas: 1–14.

Lincoln, R. J., 1979. British Marine Amphipoda: Gammaridea. British Museum Natural History, London: 658 pp.

Marchand, J., 1981. Observation de l'ecologie de *Crangon crangon* (Linné) et *Palaemon longirostris* (H. Milne Edwards) (Crustacea, Decapoda, Natantia) dans l'estuaire interne de la Loire (France). Vie et Milieu 31: 83–92.

Mauchline, J., 1980. The biology of Mysids and Euphausiids. In Advances in marine biology 18. Blaxter, J. H. S., Russell, R. S. & Younge, M. (eds). Academic Press, London: 681 pp.

McLusky, D. S., 1981. The estuarine ecosystem. Blackie, London: 150 pp.

Mees, J. & N. Fockedey, 1993. First record *Synidotea laevidorsalis* (Miers, 1881) (Crustacea: Isopoda) in Europe (Gironde estuary, France). Hydobiologia 264: 61–63.

Mees, J. & O. Hamerlynck, 1992. Spatial community structure of the winter hyperbenthos of the Schelde-estuary, The Netherlands, and adjacent coastal waters. Neth. J. Sea Res. 29: 357–370.

Mees, J., Z. Abdulkerim & O. Hamerlynck, 1994. Life history, growth and production of *Neomysis integer* in the Westerschelde estuary. Mar. Ecol. Prog. Ser. 109: 43–57.

Mees, J., A. Cattrijsse & O. Hamerlynck, 1993a. Distribution and abundance of shallow-water hyperbenthic mysids (Crustacea, Mysidacea) and euphausiids (Crustacea, Euphausiacea) in the Voordelta and the Westerschelde, south-west Netherlands. Cah. Biol. Mar. 34: 165–186.

Mees, J., A. Dewicke & O. Hamerlynck, 1993. Seasonal composition and spatial distribution of the hyperbenthic communities along the estuarine gradients in the Westerschelde. Neth. J. aquat. Ecol. 27: 359–376.

Meurs, H. G. & G. P. Zauke, 1988. Regionale und zeitliche Aspecte der Besiedlung des Elbe-, Weser- und Emsästuars mit euryhalinen Gammariden (Crustacea: Amphipoda). Arch. Hydrobiol. 113: 213–230.

Michaelis, H., H. Fock, M. Grotjahn & D. Post, 1992. The status of the intertidal zoobenthic brackish-water species in estuaries of the German bight. Neth. J. Sea Res. 30: 201–207.

Miller, P. J., 1986. Gobiidae. In Fishes of the North-eastern Atlantic and the Mediterranean, Vol. III, Whitehead, P. J. P., Bauchot, M.-L., Hureau, J.-C., Nielsen, J. & Tortonese, E. (eds) Paris: Unesco: 1019–1085.

Moreira, F, J. L. Costa, P. R. Almeida, C. P. Assis & M. J. Coasta, 1991. Age determination in *Pomatoschistus minutus* (Pallas) and *P. microps* (Krøyer) (Pisces: Gobiidae) from the upper Tagus estuary, Portugal. J. Fish Biol. 39: 433–440.

San Vincente, C. & J.-C. Sorbe, 1990. Biologia del misidaceo suprabentonico Schistomysis kervillei (Sars, 1885) en la plataforma continental aquitana (suroeste de Francia). Bentos 6: 246–267.

Sorbe, J.-C., 1978. Inventaire faunistique des Amphipodes de l'estuaire de la Gironde. Bull. Cent. Etud. Rech. sci., Biarritz 12: 369–381.

Sorbe, J.-C., 1979. Systématique et écologie des amphipodes gammarides de l'estuaire de la Gironde. Cah. Biol. mar. 20: 43–58.

Sorbe, J.-C., 1981. La macrofaune vagile de l'estuaire de la Gironde. Distribution et migration des espèces. Modes de reproduction. Régimes alimentaires. Oceanis 6: 579–592.

Sorbe, J.-C., 1983. Les Décapodes Natantia de l'estuaire de la Gironde (France). Contribution a l'étude morphologique et biologique de *Palaemon longirostris* H. MILNE EDWARDS, 1837. Crustaceana 44: 251–270.

Ter Braak, C. J. F., 1988. CANOCO – a FORTRAN program for canonical community ordination by (partial) (detrended) (canonical) correspondence analysis, principal components analysis and redundancy analysis (version 2.1). Agricultural Mat. Group. Wageningen, Ministry of Agriculture and Fisheries, 95 pp.

Van Damme, P., J. Mees & S. Maebe, 1992. Voorkomen van het zuiderzeekrabbetje *Rhithropanopeus harrisii* (Gould, 1841) in de Westerschelde. De strandvlo 12: 19–21.

Van Eck, G. T. M., N. De Pauw, N. Van de Langenbergh & G. Verreet, 1991. Emissies, gehalten, gedrag en effecten van (micro)verontreinigingen in het stroomgebied van de Schelde en Schelde-estuarium. Water 60: 164–181.

Wolff, W. J., 1973. The estuary as a habitat. An analysis of data on the soft-bottom macrofauna of the estuarine area of the rivers Rhine, Meuse, and Scheldt. Zool. Verh. Leiden 126: 242 pp.

Wolff, W. J. & A. J. J. Sandee, 1971. Distribution and ecology of the Decapoda Reptantia of the estuarine area of Rhine, Meuse and Scheldt. Neth. J. Sea Res. 5: 197–226.

Hydrobiologia **311**: 175–184, 1995.
C. H. R. Heip & P. M. J. Herman (eds), Major Biological Processes in European Tidal Estuaries.

Effects of manipulation of food supply on estuarine meiobenthos

Melanie C. Austen & Richard M. Warwick
Plymouth Marine Laboratory, Prospect Place, West Hoe, Plymouth, UK

Key words: meiofauna, mesocosm, communities, food, organic matter, nematodes

Abstract

A comparative mesocosm experiment was carried out to determine the effects of natural foods of different quality and quantity on the structure of natural meiobenthic communities collected in undisturbed sediment from the polluted Westerschelde and the comparatively undisturbed Gironde estuaries. Nematode communities are more diverse and species rich in the latter estuary. The organic matter or foods used were phytoplankton, green alga, salt marsh plant detritus and leaf litter detritus which were added at three dose rates including a high dose. There was no change in community structure in response to the treatments in either of the estuarine meiobenthic communities. Analysis of all the results from this experiment indicate that the food quantity manipulations had almost no effect on the deposit feeding meiofauna. It may be that the reserves of organic matter within the sediment were sufficient to satisfy their dietary requirements for the duration of the experiment. The abundance of diatom/epigrowth feeding nematodes which were initially dominant in the Gironde, declined substantially suggesting that they may have been food limited since diatoms were not among the sources of organic matter added to the mesocosm. There was no specific response to the five different types of organic matter added to the mesocosm

Introduction

Meiobenthos in estuaries exhibit higher diversity than macrobenthos (Warwick & Gee, 1984) and mechanisms for maintaining diversity by partitioning of resources are quite different between meiobenthos and macrobenthos. Meiobenthic species discriminate between food particles of different size shape and quality whereas for macrobenthic species spatial segregation is more important (Warwick, 1984). The wide variation in morphology of buccal structures amongst meiofaunal species, particularly amongst nematodes and the large amount of experimental evidence on individual species suggest that there is selective feeding and variation in feeding mode between species (Jensen, 1987; Austen, 1989). Most estuaries tend to be nutrient rich areas with inputs of nitrates, phosphates, silica and organic matter entering from river and sea inputs and from land runoff. These can promote high levels of local primary and secondary production in e.g. the water column, intertidal mud and sand-flats and salt marsh areas. We would expect a large amount of niche separation within the estuarine meiobenthos related to the variety and abundance of available food resources. This could explain the comparatively high abundance and diversity of the meiobenthos. However, an excess of organic inputs due to e.g. anthropogenic pollution, can create an unbalanced ecosystem with high environmental stress where high bacteria production can cause hypoxia and anoxia as well as buildup of toxic materials and by-products of decomposition.

Within the Joint European Estuaries Programme (JEEP) attention is focused on a comparison of two estuaries, the Westerschelde and the Gironde. The Westerschelde is subject to pollution from urban and industrial areas with high levels organic and inorganic waste causing seasonal anoxia (Heip, 1989). The Gironde is comparatively undisturbed by such pollution. We hypothesise that in the Westerschelde the structure of the meiobenthic communities will be most strongly influenced by the high organic and pollutant loadings but in the Gironde the availability of organic matter of different quality as well as quantity will be of more importance.

We have designed a mesocosm experiment to determine the effects of quality and quantity of organic matter on meiobenthic community structure using manipulations of food supply to field collected natural meiobenthic assemblages within laboratory mesocosms. A range of natural sources of organic matter: phytoplankton, salt marsh plant material, green macroalgae, terrigenous leaf litter and a mixture of all of these, were chosen to reflect the variety of organic matter available to estuarine meiofauna, or to associated microbial organisms on which meiofauna may feed. Our null hypothesis is that there are no significant effects of different qualities and quantities of organic matter on meiobenthic nematode community structure. Previous experimental work (Austen, 1989) indicates that the role of organic matter in structuring estuarine meiobenthic assemblages is probably more important in the higher salinity regions of an estuary. Therefore our experimental work has concentrated on assemblages from these regions of the estuaries and has been carried out at constant salinity. The experiment has been repeated using meiobenthos from locations in both estuaries which have similar salinity regimes and similar sediment structure.

Methods

Natural meiofauna assemblages were collected in undisturbed sediment cores using plastic pipes (6.4 cm diameter and 12.5 cm height) from mid-tide level on 13.6.91 at Walsoorden on the Westerschelde and on 3.6.92 at Le Verdon in the Gironde estuary. Soetart *et al.* (in press) describe the principal characteristics of these estuaries and give details of the granulometry at the sites (Walsoorden corresponds to their station WS42). Ambient conditions on the sampling dates are given in Table 1. Sediment was held in place in the core tubes by 100 μm mesh nylon attached with rubber elastic bands. Cores were transported in cool boxes containing 3–4 cm of filtered, locally collected estuarine water. To lower meiofauna metabolic rates and reduce oxygen consumption the cores were covered with layers of paper and defrosting ice packs to reduce ambient temperatures within the cool boxes by 5 °C during transport. Within 24 hours they were transferred to the mesocosm facility in Plymouth Marine Laboratory. The mesocosm consisted of four 1 m^3 plastic tanks, which were used as replicate systems, each filled with 1 μm filtered seawater (collected from the English Channel, 24 km offshore from Plymouth Marine Laboratory close to the Eddystone lighthouse and diluted with distilled water to the appropriate experimental salinity) and maintained at constant temperature. Each tank had a separate closed circulation system via an external 1 μm filter and internal circulation such that there was laminar water flow. There was a false bottom in each tank at a depth of 50 cm on which experimental sediment cores were randomly placed in four plastic mesh trays per tank. The depth of water above the surface of the sediment within the cores was 37.5 cm. The false bottom was perforated with 48 holes, 6 cm in diameter, allowing full circulation of water around the sediment cores and around the mesocosm tanks.The field collected sediment cores were divided in equal numbers between the four tanks, hereafter referred to as systems 1 to 4, and left to acclimatise. Salinity and temperature were maintained at 20 ppt and 25 ppt salinity and 15 °C and 20 °C for the Westerschelde and Gironde experiments respectively. Salinity was maintained by weekly addition of distilled water as necessary.

Table 1. Ambient salinity and temperature at estuarine sites when sediment cores with meiobenthos were collected.

	Walsoorden, Westerschelde	Le Verdon, Gironde
Collection date	13.6.91	3.6.92
Salinity (ppt)	25	25
Sediment temperature (°C)	14	23
Seawater temperature (°C)	14	19
Air temperature (°C)	16	20

After 7 days a dosing regime commenced using five different sources of organic matter to act as food either directly for the meiofauna or indirectly for bacteria in the sediment which would then be consumed by the meiofauna:

1. Phytoplankton (P) – freeze-dried phytoplankton powder which was 80% *Haematococcus*, 10% *Isochrysis* and 10% *Porphyridium* (carbon 47.9%, nitrogen 7.1%).
2. Salt marsh 1 (S) – air-dried and finely ground, one year old *Scirpus* straw collected from the Westerschelde (carbon 42.9%, nitrogen 3.7%).
3. Salt marsh 2 (E) – freeze dried and finely ground, green alga, *Enteromorpha* sp. found growing on sediment between salt marsh vegetation in the Westerschelde (carbon 26.7%, nitrogen 5.7%).

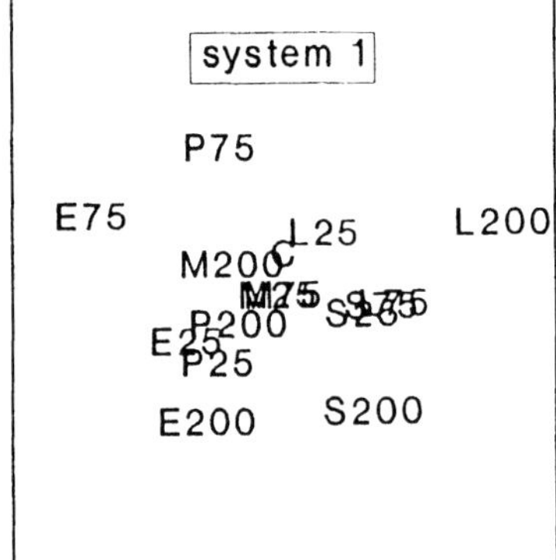

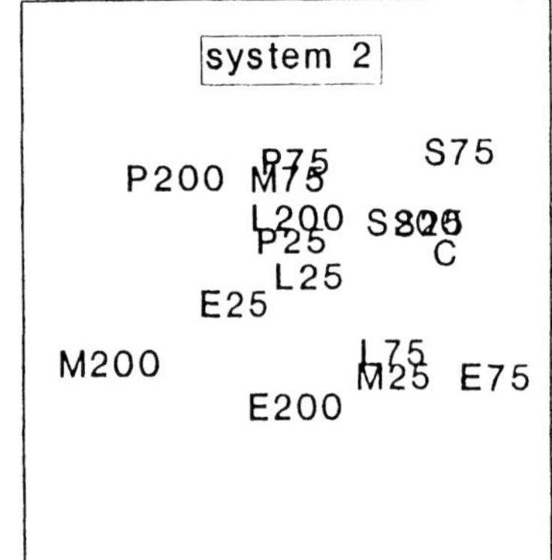

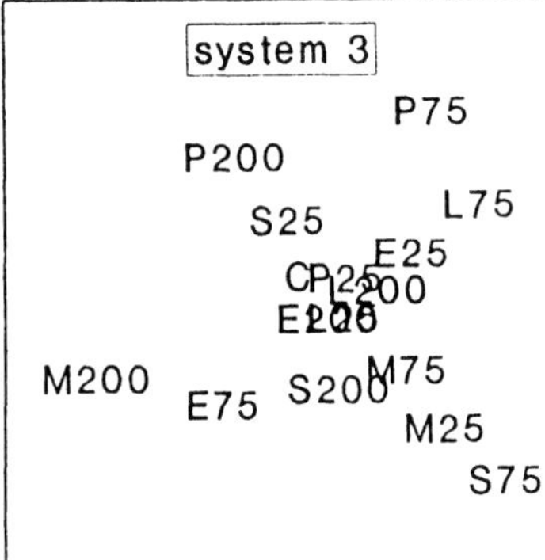

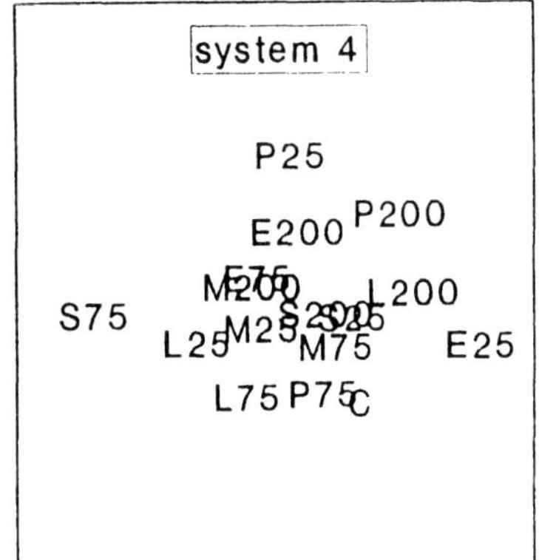

Fig. 1. Multi dimensional scaling ordinations for untransformed Westerschelde nematode assemblage data from each of four mesocosm basins (systems 1–4) after 16 weeks. C: control (0 g C m^{-2} yr^{-1}), P: phytoplankton, S: *Scirpus*, E: *Enteromorpha* sp., L: leaf litter, M: mixture at dose rates of 25, 75 and 200 g C m^2 yr^{-1}.

4. Leaf litter (L) – collection of finely ground mixed dried leaves collected from local woodland in Plymouth (carbon 47.4%, nitrogen 3.5%).
5. Mixture (M) – a mixture of 1–4 such that each contributes the same amount of carbon to the total mix (carbon 38.9%, nitrogen 5.1%).

Cores were dosed at two week intervals at rates corresponding to 25, 75 and 200 g C m^{-2} y^{-1}. These rates corresponded to low and high estimates of normal deposition of carbon in inshore areas (Gee *et al.*, 1985) but also reflect the higher inputs in estuaries which are often eutrophic. To dose the sediment cores the water levels in the mesocosm tanks were lowered to below the surface of the sediment. Organic matter was suspended in filtered mesocosm water such that the appropriate dose (0, 0.003, 0.009 and 0.025 g C/sediment core) could be added as a 10 ml aliquot by gently squirting it from a syringe. Controls were treated with a 10 ml aliquot of filtered water only. After addition of organic matter the cores were covered with fine nylon mesh and gently flooded with mesocosm water until levels within the mesocosm tanks returned to normal. Cores were dosed such that in each mesocosm system there was at least one core representing each of the 16 treatments (5 types of organic matter × 3 dose levels = 15 + 1 control).

Meiofauna from the sediment was sampled at the beginning of the experiment, before organic matter dosing commenced but after the initial seven day acclimation period – these were the day 0 samples, and again after a period of sixteen weeks of food dosing. At the beginning of the experiment one sample was taken from each mesocosm system and the four samples were treated as replicates. Similarly, at the end of the experiment, 16 cores corresponding to each experimental treatment (including controls) were removed from each of the systems to give four replicates of each treatment, one per system. For the Westerschelde experiment meiobenthos samples were taken by removing a whole core but for the Gironde meiofaunal abundance was much higher so the experiment cores were subsampled with two smaller cores (18 mm diameter) which were combined prior to further sample processing. Samples were fixed in 5% formalin. Large particles and

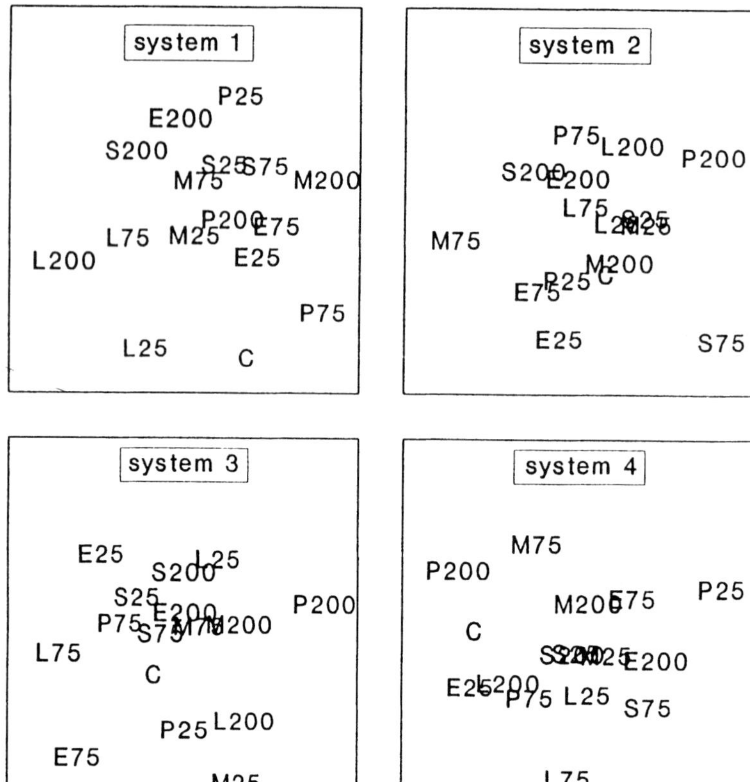

Fig. 2. Multi dimensional scaling ordinations for root-root transformed Westerschelde nematode assemblage data from each of four mesocosm basins (systems 1–4) after 16 weeks. Symbols as for Fig. 1.

Table 2. Results from modified ANOSIM significance testing of average Spearman correlation coefficient ρ_{av} for treatment and mesocosm (block) effects on nematode multivariate community structure. *P*(%): significance level, null hypothesis is rejected at *P*<5%, i.e. in no case is the null hypothesis rejected.

		ρ_{av}		*P* (%)	
	Data analysed	Treatment	Mesocosm	Treatment	Mesocosm
Westerschelde	Untransformed	0.061	−0.03	16	79
	Transformed	−0.066	−0.02	85	64
Gironde	Untransformed	−0.04	−0.06	77	100
	Transformed	−0.04	−0.04	73	92

macrobenthos were extracted by passing the sample through a 0.5 mm sieve onto a 63 μm sieve. Meiobenthos was extracted from the Westerschelde samples by 5 decantations using two 1 l measuring cylinders and three 1 hour flotations in ludox TM (specific gravity 1.15) with extracted meiobenthos and detritus retained on a 63 μm sieve (Pfannkuche & Thiel, 1988; Austen & Warwick, 1989). For the Gironde samples, which were smaller, only the ludox extractions were used. Meiobenthos subsamples were taken using a ladle or a sample splitter and these subsamples were evaporated to glycerol for mounting on microscope slides and identification.

Data analysis was carried out using standard techniques described in Clarke (1993). These include non-metric multi dimensional scaling ordination (MDS) using the Bray-Curtis similarity index for multivariate analysis (on both untransformed and double square

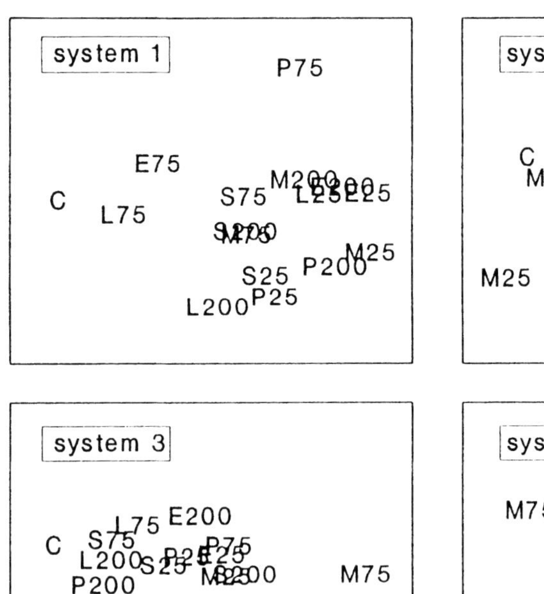

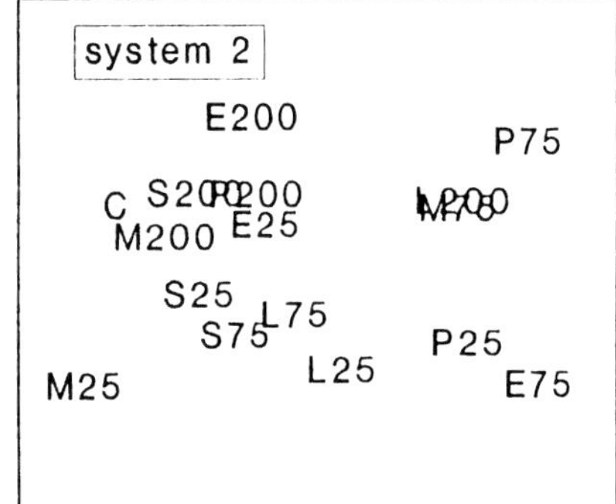

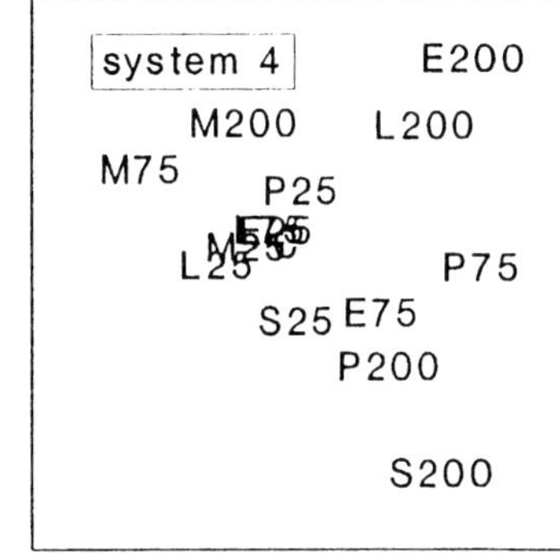

Fig. 3. Multi dimensional scaling ordinations for untransformed Gironde nematode assemblage data from each of four mesocosm basins (systems 1–4) after 16 weeks. Symbols as for Fig. 1.

root transformed data) and the computation of univariate indices: abundance, number of species, Shannon-Wiener diversity (H'), species richness (Margalef's index d) and evenness (J'). To test for significant differences between treatments a modification of the ANOSIM randomisation test was used for the multivariate data (Clarke & Warwick, in press) and ANOVA was used for the univariate indices. ANOSIM modification allows for conservative testing for treatment differences where there may be a 'block' or mesocosm system effect. The modified ANOSIM test (Clarke & Warwick, in press) is based on a measure of concordance between rank similarities of samples within each mesocosm system. Briefly, triangular matrices of rank similarities are constructed for the nematode community samples in each mesocosm system; Spearmans correlation coefficient is computed between corresponding matrix elements and then an 'average Spearman correlation coefficient' ρ_{av} is calculated between all pairs of similarity matrices, the significance of ρ_{av} is then examined by a permutation test. Unmodified ANOSIM was used to test for differences between day 0 and week 16 nematode community structure. Where significant differences were observed the program SIMPER was used to determine the relative contribution of each species to the Bray-Curtis dissimilarity term between treatments or dates.

Results

Throughout the experiment surface sediment in the cores had a 'natural' appearance – sediment colour did not change and there was no sign of anoxia (no black surface H_2S layer), there were signs of sediment disturbance by the biota e.g. mounds, tubes and holes. Small macrofaunal polychates were evident in most of the Gironde cores, particularly immediately after organic matter dosing when they would crawl about on the sediment surface.

At the beginning of the experiments there were means of 15 and 20 species per core and 3080 and 18 335 individuals per core for the Westerschelde and Gironde experiments respectively. After 16 weeks in the mesocosms the mean values were 15 and 21 species per core and 1890 and 3200 individuals per core in the Westerschelde and Gironde experiments respectively. Number of species and number of individuals per core were significantly different between Westerschelde and Gironde samples both at the beginning and end of the

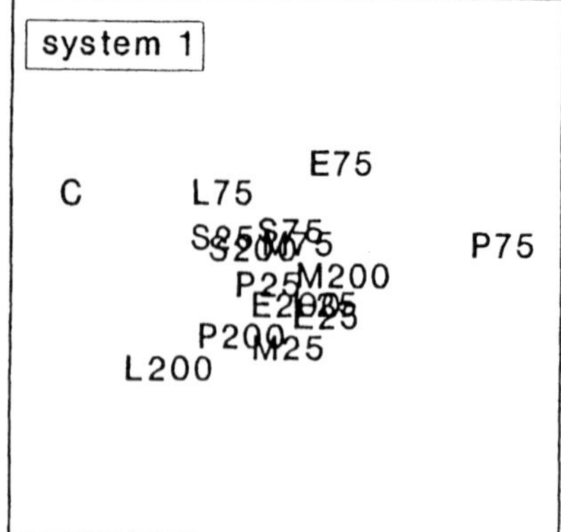

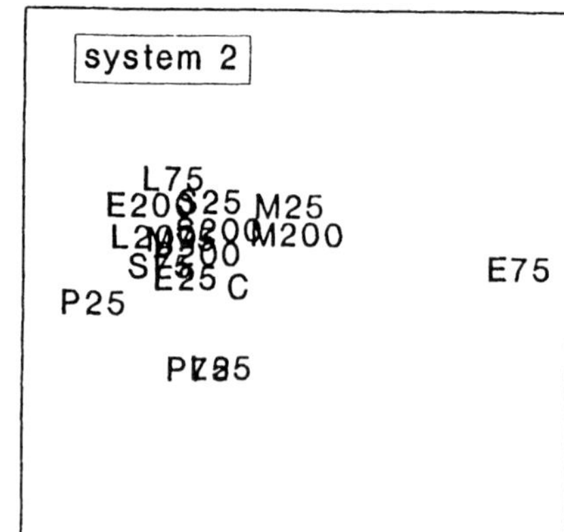

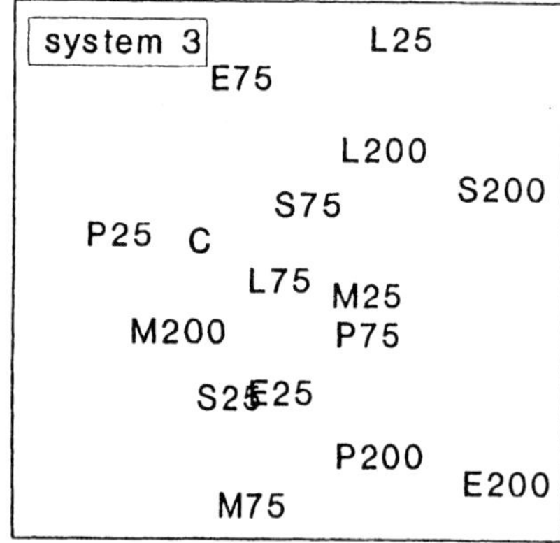

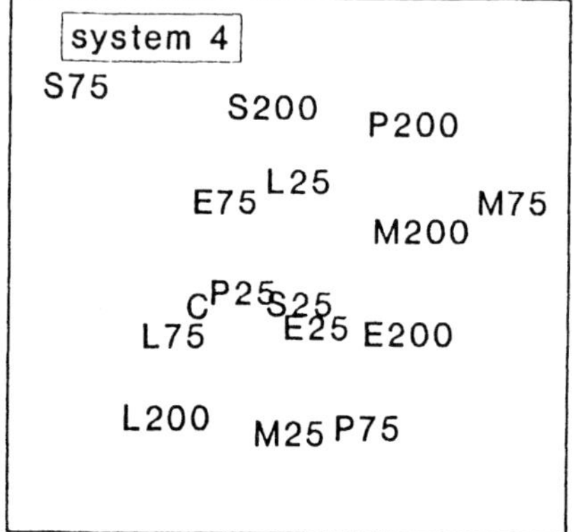

Fig. 4. Multi dimensional scaling ordinations for root-root transformed Gironde nematode assemblage data from each of four mesocosm basins (systems 1–4) after 16 weeks. Symbols as for Fig. 1.

experiment ($p<1\%$). A total of 42 and 56 nematode species were identified from the Westerschelde and Gironde experiment samples respectively.

The results of MDS analysis for the samples taken after 16 weeks are shown in Figs 1–4. There do not appear to be any similarities in the 2-dimensional configuration of the treatments between mesocosm systems. Separate modified ANOSIM tests confirm that there were no significant differences in community structure between treatments or between mesocosm systems (Table 2). The null hypothesis, that there are no significant effects of different qualities and quantities of dosed organic matter on the meiobenthic nematode community structure, cannot be rejected. In a standard ANOSIM there were significant differences between day 0 and week 16 samples for both untransformed and root-transformed data ($p<0.01\%$). SIMPER analysis (Table 3) of the Westerschelde data indicates that this was due to a decline in the abundance of the dominating genera *Viscosia* and *Halalaimus* and the less dominant *Neochromadora* and *Anaplostoma* and increases in the abundance of *Theristus* and *Geomonhystera*. In the Gironde samples *Ptycholaimellus*, *Metachromadora* and *Chromadorella* had a much lower abundance at the end of the experiment (Table 3).

There were no significant differences between treatments of any of the univariate measures of community structure in either the Westerschelde or Gironde samples at the end of the experiment ($p>5\%$; Figs 5 and 6). There was a significant reduction in abundance between day 0 and final samples but also a significant increase in evenness for both Westerschelde and Gironde samples. For the Westerschelde Shannon Wiener diversity was significantly higher at the beginning of the experiment but for the Gironde samples it was higher at the end ($p<5\%$; Figs 5 and 6). Final Gironde species richness (Margalef's d) was significantly higher than in the day 0 Gironde samples ($p<5\%$) but there was no significant difference in the number of species ($p>5\%$). In the Westerschelde samples there was no significant difference ($p>5\%$) in either species richness or number of species between the beginning and end of the experiment.

Table 3. Comparison of average nematode species abundance ($\bar{y}$) in each core between start (day 0 – 4 replicates) and end (week 16 – 64 experimental samples used as replicates) of the experiments. Species are listed in order of their average contribution ($\bar{\delta}_i$) to the average Bray Curtis dissimilarity ($\bar{\delta}$) between start and end samples, with a cut-off when the cumulative per cent contribution ($\Sigma\bar{\delta}_i\%$) to $\bar{\delta}$, reaches 80%.

Westerschelde - untransformed data
average dissimilarity between day 0 and week 16, $\bar{\delta}$ = 48.7

Species	$\bar{y}_{day\ 0}$	$\bar{y}_{week16}$	$\bar{\delta}_i$	$\Sigma\bar{\delta}_i\%$
Halalaimus sp.	979	629	10.0	20.6
Viscosia sp.	689	203	9.7	40.5
Theristus sp.	148	492	7.3	55.5
Neochromadora sp.	197	39	3.3	62.3
Tripyloides sp.	179	63	3.2	68.8
Thalassoalaimus sp.	277	124	3.1	75.2
Anaplostoma sp.	122	15	2.3	79.9

Westerschelde $\surd\surd$ - transformed data
average dissimilarity between day 0 and week 16, $\bar{\delta}$ = 34.1

Species	$\bar{y}_{day\ 0}$	$\bar{y}_{week16}$	$\bar{\delta}_i$	$\Sigma\bar{\delta}_i\%$
Geomonhystera sp.	0	47	2.7	8.0
Anaplostoma sp.	122	15	2.3	14.8
Neochromadora sp.	197	39	2.2	21.2
Daptonema sp. 1	28	32	1.7	26.1
Tripyloides sp.	179	63	1.6	30.9
Viscosia sp.	689	203	1.6	35.5
Hypodontolaimus balticus	31	1	1.6	40.1
Innocuonema sp.	28	6	1.5	44.5
Antomicron sp.	38	17	1.4	48.8
Sphaerolaimus sp.	31	16	1.4	52.8
Calyptronema sp.	21	6	1.4	56.9
Theristus sp.	148	492	1.3	60.8
Monhystera sp.	11	11	1.3	64.6
Sabatieria sp.	41	36	1.2	68.0
Chromadora sp.	27	5	1.1	71.3
Diplolaimella sp.	0	12	1.0	74.4
Halalaimus sp.	979	629	1.0	77.2
Thalassoalaimus sp.	277	124	0.9	79.7

Table 3. (Continued).

Gironde - untransformed data
average dissimilarity between day 0 and week 16, $\bar{\delta}$ = 82.2

Species	$\bar{y}_{day\ 0}$	$\bar{y}_{week16}$	$\bar{\delta}_i$	$\Sigma\bar{\delta}_i\%$
Ptycholaimellus sp.	8975	35	41.2	50.1
Metachromadora sp.	2330	172	10.4	62.8
Chromadorella sp.	1306	0	5.6	69.6
Halalaimus gracilis	1343	302	4.7	75.2
Anaplostoma sp.	847	36	3.8	79.8

Gironde $\surd\surd$ - untransformed data
average dissimilarity between day 0 and week 16, $\bar{\delta}$ = 51.5

Species	$\bar{y}_{day\ 0}$	$\bar{y}_{week16}$	$\bar{\delta}_i$	$\Sigma\bar{\delta}_i\%$
Ptycholaimellus sp.	8975	35	5.3	10.3
Chromadorella sp.	1306	0	3.7	17.5
Spilophorella sp.	309	0	2.8	23.0
Daptonema oxycerca	354	1	2.7	28.2
Anaplostoma sp.	847	36	2.6	33.2
Metachromadora sp.	2330	172	2.4	37.8
Oxystomina sp. Y	0	150	2.3	42.2
Axonolaimus sp.	245	20	1.8	45.8
Campylaimus sp.	124	0	1.7	49.1
Paracanthonchus sp.	61	1	1.5	52.0
Halalaimus gracilis	1343	302	1.4	54.7
Metalinhomoeus sp.	84	417	1.3	57.2
Aegialoalaimus sp.	127	28	1.3	59.7
Terschellingia longicaudata	556	337	1.3	62.1
Cobbia sp.	0	37	1.3	64.6
Terschellingia communis	87	24	1.1	66.8
Leptolaimus sp.	51	56	1.1	68.9
Calyptronema sp.	99	25	1.1	71.1
Praecanthonchus sp.	39	8	1.0	73.1
Microlaimus sp.	36	15	1.0	75.1
Paramonhystera sp.	0	23	1.0	77.0
Diplolaimella sp.	0	21	1.0	78.9
Daptonema procerus	0	19	0.9	80.7

Discussion

There were 20–40 and 60–75% reductions in meiofaunal abundance for Westerschelde and Gironde samples respectively from the beginning of the experiment but no reduction in the number of species. However, the abundance and diversity of the nematode community remained sufficiently high to give reasonable power to the statistical analysis of the data to determine if there were any treatment effects. The data analysis suggests that the types of organic matter and doses administered during this experiment had no effect on nematode community structure.

The Westerschelde nematode assemblages had a much lower diversity and abundance than those from the Gironde which we feel is probably a reflection of

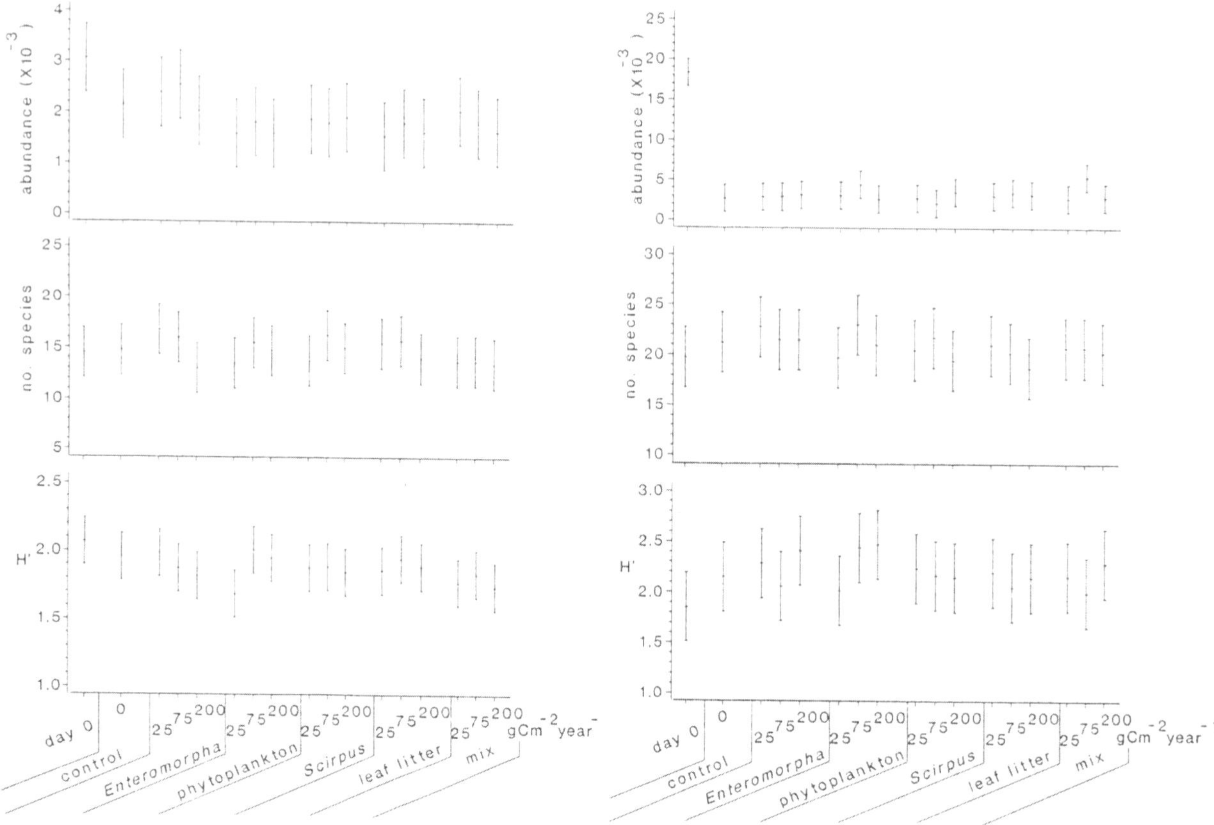

Fig. 5. Means and 95% pooled confidence intervals of univariate measures (abundance, number of species, Shannon-Wiener diversity) of Westerschelde nematode community structure at the beginning of the experiment (day 0) and after 16 weeks in the experimental mesocosms.

Fig. 6. Means and 95% pooled confidence intervals of univariate measures (abundance, number of species, Shannon-Wiener diversity) of Gironde nematode community structure at the beginning of the experiment (day 0) after 16 weeks in the experimental mesocosms.

the poor environmental conditions encountered in the former estuary. Soetart *et al.* (in press) also found higher abundances in the Gironde than the Westerschelde but in contrast with our results they found higher diversity in the Westerschelde than the Gironde. Their Westerschelde samples were taken in April 1990 and their Gironde samples in April 1992 whilst our field samples were collected in June of 1991 and 1992 respectively and the differences may be a reflection of these differences in season and sampling times. The Westerschelde nematode assemblages were dominated by non-specialist feeding genera, particularly *Halalaimus* but also *Thalassoalaimus*, *Leptolaimus*, *Tripyloides* and *Theristus* and the predator *Viscosia* with only a few diatom or epigrowth feeders. Because of the lower diversity of the original species pool in the Westerschelde experiment we did not expect the nematode assemblage to be able to respond significantly to the range of foods provided as treatments in the mesocosm and this might explain the observed lack of differences between treatments. In contrast, we hypothesised that the more diverse Gironde fauna would be able to adapt to the different food treatments so that a different response would be observed. It was expected that communities with different proportions of species reflecting feeding preferences would develop in the treatments. The experiment was run during the summer months, a time of peak reproduction, so that over the four months of the experiment there was adequate time for the community changes in response to food to occur.

The Gironde samples were initially highly dominated by chromadorids, particularly *Ptycholaimellus* and there were also large numbers of *Metachromadora*. These species were present only in much reduced numbers at the end of the experiment and were not dominant. They are classified as selective diatom/epigrowth feeders. They all have at least one tooth used either for puncturing diatoms and then sucking out the contents or for scraping diatoms and bacteria off the surface of sediment particles (Wieser, 1953; Jensen, 1987). The proportions of the other three feeding groups increased within the experimental communities i.e. non-selective deposit feeders (with large unspecialised buccal cavities, e.g. *Sabatieria, Metalinhomoeus, Theristus*) selective deposit feeders (with very tiny or no buccal cavities e.g. *Halalaimus, Oxystomina,* and *Terschellingia*) and predator omnivores (with large and heavily armoured or cuticularised buccal cavities e.g. *Viscosia,* and *Sphaerolaimus*). For species in these three groups abundances remained approximately the same or decreased slightly although some increased slightly (e.g. *Sabatieria, Metalinhomoeus, Oxystomina*). This would suggest that the food utilised by the diatom/epigrowth feeders was not available in sufficient quantity for them to survive. The mesocosms were illuminated but the light intensity was probably insufficient for vigorous diatom growth. The phytoplankton mix did not include any diatoms. It may be that this nematode feeding group requires fresh material for adequate nutrition. In addition there is some evidence that this group is quite selective in its feeding requirements (Jensen, 1986, 1987 and references therein) so they may have been food limited during the experiments.

If the diatom/epigrowth feeders were food limited the evidence in contrast for the other groups suggests that they were not food limited, but indeed that food was so abundant that the quality and quantity of the extra organic matter supplied was superfluous to requirement. The large fall in meiofaunal abundance in the Gironde experiment was due to the huge reduction in chromadorids during the experiment. The reduction in abundance of the remaining species in the Gironde samples was comparatively small and was not species specific and this was also the case for the Westerschelde samples. This smaller reduction probably occurred as a result of disturbance effects causing stress during the initial phase of collecting samples and setting up the experiments and similar effects have been observed in other mesocosm experiments (Austen unpublished). The nematodes identified and enumerated at the end of the experiment spanned the full range of life stages from juveniles through to mature adults which suggests that reproduction was occurring during the course of the experiment.

With reference to both macrobenthos and meiobenthos it has been suggested that benthic deposit feeders and subsurface feeders will not show rapid responses to fresh inputs of food since they utilise a large pool of organic matter in the sediment which is constantly being recycled (Levinton, 1972; Rudnick, 1989; Josefson, 1985). However, Rudnick (1989) demonstrated that even in just the upper 0.5 cm of surface sediment the meiobenthos were divided into 2 groups which might be part of two distinct food webs, one which utilises fresh organic matter sources and one which utilises the older organic matter in the sediment. Constituents of these two groups were determined to major taxon level and according to size range. These meiobenthic groups were divided by size and major taxa rather than by species and genera e.g. larger nematodes retained on 200 μm mesh sieves utilised older organic matter whilst small nematodes (retained on 100 μm mesh sieves) utilised fresh phytodetritus. Rudnick suggested that these two groups were divided by spatial depth in the sediment but our results suggest that trophic group may be of importance as well.

Other recent experimental work has also shown little effect of addition of food supply to meiobenthic communities. Flothmann & Werner (1992) stimulated microphytobenthos production in the field on an intertidal sand-flat but this had no effect on total meiobenthic abundance. They did not look at species composition so it is uncertain whether there was any specific response amongst the chromadorids. Gee *et al.* (1985) dosed offshore subtidal (42 m depth) meiobenthic communities in mesocosms at levels of 50 gC and 200 gC m^{-2} with powdered *Ascophyllum*; the harpacticoid copepods were considerably affected, particularly at the higher dose which went anoxic but the only effect on the nematodes was a slight reduction in abundance in the high dose treatment. In the short term, detrital food already present within estuarine sediments from the Westerschelde and the Gironde was sufficient both in terms of quantity and variety to fulfill the dietary requirements of most of the deposit feeding meiofaunal species found in these estuaries.

Acknowledgments

This work was carried out as part of the EEC funded MAST Joint European Estuaries Project (JEEP). We gratefully acknowledge the field assistance of Magda Vincx and her colleagues in the Zoology and Systematics Department at the University of Ghent, Peter Herman and Wim de Winter of CECE, Netherlands Institute of Ecology and Jacques Castel and Xabier Irigoien of the Station Biologie Marine in Arcachon.

We are grateful to Carlo Heip for his constructive criticism of the manuscript.

References

Austen, M. C., 1989. Factors affecting estuarine meiobenthic assemblage structure: a multifactorial microcosm experiment. J. exp. mar. Biol. Ecol. 130: 167–187.

Austen, M. C. & R. M. Warwick, 1989. Comparison of univariate and multivariate aspects of estuarine meiobenthic community structure. Estuar. coast. Shelf Sci. 29: 23–42.

Clarke, K. R., 1993. Non-parametric multivariate analyses of changes in community structure. Aust. J. Ecol. 18: 117–143.

Clarke, K. R. & R. M. Warwick, 1994. Similarity-based testing for community pattern: the 2-way layout with no replication. Mar. Biol. 118: 167–176.

Flothmann, S. & I. Werner, 1992. Experimental eutrophication on an intertidal sandflat: effects on microphytobenthos, meio- and macrofauna. Proceedings of the 25th European Marine Biology Symposium: 93–100.

Gee, J. M., R. M. Warwick, M. Schaaning, J. A. Berge & W. G. Ambrose Jr., 1985. Effects of organic enrichment on meiofaunal abundance and community structure in sublittoral soft sediments. J. exp. mar. Biol. Ecol. 91: 247–262.

Heip, C., 1989. The ecology of the estuaries of Rhine, Meuse and Scheldt in the Netherlands. In Ros, J. D. (ed.) Topics in marine biology. Scient. Mar. 53: 457–463.

Jensen, P., 1986. Nematode fauna in the sulphide-rich brine seep and adjacent bottoms of the East Flower Garden, NW Gulf of Mexico. Mar. Biol. 92: 489–503.

Jensen, P., 1987. Feeding ecology of free-living aquatic nematodes. Mar. Ecol. Prog. Ser. 35: 187–196.

Josefson, A. B., 1985. Distribution of diversity and functional groups of marine benthic fauna in the Skagerrak (eastern North Sea) – can larval availability affect diversity? Sarsia 70: 229–249.

Levinton, J., 1972. Stability and trophic structure in deposit-feeding and suspension-feeding communities. Am. Nat. 106: 472–486.

Pfannkuche, O. & H. Thiel, 1988. Sample processing. In Higgins, R. P. & H. Thiel (eds), Introduction to the study of meiofauna. Smithsonian Institution Press, Washington D.C.: 134–145.

Rudnick, D. T., 1989. Time lags between the deposition and meiobenthic assimilation of phytodetritus. Mar. Ecol. Prog. Ser. 50: 231–240.

Soetaert, K., M. Vincx, J. Wittoeck & M. Tulkens, 1955. Meiobenthic distribution and nematode community structure in five European estuaries. Hydrobiologia 311 (Dev. Hydrobiol. 110): 185–206.

Warwick, R. M., 1984. Species size distributions in marine benthic communities. Oecologia 61: 32–41.

Warwick, R. M. & J. M. Gee, 1984. Community structure of estuarine meiobenthos. Mar. Ecol. Prog. Ser. 18: 97–111.

Wieser, W., 1953. Die Beziehung zwischen Mundhohlengestalt, Ernahrungsweise und Vorkommen bei freilebenden marinen Nematoden. Ark. Zool., 4: 439–484.

Hydrobiologia **311**: 185–206, 1995.
C. H. R. Heip & P. M. J. Herman (eds), Major Biological Processes in European Tidal Estuaries.

Meiobenthic distribution and nematode community structure in five European estuaries

Karline Soetaert*, Magda Vincx, Jan Wittoeck & Maio Tulkens
University of Ghent, Marine Biology Section, K. L. Ledeganckstraat 35, B-9000 Gent, Belgium
**Current address: Netherlands Institute of Ecology, Centre for Estuarine and Coastal Ecology, Vierstraat 28, NL-4401 EA Yerseke, The Netherlands*

Key words: Meiobenthos, nematoda, estuary

Abstract

Meiofauna from the intertidal zone of five European estuaries (Ems, Westerschelde, Somme, Gironde, Tagus) was investigated. Samples represented a cross section of various benthic habitats from near-freshwater to marine, from pure silts to fine-sandy bottoms. The meiobenthic community comprised everywhere a fauna strongly dominated by nematodes, with meiobenthic density increasing with increasing salinity. The Ems differed from the other estuaries due to the presence of a well developed community of Copepods, Gastrotrichs, large Ciliates and/or soft-shelled Foraminiferans in some sites. The Westerschelde stood out due to the near-absence of harpacticoid copepods and, as in the Tagus, the lower meiobenthic densities in the marine part of the estuary. For nematode community analysis, we also included data from the Tamar which were obtained from the literature (Warwick & Gee, 1984). This resulted in the enumeration of 220 species, belonging to 102 genera, each with a characteristic distribution along the salinity, sedimentary and latitudinal gradients. Using the multivariate technique CANOCO, a zonation along these different physicochemical determinants was observed as well although salinity and sediment characteristics (scale of hundreds of meters to kilometers) proved to be more important in explaining community structure than latitudinal differences (scale of hundreds of kilometers). Nematode diversity was nearly entirely determined on the genus level and was positively related to salinity. Deviations from this general trend in the Gironde and the Tamar were attributed to sedimentary characteristics or to low macrobenthic predation. The presence of a typical opportunistic colonizing nematode species *Pareurodiplogaster pararmatus* in the low-salinity region of the Gironde could indicate (organic?) pollution or disturbance of the intertidal mud-flats.

Introduction

Meiobenthic communities in European estuaries have only been studied in the U.K. (Capstick, 1959; Warwick, 1971; Warwick & Price, 1979; Warwick & Gee, 1984; Moore, 1987; Austen & Warwick, 1989), Germany (Gerlach, 1953; Riemann, 1966; Skoolmun & Gerlach, 1971), the Netherlands (Van Damme *et al.*, 1980; Bouwman, 1983; Smol *et al.*, 1994) and northern France (Gourbault, 1981). No data exist on more southern estuaries. Moreover, intercomparison between estuaries is complicated by taxonomic difficulties. Especially in the past, the chaotic taxonomy of *e.g.* the nematodes made this taxon only accessible to the specialist (Gerlach, 1980). Thanks to the publication of pictorial keys (Platt & Warwick, 1983; 1988) nematode identification is now much easier – at least to the genus level. However, the identity of many species remains problematic. If different estuaries are investigated by the same researcher, as in this study, it becomes more easy to distinguish within-species variability from between-species variability and, although determinations may not always be exact, this will introduce more consistency into the results.

This study is part of a general research program, which aims at the understanding of major biological processes in European tidal estuaries (MAST CEC project, JEEP92). As part of this programme, a base-

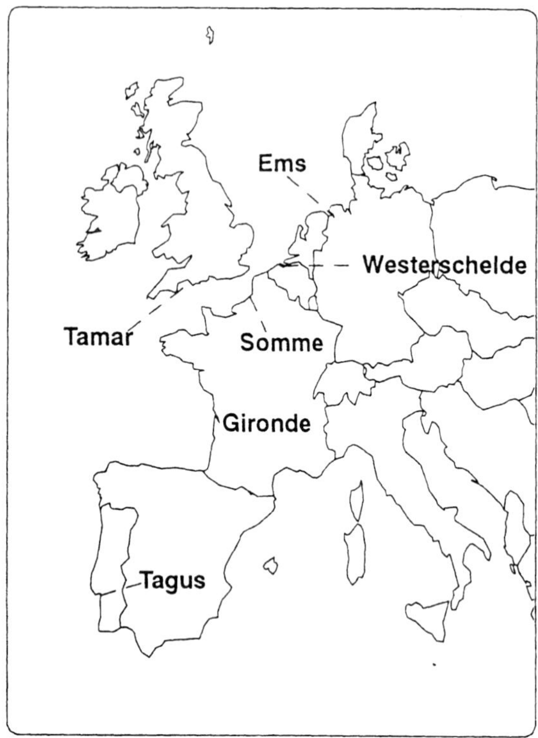

Fig. 1. Location of the various estuaries under study.

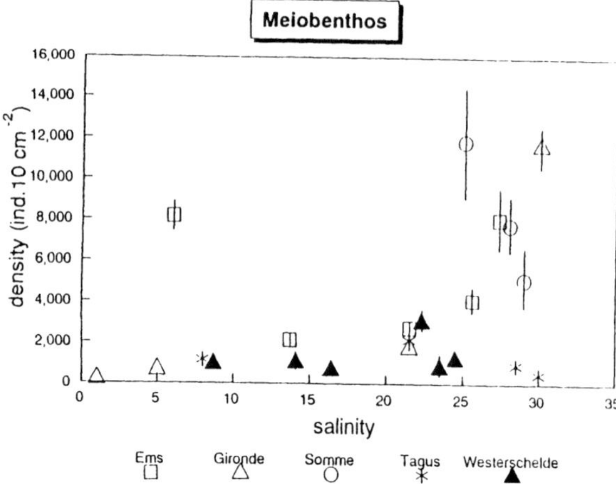

Fig. 2. Mean meiobenthos density and total range (ind 10 cm^{-2}) for the five European estuaries under study.

line study of the meiobenthos in five European estuaries was made, with emphasis on the nematode species composition. For this purpose, data on the Tamar estuary, as provided in Warwick & Gee (1984) were included. The distribution of the most important species and the influence of large-scale spatial structure on the community level are reported.

Material and methods

Samples were taken from intertidal areas of the Ems, the Tagus, the Gironde, the Somme (April 1992), and the Westerschelde (April 1990) along a salinity gradient (Fig. 1). A total of twenty five sites were sampled, and at least two replicates (10 cm^2) per station were taken with plastic cores for meiofauna and an additional one for sediment analysis. Samples from the Gironde were taken from the mean tidal level and subdivided into 0–1 and 1–5 cm slices. Samples from the Westerschelde were also vertically subdivided. However, information on the vertical distribution was not retained in this study. Vertical distribution of nematodes in the Westerschelde is discussed in Soetaert *et al.* (1994).

Salinity was measured in the water column. All sediment and meiofauna samples were treated in the same laboratory (Ghent). Sediment particle size distribution was determined using Coulter LS particle size analysis equipment.

The meiofauna was extracted using LUDOX TM as described in Heip *et al.* (1985). After colouring with Rose Bengal the meiofaunal taxa (i.e. those animals retained on a 38 μm sieve) of two replicates were enumerated after which at least 200 randomly chosen nematodes per replicate were mounted into slides for further identification. If total nematode density was less than 200 individuals, all were processed. Identification to genus – and in some cases species – level was done using the pictorial keys in Platt & Warwick (1983, 1988) and Bongers (1988). For many species further identification required consultation of the relevant literature.

The species-abundance matrix was analysed using the multivariate technique CANOCO (CANOnical COrrespondence analysis; Jongman *et al.*, 1987; Ter Braak, 1989). This method was used to relate the observed trends to the major environmental gradients. Diversity was measured as Hill's diversity numbers of order 0 (i.e. N0, the number of species present), 1 (i.e. N1, the exponential of the Shannon Wiener index), 2 and ∞ (i.e. $N\infty$, the reciprocal of the dominance of the most common species) as suggested by Heip *et al.* (1988). This spectrum of diversity indices gives a clear representation of the influence of rare (lower order indices) and more common (higher order indices) species on diversity. In order to correct for the size-dependency of diversity indices (Soetaert &

Heip, 1990) and as a means of comparing with other studies, the indices were calibrated on 100 randomly chosen individuals as described in Soetaert & Heip (1990). The life-history traits of the communities were summarized into the 'maturity index' as described in Bongers (1990) and Bongers *et al.* (1991). In short, a higher value of this index indicates a higher degree of persistence, a lower value a more opportunistic community.

The different estuarine environments

The estuaries under investigation are situated along a north-south gradient (Fig. 1). The climate changes from temperate (mild winters, relatively cool summers) in the North of the Netherlands, with a well pronounced variation in temperature and a relatively smooth precipitation curve, to warmer in Portugal, where precipitation is very variable.

The Ems estuary (the Netherlands, Germany) connects the north-eastern part of the Netherlands and the north-western part of Germany with the North Sea. It is an important shipping route with three harbours along its side. Sediments are dredged from the estuarine channels and dumped in other parts (de Jonge, 1992). The major freshwater input to the Ems estuary, the river Ems and the Westerwoldsche AA are also important sources of nutrient enrichment (de Jonge & Essink, 1992). In its upper reaches, the Ems estuary has a vast intertidal mudflat: the Dollard. The total average annual discharge varies from 80 to 180 m^3 s^{-1} with seasonal variation; tidal currents are moderate.

The Westerschelde estuary (the Netherlands) provides the link from the harbour of Antwerp towards the North Sea. It has busy shipping lanes and concurrent dredging and dumping activities of polluted sediments. Large cities on the river banks, and the high degree of industrialization stand for a high input of organics, inorganics and nutrient wastes. The high load of organic matter in the turbidity maximum zone causes strong anoxic conditions in summer (Heip, 1989). The average freshwater flow is about 100 m^3 s^{-1} and it fluctuates in a seasonal manner.

The Somme (North of France) is an embayment of the Somme and Maye river system and opens to the west in the English Channel; the northern half of the bay is subjected to strong waves from the Channel (Ducrotoy & Sylvand, 1989). The Somme is characterized by a low freshwater output (30 m^3 s^{-1}) and is thus primarily under marine influence. It consists of large intertidal areas with high biomasses of cockles (Rybarczyk *et al.*, 1992).

The Gironde estuary (South of France) is fed by the rivers Garonne and Dordogne and has the city of Bordeaux along its banks. The mean river discharge is high, varying from 600 to 1000 m^3 s^{-1} and is subjected to seasonal variations. It is a highly turbid estuary but with a distinct turbidity maximum (Castel & Feurtet, 1989).

The Tagus (Portugal, Spain) is the largest river of the Iberic peninsula and is characterized by an extensive intertidal area (>30%) (Moreira *et al.*, 1992). It is flanked by the city of Lisbon which disposes its untreated wastes into the estuary. The port of Lisbon is also an important source of industrial waste in the marine part of the estuary (Gaudencio *et al.*, 1991). Another centre of industrialization causes heavy pollution along a narrow channel in the brackish part (R. Neves, pers. comm.). The freshwater outflow of the Tagus is very variable, both annually and interannually; it varies from 30 to 18 000 m^3 s^{-1} and the salinity varies concurrently.

The Tamar estuary is situated in South-West England and is characterized by extensive tidal flats and mud banks. River discharge varies in a seasonal way from about 2 to 150 m^3 s^{-1} (Morris *et al.*, 1985). A turbidity maximum in the upper estuary acts as a trap for particles of marine, estuarine and fluviatile origin (Bale *et al.*, 1985).

Results

Sediments

The majority of intertidal stations had fine-grained sediments although some stations in the three most northerly estuaries were more coarse. Grain characteristics, salinity and name of the stations are presented in Table 1.

Meiobenthos

Meiobenthic taxa observed included Nematoda, Copepoda, Gastrotricha, Plathelminthes, soft-shelled Foraminifera, Ciliata, Polychaeta, Oligochaeta, Ostracoda, Halacarida, Cnidaria, Priapulida and Tardigrada.

Meiobenthos densities (Fig. 2) varied from 130 to 14 500 ind 10 cm^{-2} and were highest in the Somme and the marine part of the Gironde and Ems. Low meioben-

Table 1. Salinity, silt content and median grain size of the intertidal samples.

Estuary	Station	Salinity (‰)	Silt (%)	Median (μm)
Ems	E 6	6	78	18
Ems	E 5	14	16	124
Ems	E 4	22	23	107
Ems	E 2	26	45	73
Ems	E 1	27	6	136
Westerschelde	WS 1	9	0	187
Westerschelde	WS 22	14	6	133
Westerschelde	WS 32	16	1	238
Westerschelde	WS 42	22	3	167
Westerschelde	WS 61	28	40	77
Westerschelde	WS 53	29	84	15
Somme	HH	25	17	165
Somme	LC	28	17	169
Somme	LM	29	15	181
Tamar	Clifton (*)	9	94	11
Tamar	Neal point (*)	23	94	15
Tamar	West mud (*)	31	87	34
Gironde	Lamarque	2	96	9
Gironde	St Estephe	5	85	23
Gironde	Richard	22	58	60
Gironde	Le Verdon	30	92	9
Tagus	Cala do Norte	8	95	9
Tagus	Banco do Ladeiro	22	36	148
Tagus	Banco dos Cavalos	30	77	11
Tagus	Banco do Destroi	29	72	13

(*) = data from Warwick and Gee (1984).

thic densities were observed all along the transect in the Westerschelde and the Tagus and in the brackish part of the Gironde. In general, meiobenthic density increased with increasing salinity (Fig. 2).

Nematodes were always the most abundant taxon (Fig. 3) and their dominance was in the order of 81 to 99%, except for the Ems. The Ems seemed to be the only estuary where other groups were of some significance: the most upstream station had fairly large numbers of soft-shelled Foraminiferans and Copepods, the most marine station had a large Gastrotrich community (30% of all meiobenthic animals), while Turbellarians, Ciliates and soft-shelled Foraminiferans were numerous in the two most marine stations (Fig. 3).

Apart from the Ems, turbellarians and large ciliates were also present in low quantities (at most 20, resp. 75 ind 10 cm^{-2}) in the Westerschelde, while virtually absent in other estuaries. Gastrotrichs were observed in high densities in the Ems and the Somme only, while soft-shelled Foraminiferans were also numerous in the most marine station of the Gironde. Harpacticoids were present in low quantities in most estuaries; in the Westerschelde (maximum 6 ind 10 cm^{-2}) and the Somme (7 ind 10 cm^{-2}) harpacticoid densities were very low (Fig. 3).

Nematodes

A total of 220 species, belonging to 102 genera and 35 families, were recorded in the intertidal of the investigated estuaries. The majority of these species were confined to only one estuary, some were found in two or three estuaries (Appendix). Only three species were common to all estuaries: *Dichromadora cephalata*, *Halalaimus gracilis* and *Viscosia viscosa*. Another thirteen species were observed in all but one estuary: *Daptonema normandicum*, *Chromadorita tentabunda*, *Anoplostoma viviparum*, *Calyptronema maxweberi*, *Metalinhomoeus aff biformis*, *Daptonema setosa*, *Sabatieria punctata group*, *Dichromadora geophila*, *Ptycholaimellus ponticus*, *Praeacanthonchus punctatus*, *Axonolaimus paraspinosus*, *Metachromadora remanei* and *Chromadora macrolaima*.

As salinity and sediment characteristics were fairly evenly distributed (Table 1), the distribution characteristics of the most important species and genera (defined as making up at least 15% of the total community in at least one station and observed in more than two stations) were computed with respect to their salinity and sediment grain size preferences. The median distribution (50%) and the 10% and 90% occurrences as well as the total range along which the species were observed were calculated and represented as box-whisker plots (Figs 4–5). By comparing the specific distribution with the repartition of all nematodes combined (total DENSITY), the degree of selectivity for any one parameter can be evaluated. Some species exhibit broad ecological tolerances to both factors and have a distribution along the sedimentary and salinity axis that is not noteworthy different from the total nematode density: *Viscosia viscosa* and *Dichromadora cephalata*. These two species were also observed in all estuaries (see above). Most species or genera have more clear preferences.

A Canonical correspondence analysis based on species, genera or families yielded – after permutation of the X-axis in the genus and family level – very similar results. The CANOCO plot of species (or genera) and environmental variables (Fig. 6) yields

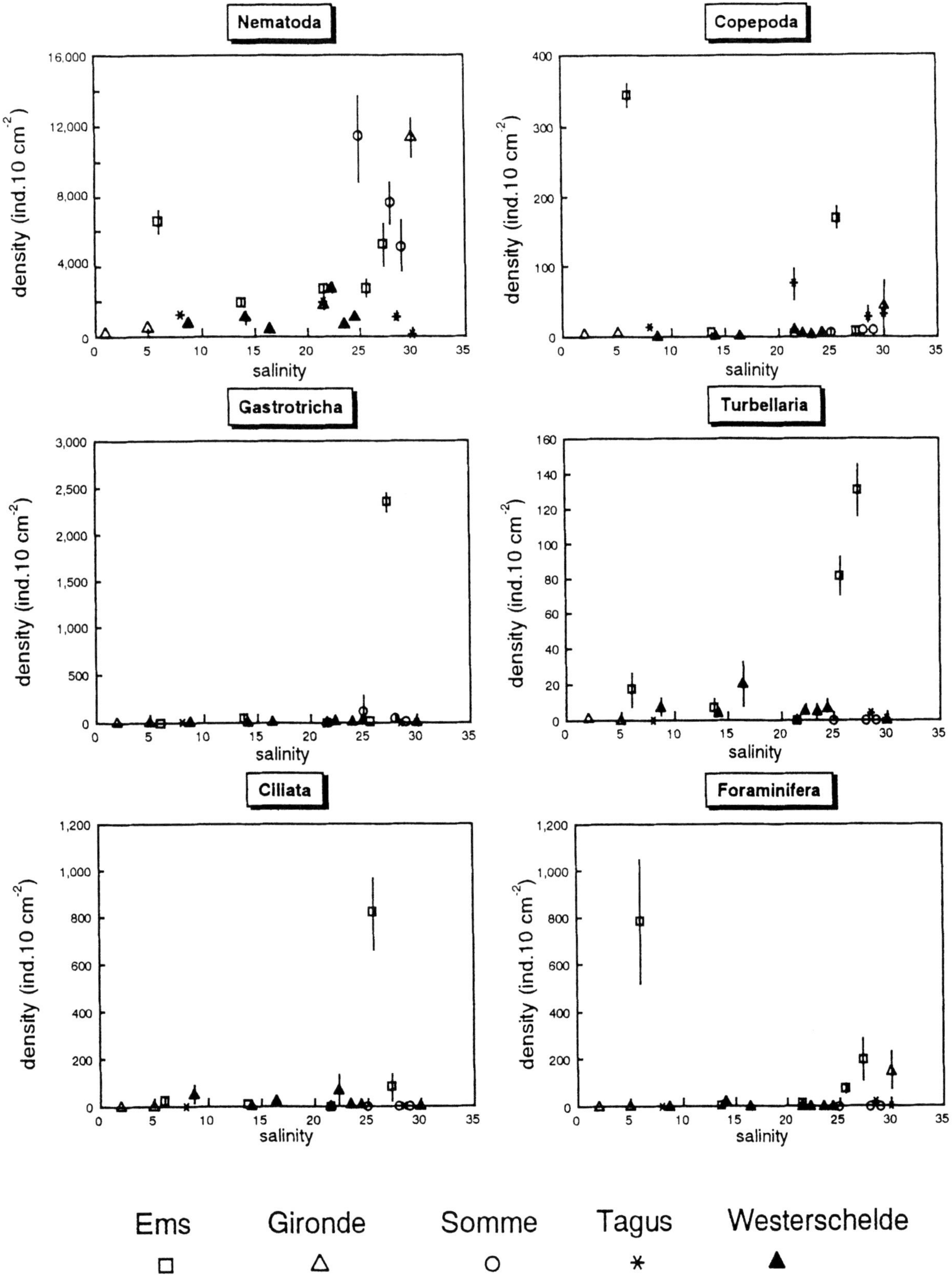

Fig. 3. Mean densities and total range (ind 10 cm^{-2}) of nematodes, copepods, gastrotrichs, turbellarians, ciliates and softshelled foraminiferans in the five European estuaries under study.

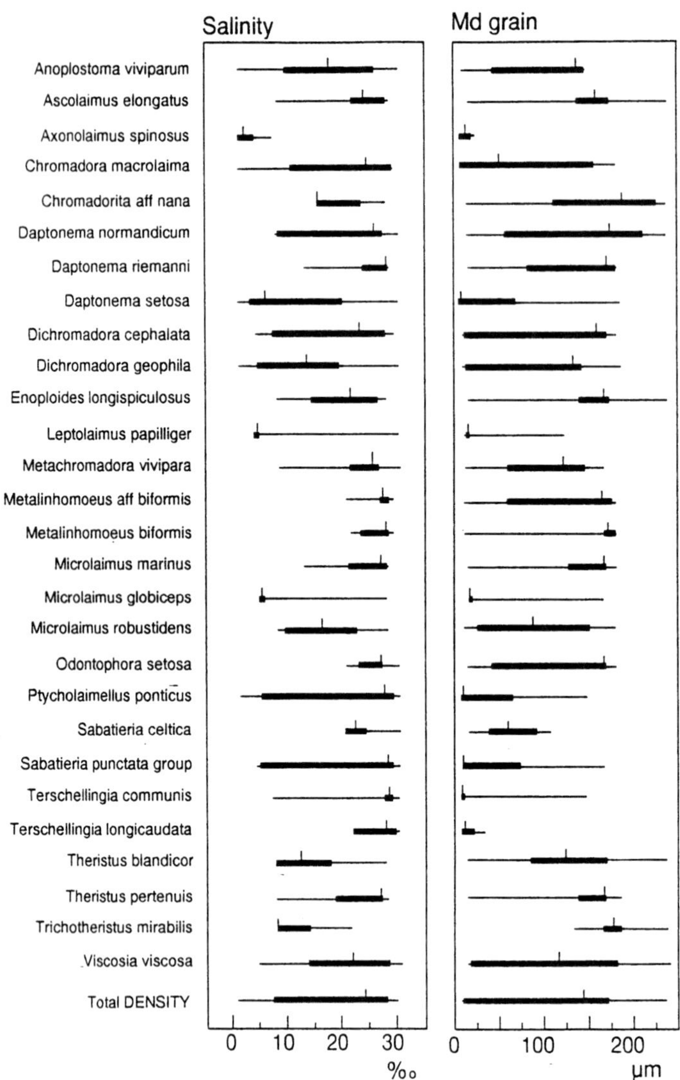

Fig. 4. Distributional characteristics of the most important species along the salinity and sediment grain gradient (all studied estuaries combined). Indicated are the total range (horizontal line), the median occurrence (vertical dash) and the 10 to 90% occurrence (horizontal bar).

similar information with respect to the figures (4–5) and (appendix): specific preferences for any variable can be 'appreciated' by orthogonally projecting the species position on the environmental axis. The positions of the species (Fig. 6) can aid in the interpretation of the station position in the same graph as they are weighted averages of the species positions. The length of the environmental arrows indicates their relative importance in explaining community structure. The relative position of the arrows reflects the relationship of the environmental variables, with orthogonality indicating no correlation, paralellism indicating positive (same direction) or negative (opposite) correlation. Thus salinity and grain characteristics (order of kilometres) were about equally important and independent factors, while latitudinal differences (correlated with grain size), although on the scale of hundreds to thousands of kilometres were much less pronounced.

The positions of the stations in the CANOCO plot with the same axes are represented in Fig. 7 for the species level. Results were very similar for genus or even family level (not depicted). There was a great overlap between stations of different estuaries. Within estuaries, the community gradually changed along the salinity gradient (arrows in Fig. 7) rather than along the sediment gradient (not depicted). This could indicate that the predictive ability of the type of sediment was overemphasized by the analysis, due to its relationship with latitude.

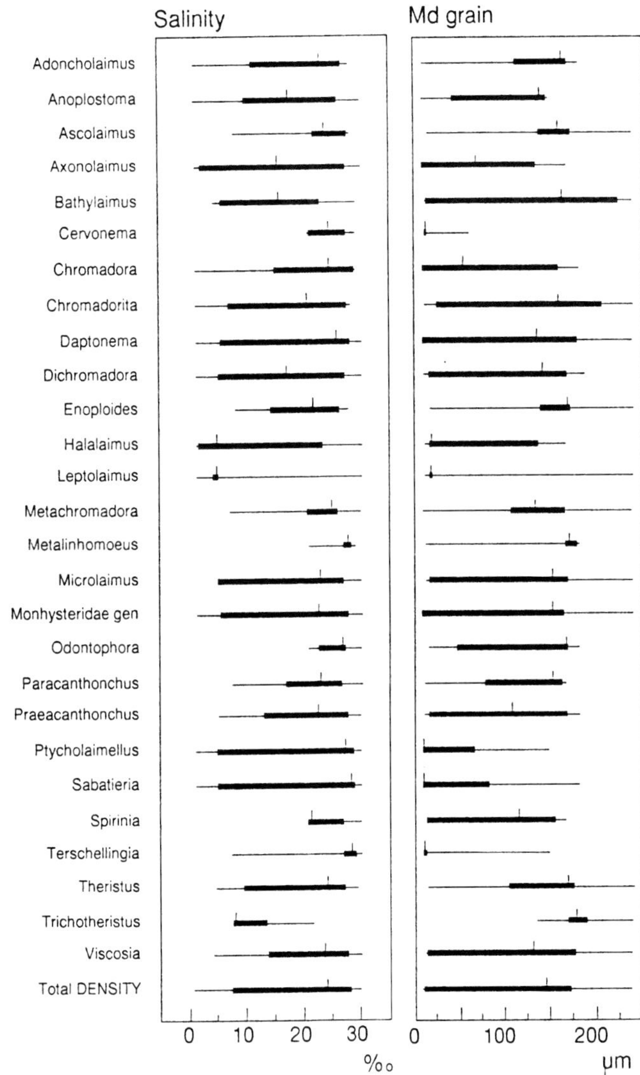

Fig. 5. Distributional characteristics of the most important genera along the salinity and sediment grain gradient. For an explanation, we refer to Fig. 4.

Nematode distributions (Figs 4–6, appendix)

The most important nematode families in terms of total density were the Xyalidae, the Axonolaimidae, the Comesomatidae, the Microlaimidae, the Linhomoeidae and the Chromadoridae.

The Comesomatidae were predominantly observed in silty sediments and preferred to some extent a more saline environment. There are two important members of this group. Species which have been described in the literature as *Sabatieria punctata*, *S. pulchra* and *S. breviseta* were present in this study both in their typical morphology as well as in all kinds of intermediate forms. This makes their systematic identity rather problematic and they were consequently grouped as one species which we called the '*Sabatieria punctata group*'. They were more strongly bound to silty sediments than the other member of the genus, *S. celtica*. Whereas the genus *Sabatieria* seemed to be more indifferent with respect to salinity and latitude, the genus *Cervonema* is a clear representative of the more saline and silty bottoms of estuaries at the lower latitudes (Gironde, Tagus).

The Xyalidae were a very abundant family encompassing three important genera: *Daptonema*, *Theristus* and *Trichotheristus*. *Trichotheristus mirabilis* and *Theristus blandicor* were only present in the most northerly estuaries (WS, resp. Ems and WS) and both prefer brackish waters. *Daptonema setosa* was found in the brackish part of all but one estuaries in silty sed-

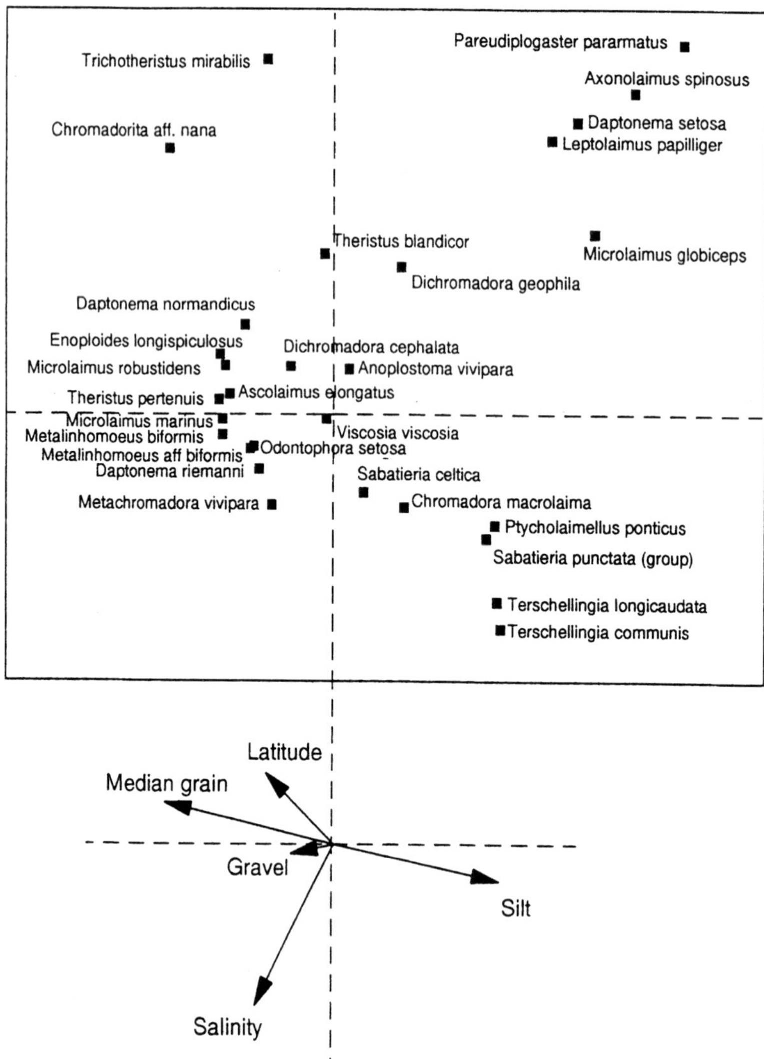

Fig. 6. Upper part: position of the most important species in two-dimensional CANOCO space (first two axes). Lower part: environmental arrows in the same CANOCO space.

iments (in the Somme the brackish area was not sampled). *Daptonema riemanni* was a common species in more marine and sandy bottoms of the northern estuaries. *Daptonema normandicum* and *Theristus pertenuis* are more northerly species with an affinity for coarser-grained and saline sediments.

The Linhomoeidae were represented by *Terschellingia longicaudata* and *Terschellingia communus* which co-occur in the silty sediments of the marine part in the Tagus, the Gironde and the Tamar and by two closely related species of the genus *Metalinhomoeus* (called *Metalinhomoeus typicus* and *M. biformis* in Bouwman, 1981, but referred to as *Metalinhomoeus biformis* resp. *M. aff biformis* in this study), which had greater preference for sandy sediments in the marine part.

The Chromadoridae were a very diverse group with important species as *Chromadora macrolaima* and *Ptycholaimellus ponticus* which are more common in silty sediments and the genus *Chromadorita* (*C. aff nana*, *C. tentabunda*) which was more frequently observed in coarser sediments and/or higher latitudes. Within the genus *Dichromadora*, we note the predominance of *D. geophila* in brackish water and the more neutral position of *D. cephalata*.

The Axonolaimidae encompassed amongst others *Ascolaimus elongatus* and *Odontophora setosa*, two northerly species which had an affinity for somewhat coarser sediments in the marine area. The

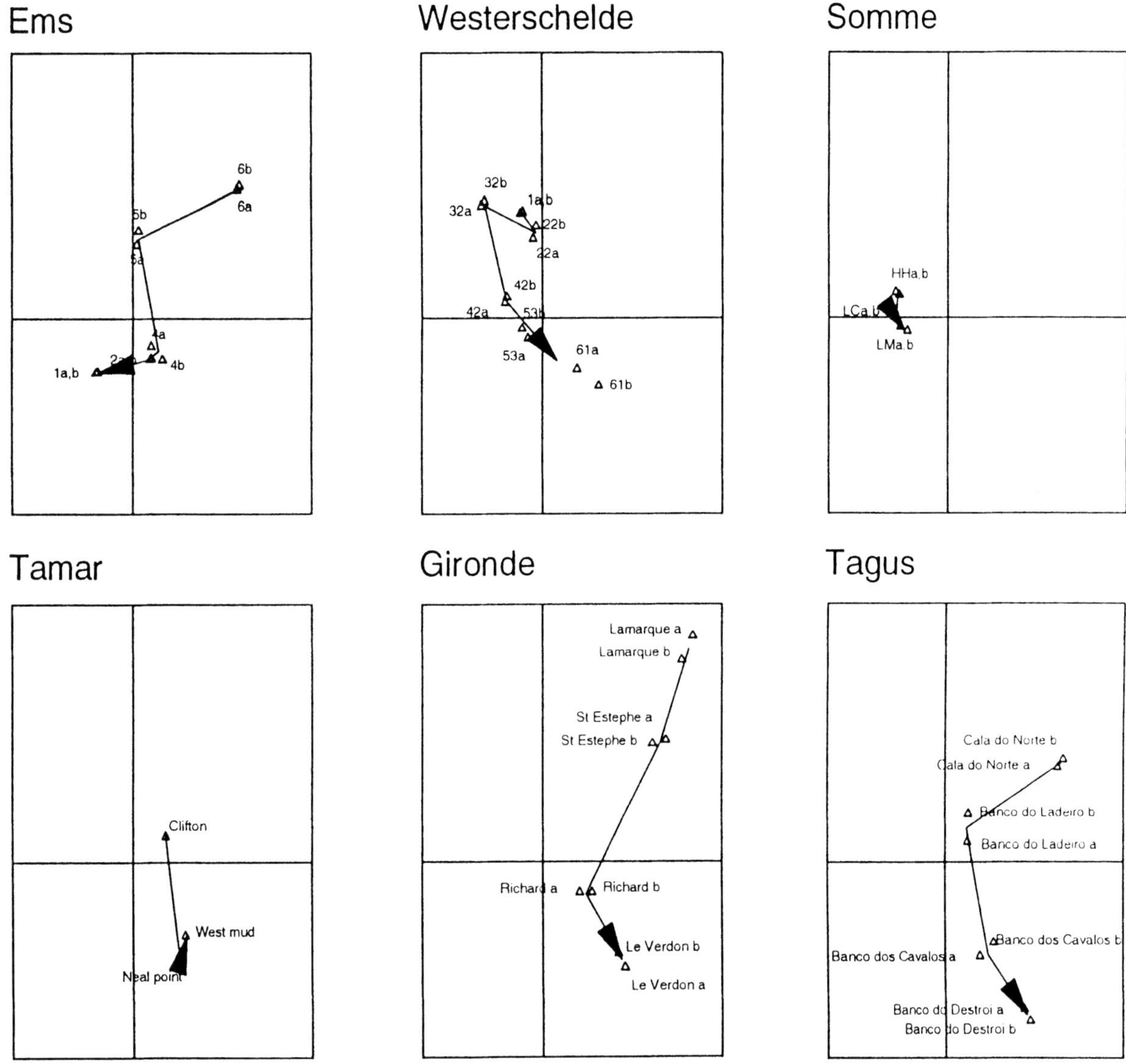

Fig. 7. Station plot of the various European estuaries in two- dimensional CANOCO space (first two axes). For convenience, the stations of one estuary were depicted separately. All plots have the same scale. The salinity gradient is depicted by the arrows.

genus *Axonolaimus* was represented by *A. spinosus*, a southern species in silty brackish-water bottoms, and *A. paraspinosus* having a broader distribution.

The Desmodoridae had as most common representatives *Metachromadora vivipara* and *Spirinia parasitifera*, which are marine species.

Amongst the Microlaimidae *Microlaimus marinus* and *M. robustidens* were more common in coarser sediment, while *M. globiceps* peaked clearly in the more freshwater part of the estuaries.

The Neodiplogasteridae – with their only representative *Pareudiplogaster pararmatus* – were only observed in a freshwater station of the Gironde, together with *Tobrilus diversipapillatus*.

Amongst the important species we also note *Anoplostoma viviparum* and *Viscosia viscosa* which had a very broad distributional range.

Nematode diversity and life-history traits

When randomly drawing 100 nematodes from the samples, a total of 8 to 29 species was found (N0, Fig. 8). Hill's diversity number of the first order (N1) varied from 3 to 22 equivalent species (Fig. 8), while N2 varied from 2 to 21, $N\infty$ from 1 to 9.

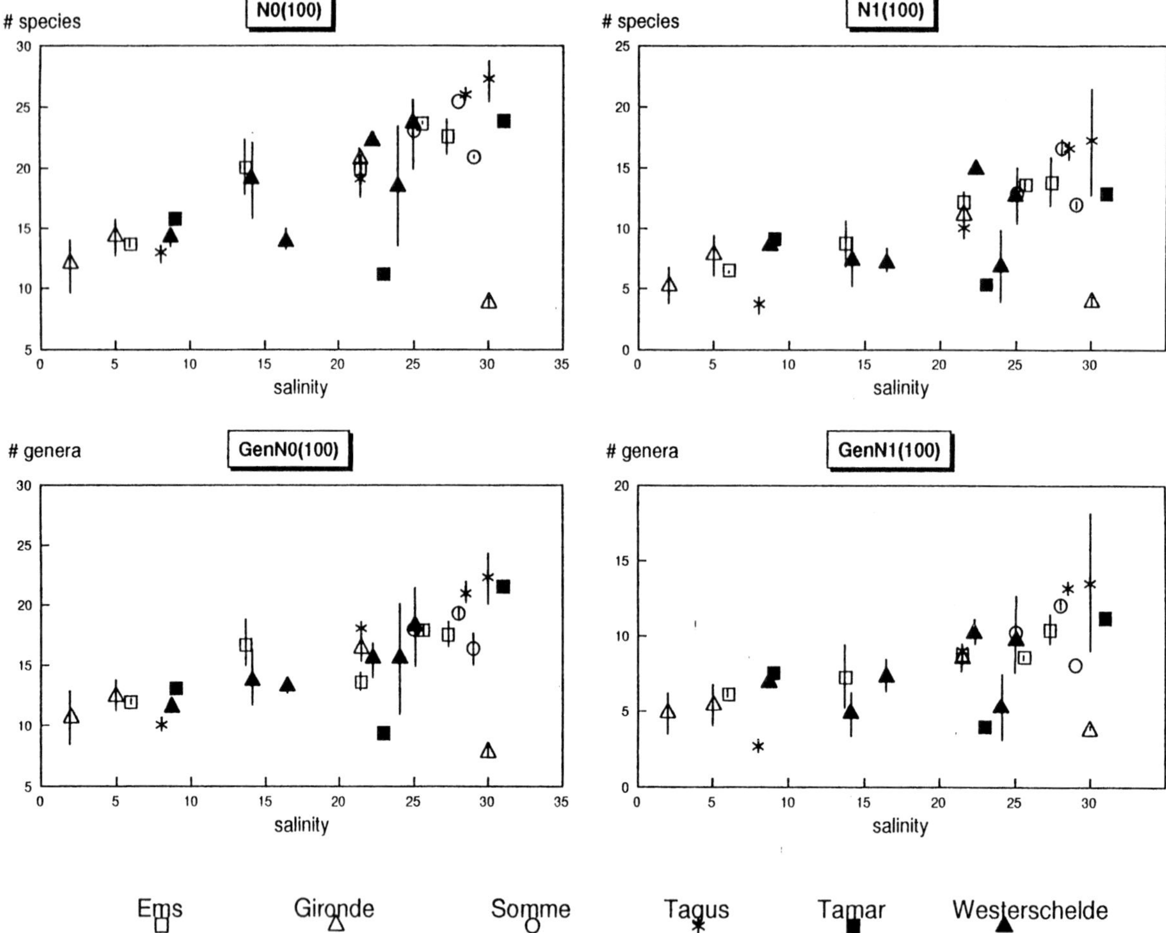

Fig. 8. Hill's diversity numbers of order 1 and 2, calibrated on 100 randomly picked nematodes for the six European estuaries, along the salinity gradient. Upper half: diversity on the species level, lower half: diversity on the genus level.

All estuaries combined, diversity increased with increasing salinity and there was no relationship with granulometry nor with latitude. No estuary could be clearly separated from the others on the basis of diversity although the Gironde and the Tamar had one station that scored relatively low.

Nematode diversity was nearly entirely expressed at the generic level: the diversity at the genus level was about 80 percent of total diversity (all Hill's numbers) and also increased with increasing salinity. On average only 1.2 species per genus were present in any one station and the vast majority were monospecific genera, even on the scale of the entire estuary. Considering the entire study area, many genera were represented by very few species (Addendum). Genera which are successful, both in number of species and density, are *Dichromadora* (5 species), *Microlaimus* (7 species), *Theristus* (9 species) and *Daptonema* (14 species).

The maturity index revealed no real trend (Fig. 9), but the fresh-water station in the Gironde had a much lower index than the other stations, which should indicate a low quality environment. This low value is caused by the relatively high concentrations of *Pareudiplogaster pararmatus* in this station, a species which has a great colonizing ability (Bongers, 1990).

Discussion

The estuarine environment shows large fluctuations of e.g. salinity, temperature and oxygen over different time scales, from a tidal cycle to a year (and longer). Meiobenthic organisms, being bound to the sediment, have to adapt to a range of these conditions and their present occurrence is based on a past set of environmental conditions. How far back one should consider the environmental history for explaining a specific distribution and abundance depends on the life-history characteristics of the species: fast-reproducing and fast-growing species will show a quick response to favourable conditions, while slow-growing and slow-reproducing nematodes have experienced much larger fluctuations. How to translate this into the values of abiotic factors (i.c. salinity) is uncertain, particularly since extreme events (ice cover, storm, dessication) could possibly be more important. We therefore chose to relate meiobenthic distributions to the abiotic parameters measured at the time of sampling.

Whereas the use of the *in situ* salinity as environmental determinant can cause concern, one could also be sceptical about the scale of this study, encompassing five (resp. 6) estuaries separated up to a thousand kilometres one from another. Moreover, only a few sites in each estuary were sampled. It has for instance been shown that the nematode and copepod fauna can be significantly different at sites as close as several metres to kilometres apart (Eskin & Coull, 1984, Phillips & Fleeger, 1985) and nematodes generally have aggregated distributions on the scale of centimetres (Heip *et al.*, 1985). Moreover, in the intertidal area some zonation along the tidal elevation gradient exists (Warwick, 1971) and for this study, samples were taken at a random spot along this gradient. Even when discarding the temporal variation, it is clear that in this study, in which we had to compromize between spatial coverage and labour intensity, a large degree of intertidal variability within any one of the estuaries has been missed. In view of all these restraints it is remarkable to find a high consistency in the nematode structure along the different estuaries. Mesoscale variability (in the order of kilometres) due to salinity changes or grain-size differences are more important than 'huge'-scale variability (hundreds of kilometres) among estuaries. Microscale variability (centimetres) seemed negligible in view of the great resemblance between subsamples. Of course the importance of such factors as salinity or grain size characteristics on nematode community structure is well documented (e.g. Ward, 1975; Warwick & Gee, 1984; Austen & Warwick, 1989; Vanreusel, 1990; Vincx *et al.*, 1990) but as this is the first study to consistently include among-estuarine variability, the similarity of these gradients in the various estuaries has been clearly demonstrated. Likewise there is nothing innovating in the salinity and sedimentary preferences reported for the species in this study, as they greatly confirm what has been observed in other areas. However, the species distributions form the basis for sophisticated multivariate techniques like CANOCO and as such they may contribute to our appreciation of and help in understanding the results from these multivariate techniques.

A striking feature of nematode assemblages is the large number of species present in any one habitat – usually an order of magnitude higher than for any other taxon (Heip *et al.*, 1985). The highest known species diversity values for nematode communities were reported from the deep sea (Soetaert *et al.*, 1991), while the lowest nematode diversity was observed in the polluted subtidal muds off the Belgian East coast (Vincx, 1990) where at some sites only one species was present. The diversity values reported here fall well in between these extremes. Diversity in the marine part of the estuaries can be compared to the ones observed in the sublittoral coastal North Sea as reported by Vincx (1990) and Vanreusel (1990).

According to Bouwman (1983) and Heip *et al.* (1985), the estuarine environment was invaded by marine species which have adapted to reduced salinities in varying degrees and these species vanish with decreasing salinity. On its upstream boundary, penetration of freshwater species (up to a salinity of about 10‰) or even of species of terrestrial origin (Bouwman, 1983) add up to those of marine origin. Hence nematode diversity usually increases from about 5‰ salinity towards both the marine and the freshwater zone (Heip *et al.*, 1985). This general trend has been confirmed in this study except for the freshwater part that was not sampled. If such a clear trend can be demonstrated, deviations from this pattern become interesting. Why was the diversity in the most marine station of the Gironde and in the mid-saline station of the Tamar (Neal point) lower than one would expect? Diversity patterns in the Tamar were discussed by Warwick & Gee (1984) and by Austen & Warwick (1989). They argue – in agreement with Huston's dynamic equilibrium hypothesis – that the lower diversity at mid-salinity could be related to the lower degree of disturbance by macrofauna, which is far less numerous here. The nematode community in the marine station of the Gironde (Le Verdon) was very similar to the low-diversity site of the Tamar: it was largely dominated by *Sabatieria punctata (group)* and *Terschellingia communis*, whereas *T. communis*, *T. longicaudata*, *Metachromadora vivipara* and *S. punctata (group)* were co-dominant in Neal point. All these species are conservative with low respiration rates and long generation times. They are typical for tidal mud flats with rather anoxic sediments (Vincx *et al.*, 1990) and this might also be the case for the sediments at Le Verdon (Castel, pers. comm.). Although the low diversity in these stations could be due to a lower degree of disturbance, as invoked by the dynamic equilibrium hypothesis, the macrofauna cannot be the cause of this at Le Verdon where macrofaunal biomass is highest (Castel, 1992) and, remarkably enough, the macrofaunal diversity is also reasonably high in this station. It may be that the low oxygen concentrations in some of the mud flats do not allow the establishment of the higher-diversity assemblage one could expect according to salinity, as only a few species have physiological tolerances suited for persisting in such a harsh environment.

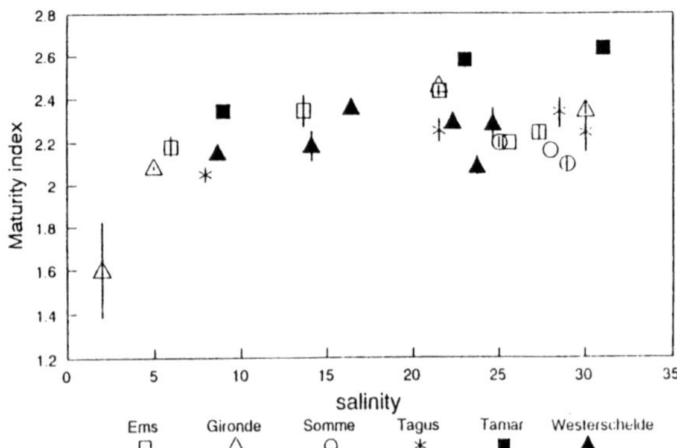

Fig. 9. Maturity index of the various stations along the salinity gradient.

Nematodes have been shown to be possible indicators of pollution or other kinds of disturbances (Heip *et al.*, 1985) and especially the influence of perturbations on diversity has been well documented (Lambshead *et al.*, 1983; Platt *et al.*, 1984) and debated (Hodda & Nicholas, 1986). Amongst the estuaries studied, organic and inorganic pollution have the highest levels in the Westerschelde, and one would expect to find significant differences in diversity in this estuary, as compared to the other estuaries. This was shown to be the case for intertidal copepods (Van Damme *et al.*, 1984), when compared to the Ems. However, the deviation of nematode diversity from the general trend in the Westerschelde is only suggestive at most and not consistent enough to establish a possible effect of pollution on nematode diversity. As for the study of copepods in Van Damme *et al.* (1984), the diversity differences between the Ems and the Westerschelde could be reflections of the density differences, because of the large dependence of diversity indices on the number of individuals (Soetaert & Heip, 1990).

Perhaps the most striking result of this study lies in the distribution of higher meiofaunal groups in the various estuaries. Whereas nematodes were overall the most abundant organisms in the intertidal zone, their dominance was much lower in the Ems compared to the other estuaries. Only in the Ems have we observed important populations of harpacticoids, turbellarians, ciliates and gastrotrichs in some stations. Although the occurrence of these high densities could very well be short-term events, this does not explain their absence in other estuaries and the causes of this are uncertain.

The paucity of harpacticoid copepods in the Westerschelde, when compared to the Ems, was already noted by Van Damme *et al.* (1984) and ascribed to pollution effects in the Westerschelde. From our study, the Westerschelde indeed appeared to be nearly devoid of intertidal populations of harpacticoids, whereas the other estuaries had, albeit weakly developed, populations in at least some of their stations. However, harpacticoids were shown to be more abundant in the subtidal area of the Westerschelde (max. 27 ind 10 cm^{-2}, Soetaert *et al.*, 1994) compared to the intertidal and it would be interesting to see whether these subtidal abundances are also consistently lower when compared to the other estuaries. Subtidal samples analyzed from the Tagus indeed suggest that Westerschelde abundances are lower since mean densities of more than 1500 ind 10 cm^{-2} were observed in one Tagus station (own unpublished data).

Total meiobenthic densities tended to increase exponentially with increasing salinities but were low in the most saline range of the Tagus and to a smaller extent of the Westerschelde. The lack of low-salinity tolerance of species of marine origin is usually invoked to explain the decreasing densities in the low salinity ranges of an estuary (e.g. Coull, 1988; Montagna & Kalke, 1992). However, whereas this could very well explain the presence or absence of species (and hence diversity), it is unclear why species that have adapted to brackish situations cannot attain large densities. Moreover, large densities were observed in the salt marshes of the brackish area of the Westerschelde (Van Damme *et al.*, 1980) and this seems to contradict the thesis of salinity tolerance.

Other explanations could be linked with food availability and both the supply and reactivity of the organic matter in the bottom is important. In the turbid regions of the estuaries primary production in the water is lowest (Kromkamp *et al.*, 1992; Soetaert *et al.*, 1994b) and this may result in a lower food supply for the benthos. Bacterial mineralization in this turbid area is high (Goosen *et al.*, 1992; Soetaert & Herman, 1995) and hence the residual organic matter of fluviatile descent, downstream from this region, will be more refractory (Soetaert & Herman, 1995) and thus less suitable as a food source. On the other hand primary pelagic productivity increases downstream as turbidity decreases (Soetaert *et al.*, 1994b). The closer to the marine zone, the more important the input of organics from marine origin will become (Soetaert & Herman, 1995) and this higher quality food source could allow the benthos to increase in abundance. Also of possible importance in explaining meiobenthic density is macrofaunal predation (e.g. Warwick & Gee, 1984). However, the magnitude of macrofaunal interaction with the meiofauna along the salinity gradient is difficult to assess. In the Westerschelde for instance, macrofaunal biomass increases with increasing salinities but the trophic structure of the macrofauna changes concurrently from detritus-feeding (and thus meiofauna consuming) in the brackish part to filter feeding in the marine part (Meire *et al.*, 1991). Van Damme *et al.* (1980) even invoked pollution effects to explain the lower densities of the Westerschelde meiobenthos compared to the Ems. However, as may be apparent from this study, meiobenthic densities of the Westerschelde are not consistently lower, except perhaps in the marine part. Moreover, the Ems was not so well chosen as a reference estuary as from our study it appears to be the only estuary where large densities in the brackish part were observed. Nevertheless, in many instances pollution decreases in parallel with increasing salinity and thus cannot be ruled out as a possible cause of benthic depauperation. Another factor which could be of possible importance is the degree of environmental fluctuation, with brackish areas that are usually more unpredictable (varying current velocities, steeper gradients of several solutes and particulates), while the human impact (dredging) usually is more intense in this region.

Pollution is expected to be high in the Tagus, which receives high amounts of untreated waste from the city of Lisbon, and many industrial effluents are pumped into the estuary. This estuary has low numbers of meiobenthic animals all along its banks, and one could be tempted to conclude that this is a pollution effect. But seasonal environmental fluctuations in the Tagus are much more pronounced than in the other estuaries and this too could cause low densities. However, too little is known about the abiotic and biotic factors from this area to be conclusive about the causes of meiobenthic distributions.

Another way of looking for anomalies among stations is by comparing the life-history traits of nematodes: are they predominantly *r* or *K* strategists? Very recently, an index has been proposed that summarizes this information into one number: the maturity index (Bongers, 1990; Bongers *et al.*, 1991). On the whole, not much information could be extracted from this measure and it appeared to be unable to reveal gradients. However, it did point to a spuriously low 'maturity' of the station located in the most freshwater part of the Gironde (Lamarque). This muddy intertidal flat was also very poor in macrobenthos, which was dominated by oligochaetes (Castel, 1992) and it is situated at about 30–40 km from the city of Bordeaux.

The low maturity of the nematode community could nearly entirely be accounted for by the dominance of the freshwater nematode *Pareudiplogaster pararmatus*. This nematode is known to be successful in waste-water exposed intertidal mud flats in the low-salinity range of estuaries (Romeyn *et al.*, 1983) and belongs to a family which has high colonization capabilities (Bongers, 1990). Being ovoviviparous, it can survive under extreme conditions and become dominant due to the disappearance of less resistant competitors. Its dominance at Lamarque could thus be indicative for some kind of organic enrichment in this area although the total densities were relatively low. Another possibility is that we have observed a pioneer community after some kind of 'catastrophic event' (e.g. storm) has taken place.

In 1980 *Pareudiplogaster pararmatus* was the most important member of the nematode fauna in an area exposed to organic pollution in the Ems (Bouwman *et al.*, 1983). The fact that it was not observed in the Ems during this study could indicate the improved conditions in this part of the estuary as was shown from the macrobenthos (Esselink *et al.*, 1989).

Although ecological interpretations are difficult to make without an extensive background environmental data set, this study has provided some insight in the structure of estuarine meiobenthic communities. The results of this work could serve as a base-line for a more intensive study where smaller spatial scales and temporal variation should be taken into account and a more extensive environmental data set should be procured. If one wants to spot pollution or produce conclusive evidence for the causes of benthic distributions, a global research project as this one is not appropriate. Yet it can provide a reference frame against which other results can be put into perspective.

Acknowledgments

This work was supported by a grant from the European Community in the framework of the MAST-JEEP project (first author). The authors are grateful to all investigators who were responsible for the sampling of the meiobenthos in the various estuaries and they regret the loss of samples from the Shannon, which were retained somewhere at the – as yet still existing – internal borders of Europe. We acknowledge the help of Danielle Schram concerning the analysis of sediment samples. The first author is grateful for the working facilities offered in the NIOO-CEMO. We also thank Johan Craeymeersch and Carlo Heip for a critical evaluation of the manuscript. This is article number 739 from the NIOO-CEMO. The extensive data set was included into the JEEP data base and is available on request.

References

Austen, M. C. & R. M. Warwick, 1989. Comparison of univariate and multivariate aspects of estuarine meiobenthic community structure. Estuar. coast. Shelf Sci. 29: 23–42.

Bale, A. J., A. W. Morris & R. J. M. Howland, 1985. Seasonal sediment movement in the Tamar estuary. Oceanol. Acta 8: 1–16.

Bongers, T., 1988. De Nematoden van Nederland. Natuurhistorische bibliotheek KNNV nr 46. Pirola, Schoorl.

Bongers, T., 1990. The maturity index: an ecological measure of environmental disturbance based on nematode species composition. Oecologia 83: 14–19.

Bongers, T., R. Alkemade & G. W. Yeates, 1991. Interpretation of disturbance-induced maturity decrease in marine nematode assemblages by means of the Maturity Index. Mar. Ecol. Progr. Ser. 76: 135–142.

Bouwman, L. A., 1981. A survey of nematodes from the Ems estuary. Part I. Systematics. Zool Jahrbuche (Systematik, Okologie, Geographie) 108: 335–385.

Bouwman, L. A., 1983. A survey of nematodes from the Ems estuary: species assemblages and associations. Zool. Jahrbuche (Systematik, Okologie, Geographie) 110: 345–376.

Bouwman, L. A., K. Romeyn, D. R. Kremer & F B. van Es, 1983. Occurrence and feeding biology of some nematode species in Aufwuchscommunities. In Systematics, ecology and feeding biology of estuarine nematodes. BOEDE publicaties en verslagen nr 3: 107–127.

Capstick, C. K., 1959. The distribution of free-living nematodes in relation to salinity in the middle and upper reaches of the river Blyth estuary. J. anim Ecol. 28: 189–210.

Castel, J. & A. Feurtet, 1989. Dynamics of the copepod *Eurytemora affinis hirundoides* in the Gironde estuary: origin and fate of its production. In Ros, J. D. (ed.), Topics in marine biology. Scient. Mar. 53: 577–584.

Castel, J., 1992. Comparative field study of the ecological structure of major european tidal estuaries: the Gironde estuary. In P. Herman (ed.). JEEP 92: Major biological processes in European tidal estuaries – report of the workshop held in Plymouth: 55–64.

Coull, B. C., 1988. Ecology of the marine meiofauna. In Higgins, R. P. & H. Thiel (eds). Introduction to the study of the meiofauna. Smithsonian Institution Press, Washington D.C. London: 18–38.

Ducrotoy, J. P. & B. Sylvand, 1989. Baie des Veys and baie de Somme (English Channel): comparison of two macrotidal ecosystems. In M. Elliot & J. P. Ducrotoy (eds), Estuaries and coasts: spatial and temporal intercomparisons. ECSA 19 Symposium: 207–210.

Eskin, R. A. & B. C. Coull, 1984. A priori determination of valid control sites: an example using marine meiobenthic nematodes. Mar. envir. Res. 12: 161–172.

Esselink, P., J. van Belkum & K. Essink, 1989. The effect of organic pollution on local distribution of Nereis diversicolor and Corophium volutator. Neth. J. Sea Res. 23: 323–332.

Gaudencio, M. J., M. T. Guerra & M. Glemarec, 1991. Recherches biosedimentaires sur la zone maritime de l'estuaire du Tage, Portugal: donnees sedimentaires preliminaires. In M. Elliot & J. P. Ducrotoy (eds), Estuaries and coasts: spatial and temporal intercomparisons. ECSA 19 Symposium: 11–16.

Gerlach, S. A., 1953. Die Biozonotische gliederung der Nematodenfauna an den Deutschen Kusten. Zeitschrift für Morfologie und Okologie der Tiere 41: 411–512.

Gerlach, S. A., 1980. Development of marine nematode taxonomy up to 1979. Veroff. Inst. Meeresforsch. Bremerh. 18: 249–255.

Gourbault, N., 1981. Les peuplements de nematodes du chenal de la baie de Morlaix (premieres donnees). Cah. Biol. Mar. 22: 65–82.

Heip, C., M. Vincx & G. Vranken, 1985. The ecology of marine nematodes. Oceanogr. Mar. Biol. Ann. Rev. 23: 399–489.

Heip, C., P. Herman & K. Soetaert, 1988. Data processing, evaluation and analysis. In Higgins, R. P. & H. Thiel (eds), Introduction to the study of the meiofauna. Smithsonian Institution Press, Washington D.C. London: 197–231.

Heip, C., 1989. The ecology of the estuaries of Rhine, Meuse and Scheldt in the Netherlands. In Ros, J. D. (ed.), Topics in marine biology. Scient. Mar. 53: 457–463.

Hodda, M. & W.L . Nicholas, 1986. Temporal changes in littoral meiofauna from the Hunter River estuary. Aust. J. mar. Freshwat. Res. 37: 729–741.

Jonge, V. N. de, 1992. Tidal flow and residual flow in the Ems Estuary. Estuar. coast. Shelf Sci. 34: 1–22.

Jonge, V. N. de & K. Essink, 1992. The Ems estuary: water circulation, sediment dynamics and nutrient enrichment. In P. Herman (ed.). JEEP 92: Major biological processes in European tidal estuaries – report of the workshop held in Plymouth: 23–38.

Jongman, Q., C. Ter Braak & O. Van Tongeren, 1987. Data analysis in community and landscape ecology. PUDOC Wageningen, 299 pp.

Kramer, K. J. M., R. M. Warwick & U. H. Brockmann, 1992. Manual of sampling and analytical procedures for tidal estuaries. JEEP 92.

Lambshead, P. J. D., H. M. Platt & K. M. Shaw, 1983. The detection of differences among assemblages of marine benthic species based on an assessment of dominance and diversity. J. nat. Hist. 17: 859–874.

Meire, P. M., J. J. Seys, T. J. Ysebaert & J. Coosen, 1991. A comparison of the macrobenthic distribution and community structure between two estuaries in SW Netherlands. In M. Elliot & J. P. Ducrotoy (eds), Estuaries and coasts: spatial and temporal intercomparisons. ECSA 19 Symposium: 221–230.

Montagna, P. A. & R. D. Kalke, 1992. The effect of freshwater inflow on meiofaunal and macrofaunal populations in the Guadalupe and Nueces Estuaries, Texas. Estuaries 15: 307–326.

Moore, C. G., 1987. Meiofauna of an industrialized estuary and Firth of Forth, Scotland. Proc. r. Soc. Edinb. 93B: 415–430.

Moreira, F., C. A. Assis, P. R. Almeida, J. L. Costa & M. J. Costa, 1992. Trophic relationships in the community of the Upper Tagus Estuary (Portugal): a preliminary Approach. Estuar. coast. Shelf Sci. 34: 617–623.

Morris, A. W., R. J. M. Howland, E. M. S. Woodward, A. J. Bale & R. F. C. Mantoura, 1985. Nitrite and Ammonia in the Tamar estuary. Neth. J. Sea Res. 19: 217–222.

Phillips, F. E. & J. W. Fleeger, 1985. Meiofauna meso-scale variability in two estuarine habitats. Estuar. coast. Shelf Sci. 21: 745–756.

Platt, H. M. & R. M. Warwick, 1983. Freeliving marine nematodes. Part I: British Enoplids. Synopses of the British Fauna (New series) no 28, Cambridge University Press.

Platt, H. M. & R. M. Warwick, 1988. Freeliving marine nematodes. Part II: British Chromadorids. Synopses of the British Fauna (New series) no 38, E. J. Brill, Leiden.

Platt, H. M., K. M. Shaw & P. J. D. Lambshead, 1984. Nematode species abundance patterns and their use in the detection of environmental perturbation. Hydrobiologia 118: 59–66.

Riemann, F., 1966. Die interstitielle Fauna im Elbe Estuar. Verbreitung und Systematik. Arch. Hydrobiol. Suppl. 31: 1–279.

Romeyn, K., L. A. Bouwman & W. Admiraal, 1983. Ecology and cultivation of the herbivorous brackish-water nematode *Eudiplogaster pararmatus*. Mar. Ecol. Progr. Ser. 12: 145–153.

Rybarczyk, H., F. Jamet & M. Desprez, 1992. Baie de Somme: Bilan 1991. In P. Herman (ed.), JEEP 92: Major biological processes in European tidal estuaries – report of the workshop held in Plymouth: 49–53.

Skoolmun, P. & S. A. Gerlach, 1971. Jahreszeitliche fluktuationen der Nematodenfauna im Gezeitenbereich des Weseraestuar. Veroff. Inst. Meeresforsch. Bremerh. 13: 119–138.

Smol, N., K. A. Willems, J. C. R. Govaere & A. J. J. Sandee, 1994. Composition, distribution and biomass of the meiobenthos in the Oosterschelde estuary (SW Netherlands). Hydrobiologia 282/283 (Dev. Hydrobiol. 97): 197–217.

Soetaert, K. & C. Heip, 1990. Sample-size dependence of diversity indices and the determination of sufficient sample size in a high-diversity deep-sea environment. Mar. Ecol. Progr. Ser. 59: 305–307.

Soetaert, K., C. Heip & M. Vincx, 1991. Diversity of nematode assemblages along a mediterranean deep-sea transect. Mar. Ecol. Progr. Ser. 75: 275–282.

Soetaert, K., M. Vincx, J. Wittoeck, M. Tulkens & D. Vangansbeke, 1994. Spatial pattern of Westerschelde meiobenthos. Estuar. coast. Shelf Sci. 39: 367–388.

Soetaert, K. & P. M. J. Herman, 1995. Carbon flows in the Westerschelde estuary (The Netherlands) evaluated by means of an ecosystem model (MOSES). Hydrobiologia 311 (Dev. Hydrobiol. 110): 247–266.

Soetaert, K., P. M. J. Herman & J. Kromkamp, 1994. Living in the twilight: estimating net phytoplankton growth in the Westerschelde estuary (The Netherlands) by means of an ecosystem model (MOSES). J. Plankt. Res. 16: 1277–1301.

Ter Braak, C. J. F., 1989. CANOCO – an extension of DECORANA to analyze species-environment relationships. Hydrobiologia 184: 169–170.

Van Damme, D., C. Heip & K. A. Willems, 1984. Influence of pollution on the harpacticoid copepods of two North Sea estuaries. Hydrobiologia 112: 143–160.

Van Damme, D., R. Herman, Y. Sharma, M. Holvoet & P. Martens, 1980. Fluctuations of the meiobenthos communities in the Westerschelde estuary. Ices-report CM/L 23: 131–170.

Vanreusel, A., 1990. Ecology of free-living marine nematodes from the Voordelta (Southern Bight of the North Sea). I. Species composition and structure of the nematode communities. Cah. Biol. mar. 31: 439–462.

Vincx, M., P. Meire & C. Heip, 1990. The distribution of nematodes communities in the Southern Bight of the North Sea. Cah. Biol. mar. 31: 107–129.

Vincx, M., 1990. Diversity of the nematode communities in the southern bight of the north sea. Neth. J. Sea Res. 25: 181–188.

Ward, A. R., 1975. Studies on the sublittoral free-living nematodes of the Liverpool Bay. II. Influence of sediment composition on the distribution of marine nematodes. Mar. Biol. 30: 217–225.

Warwick, R. M., 1971. Nematode associations in the Exe estuary. J. Mar. Biol. Assoc. U.K. 51: 439–454.

Warwick, R. M. & R. Price, 1979. Ecological and metabolic studies on free-living nematodes from an estuarine mud flat. Estuar. coast. mar. Sci. 9: 257–271.

Warwick, R. M. & J. M. Gee, 1984. Community structure of estuarine meiobenthos. Mar. Ecol. Progr. Ser. 18: 97–111.

Appendix. Species list, indicating the maximum density (10 cm^{-2}) of each species in the five estuaries of the current study and the Tamar (data derived from Warwick & Gee, 1984).

	Ems	Wester-schelde	Somme	Tamar	Gironde	Tagus
Enoplida						
Enoplidae						
Enoplus brevis	–	–	**	–	–	–
Enoplus sp	*	*	–	–	–	–
Thoracostomopsidae						
Enoploides longispiculosus	**	***	***	–	–	–
Enoplolaimus litoralis	–	–	**	–	–	–
Enoplolaimus propinquus	–	**	–	–	–	–
Anoplostomatidae						
Anoplostoma viviparum	***	**	–	**	**	***
Chaetonema riemanni	**	–	–	–	–	–
Ironidae						
Syringolaimus sp	–	*	–	–	*	*
Leptosomatidae						
Leptosomatidae T1	–	–	–	–	–	**
Oxystominidae						
Halalaimus gracilis	***	**	**	*	**	**
Halalaimus sp	–	–	–	*	**	–
Halalaimus T1	–	–	–	–	–	**
Halalaimus T2	–	–	–	–	–	**
Halalaimus T3	–	–	–	–	–	*
Nemanema cylindricaudatum	**	–	–	*	–	–
Oxystomina elongata	–	–	**	*	–	–
Oxystomina sp	**	–	–	–	–	–
Thalassoalaimus septentrionalis	**	*	–	–	–	**
Oncholaimidae						
Adoncholaimus fuscus	***	***	***	–	–	–
Adoncholaimus thalassophygas	–	–	–	–	*	–
Oncholaimellus mediterraneus	–	*	–	–	–	–
Oncholaimidae sp	–	–	–	–	–	**
Oncholaimus oxyuris	–	*	***	–	–	–
Oncholaimus sp	**	–	–	–	*	–
Viscosia abyssorum	–	–	–	*	–	–
Viscosia aff rustica	***	**	–	–	–	–
Viscosia glabra	**	–	–	–	–	–
Viscosia viscosa	***	***	***	**	***	***
Enchelidiidae						
Calyptronema maxweberi	**	**	***	*	**	–
Eurystomina sp	–	–	–	–	–	**

Appendix (Continued).

	Ems	Wester-schelde	Somme	Tamar	Gironde	Tagus
Tripyloididae						
Bathylaimus australis	–	**	–	–	–	–
Bathylaimus sp	–	–	–	–	*	**
Bathylaimus stenolaimus	–	*	–	–	–	–
Tripyloides gracilis	–	**	***	*	**	–
Tobrilidae						
Tobrilus diversipapillatus	–	–	–	–	*	–
Trefusiida						
Trefusiidae						
Trefusia Longicaudata	–	***	–	–	–	–
Trefusia multipapillatus	***	–	–	–	–	–
Trefusia S1	–	–	***	–	–	–
Trefusia S2	–	–	**	–	–	–
Trefusia W2	–	*	–	–	–	–
Chromadorida						
Chromadoridae						
Atrochromadora microlaima	*	*	–	**	–	–
Chromadora aff nudicapitata	–	–	–	–	–	**
Chromadora axi	–	*	–	–	–	**
Chromadora macrolaima	***	**	***	–	****	***
Chromadorida sp	**	–	–	–	*	**
Chromadorina germanica	–	–	–	–	*	–
Chromadorita aff nana	**	***	–	–	–	–
Chromadorita tentabunda	**	***	***	–	**	**
Dichromadora cephalata	**	*	***	**	**	**
Dichromadora cucullata	–	*	–	–	–	–
Dichromadora geophila	***	**	–	**	**	***
Dichromadora hyalocheile	–	–	***	–	–	–
Dichromadora sp	–	–	**	–	–	**
Dichromadora T3	–	–	–	–	–	*
Hypodontolaimus balticus	–	–	–	*	–	–
Hypodontolaimus W1	–	**	–	–	–	–
Neochromadora sp	–	–	–	*	–	–
Prochromadorella ditlevseni	–	**	***	–	–	–
Ptycholaimellus ponticus	***	*	–	**	****	**
Spilophorella candida	–	*	–	–	–	*
Spilophorella paradoxa	–	*	**	*	–	*

Appendix (Continued).

	Ems	Wester-schelde	Somme	Tamar	Gironde	Tagus
Comesomatidae						
Cervonema G1	–	–	–	–	**	–
Cervonema T1	–	–	–	–	–	***
Paracomesoma sp	–	–	–	*	–	–
Sabatieria celtica	***	*	–	*	–	–
Sabatieria longisetosa	–	–	–	–	–	**
Sabatieria longispinosa	***	–	–	–	–	–
Sabatieria punctata group	****	***	–	**	****	***
Sabatieria sp	**	–	**	–	*	**
Setosabatieria hilarula	–	–	–	*	–	**
Ethmolaimidae						
Neotonchus aff cupulatus	–	*	–	–	–	–
Neotonchus sp	–	–	–	*	–	–
Cyatholaimidae						
Cyatholaimidae sp	–	–	***	–	–	–
Paracanthonchus aff caecus	–	–	–	*	–	–
Paracanthonchus aff heterodontus	–	–	–	–	–	***
Paracanthonchus aff thaumasius	–	–	****	–	–	–
Paracanthonchus caecus	**	–	–	–	–	–
Paracanthonchus heterodontus	***	–	–	–	–	**
Paracanthonchus thaumasius	–	**	–	–	–	–
Paracyatholaimoides W1	–	*	–	–	–	–
Paracyatholaimus W1	–	**	–	–	–	–
Praeacanthonchus punctatus	***	*	***	*	**	–
Selachinematidae						
Halichoanolaimus robustus	**	–	–	–	–	*
Desmodoridae						
Desmodora A	–	–	–	*	–	–
Desmodora B	–	–	–	*	–	–
Leptonemella sp	–	–	**	–	–	–
Metachromadora aff suecica	***	**	–	–	–	–
Metachromadora remanei	**	***	***	–	**	**
Metachromadora sp	–	–	**	–	–	–
Metachromadora vivipara	****	***	–	***	–	–
Molgolaimus cuanensis	***	–	–	–	–	–
Molgolaimus S1	–	–	***	–	–	–
Molgolaimus S2	–	–	**	–	–	–
Molgolaimus sp	–	–	***	–	–	*
Molgolaimus tenuispiculum	–	–	–	*	–	–
Molgolaimus turgofrons	–	*	–	–	–	–
Onyx sagittarius	–	*	–	–	–	–
Sigmaphoranema aff rufus	–	**	–	–	–	–
Spirinia parasitifera	**	***	–	*	–	–

Appendix (Continued).

	Ems	Wester-schelde	Somme	Tamar	Gironde	Tagus
Microlaimidae						
Aponema torosa	–	–	–	*	–	–
Calomicrolaimus S1	–	–	****	–	–	–
Calomicrolaimus S3	–	–	****	–	–	–
Microlaimus arenicola	–	**	***	–	–	–
Microlaimus globiceps	****	**	***	–	–	–
Microlaimus marinus	***	***	****	–	–	–
Microlaimus parahonestus	***	***	–	**	–	–
Microlaimus robustidens	–	–	****	*	–	–
Microlaimus sp	**	–	–	–	–	–
Microlaimus W1	–	**	–	–	–	–
Microlaimus W4	–	*	–	–	–	–
Monoposthiidae						
Monoposthia mirabilis	–	**	–	–	–	–
Monoposthia sp	**	–	–	–	–	–
Nudora bipapillata	–	–	–	**	–	–
Leptolaimidae						
Antomicron elegans	**	–	**	–	*	*
Camacolaimus tardus	–	**	***	–	–	–
Dagda bipapillata	–	*	–	–	–	–
Deontolaimus papillatus	–	*	–	–	–	–
Leptolaimus acicula	–	*	–	–	–	–
Leptolaimus ampullaceus	–	*	–	–	–	–
Leptolaimus elegans	–	**	–	–	–	–
Leptolaimus luridus	–	–	–	–	–	*
Leptolaimus papilliger	****	–	–	*	**	–
Leptolaimus S1	–	–	***	–	–	–
Leptolaimus sp	–	*	–	–	*	*
Stephanolaimus aff spartinae	–	–	***	–	–	–
Haliplectidae						
Haliplectus wheeleri	–	–	–	–	–	**
Aegialoalaimidae						
Aegialoalaimus aff tenuicaudatus	–	–	–	–	–	**
Aegialoalaimus elegans	**	–	–	–	**	*
Cyarthonema E1	**	–	–	–	–	–
Cyarthonema germanica	***	–	–	–	**	–
Cyarthonema W1	–	*	–	–	–	–
Southernia zosterae	**	**	–	–	–	–

Appendix (Continued).

	Ems	Wester-schelde	Somme	Tamar	Gironde	Tagus
Tubolaimoididae						
Chitwoodia warwicki	–	**	–	–	–	–
Meyliidae						
Meyliidae T1	–	–	–	–	–	**
Desmoscolecidae						
Calligyrus sp	–	–	–	*	–	–
Desmoscolex falcatus	–	–	–	**	–	–
Tricoma sp	–	–	–	*	–	–
Monhysterida						
Monhysteridae						
Diplolaimella sp	–	–	***	–	*	*
Monhysteridae sp	**	**	***	*	**	**
Monhysteridae T1	–	–	–	–	–	**
Xyalidae						
Daptonema acc Bouwman	***	*	***	–	–	–
Daptonema cfr biggi	–	–	–	–	**	–
Daptonema G1	–	–	–	–	***	–
Daptonema kornoense	–	–	–	–	–	**
Daptonema normandicum	**	***	***	*	**	–
Daptonema oxycerca	***	–	–	*	***	*
Daptonema procera	–	–	–	*	–	–
Daptonema riemanni	****	***	****	–	–	–
Daptonema setosa	***	**	–	*	***	***
Daptonema sp	**	**	***	–	***	***
Daptonema T1	–	–	–	–	–	**
Daptonema T2	–	–	–	–	–	*
Daptonema tenuispiculum	–	**	–	–	–	–
Daptonema W1	–	**	–	–	–	–
Daptonema xyaliforme	**	–	–	–	**	**
Metadesmolaimus 2	–	*	–	–	–	–
Metadesmolaimus E1	**	–	–	–	–	–
Metadesmolaimus gaelicus	–	*	–	–	–	–
Paramonohystera E1	***	–	–	–	**	**
Paramonohystera sp	–	*	–	–	–	–
Pseudotheristus furcatus	–	–	–	–	–	*
Theristus 1	–	**	–	–	–	–
Theristus acer	**	–	***	*	–	**
Theristus aff profundus	–	*	–	–	–	–
Theristus blandicor	****	***	–	–	–	–
Theristus cfr subcurvatus	**	**	–	–	–	–

Appendix (Continued).

	Ems	Wester-schelde	Somme	Tamar	Gironde	Tagus
Theristus ensifer	–	*	***	–	–	–
Theristus G1	–	–	–	–	*	–
Theristus longus	***	**	***	–	–	–
Theristus pertenuis	***	**	****	–	–	–
Theristus sp	**	–	***	–	–	–
Trichotheristus mirabilis	–	***	–	–	–	–
Xyala striata	–	*	–	–	–	–
Sphaerolaimidae						
Sphaerolaimus balticus	–	–	–	*	–	–
Sphaerolaimus gracilis	–	–	–	*	**	**
Sphaerolaimus hirsutus	**	–	–	*	**	*
Sphaerolaimus sp	–	**	–	–	–	–
Siphonolaimidae						
Siphonolaimus sp	–	*	–	–	–	–
Linhomoeidae						
Desmolaimus S1	–	–	***	–	–	–
Desmolaimus T1	–	–	–	–	–	*
Desmolaimus zeelandicus	***	–	–	**	–	–
Eleutherolaimus aff stenosoma	–	–	–	–	–	**
Eleutherolaimus amasi	**	–	–	–	–	–
Eleutherolaimus sp	**	–	–	–	–	–
Eleutherolaimus stenosoma	**	**	–	*	–	–
Linhomoeidae T1	–	–	–	–	–	**
Linhomoeidae sp	**	*	***	–	**	*
Linhomoeidae W1	–	*	–	–	–	–
Linhomoeidae W2	–	*	–	–	–	–
Linhomoeidae W4	–	*	–	–	–	–
Linhomoeidae W5	–	**	–	–	–	–
Linhomoeus S1	–	–	***	–	–	–
Megadesmolaimus W1	–	**	–	–	–	–
Metalinhomoeus aff biformis	**	**	****	–	**	***
Metalinhomoeus biformis	–	**	****	–	–	**
Paralinhomoeus ilensis	**	–	–	–	–	–
Paralinhomoeus sp	**	–	–	–	–	**
Paralinhomoeus T1	–	–	–	–	–	**
Terschellingia communis	–	*	–	**	****	**
Terschellingia longicaudata	–	–	–	***	***	**

Appendix (Continued).

	Ems	Wester-schelde	Somme	Tamar	Gironde	Tagus
Axonolaimidae						
Ascolaimus elongatus	***	***	****	–	–	–
Axonolaimus cfr orus	–	–	–	–	–	**
Axonolaimus paraspinosus	***	*	–	**	**	**
Axonolaimus spinosus	–	–	–	–	**	*
Odontophora aff. paravilloti	–	*	–	–	–	–
Odontophora rectangula	***	***	–	–	–	–
Odontophora setosa	***	***	****	*	–	–
Odontophora sp	**	–	***	–	–	–
Odontophora W4	–	**	–	–	–	–
Pseudolella granulifera	–	–	–	–	***	***
Diplopeltidae						
Campylaimus gerlachi	–	–	–	–	–	**
Diplopeltis incisus	–	–	–	*	–	–
Diplopeltula asetosa	**	–	–	–	–	–
Rhabditida						
Diploscapteridae						
Diploscapter sp	**	*	–	–	–	–
Neodiplogasteridae						
Pareudiplogaster pararmatus	–	–	–	–	***	–
Dorylaimida						
Dorylaimida						
Dorylaimida W2	–	*	–	–	–	–

(–) absent, (*) 0–10 ind, (**), 10–100, (***) 100–1000, (****) >1.000

Hydrobiologia **311**: 207–214, 1995.
C. H. R. Heip & P. M. J. Herman (eds), Major Biological Processes in European Tidal Estuaries.

The response of two estuarine benthic communities to the quantity and quality of food

Michael A. Kendall, John T. Davey & Steve Widdicombe
Natural Environment Research Council, Plymouth Marine Laboratory, 1 Prospect Place, The Hoe, Plymouth PL1 3DH, UK

Key words: estuaries, benthos, soft sediments, food, experimental manipulation, community response

Abstract

Experimental manipulations of food supply were performed on soft sediment cores from two European estuaries, the Westerscheldt and the Gironde, with a view to determining benthic macrofaunal community response. Over a period of twenty weeks in a laboratory mesocosm system, both communities showed losses in terms of numbers of individuals and small, but non-significant, losses in terms of numbers of species. Whereas no effect of the different types of foods or the dose levels at which they were supplied was detected for the Westerscheldt benthic community, that of the Gironde showed some significant response. This was largely attributed to the differential mortality of spionid polychaetes across the dose levels used, with the highest dose, equivalent to 200 g C m^{-2} yr^{-1}, only just maintaining their initial population densities. The results are discussed in terms of the importance of lateral advection of food particles at the benthic boundary layer and the general insufficiency of many estimates of carbon input to shallow benthic systems.

Introduction

The relationship between the quantity of organic matter potentially available to benthic organisms and the characteristics of the faunal assemblages in which they live is complex. Studies in the field and the laboratory have shown that the number of species and the number of individuals can be influenced by both the quantity and quality of the organic matter on which they feed. Whitlach (1980, 1981) demonstrated correlations between species richness and the total amount of organic carbon in surface sediments and showed a relationship between faunal diversity and that of potential food particles. Other authors (e.g. Beukema & Cadee, 1986; Creutzberg *et al.*, 1984) have demonstrated increases in biomass or abundance of individuals as a result of an increased food supply. Laane *et al.* (1987) have suggested that the caloric content of suspended matter can be used as an indicator of differences in biomass and production of heterotrophic organisms between estuaries. In inshore waters, the rate of deposition of organic matter is generally in the region of 25–75 g C m^{-2} (Gee *et al.*, 1985 and references therein) but in many areas this is substantially supplemented by anthropogenic material. As the organic loading to the benthos increases, the number of species present declines (although initially the number of individuals might rise) until in the most badly affected areas macrofauna is totally absent (Pearson & Rosenberg, 1978). The effects of gross organic enrichment in the marine environment are well documented but the effects of more moderate variations in food supply are less clear.

Benthic animals utilise organic matter from many sources including *in situ* production of bacteria and algae, sedimentation of phytoplankton and advection of terrigenous material. The availability and nutritive value of these different foods varies greatly (Lopez & Levinton, 1987; Levinton & Stewart, 1988). In coastal waters, the relative contribution of each will differ from place to place depending on topography and land drainage patterns, but at any single location it is difficult to assess the relative importance to the benthos of the different constituents of the pool of organic matter.

The Gironde and Scheldt estuaries contrast sharply in the quantity and composition of the organic matter which they carry and have clearly different intertidal macrofaunal assemblages. While the former is subject to both organic and inorganic pollution (Heip, 1989) the latter has a rural catchment area and is comparatively clean. The experiments described in this paper were designed to examine the relationship between the pollutant load of the estuaries and their fauna. The hypothesis examined, the same as for the meiofaunal manipulations (Austen & Warwick, 1995), was that while the fauna of the Westerscheldt has already been influenced by high organic pollution loadings that of the cleaner Gironde will be more responsive to the quality and quantity of organic matter.

Methods

Two hundred fresh sediment cores (6.4 cm diameter and 12.5 cm height) containing natural local faunal assemblages were collected from mid-tide level at Walsoorden on the Westerschelde in June 1991 and a further 200 from Le Verdon in the Gironde estuary in June 1992. They were transported to Plymouth where they were transferred to the Plymouth Marine Laboratory's mesocosm and held under conditions as described by Austen & Warwick (1995). They were fed at weekly intervals with suspensions of five different detrital foods (Table 1). Three dose levels were used equivalent to 25, 75 and 200 g C m^{-2} yr^{-1}. Cores representative of all treatments including unfed controls were harvested at 4 and 20 weeks, the former hereafter referred to as the 'start' condition and the latter as the 'end' condition. The first four weeks in the mesocosm were to allow acclimation of the cores to experimental conditions. At harvest, the cores were sieved at 0.5 mm to recover macrofauna which were fixed in 10% seawater formalin, stained in rose bengal, counted and identified to species.

The data were analysed using the PRIMER statistical package developed at PML, chiefly by applying Multidimensional Scaling (MDS) to Bray Curtis similarity matrices of both raw and $\sqrt{\sqrt{}}$ transformed data. In this way, clustering in the data could be discerned and the species responsible for major differences between samples and treatments identified. Tests for faunal differences between groups of cores were tested for using ANOSIM and the species responsible for those differences identified using similarity terms

Table 1. Constitution of detrital foods supplied to cores in mesocosm experiments.

Food type	Ingredients	Carbon content	Nitrogen content
Phytoplankton	80% Haematococcus 10% Isochrysis 10% Porphyrium freeze-dried powder	47.9%	7.1%
Scirpus	Year-old finely ground Scirpus straw	42.9%	3.7%
Enteromorpha	Freeze-dried, finely ground green alga	26.7%	5.7%
Leaf Litter	Finely ground dried woodland leaves	47.4%	3.5%
Mixture	All the above to give equal C contributions	38.9%	5.1%

analysis (SIMPER): both methods of analysis have been discussed by Clarke (1993).

Results

Table 2 summarises the main features of the macrofaunal communities recorded from the cores in the two experiments. In the Westerscheldt cores, the numerical dominants among the 15 species recorded were *Heteromastus filiformis* and *Pygospio elegans*. The biomass dominant was *Nereis diversicolor*. In the Gironde, the 14 species recovered were numerically dominated by the spionid, *Streblospio shrubsolii*, with the biomass dominated by *Nereis diversicolor* and *Macoma balthica*.

In the Westerscheldt cores, although the mean number of species remained constant throughout the experiment, mean abundances fell. In the Gironde cores, abundances of individuals were much lower from the outset, but these also declined, although they did so least in certain high dose-level treatments.

Multivariate analysis

Westerscheldt experiment

In the MDS plots, the cores from the start condition were found to cluster closely, indicating a consistent community structure. But the cores from the final condition were more widely dispersed, with no patterns to

Table 2. Mesocosm experiments in the Westerscheldt and Gironde: Numbers of individuals (NI) and of species (NS) in cores from the Start condition and from the High, Medium and Low dose treatments and controls (unfed). Data given ±95% confidence interval (CI).

	Start	High	Mid	Low	Control
Westerscheldt					
NI + 95% CI	113.12 + 41.23	73.04 + 39.72	64.00 + 20.38	77.44 + 24.5	559.00 + 19.71
NS + 95% CI	5.56 + 1.21	4.60 + 1.01	4.67 + 1.15	4.62 + 0.88	4.25 + 0.96
Gironde					
NI + 95% CI	31.38 + 11.02	27.80 + 7.60	16.40 + 4.43	10.55 + 3.20	6.75 + 3.77
NS + 95% CI	5.50 + 0.60	5.55 + 0.53	4.75 + 0.51	4.40 + 0.73	3.25 + 0.63

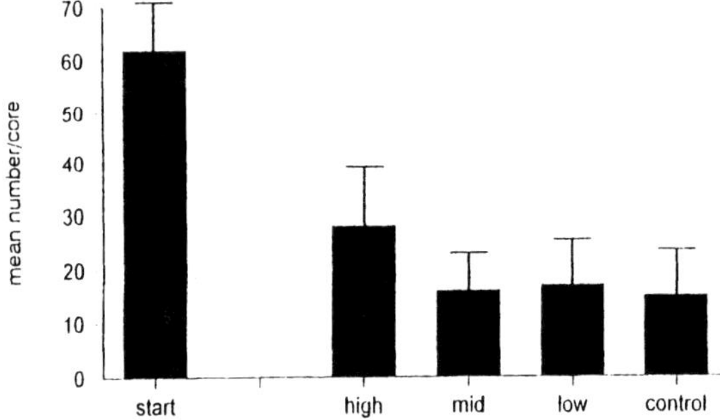

Fig. 1. Pygospio elegans: Mean number per core in Westerscheldt mesocosm experiment, with 95% confidence interval.

suggest specific treatment effects. This was only slightly less so for untransformed data than for the double square root transformed data.

Similarity Terms Analysis (SIMPER) showed that this increased variability could be related to changes in the abundance of only a small number of species. In the analysis of transformed data, 80% of this dissimilarity could be related to declines in the abundance of the bivalves and the polychaetes *Polydora ciliata*, *Pygospio elegans* and *Nereis diversicolor*. There was, however, a small increase in the abundance of *Manayunkia aestuarina*. SIMPER analysis of the untransformed data showed that 80% of the dissimilarity between sampling dates related to the decline of *Pygospio elegans* (Fig. 1) and *Heteromastus filiformis*. There was no significant difference between the mean number of species present at the start and end of the experiment. Despite the overall pattern of mortality, it was nevertheless evident that some reproduction and recruitment had occurred. Juveniles were recorded of *Pygospio*, *Heteromastus*, *Polydora*, *Manayunkia* and the isopod *Cyathura carinata*.

Data from the end condition cores were also analysed separately using MDS on both transformed and untransformed data. In neither case did the resulting plots show any relationship between the dose of food supplied and the fauna of a particular core. Multivariate analysis of variance (Clarke & Warwick, 1994) showed that there was a significant block (system) effect within the experiment, so that tests for differences between foods and between doses using ANOSIM could not be performed.

Gironde experiment

As with the Westerscheldt experiment, MDS plots of both untransformed and transformed data (Fig. 2) showed a tight cluster for all the start-condition cores with a much more variable cloud of points for the end-condition cores. There was no clear separation of the two data sets but in the untransformed plot, the data points for the end-condition cores tended to lie above and to the left of those representing the start condition with the high dose cores closest to the start condition. In an MDS plot of the untransformed end-condition data only, the tendency was detected for the cores to be arranged from left to right in the order of increasing dose, with sample variability decreasing as dose increased. No such clear pattern was evident in the analysis based on transformed data.

These results suggest that while there were no significant changes in species composition as a result of manipulating the food supplied, there were changes in patterns of dominance. Since multivariate analysis of variance demonstrated that for untransformed data there was no significant difference between blocks, the treatments were compared using ANOSIM.

To demonstrate more clearly the relationship between food type, dose, and faunal response, sepa-

a)

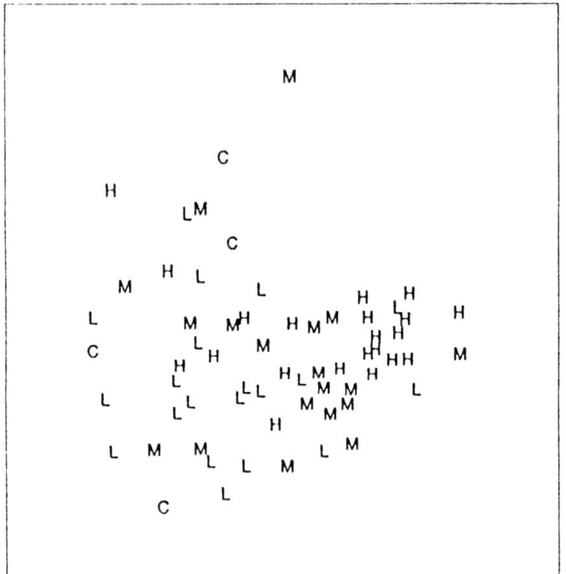

b)

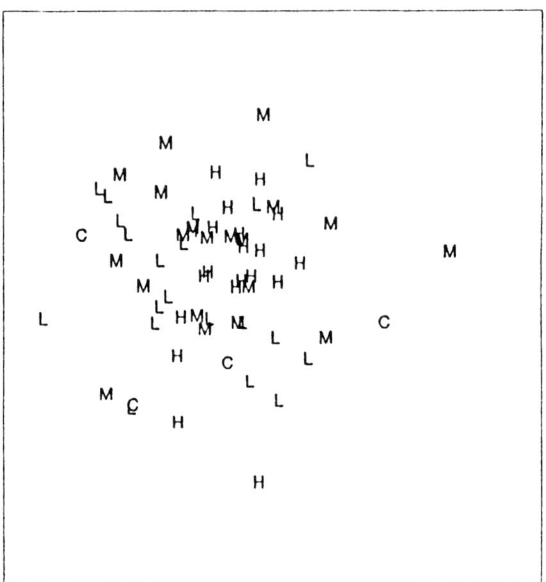

Fig. 2. Gironde mesocosm experiment: MDS plots of end-condition cores by dose level,(a) Untransformed data, (b) transformed data. C = Control, L = Low, M = Medium and H = High dose.

rate MDS plots were produced for each, using untransformed data. Figure 3 shows that for foods other than *Scirpus* and *Enteromorpha*, the four high dose treatments were closely clustered together on the right hand side of the plot while the remaining treatments showed no consistent pattern. ANOSIM results are summarised in Fig. 4. They indicate significant responses ($p<0.05$) to Leaf at high dose treatments over the other levels, and to 'Mixture' and Phytoplankton at high and medium dose over the low dose and controls. Cores fed on *Enteromorpha* and *Scirpus* showed no significant differences between treatments.

SIMPER analysis on untransformed data showed that almost 80% of the difference between start and end data related to the level of the decline in the abundance of bivalves, *Streblospio*, *Polydora ciliata* and *Pygospio*. *Streblospio* was the single most influential species within these analyses. Figure 5 compares the abundance of this polychaete at the start of the experiment with its numbers in cores treated with the various food doses. This species was significantly more abundant at high food doses than at the lower levels. High variability in the initial samples prevented any significance being attained in comparisons between the start and end of the experiment.

Discussion

By their very nature, laboratory experiments cannot fully mimic the natural situation, but by holding conditions constant and manipulating single factors they can be used to elucidate effects which cannot be determined by field-sampling alone. One cannot expect the fauna of experimental cores maintained in the laboratory to mirror exactly the community from which they were taken; some divergence must be expected. In interpreting the results it is important to evaluate the significance of any differences between real and experimental conditions. In the experiments described in this paper we believe that we achieved a reasonable simulation of a natural community. In the high dose treatments, the fauna remained close to that at the start of the experiment. This alone must indicate that the mesocosm system was not inimical to the fauna which we were manipulating. In preliminary experiments, with the addition of higher doses of organic matter, we observed the development of bacterial mats and the presence of reducing conditions at the sediment surface but such conditions were never encountered during the experiment which we describe here.

The mortality of the fauna in the mesocosm experiments was not uniform and cannot be attributed to an adverse effect of the mesocosm itself. The similarity of

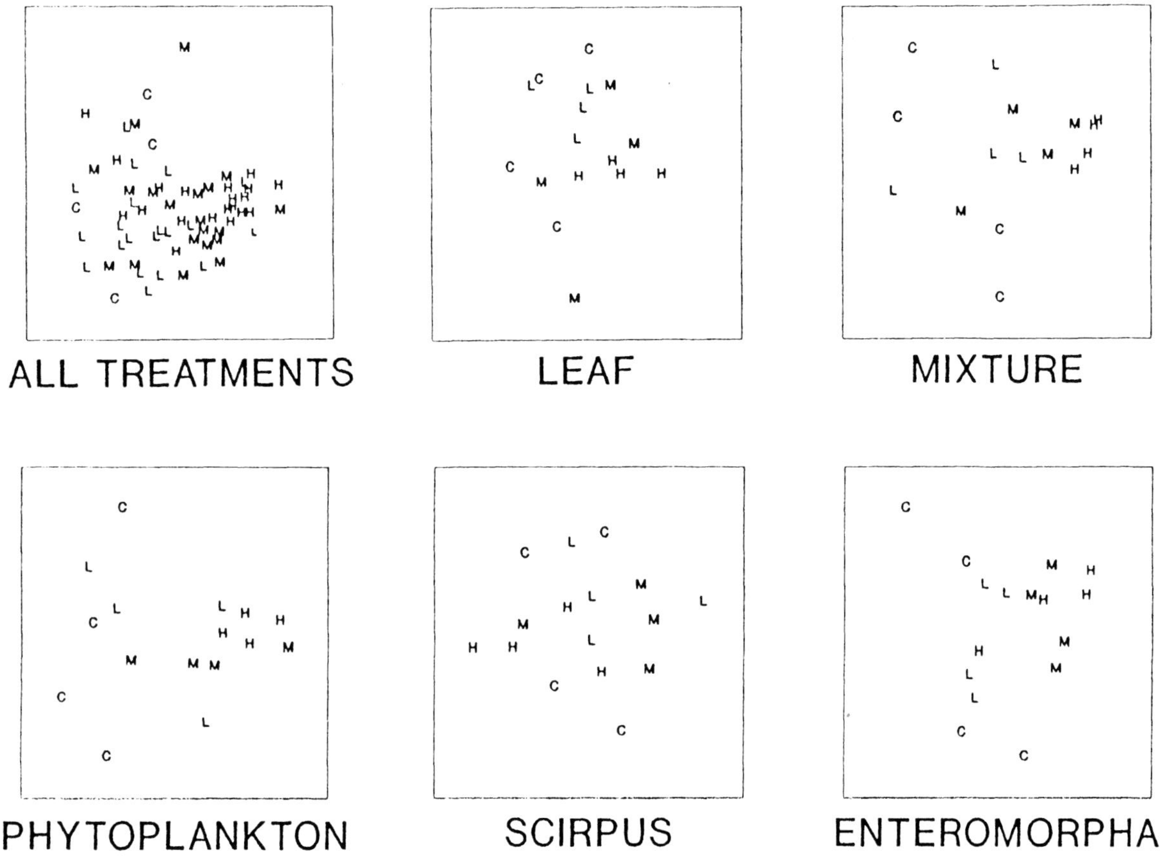

ALL TREATMENTS
LEAF
MIXTURE
PHYTOPLANKTON
SCIRPUS
ENTEROMORPHA

Fig. 3. Gironde mesocosm experiment separate MDS plots of untransformed data from end condition cores according to type of food. L = Low, M = Medium and H = High dose.

the end-condition fauna in Gironde cores at high dose treatments of leaf, phytoplankton and 'mixture' when compared with the start condition of all Gironde cores supports the contention that, given a sufficient quantity of the right food, the system was capable of supporting a normal estuarine assemblage. While it appears that our hypothesis that the faunas of the two estuaries would perform differently under variable organic loadings is supported, the nature of the changes are not those which we might have predicted. Our intention was to feed the sediment cores at rates which would vary between underfeeding to substantial overfeeding but in the Gironde, only the high dose treatments did not show a decline in the abundance of the fauna. This must call into question the validity of literature estimates (Gee *et al.*, 1985) for 'typical' or 'normal' rates of carbon flux to the benthos in estuarine and coastal waters.

Most estimates of the benthic carbon flux in marine environments are based on the collection of downward sinking particles in sediment traps and thus make little

H
M
L
C
Leaf
Mixture
Phytoplankton
Scirpus
Enteromorpha

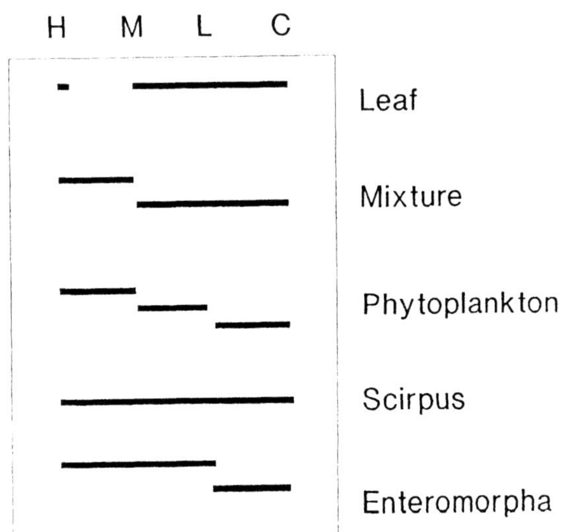

Fig. 4. Gironde mesocosm experiment: separation of dose level results from end-condition cores using ANOSIM. Lines under dose types (H, M, L, C as in Fig. 2) link results not significantly different at the $p = 0.05$ level.

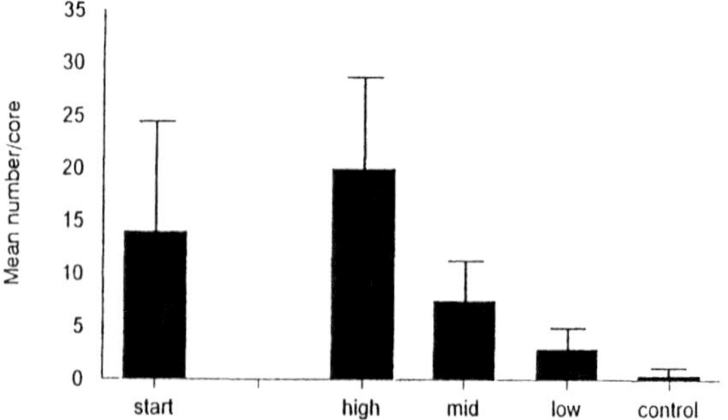

Fig. 5. Streblospio shrubsolii: mean number per core in Gironde mesocosm experiment with 95% confidence interval.

or inadequate allowance for lateral advection of particles, especially in the benthic boundary layer (Graf, 1992). Estuarine animals feeding at the sediment/water interface primarily experience a horizontal flux of particles, many of a size and density such that they have never have been in suspension and hence which will have been underestimated by sediment traps. Moreover, the labile proportion of the carbon available to the animals is not well documented and may bear little relation to analyses based only on sediment trap data. The lack of lateral currents in the mesocosm could have been particularly inimical to spionids which respond facultatively to variable flow conditions by switching between suspension and deposit feeding (Taghon *et al.*, 1980). This may have obliged the spionids to deposit feed on a restricted food source with a resultant increase in intra-specific competition. Although there is no direct evidence, it may be that the decline from high numbers of individuals in Scheldt cores reflected this type of competition given that the same conditions sustained the smaller numbers in the Gironde cores with high dose treatments of leaf, phytoplankton or 'mixture'.

However, it remains possible that the amount of labile carbon present in our feeding regimes was not available to the animals. This could have been due to either a rapid release and loss from the system of dissolved organic matter or by a proportion of the carbon being in a refractory state. It is also possible that the particle size may have been too large, leading to insufficient digestion during gut passage. Such speculations and the nature of any interaction with the meiofauna might form the basis for further experimentation using the mesocosm system. The results themselves, underline the difficulties of understanding, let alone simulating, the total pattern of supply of utilizable carbon to the benthic communities of shallow marine habitats.

Acknowledgments

We thank Melanie Austen for her co-operation in enabling us to obtain macrofaunal material from her mesocosm experiments, and Andrea McEvoy for technical assistance.

References

Austen, M. A. & R. M. Warwick, 1995. Effects of manipulation of food supply on estuarine meiobenthos. Hydrobiologia 311 (Dev. Hydrobiol. 110): 175–184.

Beukema, J. J. & G. C. Cadee, 1986. Zoobenthos responses to eutrophication of the Dutch Wadden Sea. Ophelia 26: 55–64.

Clarke, K. R., 1993. Non parametric multivariate analysis of changes in community structure. Aust. J. Ecol. 18: 117–143.

Clarke, K. R. & R. M. Warwick, 1994. Similarity based testing for community pattern: the two-way layout with no replication. Mar. Biol. 118: 167–176.

Creutzberg, F., P. Wapenaar, G. Duineveldt & N. L. Lopez, 1984. Distribution and density of the benthic fauna in the southern North Sea in relation to bottom characteristics and hydrographic conditions. P-V. Réun. Cons. Int. Explor. Mer 183: 101–110.

Gee, J. M., R. M. Warwick, M. Schanning, J. A. Berge & W. G. Ambrose, 1985. Effects of organic enrichment on meiofaunal abundance and community structure in sublittoral soft sediments. J. exp. mar. Biol. Ecol. 91: 247–262.

Graf, G., 1992. Benthic-Pelagic coupling: a benthic view. Oceanogr. Mar. Biol. annu. Rev. 30: 149–190.

Heip, C., 1989; The ecology of the estuaries of the Rhine. In Ros, J. D. (ed.), Meuse and Scheldt in the Netherlands. Topics in Marine Biology. Scient. Mar. 53: 457–463.

Laane, R. W. P. M., H. Etcheber & J. C. Relexans, 1987. Particulate organic matter in estuaries and its ecological implications for macrobenthos. SCOPE/UNEP Sonderband 64: 71–91.

Levin, L. A. & E. L. Creed, 1986. Effect of temperature and food availability on reproductive responses of *Streblospio benedicti* (Polychaeta: Spionidae) with planktotrophic or lecithotrophic development. Mar. Biol. 92: 103–113.

Levinton, J. S. & S. Stewart, 1988. Effects of sediment organics, detrital input and temperature on demography, production and body size of a deposit feeder. Mar. Ecol. Prog. Ser. 49: 259–266.

Lopez, G. R. & J. S. Levinton, 1987. Ecology of deposit feeding animals in marine sediments. Quart. Rev. Biol. 62: 235–260.

Pearson, T. H. & R. Rosenberg, 1978. Macrobenthic succession in relation to organic enrichment and pollution of the marine environment. Oceanogr. Mar. Biol. Annu. Rev. 16: 229–311.

Taghon, G. L., A. R. M. Nowell & P. A. Jumars, 1980. Induction of suspension feeding in spionid polychaetes by high particulate fluxes Science, N.Y. 210: 562–564.

Whitlach, R. B., 1980. Patterns of resource utilization and coexistence in marine intertidal deposit feeding communities. J. mar. Res. 38: 743–765.

Whitlach, R. B., 1981. Animal-sediment relationships in intertidal marine benthic habitats: some determinants of deposit-feeding species diversity. J. exp. mar. Biol. Ecol. 53: 31–45.

Appendix 1. Kendall, Davey & Widdicombe, Plymouth Marine Laboratory. Macrofaunal recoveries at week 20 from 64 cores in the Gironde mesocosm experiment: 16 cores were analysed from each of four mesocosm basins ('System'). Treatment codes: C = unfed Controls; L = Leaf, M = Mixture, P = Phytoplankton, S = *Scirpus* and E = *Enteromorpha*; 25, 75 and 200 = dose levels in g C m^{-2} Yr^{-1} equivalent.

System 1	C	L 25	L 75	L 200	M 25	M 75	M 200	P 25	P 75	P 200	S 25	S 75	S 200	E 25	E 75	E 200
Nepthys	0	1	0	0	0	1	0	0	0	0	0	0	1	0	0	0
Nereis	0	1	0	1	0	0	1	1	0	1	1	0	0	0	1	1
Polydora	0	2	1	5	4	4	12	0	2	3	0	0	0	0	0	0
Pygospio	6	1	2	1	4	1	10	1	1	0	0	2	4	1	1	8
Heteromastus	3	1	0	0	0	1	0	0	0	0	0	0	3	0	0	2
Streblospio	1	1	7	4	6	3	13	20	10	23	15	4	0	2	9	42
Bivalve	1	2	1	3	3	8	3	1	6	3	0	0	2	0	0	4
T. pseudogaster	7	0	0	1	0	0	0	8	0	0	1	0	0	0	0	1
Cyathura	0	0	0	0	0	0	0	0	0	1	0	1	0	0	1	0
Pseudopolydora I	0	0	1	0	0	0	1	0	1	0	0	1	0	1	0	1
Eteone longa	0	0	0	0	0	0	0	0	0	0	0	0	0	0	0	0
Pseudopolydora II	0	0	0	0	0	0	0	0	0	0	0	0	0	0	0	0
T. benedi	0	0	0	0	0	0	0	0	0	0	0	0	0	0	1	0

System 2	C	L 25	L 75	L 200	M 25	M 75	M 200	P 25	P 75	P 200	S 25	S 75	S 200	E 25	E 75	E 200
Nepthys	0	0	0	0	0	0	0	0	0	0	0	0	0	0	0	0
Nereis	1	1	1	2	0	1	1	1	1	1	0	1	2	0	0	0
Polydora	0	0	1	4	3	1	2	2	0	4	1	1	5	8	7	6
Pygospio	1	2	0	1	0	1	2	1	0	1	0	0	3	6	9	5
Heteromastus	1	0	1	0	2	3	4	3	0	0	1	0	0	1	5	0
Streblospio	0	2	0	22	2	12	24	0	30	61	1	2	1	3	19	11
Bivalve	0	0	1	2	0	0	0	2	0	0	0	0	0	0	0	3
T. pseudogaster	0	0	3	0	0	3	1	1	0	0	0	1	2	0	2	0
Cyathura	0	0	0	0	0	0	0	0	0	0	0	0	0	0	0	0
Pseudopolydora I	0	0	0	0	0	2	2	0	0	0	0	0	0	1	0	1
Eteone longa	0	0	0	1	0	0	0	1	0	1	0	0	0	0	0	0
Pseudopolydora II	0	0	0	0	0	0	0	0	0	0	0	0	0	0	0	0
T. benedi	0	0	0	0	0	0	0	0	0	0	1	1	0	0	0	0

System 3	C	L 25	L 75	L 200	M 25	M 75	M 200	P 25	P 75	P 200	S 25	S 75	S 200	E 25	E 75	E 200
Nepthys	0	0	0	0	0	2	0	0	0	0	0	0	1	0	0	1
Nereis	0	0	0	1	0	0	1	0	1	1	0	1	0	0	0	0
Polydora	1	1	1	2	1	1	0	1	6	4	2	0	0	0	3	0
Pygospio	0	1	1	0	2	6	4	0	6	0	2	3	0	1	8	1
Heteromastus	0	1	0	0	0	0	1	1	0	2	0	5	2	0	0	3
Streblospio	0	2	1	3	0	1	15	1	8	31	1	0	0	3	7	4
Bivalve	1	0	0	4	0	1	10	0	1	2	1	0	1	1	0	1
T. pseudogaster	2	0	0	0	0	0	0	4	0	0	0	0	0	0	0	0
Cyathura	0	0	0	0	0	0	0	0	0	0	0	1	0	0	1	0
Pseudopolydora I	0	0	0	0	0	0	1	0	0	0	0	0	0	0	3	0
Eteone longa	0	0	0	0	0	0	0	0	0	0	0	0	0	0	0	0
Pseudopolydora II	0	0	0	0	1	0	0	0	0	0	0	0	0	0	0	0
T. benedi	0	0	0	0	0	0	0	0	0	0	0	0	0	0	0	0

System 4	C	L 25	L 75	L 200	M 25	M 75	M 200	P 25	P 75	P 200	S 25	S 75	S 200	E 25	E 75	E 200
Nepthys	0	0	1	0	0	0	0	0	0	0	0	0	0	1	0	0
Nereis	0	1	0	1	1	0	1	0	0	1	1	0	1	0	1	1
Polydora	0	0	0	1	4	6	0	0	6	5	0	2	2	1	1	0
Pygospio	1	1	0	4	2	1	0	3	6	2	2	1	2	6	0	4
Heteromastus	0	2	0	5	0	0	1	0	1	0	2	0	2	1	0	1
Streblospio	1	0	0	7	1	7	15	7	1	16	4	6	1	3	8	29
Bivalve	0	0	0	1	2	1	3	0	3	10	0	3	0	1	3	4
T. pseudogaster	0	0	1	2	0	1	1	0	0	1	0	6	0	4	0	0
Cyathura	0	0	0	0	1	0	0	1	0	0	0	0	0	0	0	0
Pseudopolydora I	0	0	1	1	3	3	1	0	0	2	1	2	0	4	0	2
Eteone longa	0	0	0	0	0	0	0	0	0	0	0	0	0	0	0	0
Pseudopolydora II	0	0	0	0	0	0	0	0	0	0	0	0	0	0	0	0
T. benedi	0	0	0	0	0	0	0	0	0	0	0	0	0	0	0	0

Appendix 2. Gironde Mesocosm Experiment: Numbers of Individuals (NI) and numbers of species (NS) of macrofauna recovered from cores at the start and, from four treatment conditions, at the end of the experiment 'High', 'Mid', 'Low' = dose levels of feeding equivalent to 200, 75 and 25 g C m^{-2} Yr^{-1} respectively; control = unfed controls. CI = confidence interval. Mean and standard deviation etc not calculated for the four control cores.

	Start		High		Mid		Low		Control	
	NI	NS	NI	NS	NI	NS	NI	sp.	NI	NS
	20	5	15	6	12	5	9	7	18	5
	17	4	40	6	18	6	17	4	3	3
	19	4	31	5	8	4	31	5	4	3
	45	4	10	1	20	5	17	3	2	2
	16	7	59	7	13	5	4	3		
	77	5	32	6	7	5	5	3		
	30	6	36	7	23	7	7	3		
	28	6	68	5	31	2	11	7		
	22	6	13	5	6	5	4	4		
	11	6	26	5	42	5	19	5		
	25	5	10	4	3	3	5	4		
	22	6	32	6	11	5	4	3		
	29	7	10	5	22	5	7	4		
	91	8	4	3	1	4	6	4		
	27	4	10	5	22	5	5	3		
	23	5	22	8	3	3	4	3		
			22	6	19	6	14	7		
			37	7	17	5	11	3		
			8	5	28	6	10	5		
			41	6	13	1	21	8		
mean	31.38	5.5	27.80	5.55	16.10	4.75	10.55	4.40		
ρ	22.04	1.21	17.20	1.19	9.92	1.10	7.28	1.64		
95% CI	11.02	0.60	7.60	0.53	4.43	0.51	3.20	0.73		

Appendix 3. Westerscheldt Mesocosm Experiment: Numbers of Individuals (NI) and numbers of species (NS) of macrofauna recovered from cores at the start and, from four treatment conditions, at the end of the experiment 'High', 'Mid', 'Low' = dose levels of feeding equivalent to 200, 75 and 25 g C m^{-2} Yr^{-1} respectively; control = unfed controls.

	Start		High		Mid		Low		Control	
	NI	NS	NI	NS	NI	NS	NI	sp.	NI	NS
	67	7	56	4	70	5	41	3	56	4
	184	6	98	5	48	4	51	4	87	5
	107	6	128	6	74	4	101	5	41	5
	110	5	184	6	89	4	62	4	52	3
	96	5	57	6	82	4	62	4		
	136	5	81	4	52	7	115	6		
	98	7	87	5	92	4	101	4		
	174	8	56	4	64	4	105	4		
	145	6	63	3	62	5	53	4		
	107	6	68	4	71	4	84	5		
	90	5	176	5	12	7	74	6		
	121	4	35	4	22	4	59	4		
	15	3	41	5			103	5		
	123	5	66	4			54	5		
	151	5	54	4			108	6		
	86	6	45	4			66	5		
			46	4						
			40	4						
			52	3						
			30	6						
			77	5						
			52	5						
			72	7						
			89	4						
Mean	113.12	5.56	73.01	1.6	64	4.67	77.44	4.62	59	4.25
ρ	41.23	1.21	39.72	1.01	20.38	1.15	24.55		19.71	0.96

Hydrobiologia **311**: 215–224, 1995.
C. H. R. Heip & P. M. J. Herman (eds), Major Biological Processes in European Tidal Estuaries.

Estimating estuarine residence times in the Westerschelde (The Netherlands) using a box model with fixed dispersion coefficients

Karline Soetaert & Peter M. J. Herman
Netherlands Institute of Ecology, Centre for Estuarine and Coastal ecology, Vierstraat 28, NL-4401 EA Yerseke, The Netherlands

Key words: residence times, Westerschelde, estuary

Abstract

The residence time of the water masses in the Westerschelde estuary was determined using a simple compartment-model that simulates the advective-diffusive transport of a conservative dissolved substance (chlorinity). The residence time of a water parcel in the upstream part of the estuary (i.e. the time needed for this water parcel to leave the estuary) varied from about 50 days in winter to about 70 days in summer. The most seaward compartment had residence times of about 10–15 days.

Dispersive coefficients that are fixed in time were able to reproduce the observed salinity distributions very well in the Westerschelde. They were obtained by calibration on observed chlorinities. It is argued that the apparent relationship of dispersive coefficients with freshwater flow, which is observed in certain studies, could (partly) reflect the deviation from steady state conditions which are required assumptions to calculate these dispersive coefficients directly from salinity profiles.

Introduction

Flushing rates are an important characteristic of estuaries as this greatly affects the speciation of various dissolved substances and ultimately determines the amount of constituents that reaches the sea (Wollast, 1983). As detailed physical models are difficult to develop and are not always available, one can resort to tidally-averaged advective-dispersive models that describe the flow of matter across supposedly homogeneous compartments. These models can then be used to estimate the characteristic time scales of estuaries or they can be incorporated into global ecosystem models (Helder & Ruardij, 1982; Uncles & Radford, 1980).

In some of these box models, dispersion is allowed to fluctuate in time and hence the associated coefficients have to be calculated on several occasions directly on observed salinity profiles (Radford, 1978; Ruardij & Baretta, 1988). However, these calculations require the system to be in steady state, a prerequisite which is not always fulfilled (Loder & Reichard, 1981). Moreover, in these calculations the uncertainty in the salinity data set is translated into the dispersive coefficients.

In what follows we show that constant dispersive coefficients can reproduce the transport of a dissolved substance very well in the Westerschelde. These (constant) dispersion coefficients were calibrated on an observed data set of salinity using a simple dynamic transport model. The advantage of calibration in a dynamic model over direct calculation on a salinity profile is that steady state conditions need not be assumed. The resulting transport model was then used to calculate the characteristic time scales of the estuary.

This work is part of a multidisciplinary research aiming at the construction of a global ecosystem model of the Westerschelde (SW Netherlands). This is the only remaining true estuary of the Dutch delta. Freshwater discharge varies seasonally (mean 100 m^3 s^{-1}) (Heip, 1988) and is several orders of magnitude smaller than tidal exchange. Salinity intrudes for about 100 km inland and a freshwater tidal zone reaches another 50 kilometres more upstream. The salinity zones are relatively stable.

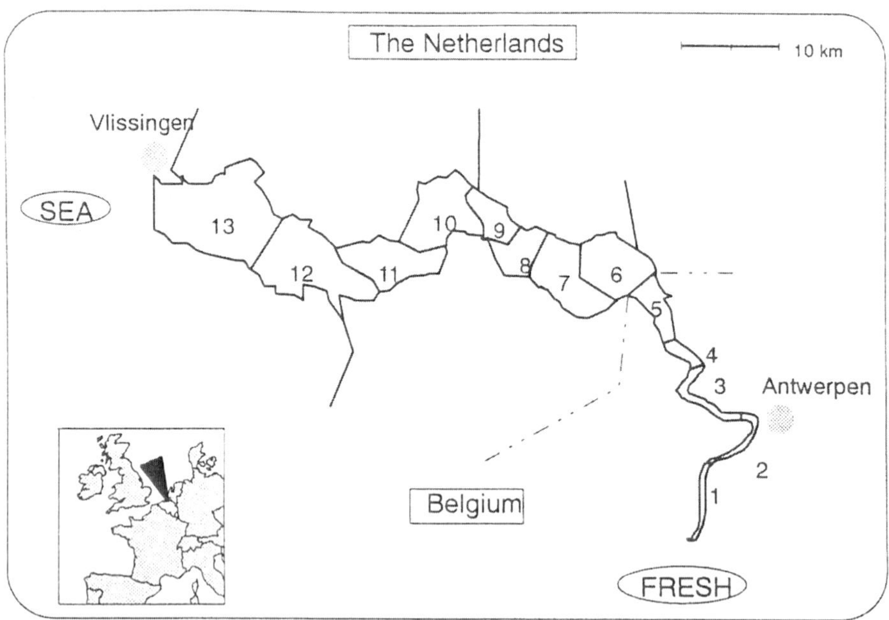

Fig. 1. Compartimentalisation of the Westerschelde pelagic.

Table 1a. Morphological characteristics of the model compartments at N.A.P.

Comp. no.	Volume (10^3 m^3)	Surface (10^3 m^2)	Depth (m)	Cross surface (10^3 m^2)	Total length (m)
1	26649	2973	9.0	4	7950
2	31957	3075	10.4	4	8300
3	59460	6387	9.3	6	9600
4	28807	2854	10.1	8	5100
5	69830	7772	9.0	14	9700
6	131128	16420	8.0	17	5950
7	86636	14380	6.0	32	5700
8	106008	14380	7.4	35	5300
9	158476	13360	11.9	39	5900
10	312588	34600	9.0	45	6900
11	300120	30300	9.9	50	6200
12	593684	49360	12.0	75	12100
13	873080	63620	13.7	80	13300

Table 1b. Bulk dispersion coefficients (E') and tidal dispersion coefficients (E) at the compartment interfaces.

From Comp. no.	to	Bulk disp. coeff E' (m^3 s^{-1})	Tidal disp. coeff E (m^2 s^{-1})
fresh	1	0	0
1	2	45	91
2	3	9	16
3	4	213	224
4	5	280	188
5	6	262	130
6	7	652	153
7	8	455	75
8	9	1226	186
9	10	960	146
10	11	1921	265
11	12	1161	170
12	13	2325	381
13	sea	2120	352

Material and methods

Transport of dissolved substances in the Westerschelde was simulated by means of a box-model. It consists of an advective-dispersive finite difference equation, described in a constant volume reference frame that moves with the tides (O'Kane, 1980; Thomann & Mueller, 1987). The compartimentalisation of the model area at mid-tide is as in Fig. 1. Some morphological characteristics of the model compartments are in Table

1. The monthly data set of the advective flows across the compartment interfaces and the extensive chlorinity data set used for calibration were kindly provided from the SAWES model (SAWES, 1991; Van Eck & de Rooiy, 1990). Each salinity measurement was first transposed to the location with respect to the fixed volume reference frame (mean tidal level) and then assigned to one of the modeled compartments, using data obtained from SAWES (1991).

Implementation of the advective-dispersive transport equation was done in the modelling environment SENECA (de Hoop *et al.*, 1993). This software package takes care of the mathematical routines that are common to most modelling exercises. Numerical integration can be performed by means of a 4th or 5th order Runge-Kutta or adaptive Euler method. During the calibration process the SENECA algorithm searches, within the given range of parameter values, for the parameter set that minimizes the weighted sum of squared deviations of modelled to observed time series (Klepper *et al.*, 1991; de Hoop *et al.*, 1993). SENECA also provides easy input (e.g. waste loads, boundary conditions, forcing functions, initial conditions) and output management (extensive graphical and listing facilities). When using this model environment, the modeler simply has to define parametres, variables, state variables, forcing functions and waste loads, starting conditions and implement the advective-dispersive equation.

The advective-diffusive transport equation

The differential equation which describes the concentration (s) of a conservative substance as a function of time (t) and space (x) in a reference frame that moves with the tides is given by:

$$\frac{\partial s}{\partial t} = -\frac{1\partial}{A\partial x}(Qs) + \frac{1\partial}{A\partial x}(EA\frac{\partial s}{\partial x}) \qquad (1)$$

(Thomann & Mueller, 1987; O'Kane, 1980).

Mass transport is a function of the freshwater flow (advective transport, first term) and a transport caused by heterogeneities introduced by the tides (dispersive transport, second term).

This differential equation can be replaced by an approximate algebraic equation, which is solved numerically by computer. In practice we used a 'finite difference' approximation, which calculates concentrations (s_i) at the centre of supposedly homogeneous compartments (i). In the case of a one-dimensional estuary one obtains a set of N first-order ordinary linear differential equations:

$$V_i\frac{ds_i}{dt} = Q_{i-1,i}(\alpha_{i-1,i}s_{i-1} + (1-\alpha_{i-1,i})s_i) - \qquad (2)$$
$$Q_{i,i+1}(\alpha_{i,i+1}s_i + (1-\alpha_{i,i+1})s_{i+1}) +$$
$$E'_{i,i+1}(s_{i+1} - s_1) - E'_{i-1,i}(s_i - s_{i-1}),$$

i = 1,.. N (Thomann & Mueller, 1987), with i = 1 denoting the spatial compartment most upstream, i = N the most downstream compartment.

$E'_{i,i+1} = E_{i,i+1} * A_{i,i+1}/\Delta x$, the 'bulk' dispersion or mixing coefficient ($m^3\ s^{-1}$) and $Q_{i,i+1}$ the advective flow ($m^3\ s^{-1}$) between compartments i and i+1. $A_{i,i+1}$ is the area of the interface between compartments i and i+1 (m^2), Δx the dispersion length (m). $\alpha_{i,i+1}$ is a weighting coefficient, used to calculate the concentration at the interface between boxes i and i+1, based on known values at the centre of both boxes. s_0 and s_{N+1} are then the concentrations of the conservative substance at the upstream, resp. downstream boundary.

The advective flows across the compartment interfaces ($Q_{i,i+1}$) are not always available and frequently one adapts an approximate finite difference equation which uses only freshwater flows at the upstream boundary (Q_r):

$$V_i\frac{ds_i}{dt} = r(s_{i-1} - s_i) + E'_{i,i+1}(s_{i+1} - s_i)$$
$$- E'_{i-1,i}(s_i - s_{i-1}), \qquad (3)$$

(i=1,.. N) when using backward differences (α = 1), (e.g. Miller & McPherson, 1991), or

$$V_i\frac{ds_i}{dt} = Q_r\frac{1}{2}(s_{i-1} - s_{i+1}) + E'_{i,i+1}(s_{i+1} - s_i)$$
$$- E'_{i-1,i}(s_i - s_{i-1}), \qquad (4)$$

(i =1,.. N), in the case of centered differences (α = .5), (e.g. Helder & Ruardij, 1982; Ruardij & Baretta, 1988).

In many studies the bulk mixing coefficients (E') are determined empirically from an observed data set of a conservative substance (salinity). It follows (for backward differences), from formula (3) that

$$E'_{i,i+1} = \frac{V_i\frac{ds_i}{dt} + (s_i - s_{i-1})(Q_r + E'_{i-1,i})}{(s_{i+1} - s_i)}, \qquad (5)$$

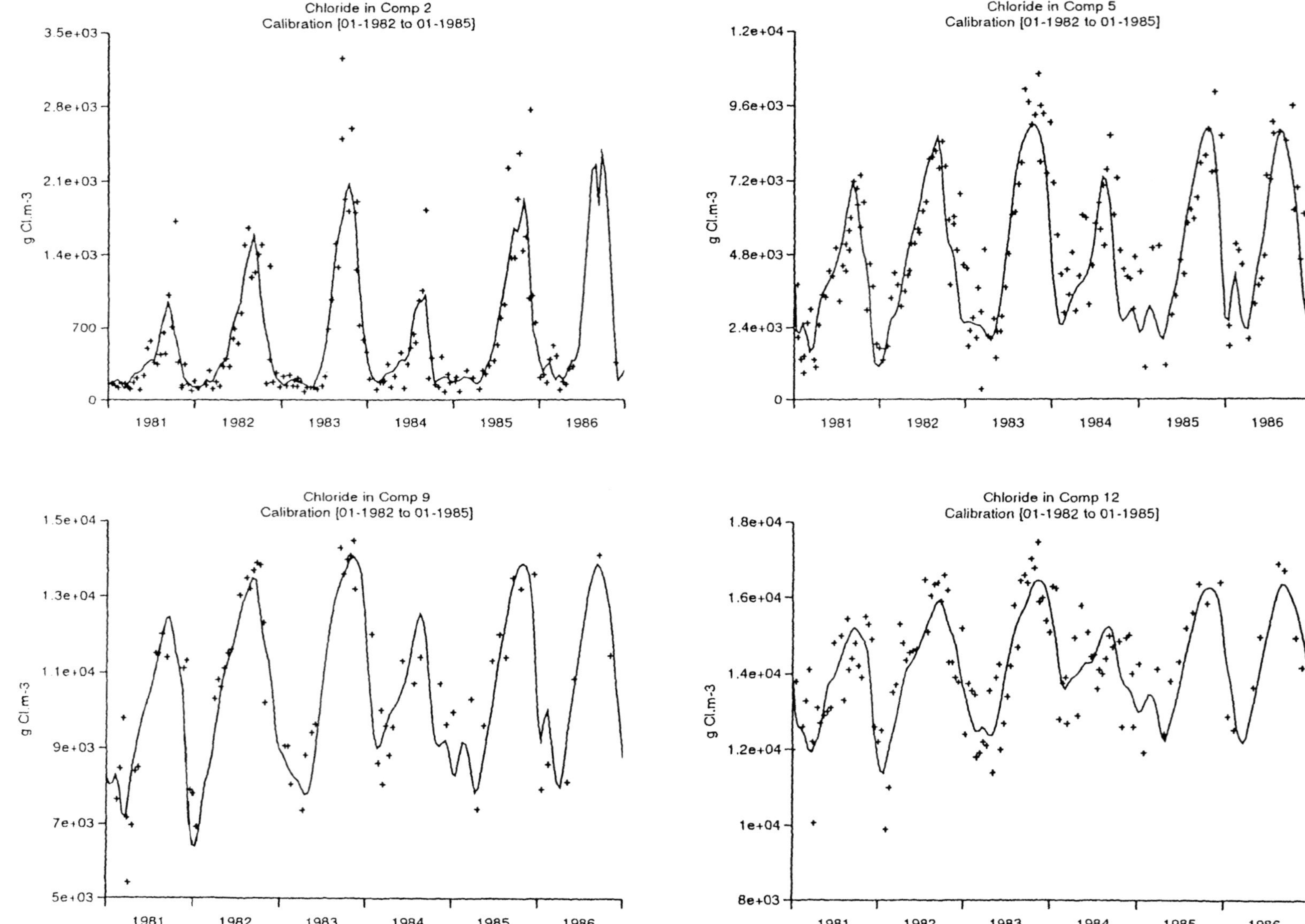

Fig. 2. Simulated and observed chlorinity in compartments 2, 5, 9 and 12. Calibration was on data from [01-1982 to 01-1985]. Verification of the model: [01-1981 to 01-1982] and [01-1985 to 01-1987].

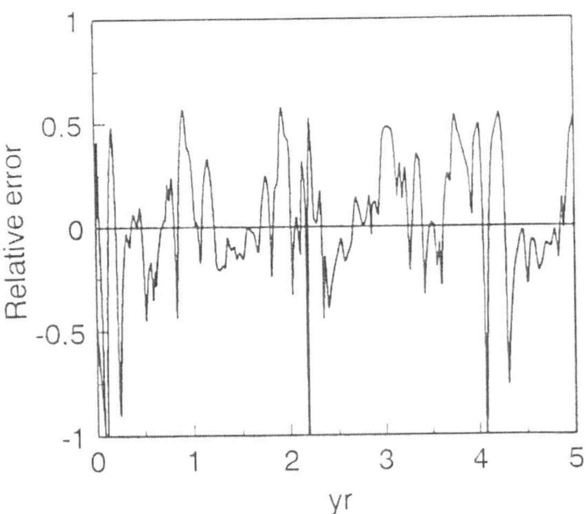

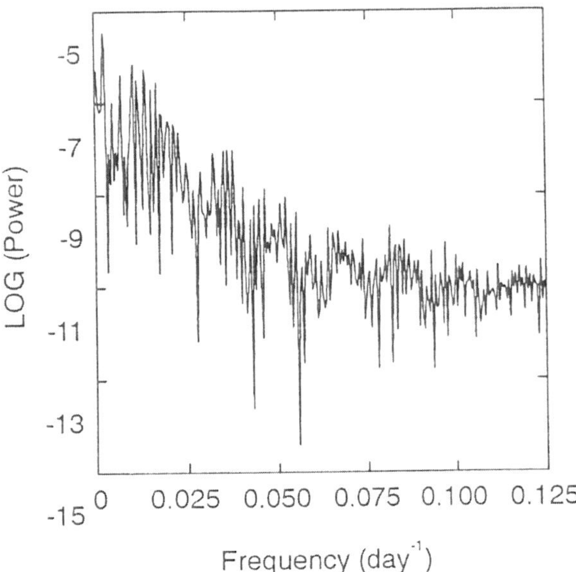

Fig. 3. Above: Residuals of modeled versus observed chlorinities in model compartment 5. Relative error = (observed concentration − simulated concentration) / observed concentration. Below: Spectral analysis of the residual errors of modeled versus observed chlorinities in compartment 5.

(i = 1,.. N), which, under the assumption of steady state ($V_i \frac{ds_i}{dt} = 0.$) becomes:

$$\overline{E'_{i,i+1}} = \frac{(s_i - s_{i-1})(Q_r + E'_{i-1,i})}{(s_{i+1} - s_i)} \qquad (6)$$

(Miller & McPherson, 1991). When dispersion at the most upstream boundary ($E'_{0,1}$) is known one can calculate all dispersion coefficients on the observed profile of salinity, working from up- to downstream, provided (s_{i+1} - s_i) is not equal to 0. Advection at the upstream boundary is the most important process and $E'_{0,1}$ is usually set to 0 or some very small number.

When the cross-sectional area, water flow and dispersion coefficients are constant along the estuary, which is in steady state, formula (1) becomes

$$\frac{Q}{A}\frac{ds}{dx} = E\frac{d^2s}{dx^2},$$

from which

$$\frac{EA}{\Delta x} = E' = \frac{Qs}{\Delta s}.$$

Thus in the case of centered differences the bulk dispersion coefficients can be estimated as:

$$\overline{E'_{i,i+1}} = \frac{Q_r(s_i + s_{i+1})}{2(s_{i+1} - s_i)} \qquad (7)$$

(Helder & Ruardij, 1982; Ruardij & Baretta, 1988).

Apart from being based on the (usually false) assumption of steady state, the inherent variability in the salinity profile requires that the data set should be smoothed carefully before attempting to estimate dispersion coefficients with formula (6) or (7). Formula (7) rests on the assumption of a constant cross-sectional area, which is never true in alluvial estuaries. Although formula (6) requires less assumptions than formula (7), a major drawback of (6) is that errors from the more upstream compartments are propagated downstream.

Dissolved transport in the Westerschelde

In order to circumvent the problems of direct estimation of dispersion coefficients on salinity profiles, we chose to estimate fixed dispersion coefficients by means of calibration in a dynamic simulation model. During the calibration procedure in SENECA (de Hoop *et al.*, 1993) a large number of simulations is run, starting with initial guesses of the dispersion coefficients (within preset limits). An algorithm then iteratively searches for this set of dispersion coefficients for which the weighted sum of squared deviations of modeled to observed time series of chlorinity is minimized.

For the model of the Scheldt estuary, the formula (2) was implemented, using advective flows across the compartment interfaces. A calibration on both the (fixed) dispersion coefficients and the weighting coefficients was then run based on chlorinity data of 1982 to 1985 (Fig. 2). The resulting fit of the data (1981–1986) provides evidence that dissolved transport in the Westerschelde can be adequately represented by means of fixed dispersion coefficients. Best goodness of fit was obtained using backward differences ($\alpha = 1$). The 'best' dispersion coefficients obtained are in Table 1.

As the dispersion coefficients were estimated during dynamic simulation, our method does not rely on the assumption of steady state. However, we assumed that constant, rather than fluctuating, dispersion coefficients were able to represent the transport of dissolved substances in the Westerschelde estuary. To see whether we had not missed any strong temporal signals (e.g. fortnightly, seasonally, yearly) we analysed the residuals of simulated versus observed chlorinities in the Westerschelde model using spectral analysis. A strong temporal cycle results in a high and delineated peak in the spectral plot, the position of which indicates the periodicity. The residuals and the resulting spectral plot is depicted in Fig. 3 for model compartment 5, which gave the worst fit of observed versus modeled chlorinities. It is much the same for the other compartments. There were no indications of strong periodic signals (no delineated peaks) suggesting that there were no important temporal variations in dispersive flows. This justifies the use of constant dispersion coefficients in the Westerschelde.

Estimating fluctuating dispersion coefficients from salinity profiles

We then used the transport model of the Westerschelde to investigate the magnitude of the error introduced when estimating dispersion coefficients directly from chlorinity profiles. We used the calculated chlorinity in the model as 'observed data' and estimated the dispersive coefficients with formula (6) and (7) for each simulated day, based on these generated profiles of 'observed chlorinities'.

It should be noted that by using the modeled chlorinities as observed data, one source of error in the estimation procedure (related to the uncertainty of the salinity profile) has been omitted. Furthermore as the chlorinity of the freshwater boundary was not modeled, and in view of the downstream propagation of errors for formula (6), we started the estimation beginning with $E'_{2,3}$ using the real value of $E'_{1,2}$ as a start (for formula (6)). The results of formula (6) from model compartment 5 are represented against time in Fig. 4. The other compartments gave similar results. Also results using formula (7) were comparable and are not depicted. One of the assumptions for using the formulae was a steady state condition, i.e. $ds_i/dt = 0$. From formulae (5) and (6) it follows that the error introduced by the steady state assumption is given by

$$\overline{E'_{i,i+1}} - E'_{i,i+1} = -\frac{V_i \frac{ds_i}{dt}}{s_{i+1} - s_i}, \qquad (8)$$

where $E'_{i,i+1}$ is the true and $\overline{E'_{i,i+1}}$ is the estimated bulk dispersion coefficient. In Fig. 4 (upper part), the horizontal line gives both the correct value of the dispersion coefficient (which is constant) and a zero rate of change of chlorinity (or steady state). The estimated bulk dispersion coefficient fluctuates around the dispersion coefficient that was actually used in the simulation (280 $m^3\,s^{-1}$) and varies from about 200 to 320 $m^3\,s^{-1}$. As was shown in formula (8), the estimated dispersive coefficients merely mirror ds_i/dt (or dCl/dt) and the larger the derivative ds_i/dt, the larger the estimation error. Due to the relationship between freshwater flow and this rate of change (Fig. 4, middle part), this is translated into a relationship between estimated dispersion coefficients and freshwater flow (Fig. 4 lower part).

Thus, because of the non-steady state of the system, estimating dispersion coefficients on salinity profiles generates seasonally fluctuating dispersion coefficients, even if the true dispersion is constant and the estimates deteriorate with increasing freshwater flow. It was amongst others this apparent relationship of dispersion and freshwater flow that led to the inclusion of a fluctuating dispersion coefficient in some ecological models (Helder & Ruardij, 1982; Radford, 1979 and Ruardij & Baretta, 1988). Although we are not claiming that constant dispersion coefficients could represent dissolved transport in these ecological models equally well as in the Westerschelde, the possible effect of non-steady state conditions on their estimates could explain part of the apparent fluctuations in dispersion coefficients.

Calculating characteristic time scales

Finally our model was used to calculate the residence time and the age of the sea water of water parcels in

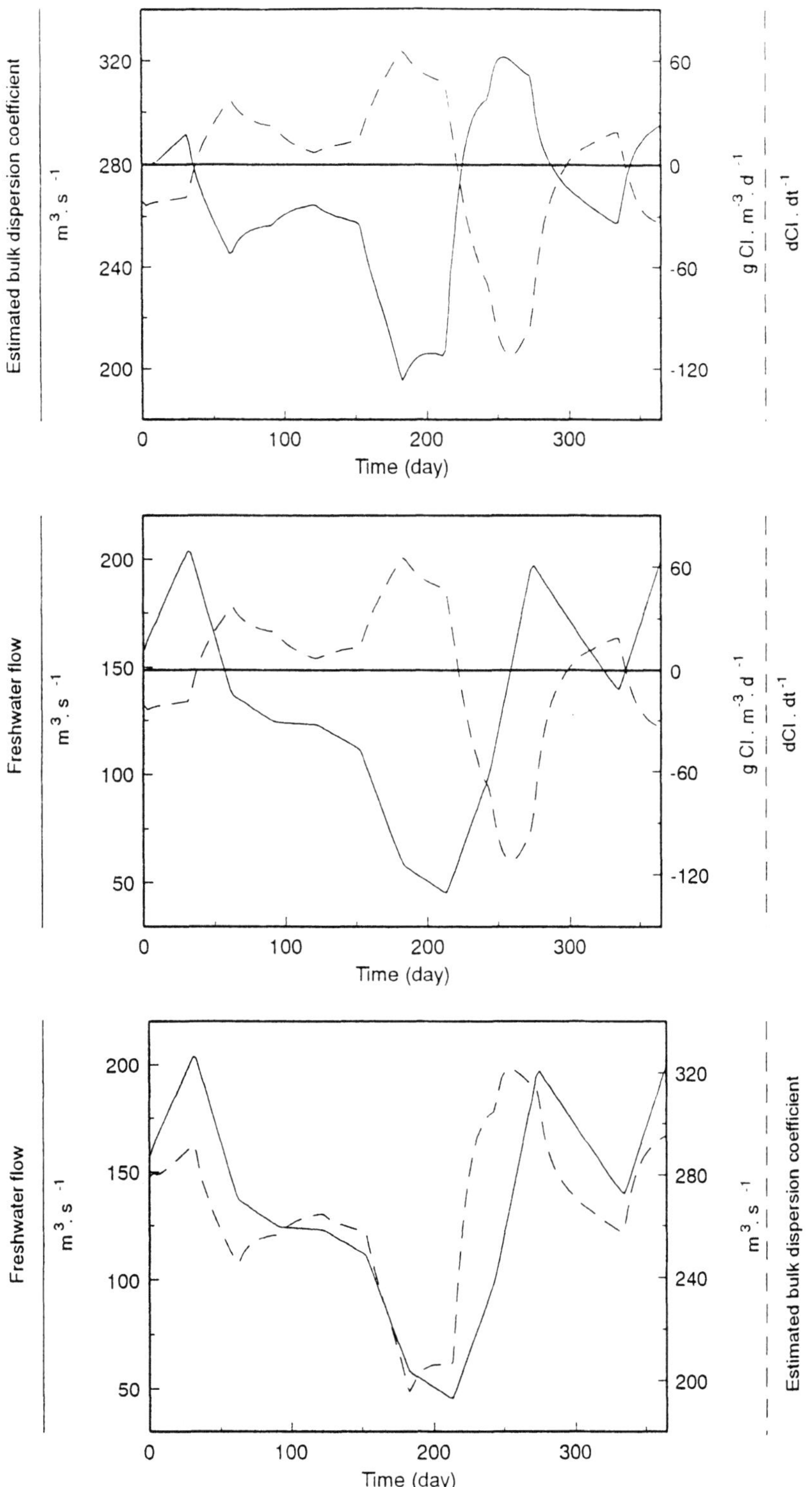

Fig. 4. Above: Estimated bulk dispersion coefficient $\overline{E'_{4,5}}$ at the interface of model compartment 4 and 5, and the rate of change for model compartment 5 (*dCL/dt*) during the simulation. Estimates were based on simulated chlorinity profiles, using a constant dispersion coefficient. The solid line indicates a constant value of Cl (*dCL/dt* = 0.) and the 'real' (constant) dispersion coeficient.
Middle: Freshwater flow and rate of change for compartment 5. The solid line indicates a constant value of Cl (*dCl/dt* = 0.)
Below: Freshwater flow and estimated bulk dispersion coefficient at the interface of model compartment 4 and 5.

the Westerschelde estuary as in Zimmerman (1976) and Takeoka (1984).

The residence time of a water parcel is defined as the average time a parcel needs to leave the estuary. It can be calculated for water parcels in the centre of each compartment (*j*) by setting as an initial condition the concentration of a conservative substance in this compartment to 1 (g m^{-3}) while the other compartments are empty. Next the relative decline of total mass is followed in time by the simulation model, keeping the concentration at the boundaries 0. If $\gamma(j,t)$ represents the relative amount still present in the estuary at time *t*, i.e.

$$\gamma(j,t) = \frac{1}{S(j,0) * V(j)} \sum_{i=1}^{N\mathrm{comp}} S(i,t) * V(i),$$

where N_{comp} is the number of compartments, $S(i,t)$ is the concentration of the substance in compartment i at time *t* and *V*(i) the total volume of this compartment. $S(i,t) \times V(\mathrm{i})$ is then the total mass of the substance present in compartment i at time *t*. Then

$$\text{residence time}(j) = -\int_0^{\infty} t\frac{d_{\gamma}(j,t)}{dt}dt$$

represents the average residence time of this water parcel (Takeoka, 1984; Zimmerman, 1976).

After integration by parts it can be shown that this equals

$$\int_0^{\infty} \gamma(j,t)dt$$

(Takeoka, 1984). This integral is easily solved in SENECA.

In Fig. 5 the residence time for water masses in the middle of the model compartments is calculated for both a winter (start 01-01-1984) and a summer situation (start 01-06-1984). Freshwater flows were allowed to change seasonally. The residence time of a water parcel in the most upstream compartments is in the order of 50 to 70 days, the most seaward compartments have residence times of about 10–15 days. Differences between summer and winter conditions are especially apparent in the most upstream compartments as these are mainly dominated by advective processes. As the system becomes more dispersive towards the sea side, the difference in residence times between summer and winter conditions is much less pronounced.

Under steady state conditions, setting the initial concentrations ($S(i,0)$, for i = 1.. Ncomp) to 0 and assuming a constant concentration at the seaward

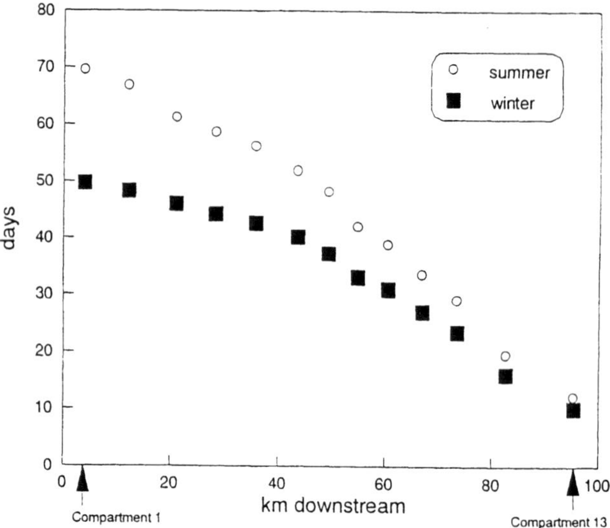

Fig. 5. Residence times of water in the centre of the model compartments for a typical winter and summer condition.

boundary ($c(\mathrm{sea},t)$ = 1), the age of the sea water in compartment *j* can be estimated as

$$\text{age of sea water } (j) = \frac{1}{V(j) * S(j,\infty)} \int_0^{\infty} \frac{t * d(S(j,t) * V(j))}{dt}dt,$$

(Zimmerman, 1976; Takeoka, 1984), where $V(j)$ is the volume of compartment *j*, $S(j,t)$ is the concentration in compartment *j* at time *t*, $S(j,\infty)$ is the concentration at the end of the simulation ($t = \infty$). In practice the 'end of the simulation' is achieved when the concentrations remain constant, i.e.

$$\frac{d(S(j,t) * V(j))}{dt} = 0.$$

This integral is easily solved in SENECA. Since the steady state is assumed, the age of seawater in a compartment and the residence time in this compartment (calculated for steady state conditions) must be equal (Takeoka, 1984).

Numerical dispersion

In addition to the true dispersion, an undesired dispersion can be introduced (numerical dispersion) as, due to the crude spatial resolution, material is shunted too rapidly from one part of the estuary to the other. The problem is especially critical if the advective transport and the dispersive mixing work in the same direction. For the calculation of residence times a sharp concentration gradient is initially created which is then tempered by both dispersion and advection which work in the same direction. As a first test on the degree of numerical dispersion, the magnitude of this initial gradient was increased 100-fold. This had no effect on the estimated residence time.

Next the calculated 'age of the sea water' was compared with the residence times determined under similar steady-state conditions. When estimating the age of seawater the concentration gradient is such that the dispersive and the advective term work in opposite direction, whereas for the calculation of the residence time both act in the same sense. Thus the difference in both estimation methods can also provide a means of assessing the magnitude of numerical dispersion. For the model of the Westerschelde, there was no difference whatsoever between the two methods for the 11 most downstream compartments. A difference of 0.14 and 0.07% was apparent for the most upstream and the second most upstream compartment. These differences were considered to be negligible.

Conclusions

The model presented here does not intend to study mixing processes in the Westerschelde in much detail. This is the scope of advanced physical models which take into account the influence of complex bottom topography in the area on the tidal currents. Instead we wanted to show that a simple mixing equation with fixed coefficients, used in a compartment model can reproduce convincingly real-world observations in the Westerschelde. This transport equation has been incorporated into an ecosystem model of the Westerschelde (Soetaert & Herman, 1995 a, b). The time scales calculated from this model can increase our awareness of the chemical and biological alterations that can take place in dissolved species that reside in the estuary.

Although it is likely that mixing intensity will change with tidal range through the Spring/Neap cycle (Radford, 1978), or with magnitude and direction of wind speed (Zimmerman, 1976) these factors did not seem very important for the Westerschelde.

An apparent relationship between dispersion coefficients, estimated from salinity profiles, and freshwater flow was mentioned by Helder & Ruardij (1982) and Uncles and Radford (1980) who ascribed it to the dependence of dispersion on buoyancy-driven horizontal circulation. Also the higher wind speeds that coincide with periods of higher run-off could provide an additional seasonal variability. However, we showed that dispersion coefficients, when estimated based on chlorinity profiles generated with a constant-dispersion model, fluctuated temporally around the true (constant) value because of non-steady state conditions in the estuary. The estimate of the dispersion coefficient deteriorated with increasing freshwater flow, and this may be an additional explanation for the observed relationship between freshwater flow and dispersion coefficients.

We conclude that merely calculating dispersion coefficients from observed salinity profiles provides no evidence for fluctuating dispersion in an estuary. A more thorough proof for the existence or absence of temporally variable dispersive effects consists in the description of advective-dispersive transport of conservative substances using fixed coefficients and a consequent investigation of the residuals of modeled and observed concentrations for strong temporal signals.

Acknowledgements

This study was supported by the ECOLMOD project of Rijkswaterstaat (The Netherlands) and by the MAST-JEEP project of the CEC. The authors are grateful to dr. Van Eck for providing us with the SAWES data set. We also thank prof.dr. R. Neves for critically reading the manuscript. This is manuscript number 729 from the NIOO-CEMO.

References

de Hoop, B. J., P. M. J. Herman, H. Scholten & K. Soetaert, 1993. SENECA 2.0. A Simulation ENvironment for ECological Application. Manual.

Heip, C., 1988. Biota and abiotic environment in the Westerschelde estuary. Hydrobiol. Bull. 22: 31–34.

Helder, W. & P. Ruardij, 1982. A one-dimensional mixing and flushing model of the Ems-Dollard estuary: calculation of time scales at different river discharges. Net. J. Sea Res. 15: 293–312.

Klepper, O., H. Scholten & J. P. G. Van de Kamer, 1991. Prediction uncertainty in an ecological model of the Oosterscheldt estuary. J. forecasting 10: 191–209.

Loder, T. C. & R. P. Reichard, 1981. The dynamics of conservative mixing in estuaries. Estuaries 41: 64–69.

Miller, R. L. & B. F. McPherson, 1991. Estimating estuarine flushing and residence times in Charlotte Harbor, Florida, via salt balance in a box model. Limnol. Oceanogr. 36: 602–612.

O'Kane, J. P., 1980. Estuarine water-quality manaqement. Pitman, Boston.

Radford, P. J., 1978. Some aspects of an estuarine ecosystem model - GEMBASE. In Jorgensen (ed.), State of the art in ecological modelling, Pergamon Press, Oxford, 301–322.

Ruardij, P. & J. W. Baretta, 1988. The construction of the transport model. In Baretta & Ruardij (eds), Tidal flat estuaries. Simulation and analysis of the Ems estuary. Springer-Verlag, Berlin: 65–76.

SAWES, 1992. Waterkwaliteitsmodel Westerschelde. WL-rapport T257.

Soetaert, K. & P. M. J. Herman, 1995a. Nitrogen dynamics in the Westerschelde estuary (SW Netherlands) estimated by means of the ecosystem model MOSES. Hydrobiologia 311 (Dev. Hydrobiol. 110): 225–246.

Soetaert, K. & P. M. J. Herman, 1995b. Carbon flows in the Westerschelde estuary (The Netherlands) evaluated by means of an ecosystem model (MOSES). Hydrobiologia 311 (Dev. Hydrobiol. 110): 247–266.

Takeoka, H., 1984. Fundamental concepts of exchange and transport time scales in a coastal sea. Cont. Shelf Res. 3: 311–326.

Thomann, R. V. & J. A. Mueller, 1987. Principles of surface water quality modelling and control. New York, Harper & Row.

Uncles, R. J. & P. J. Radford, 1980. Seasonal and spring-neap tidal dependence of axial dispersion coefficients in the Severn - a wide, vertically mixed estuary. J. Fluid Mech. 98: 703–726.

Van Eck, G. Th. M. & N. M. de Rooij, 1990. Development of a water quality and bio-accumulation model for the Scheldt estuary. In W. Michaelis (ed.), Coastal and Estuarine Studies. Springer Verlag. Berlin, Heidelberg, etc. 95–104.

Wollast, R., 1983. Interactions in estuaries and coastal waters. In Bolin, B. & R. B. Cook (eds), The major biogeochemial cycles and their interactions. Scope.

Zimmerman, J. T. F., 1976. Mixing and flushing of tidal embayments in the western Dutch Wadden Sea. Part I.: distribution of salinity and calculation of characteristic mixing time scales. Neth. J. Sea Res. 10: 149–191.

Hydrobiologia **311**: 225–246, 1995.
C. H. R. Heip & P. M. J. Herman (eds), Major Biological Processes in European Tidal Estuaries.

Nitrogen dynamics in the Westerschelde estuary (SW Netherlands) estimated by means of the ecosystem model MOSES

Karline Soetaert & Peter M. J. Herman
Netherlands Institute of Ecology, Vierstraat 28, NL-4401 EA Yerseke, The Netherlands

Key words: Nitrogen, budget, nitrification, denitrification, Westerschelde, estuary

Abstract

A tentative nitrogen budget for the Westerschelde (SW Netherlands) is constructed by means of a simulation model with thirteen spatial compartments. Biochemical and chemical processes in the water column are dynamically modeled; fluxes of dissolved constituents across the water-bottom interface are expressed by means of diagenetic equations.

The model is calibrated on a large amount of observed variables in the estuary (1980–1986) with relatively fine temporal and spatial detail. Additional constraints are imposed by the stoichiometric coupling of carbon, nitrogen and oxygen flows and the required conservation of mass. The model is able to reproduce rather well the observed distributions of nitrate, ammonium, oxygen and Kjeldahl nitrogen both in time and space. Also, model output of biochemical oxygen demand and total organic carbon falls within observed ranges.

By far the most pervasive process in the nitrogen cycle of the estuary is nitrification which mainly takes place in the water column of the upper estuarine part. On average about three times as much nitrate is leaving the estuary at the sea side compared to what enters from the river and from waste discharges. Ammonium on the other hand is consumed much faster (nitrification) than it is regenerated and only about one third of the total import leaves the estuary at the sea side. The budget for detrital nitrogen reveals import from the river, from wastes and from the sea. Phytoplankton uptake of inorganic nitrogen is negligible in the model.

About 21% of total nitrogen, 33% of inorganic nitrogen, is removed from the estuary (mainly to the atmosphere through denitrification) and the load of nitrogen net exported to the sea amounts to about 51 000 tonnes per year. Total denitrification in our model is lower than what was estimated in the literature from the late seventies, where a nitrogen removal up to 40–50% of the total inorganic load was reported. Part of the differences could be methodological, but inspection of the nutrient profiles that led to these conclusions show them to be different to the ones used in our study. The oxygen deficient zone has moved upstream since the late seventies, entrailing the zone of denitrification into the riverine part of the Schelde. The nitrification process now starts immediately upon entering the estuary.

Introduction

By far the major input of land-derived nutrients into the coastal sea occurs through riverine transport (Billen *et al.*, 1991). On their way to the coastal zone, the nutrients have to pass through estuaries where both the quantity and the quality of the constituents is altered. The Westerschelde estuary (SW Netherlands) is the last remaining true estuary from the Dutch Delta region and discharges on average about 100 m^3 freshwater per second into the Southern Bight of the North Sea (Van Eck & de Rooij, 1990). At the scale of the entire North Sea ecosystem, the Westerschelde contributes only marginally to nutrient load (Brockmann *et al.*, 1988) but its effect locally is enhanced due to the presence of a residual gyre which elongates the residence time of the water masses near the estuarine mouth (Nihoul & Ronday, 1975).

The Westerschelde is one of the most polluted estuaries in the world as it receives a substantial amount of largely untreated industrial and domestic wastes (Heip, 1988). Thus the estuary has an unusually high load of organic and inorganic substances. Although nitrogen is never limiting phytoplankton growth in the Westerschelde estuary (Kromkamp *et al.*, 1992; Soetaert *et al.*, 1994), it is the limiting macronutrient over most of the North Sea (Brockmann *et al.*, 1991) and the most common blooming species in the North Sea, the nuisant *Phaeocystis*, is probably nitrogen limited in Belgian coastal waters (Lancelot & Billen, 1984). As such, the amount of outwelling of nitrogen from the estuary could be important for the degree of coastal eutrophication.

A decade ago, a nitrogen budget for the Westerschelde was constructed by Billen *et al.* (1985, 1986, 1988). This budget was based on data from 1975 to 1983. They found that about 40–50% of the total nitrogen load was removed within the estuary, partly due to denitrification in the anaerobic water column in the most upstream estuarine stretch. Moreover, the authors suggested that restoring higher oxygen content in the river or estuary, without eliminating the inorganic nitrogen load, would reduce denitrification in the water column and thus – paradoxically – tend to increase the nitrogen discharge to the sea.

We have studied the flow and speciation of the various nitrogen species in the Westerschelde estuary by means of an ecosystem simulation model. The aim was to determine the fate and turnover of nutrients entering the estuary and to describe the temporal and spatial patterns of nutrient concentrations and fluxes in the estuary. Estimates are based on data from the years 1980–1986. Our – independently derived – budget is compared to that of Billen *et al.* (1985, 1986, 1988) and the results are discussed in the light of estuarine eutrophication.

The modelling effort that is presented here is achieved as part of a project financed by the ministry of transport and public works (The Hague) and as part of the Joint Estuarine Ecosystem Programme sponsored by the European community. This is one of a series of papers describing the ecosystem simulation model of the Westerschelde.

Material and methods

The ecological MOdel of the Schelde Estuary (hereafter denoted as MOSES) was implemented in the simulation environment SENECA (de Hoop *et al.*, 1993). This modelling package takes care of most routines common to modelling exercises (automatic calibration, sensitivity analysis, numerical integration) and provides easy input-output management. MOSES is a compartmentalized model, describing processes in the pelagic, in the intertidal and in the sublittoral. These zones are subdivided into thirteen spatial compartments. The sublittoral is defined as that part of the bottom in a pelagic compartment that is deeper than the intertidal and less deep than 10 m. The intertidal consists of thirteen distinct morphological units (see Fig. 1).

Transport of dissolved matter is modeled by a constant-volume advective-dispersive finite difference approximation (Thomann & Mueller, 1987) which has been calibrated on chlorinity data (Soetaert & Herman, 1995a). Particulate matter is also transported by means of advective and dispersive processes, but net advection at the most seaward compartments is directed into the estuary. This allows for the creation of a turbidity maximum and a seasonal pattern in the concentration of particulate matter (Soetaert & Herman, 1993). Organic matter is then moved in the model by assuming that it behaves partly dissolved, partly particulate.

The biological processes described in the model are (pelagic and benthic) primary production and feeding, respiration, excretion and physiological and predatory mortality of secondary producers (micro- and mesozooplankton, hyperbenthos, benthic deposit and suspension feeders). MOSES is essentially a carbon-based model with the oxygen and nitrogen cycles linked via stoichiometric equations. For a rationale on the modelling strategy and the other modeled processes, we refer to Soetaert & Herman (1993), Soetaert *et al.* (1994) and Soetaert & Herman (1995b). Part of the data set (waste loads, nutrient concentrations, BOD) was kindly provided by Dr Van Eck and Mr A. Schouwenaar (SAWES, 1991).

Biochemical and physical processes in the water phase

In what follows, STATE VARIABLES (e.g. concentrations of the various nutrients) are denoted by capital letters, parameters (e.g. rate constants, coefficients) are denoted in italics.

In short, the model simulates the temporal evolution of state variables by integrating their rate of change (denoted as dX/dt) in time. Numerical integration is performed within the simulation package SENECA (de Hoop *et al.*, 1993) by Euler integration with adaptive time step.

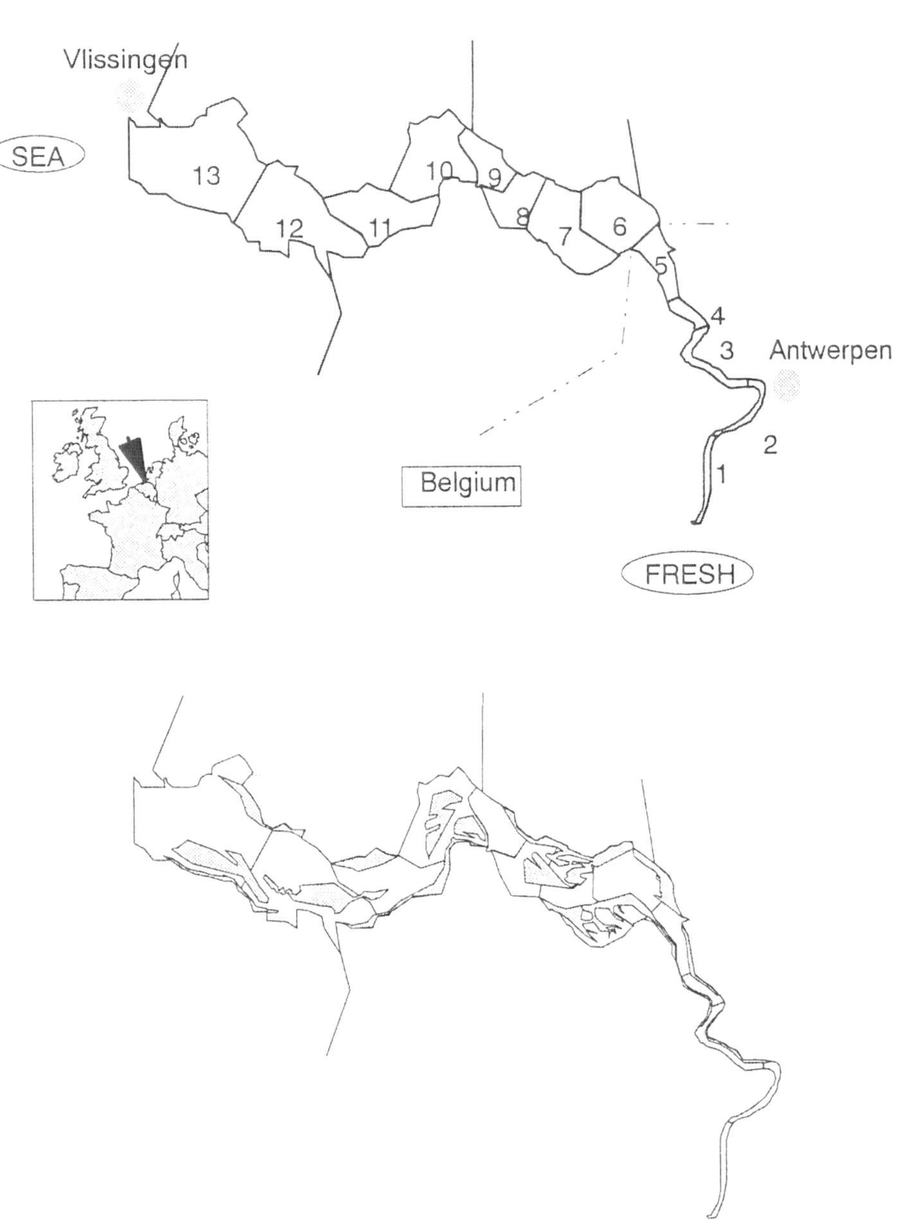

Fig. 1. Spatial compartimentalisation of the Westerschelde estuary. Upper half: pelagic compartments, lower half: intertidal compartments.

State variables (Table 1)
Two dissolved inorganic nitrogen state variables are considered: ammonium (NH_4) and nitrate + nitrite (NITR). Nitrite and nitrate are combined into one state variable since nitrite is generally rapidly converted to nitrate. Moreover, in the Westerschelde, the nitrite concentration usually is less than 5% of the total pool. Nitrous oxide and dinitrogen gas, both products of denitrification, are not modeled as these are lost to the atmosphere.

Dissolved oxygen (OX) has a strong impact on the nitrogen cycle as it determines the type of processes (nitrification-denitrification, aerobic degradation) that will take place in the estuary.

The detritus in rivers and estuaries is a complex mixture of biochemicals with oxidation times ranging

Table 1. State variables of MOSES relevant to the biochemical submodel.

State variable	Units	Acronym
Oxygen	$g\,O_2\,m^{-3}$	OX
Ammonium	$g\,N\,m^{-3}$	NH4
Nitrate+Nitrite	$g\,N\,m^{-3}$	NITR
Fast-decay detritus	$g\,C\,m^{-3}$	FD
Slow-decay detritus	$g\,C\,m^{-3}$	SD

Table 2. Effect of the modeled process on the state variables. '++' indicates increase, '- -' indicates decrease, '/' = no effect.

State variable: Process:	NITR	NH4	FD	SD	OX
Aerobic mineralisation	/	++	- -	- -	- -
Denitrification/nitrate reduction	- -	++	- -	- -	/
Nitrification	++	- -	++	/	- -
Aeration	/	/	/	/	++
Waste import	++	++	++	++	++

from hours to thousands of years (Spitzy & Ittekot, 1991). Although the labile organic matter in natural water is present as a continuum of biodegradabilities and C/N content (Garber, 1984), such precision cannot be attained by a simple ecosystem model. Thus, following Lancelot & Billen (1985) and Billen & Lancelot (1988), in MOSES two detritus state variables are distinguished, namely slowly and fast decaying detritus (SD and FD) each with its own biodegradabilities (*cSDMin* resp. *cFDMin*) and N/C content (*NCrSD* resp. *NCrFD*). The (arbitrary) distinction between dissolved and particulate organic matter is not made here. Both the slow and fast decaying detritus fractions consist of a dissolved and a particulate phase.

As in many other models (e.g. Klepper, 1989; Di Toro *et al.*, 1971), bacteria are not considered separately in MOSES, but they are included in the detritus fraction.

The modeled processes in the water column are aerobic mineralisation, denitrification, nitrification, aeration and waste import. Their effect on the state variables is summarized in Table 2. The equations used are given in Table 3. Table 4 explains the variables used to describe rates in these equations, whereas Table 5 gives the value and meaning of the parameters.

'Aerobic mineralization' is modeled as a first-order process with respect to the organic load (Streeter & Phelps, 1925) (Eqs. 1 and 2, Table 3). The influence of temperature on mineralisation is represented by means of a Q10 function. Oxygen limitation is modeled by means of a hyberbolic relation. The amount of ammonium produced by means of mineralisation relates to the amount of CO_2 production through the (fixed) Nitrogen to Carbon ratio of the detritus fraction (Eq. 3, Table 3). Per amount of carbon that is thus converted an amount *OCr* of oxygen is consumed (Eq. 4, Table 3).

'Denitrification' is an anaerobic and heterotrophic process, using nitrate and nitrite as an electron acceptor. The products of denitrification are N_2 or N_2O which are lost to the atmosphere. Nitrate reduction can also lead to NH_4^+ formation, especially at high organic matter content and low nitrate concentration. Denitrification is modeled as a first-order kinetic equation with respect to nitrate content. It depends on temperature (Q10 formulation) and is depressed at high oxygen concentrations (hyperbolic relation). The detritus fraction can also become limiting (Dlim) (Seitzinger, 1988) and both the availability and the reactivity of the two detritus fractions has to be considered. Dlim is expressed as a Monod equation with half saturation constant *ksDenD* (Eq. 5, Table 3). Denitrification rate is given in Eq. 6 (Table 3). For each gram of nitrate reduced, a (fixed) amount *CNrDenit* of detritus is converted to CO_2. The part of fast and slow detritus that is denitrified is both function of total load and reactivity (Eqs. 7 and 8, Table 3). Carbon to nitrogen stoichiometry is assured with ammonium. During nitrate reduction, nitrate is partly converted into ammonium. The resulting rate of change in ammonium is given in Eq. 9 (Table 3).

'Nitrification' is an aerobic process, usually mediated by two different types of autotrophic bacteria, which obtain energy from oxidation of ammonium to nitrite, followed by the oxidation of nitrite to nitrate. They obtain cell carbon from carbon dioxide. The process depends on temperature (Q10 formulation) and is inhibited in anaerobic conditions (hyperbolic relation). Owing to the slow development of nitrifying bacteria (Billen *et al.*, 1986) there is an upper limit to the rate of nitrification. This was modeled as a hyperbolic relationship with ammonium (Helder & De Vries, 1983; Somville *et al.*, 1982). The nitrification rate *cNITR* is then given by Eq. 10 (Table 3). Being autotrophic, the nitrifying bacteria produce organic carbon from CO_2, with a low efficiency (Helder & de Vries, 1983;

Table 3. Formulations used for the processes affecting nitrogen in the model. See Table 1 for a list of state variables, Table 4 for a list of variables used in these expressions, Table 5 for a list of parameters. The symbol @ represents all other processes (e.g. transport, uptake or excretion by zooplankton, etc...) affecting the state variables, but not explicitly discussed in this paper. Terms ending on 'Imp' in the differential equations of the state variables denote imports from (lateral) waste loads into the estuary. In several expressions, T denotes ambient temperature, T_0 denotes base temperature at which the basic rate is defined. Expressions are explained in the text.

MinFD	= cFDMin * exp ((T-T_0) * ln(Q10Min)/10) * (Oxsat/ (Oxsat + ksMinOX)) * FD	(1)
MinSD	= cSDMin * exp ((T-T_0) * ln(Q10Min)/10) * (Oxsat/ (Oxsat + ksMinOX)) * SD	(2)
NH4Min	= MinFD * NCrFD + MinSD *NCrSD	(3)
OXMin	= (MinFD + MinSD) * OCr	(4)
DLim	= (cFDMin * FD + cSDMin * SD) / (ksDenD + cFDMin * FD + cSDMin * SD)	(5)
Denitri	= cDenitr * exp ((T-T_0) * ln (Q10Denit)/10) * (ksDenOX/ (OX + ksDenOX)) * Dlim * NITR	(6)
fDenit	= Denitri * CNrDenit * cFDMin * FD / (cFDMin * FD + cSDMin * SD)	(7)
sDenit	= Denitri * CNrDenit * cSDMin * SD / (cFDMin * FD + cSDMin * SD)	(8)
NH4Denit	= Denitri * pNH4Denit + fDenit * NCrFD + sDenit * NCrSD	(9)
Nitri	= cNitr * (NH4 / (NH4 + ksNH4Nit)) * exp ((T-T_0) * ln (Q10Nit) / 10) * OX/ (OX + ksNitOX))	(10)
fNit	= CNrNit * Nitri	(11)
NH4Nit	= fNit * NCrFD	(12)
OXNit	= Nitri * (16 * 5 / 14-CNrNit * OCr)	(13)
Aer	= cOXex * (satOX-OX) * exp (0.023 * T) / depth	(14)
dFD/dt	= @ - MinFD - fDenit + fNit + fIMP	(15)
dSD/dt	= @ - MinSD - sDenit + sIMP	(16)
dNITR/dt	= @ - Denitri + Nitri + NitrImp	(17)
dNH4/dt	= @ + NH4Min + NH4Denit - Nitri - NH4Nit + NH4Imp	(18)
dOX/dt	= @ - OXMin - OXNit + Aer + OXImp	(19)

Table 4. List of variables used in the expressions of Table 3.

Variable	Units	Comments
MinFD	$g\,C\,m^{-3}\,d^{-1}$	Aerobic mineralisation rate of fast detritus
MinSD	$g\,C\,m^{-3}\,d^{-1}$	Aerobic mineralisation rate of slow detritus
NH4Min	$g\,N\,m^{-3}\,d^{-1}$	NH4 production rate in aerobic mineralisation
OXMin	$g\,O_2\,m^{-3}\,d^{-1}$	OX consumption rate in aerobic mineralisation
DLim	-	Limitation of denitrification by detritus
Denitri	$g\,N\,m^{-3}\,d^{-1}$	Denitrification rate
fDenit	$g\,C\,m^{-3}\,d^{-1}$	Loss rate of fast detritus in denitrificaiton
sDenit	$g\,C\,m^{-3}\,d^{-1}$	Loss rate of slow detritus in denitrification
NH4Denit	$g\,N\,m^{-3}\,d^{-1}$	Rate of change in NH4 due to denitrification
Nitri	$g\,N\,m^{-3}\,d^{-1}$	Loss rate of NH4 due to nitrification
fNit	$g\,C\,m^{-3}\,d^{-1}$	Production rate of fast detritus in nitrification
NH4Nit	$g\,N\,m^{-3}\,d^{-1}$	Uptake rate of NH4 for growth of nitrifiers
OXNit	$g\,O_2\,m^{-3}\,d^{-1}$	Oxygen consumption rate in nitrification
Aer	$g\,O_2\,m^{-3}\,d^{-1}$	Reaeration rate
SatOX	$g\,O_2\,m^{-3}\,d^{-1}$	Saturation concentration of oxygen at ambient temperature and salinity

Table 5. Best parameter values.

Parameter	Value	Units	Comment
BOD20	0.05	d^{-1}	Biochemical oxygen demand of wastes at 20 dg
fdFRSH	0.7	-	part of riverine detritus that is fast decaying
fdSEA	0.4	-	part of seawater detritus that is fast decaying
Q10Min	1.6	-	Q10 value for aerobic mineralisation
KsMinOX	0.54	$g O m^{-3}$	monod ct of oxygen limitation aerobic degrad
cSDMin	2.E-3	d^{-1}	Aerobic degradation of slow detritus at 20 dg
cFDMin	0.11	d^{-1}	Aerobic degradation of fast detritus at 20dg
NCrFD	0.3	$g N (g C)^{-1}$	nitrogen-carbon ratio of fast detritus
NCrSD	0.1	$g N (g C)^{-1}$	nitrogen-carbon ratio of slow detritus
OCr	2.7	$g O_2 (g C)^{-1}$	gr Oxygen respired or produced per gram carbon
Q10Denit	1.65	-	Q10 value for denitrification
ksDenOX	0.7	$g O_2 m^{-3}$	monod ct in oxygen limitation denitrification
ksDenD	3.75	$g C m^{-3}$	monod ct in detritus limitation of Denitrif
cDenitr	1.	$g N (g N)^{-1} d^{-1}$	denitrification rate at 10 degr Celsius
CNrDenit	1.04	$g C (g N)^{-1}$	g C oxidized per gram N during denitrification
pNH4Den	0.2	-	part of NH4 produced during nitrate reduction
Q10Nit	3.37	-	Q10 value for nitrification
ksNitOX	2.5	$g O_2 m^{-3}$	monod ct in oxygen limitation nitrification
ksNH4Nit	9.	$g N m^{-3}$	monod ct in ammonium limitation nitrification
cNitr	1.	$g N d^{-1}$	nitrification coefficient at 10 degr Celsius
CNrNit	0.033	$g C (g N)^{-1}$	gr C fixed per g N during nitrification process
cOXex	1.75	$m d^{-1}$	coefficient for oxygen exchange water-atmosphere

Brock & Madigan, 1991). As bacterial biomass is not modeled as such, the product of nitrification joins the fast decaying detritus part (Eq. 11, Table 3). Stoichiometric equilibrium is assured with ammonium (Eq. 12, Table 3). Nitrification is an oxygen demanding process: per gram nitrogen an amount ONrNit of oxygen is converted. The products of the oxidation of ammonium are water and nitrate and this requires 5 mol (16 × 5 gram) oxygen per mole (14 gram) nitrogen (Brock & Madigan, 1991). Incorporating CO_2 into fast detritus releases oxygen, according to the fixed factor *OCr*. Thus the rate of change in oxygen, due to nitrification, is given by Eq. 13 (Table 3).

The 'oxygen exchange' between water column and atmosphere is one of the most important sources of oxygen to the water column. It is modelled as in O'Kane (1980) (Eq. 14, Table 3; where satOX is oxygen saturation concentration, calculated from temperature and salinity).

'Waste discharges' of inorganic nitrogen were obtained from the SAWES database (SAWES, 1991). Waste input of organic nitrogen is estimated based on Kjeldahl nitrogen, ammonium and BOD values discharged per day (Biochemical oxygen demand), obtained from SAWES (1991). The total load of carbon was calculated from dayly BOD values according to Thomann & Mueller (1987). Total organic nitrogen was obtained by substracting ammonium from Kjeldahl nitrogen waste load data. This organic nitrogen waste load was used to partition the carbon wastes into a fast and slow detrital fraction by requiring that the fast and slow detrital fractions in the carbon waste loads have their (fixed) model values *NCrFD* and *NCrSD*. Import values ($g d^{-1}$) were divided by the volume of the compartment into which the discharge took place, to calculate rates of change of state variables due to import.

'Boundary conditions' at the freshwater and seawater boundaries have to be specified for those constituents that are transported. For the dissolved nutrients and oxygen, boundary concentrations were available as such and obtained from the SAWES database (SAWES, 1991). The boundary conditions of fast and slow decaying detritus were calculated using Biochemical Oxygen Demand values at the fresh and marine boundary. Calculations were based on the mineralization rates of the detritus fractions in the estuary (*cFDMin, cSDMin*) and assuming that a fixed part of imported detritus is fast decaying (*fdFRSH*, resp. *fdSEA*). The amount of phytoplankton carbon was substracted to give the total detrital carbon at the boundary.

Benthic-pelagic exchange processes

As organic matter sediments to the bottom it is degraded by a sequence of oxidants. These reactions promote a flow of dissolved chemicals across the sediment-water interface (Malcolm & Stanley, 1982). It has been shown that even extremely idealized diagenetic models can account for the major trends of the behavior of various substances in the sedimentary column (Billen, 1982). For the Westerschelde, we implemented the idealized model of nitrogen recycling as proposed by Lancelot & Billen (1985). This model was extended to include the exchange of oxygen across the sediment-water interface (Soetaert & Herman, 1993).

In the diagenetic model of Lancelot & Billen (1985) the rate of denitrification, and the release or uptake of nitrate and ammonium from the bottom is calculated from the input of organic matter (consisting of two degradable fractions) to the bottom, from the overlying nitrate and oxygen concentration and from the mixing coefficient of the sediment interstitial and solid phases. The model assumes that nutrient advection due to sediment accumulation is negligible. It considers an oxic and an anoxic bottom layer. In the oxic layer organic matter is mineralized, thus producing ammonium. Nitrification (oxidation of ammonium) proceeds at a rate proportional to this aerobic degradation and is also restricted to the oxic layer. The anoxic layer is the site of denitrification (reduction of nitrate) which is defined as a first-rate process with respect to nitrate concentration.

Calibration

Calibration of the model proceeded by fitting model output to observed data of nitrate + nitrite, ammonium, oxygen, Kjeldahl nitrogen, Biochemical oxygen demand and total organic carbon. The calibration process consists of varying the parameter values, within predefined ranges obtained from the literature, until a good fit of model versus observed data is found.

An initial set of 'good' parameter ranges was obtained using the automatic calibration procedure of SENECA (de Hoop *et al.*, 1993). Within these ranges additional fine-tuning was performed by manually adjusting the parameter values, until a fit was found that was satisfying for all variables.

Parameter values were calibrated for the period 1984–1985. For verification, the model was subsequently run for the period 1980 till 1985 (Figs 2–4), where the years 1980–1983 constitute the verification data base. In order to assess conservative behavior of the constituents, a separate model was run allowing only advective-dispersive transport and waste import (but no sedimentation). Comparing the 'conservative' mixing of the nutrients with observed data then indicates whether a variable is consumed or produced in the estuary.

For a list of best parameter values we refer to Table 5.

Results

Calibration results

Ammonium (Fig. 2-left)
Observed data and model output for MOSES spatial compartments 1 and 2 were very erratic. As these compartments are almost not influenced by tidal mixing their concentration is mainly determined by the ammonium load at the freshwater boundary. From spatial compartment 3 on a consistent seasonal cycle was apparent. Both the spatial and temporal gradients of ammonium were well represented in the model. The conservative mixing plot (compartment 8) shows ammonium to be consumed in the estuary.

Nitrate (Fig. 2-right)
There were not many observed data for nitrate from the three most upstream compartments, but those that were available were reasonably fitted. Nitrate in the most seaward compartments appeared to be slightly overestimated in the fall, underestimated in spring. The behavior of 'conservative' nitrate shows that nitrate is produced in the estuary.

Oxygen (Fig. 3-left)
Observed data of oxygen showed a large variability up to compartment 6. More downstream a clear seasonal cycle was apparent. Oxygen was somewhat overestimated in spatial compartments 1 to 3. The mean concentration was well represented from compartments 4 to 7 although model output was much less variable. Downstream from compartment 8, the oxygen concen-

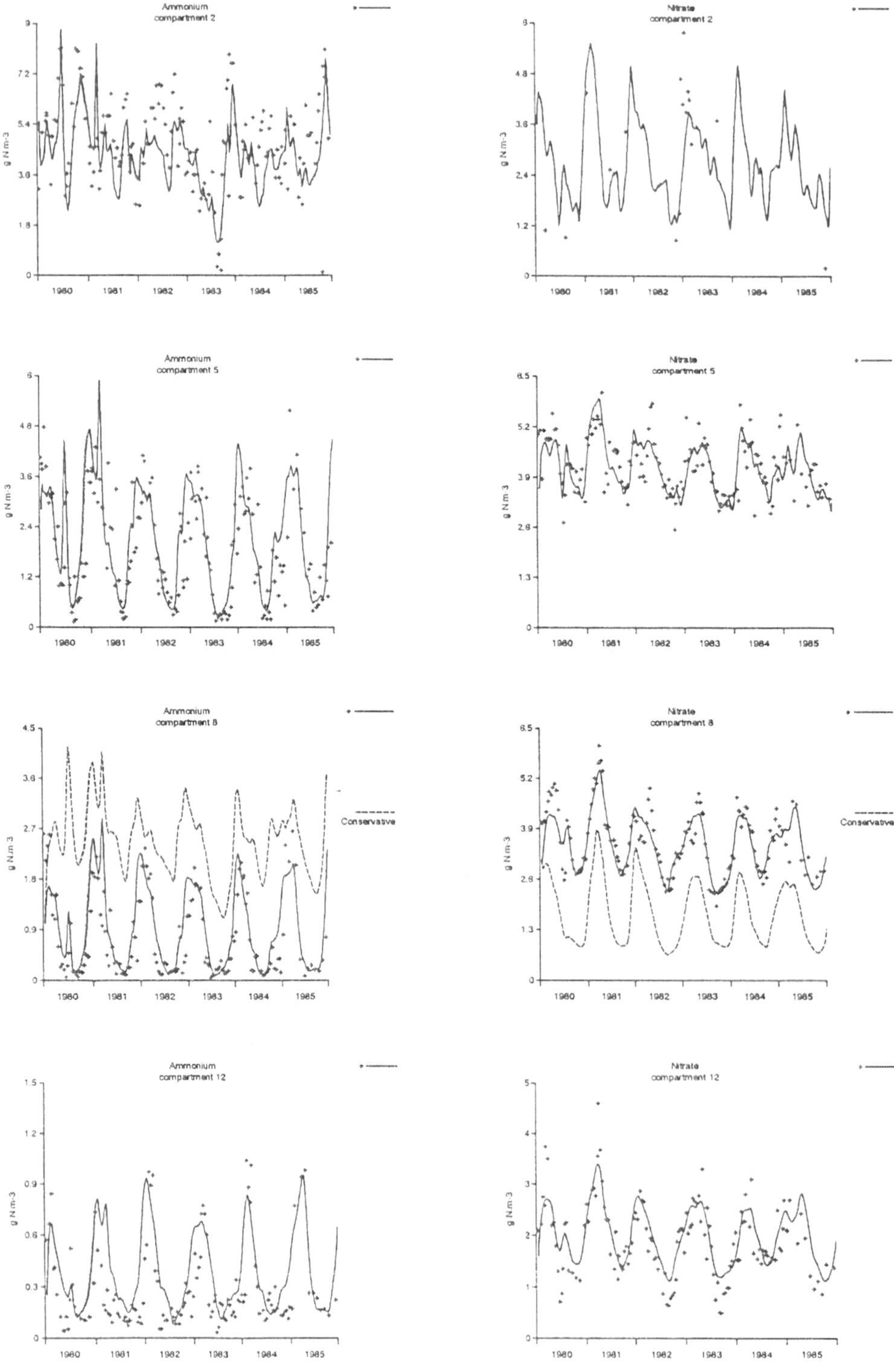

Fig. 2. Observed data (+) and model output of ammonium (left) and nitrate (right) for MOSES spatial pelagic compartments 2, 5, 8 and 12. The conservative concentration (only transport and waste input) of the nutrients is represented for model compartment 8. Calibration occurred on data of 1984–1985.

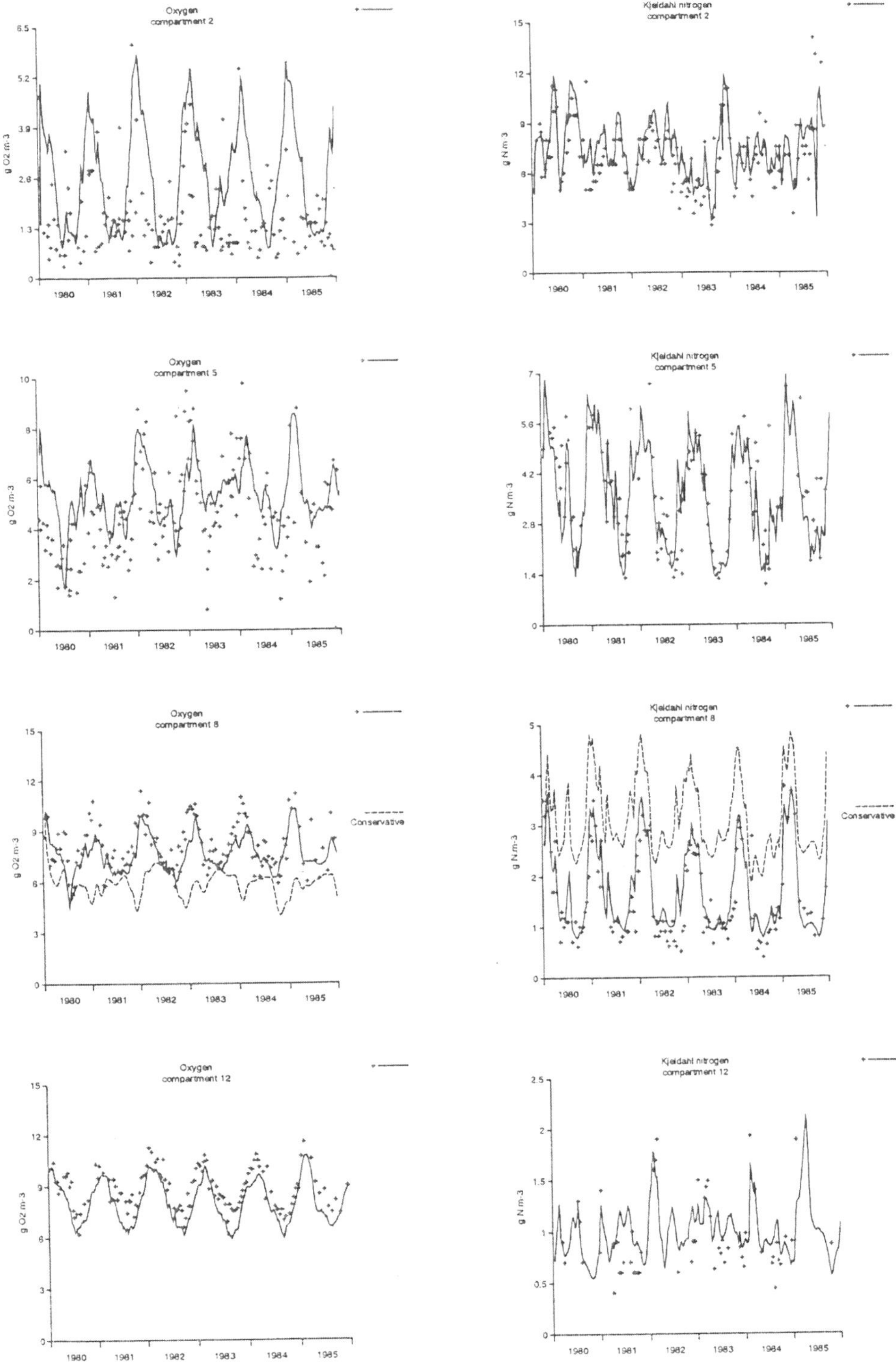

Fig. 3. Observed data (+) and model output of oxygen (left) and Kjeldahl nitrogen (right) for MOSES pelagic compartments 2, 5, 8 and 12. The conservative concentration (only transport and waste input) of the nutrients is represented for model compartment 8. Calibration occurred on data of 1984–1985.

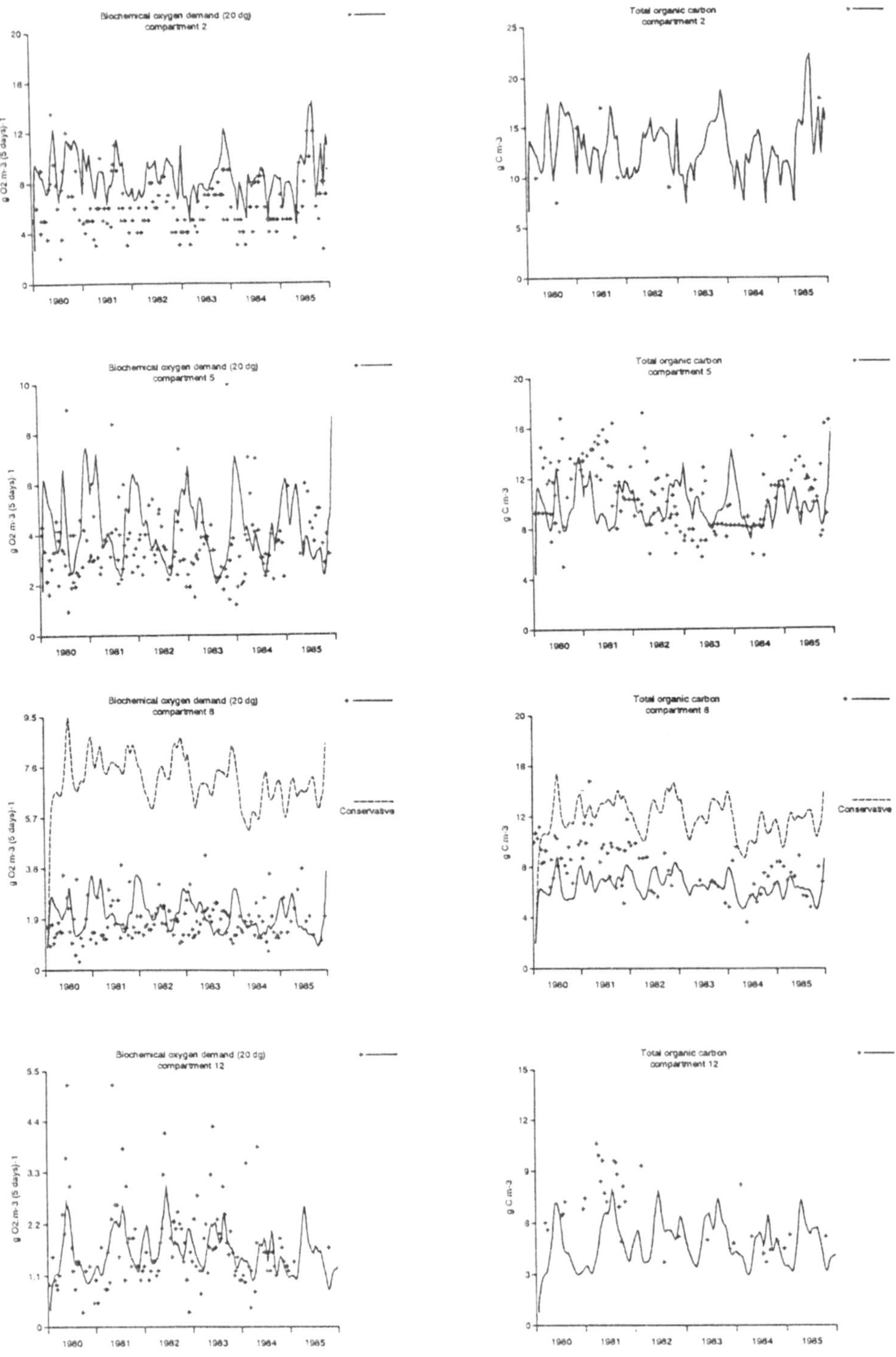

Fig. 4. Observed data (+) and model output of Biochemical oxygen demand (left) and total organic carbon (right) for MOSES pelagic compartments 2, 5, 8 and 12. The conservative concentration (only transport and waste input) of the nutrients is represented for model compartment 8. Calibration occurred on data of 1984–1985.

tration was slightly underestimated but the seasonality was well represented. The conservative mixing plot shows the extent to which oxygen is added to the estuarine waters.

Kjeldahl nitrogen (Fig. 3-right)
Kjeldahl nitrogen is the sum of ammonium and total organic nitrogen. In the Westerschelde it consists for more than half of ammonium. Both the temporal and spatial pattern is well represented by the model. Comparison with conservative Kjeldahl-N shows the consuming nature of the estuary with respect to this constituent.

Biochemical oxygen demand (Fig. 4-left)
BOD values were calculated by adding all fast decaying and slow decaying pelagic components (detritus, phyto- and zooplankton) in the model and calculating the amount of oxygen necessary for the oxidation of this load during 5 days at 20 degrees. The observed data were very variable and unpredictable, whereas model output showed a seasonal trend. However, modeled values fell well within observed ranges. The conservative mixing plot reveals that biochemical oxygen demand decreases substantially in the estuary.

Total organic carbon (Fig. 4-right)
As for BOD, total organic carbon observed data showed much variability which was less pronounced in model output. There was a decrease in TOC from 1980 towards 1986 which was not reproduced by the model. The estuary acts as a consumer of TOC.

Spatial and temporal patterns of organic matter and nutrients in the model (Fig. 5)

As the model adequately fits temporal and spatial patterns of the various nutrient concentrations and organic nitrogen, we used model output to describe general trends in standing stock. A synthetic year was created by averaging over six years (1980–1985) and an additional distance weighted least squares smoothing (Systat, 1992) was performed. By doing so, much of the short-term variability in the model output has been removed such as to reveal the inherent large-scale spatial and temporal signature. Moreover, these plots directly result from the calculated rates which facilitates their interpretation.

Nitrate concentrations generally increase from about 2 gN m^{-3} at the riverside to about 5 g N m^{-3} in the middle of the estuary, after which they steadily decline towards the sea. Nitrate concentrations peak in spring and are lowest in late summer. An additional small peak in summer is the result of higher nitrification rates in the estuary then. Ammonium concentrations fall drastically from about 6 gN m^{-3} near the river to low levels at the sea side. In summer, ammonium concentrations are minimal, highest concentrations are attained in winter. Oxygen concentrations steadily increase towards the sea and from summer to winter. A small increase in oxygen concentration, due to primary production, is visible in mid-summer.

Organic nitrogen peaks in the model in the maximum turbidity zone (model compartment 2) in winter and fall. In summer, a smaller maximum is present more upstream (mainly freshwater algae). Organic nitrogen generally decreases from the freshwater boundary towards the sea, except in summer when concentrations are rising slightly from the midestuary downstream.

Spatial and temporal patterns of the various processes in the model (Fig. 6)

Aerobic mineralisation is nearly entirely patterned to the organic nitrogen load, but is more pronounced in summer (temperature effect). Also the increase of mineralisation rates towards the sea is more pronounced than for organic nitrogen content. This reflects the reactivity of the organic matter which is higher near the sea due to import of relatively labile marine organics.

Denitrification rate in the water column is highest in the badly aerated water masses. It peaks in summer and is highest in spatial compartment 2, which has the best mix of (high) nitrate and (low) oxygen concentration. A smaller amount of nitrogen is lost due to denitrification in the bottom. Sediment denitrification is most pronounced in pelagic spatial compartment 7 which has the most extensive tidal flats.

Nitrification in the water column is highest in summer and in the most upstream compartment, where ammonium concentration is highest. The amount of nitrogen incorporated into bacterial biomass by this process is small.

Nitrogen uptake of the phytoplankton is function of the net primary production, which in turn is determined by light intensity, turbidity and standing stock. In summer net production in the model is positive and nitrogen is taken up by algal cells. There is a sharp decline of net productivity towards the maximum turbidity zone (spatial compartment 2) after which N uptake by phy-

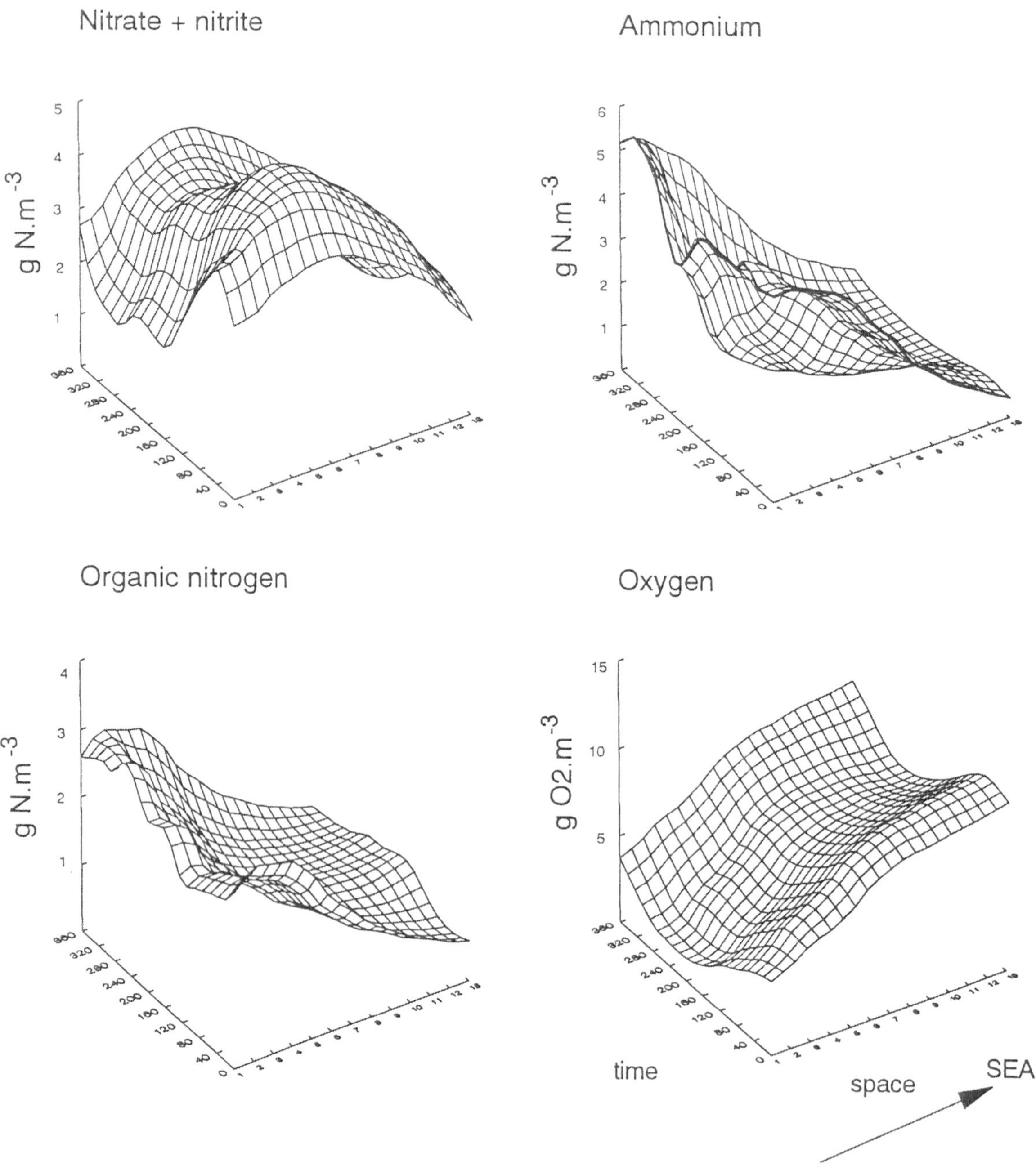

Fig. 5. Temporal and spatial patterns of the various nitrogen species and oxygen. (six-years average of model output). The X axis represents the number of the pelagic model compartment. The Y axis gives the day number (starting at 1 jan).

toplankton gradually increases towards the sea. The slight dip in compartment 9 is due to the unfavourable proportion of shallow and deep regions in this compartment. In winter, light intensity on the surface is lower and turbidity in the estuary is higher. As a result phytoplankton respiration exceeds gross primary production in the model which implies a net production of inorganic nitrogen. For more information about the implementation and description of phytoplankton production we refer to Soetaert *et al.* (1994).

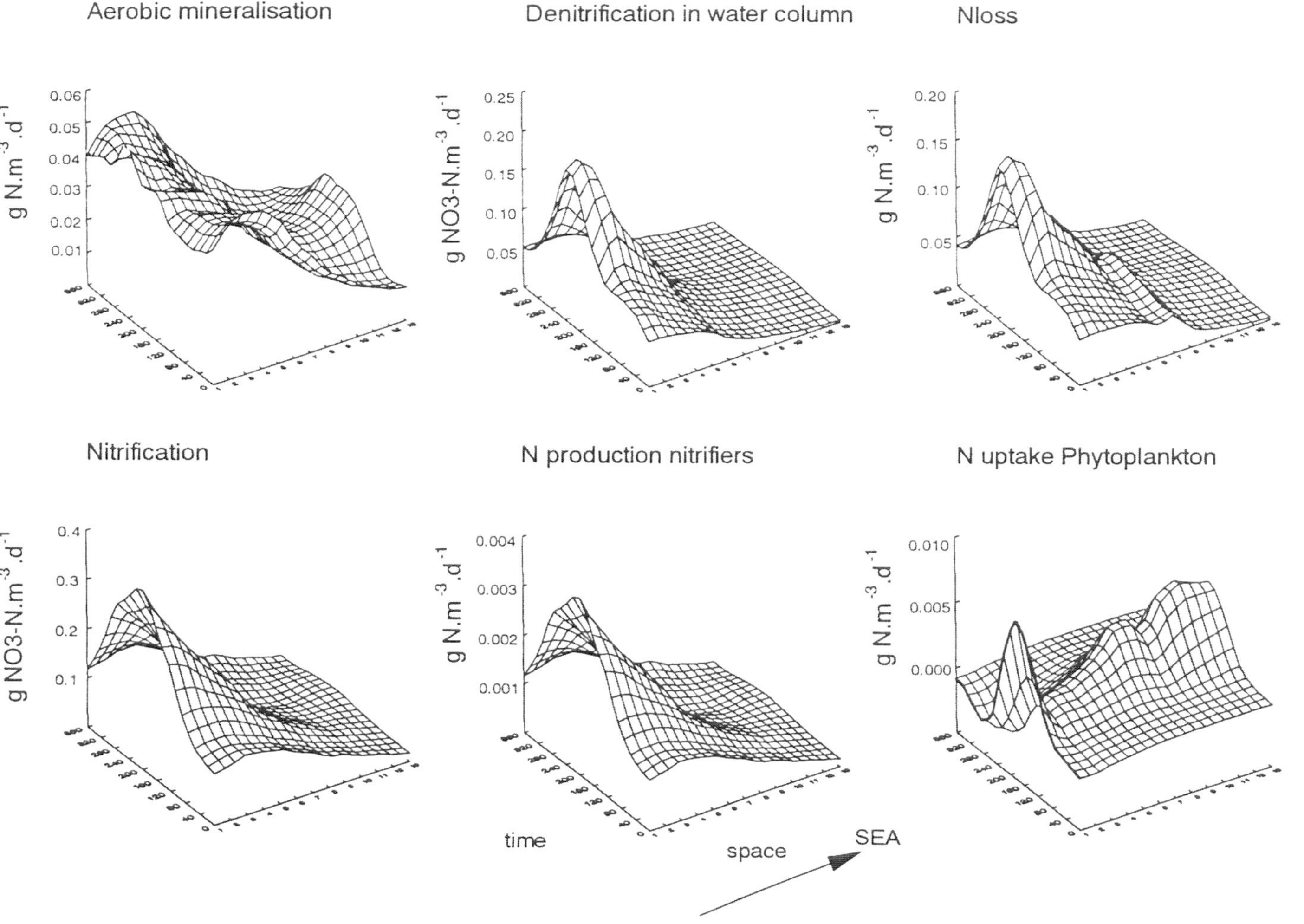

Fig. 6. Temporal and spatial patterns of the various processes. (six-years average). Aerobic mineralisation, denitrification, nitrification and N-uptake of phytoplankton are all in the water column. Nloss also includes benthic nitrogen loss. The X axis represents the number of the pelagic model compartment, the Y axis gives the day number (starting at 1 jan).

A yearly averaged nitrogen budget (Figs 7–9)

The estuary is mainly funnel shaped and the cross sectional surfaces at the freshwater and the marine boundary are in the proportion of 1 to 20. Thus certain processes, although occurring at much higher rates in the most upstream part of the estuary could prove to be unimportant at the scale of the entire estuary.

Yearly averaged rates are represented in Fig. 7. This figure indicates the importance of each process with respect to nitrogen concentration in a certain spatial compartment. The total amount of nitrogen that is transformed in a transverse cross-section of 1 m length is represented in Fig. 8. It reflects the impact of the process at a certain point along the estuary with respect to total nitrogen load.

Except for phytoplankton uptake, the rates of pelagic processes decrease towards the sea (Fig. 7). Benthic rates (expressed per pelagic units) are most pronounced in spatial compartment 7, which has the most extensive tidal flats (Fig. 7). With the exception of (pelagic and benthic) denitrification, the impact of all processes on total nitrogen load increases towards the sea (Fig. 8).

Nitrification has the greatest impact on the nitrogen load and on average 0.22 g NO3-N m^{-3} d^{-1} is produced in the most upstream compartment (Fig. 7). Regeneration of ammonium by means of aerobic mineralisation proceeds at a much lower rate than its consumption by the nitrification process. Denitrification is the second most important process in the brackish part of the estuary. It is mainly located in the water column. In the most aerated water masses downstream, pelagic denitrification rates are very small (less than 0.01 g NO3-N m^{-3} d^{-1}) (Fig. 7). But here too more than 60 percent of total nitrogen loss is occurring in the water column due the unfavourable volume/surface ratio in these compartments (Fig. 8). Only in MOSES spatial compartment 7, with its extensive tidal flats, more than 50% of total nitrogen loss occurs in the bottom. Phytoplankton uptake is unimportant in the Westerschelde nitrogen budget. On a yearly basis, net nitrogen production by algae (due to mortality and lysis of imported algal cells) is higher than total algal nitrogen uptake in MOSES spatial compartments 2, 3 and 4. The phytoplankton in the other compartments is a net consumer of ammonium.

A yearly integrated budget (average of the years 1980–1985) is given in Fig. 9. Only about 30% of total ammonium entering the estuary leaves at the mouth. Nitrate discharge into the sea is about three times higher than what comes in. Organic nitrogen enters the estuary as part of the river load, in waste discharges and from the sea. 26 000 tonnes of organic nitrogen is lost within the estuary per year, mainly by means of (aerobic and anoxic) pelagic degradation. The remainder sediments to the bottom.

Of the 71 000 tonnes of nitrogen that enters the estuary per year, only 56 000 tonnes are exported to the sea. Thus about 21% of total nitrogen import is lost in the estuary per year.

Discussion

The Westerschelde estuary is a passage of waters from the river Schelde watershed to the southern Bight of the North Sea. Tidal currents result in the mixing of fresh and sea water and the Westerschelde nutrient composition to some degree reflects both endmember concentrations. Superimposed on these hydrodynamical forces are the various biological and chemical processes that react in concert and further change the nutrient load.

Whereas concentrations of nutrients in the watercolumn are relatively easy to measure, their interpretation is more difficult. Frequently the concentration is plotted against a conservative index of mixing (chlorinity) and the resulting mixing plots are interpreted as indicative of removal, addition or conservative behavior of the nutrient. However, seasonal variations in the freshwater and marine nutrient concentrations and chlorinity may complicate the interpretation as they result in 'pulses' of nutrients that propagate through the estuary in time (Loder & Reichard, 1981).

A mathematical model is an ideal tool to evaluate the different processes that change the nutrient composition in estuaries. Being based on a well-calibrated transport model (Soetaert & Herman, 1995a), the effect of the ever changing nutrient concentration at the boundaries is now included into the model and this should make the influence of biochemical and biological processes more obvious.

The Westerschelde estuary is especially interesting due to the high load of detritus and nutrients. This, combined with the relatively high residence times in the estuary (50 to 70 days in Soetaert & Herman, 1995a) results in clear and temporally consistent patterns of the various nutrients (Nixon & Pilson, 1983), which makes the Westerschelde an ideal area for modelling.

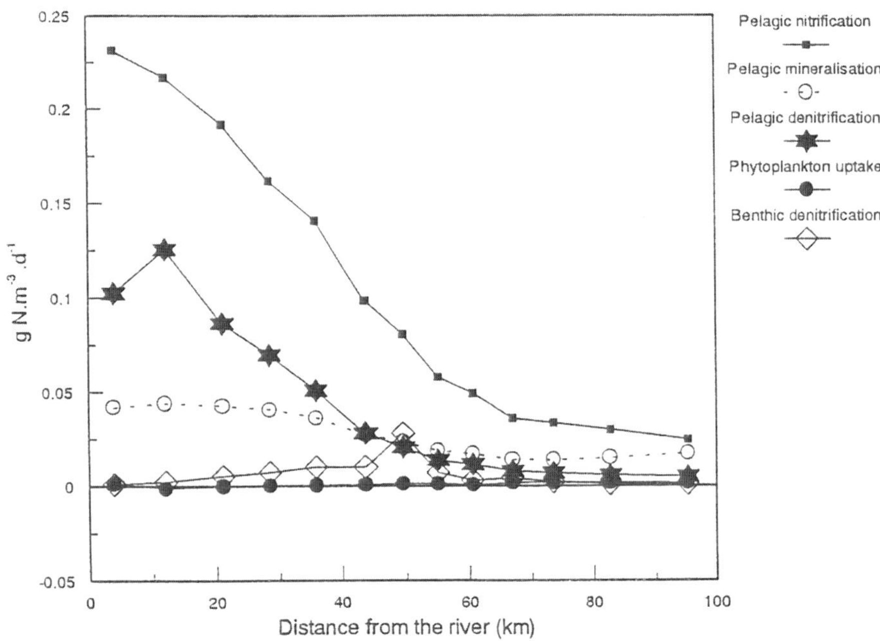

Fig. 7. Yearly averaged process rates (g N m^{-3} d^{-1}) for the different compartments. Pelagic nitrification and pelagic denitrification is in g NO_3-N d^{-1}, benthic denitrification is in g N d^{-1} lost to the atmosphere.

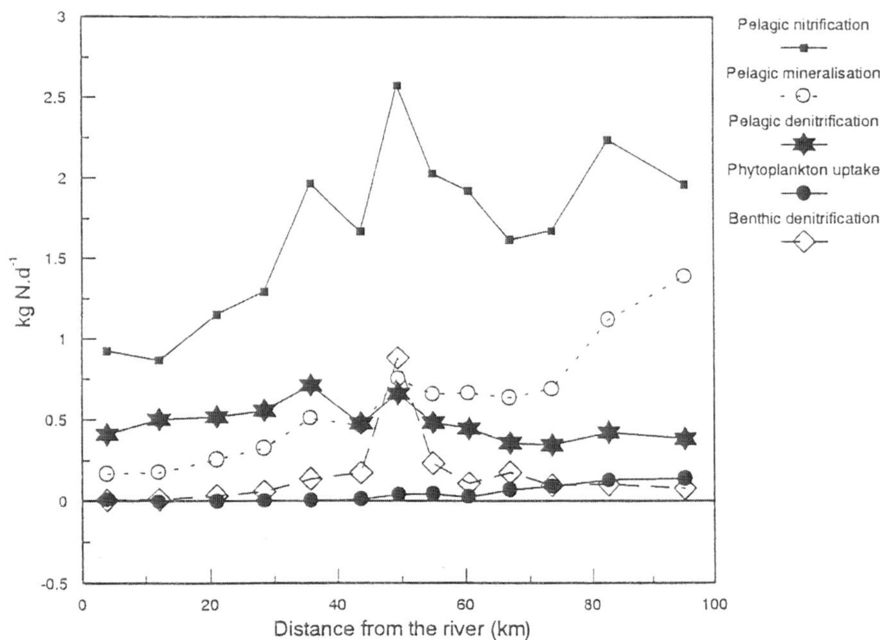

Fig. 8. Yearly averaged nitrogen budget (in kg N d^{-1}) for a transverse cross-section of 1 m length. Pelagic denitrification and pelagic denitrification is in kg NO_3-N d^{-1}, benthic denitrification is in kg N d^{-1} lost to the atmosphere.

The major sources of nitrogen to the estuary are the river and waste discharges (Wattel & Schouwenaar, 1991). The majority of this supply (about 70%) is as inorganic nitrogen, the remainder consists of detritus.

Among the various biochemical processes that act on this nitrogen load, nitrification has the largest effect

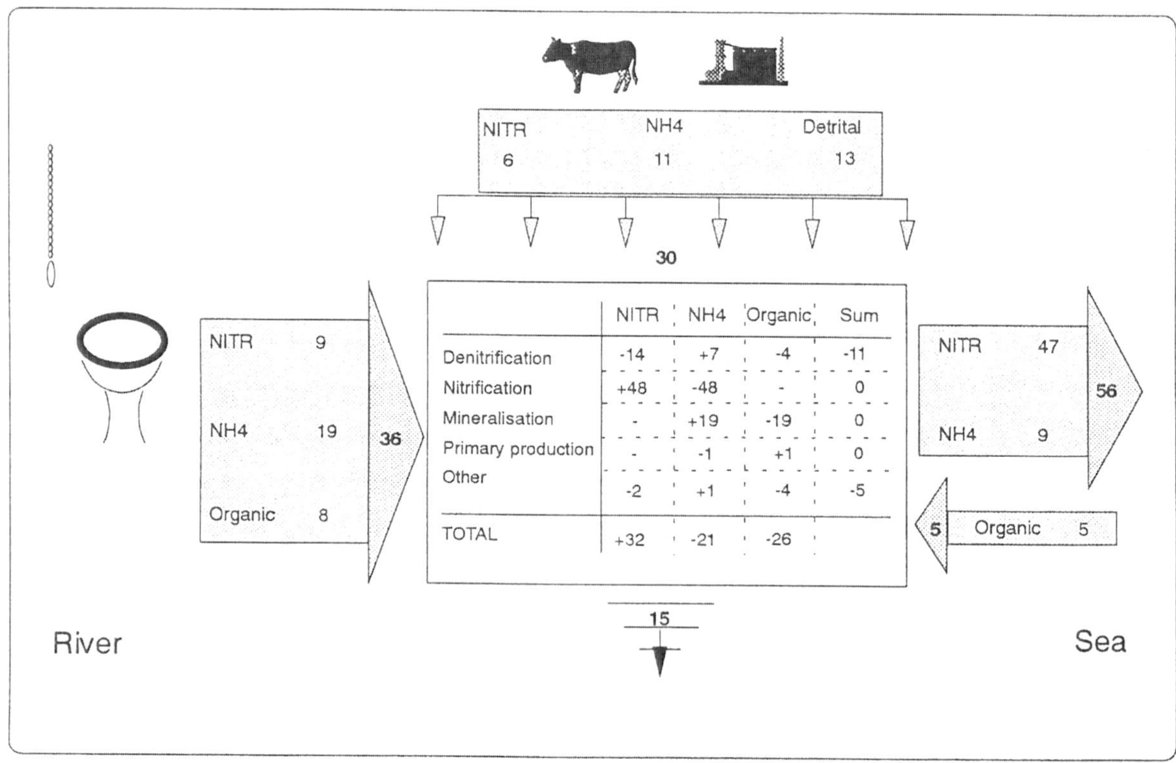

Fig. 9. Global nitrogen budget in the Westerschelde in 1000 tonnes N yr^{-1}. Denitrification, nitrification, aerobic mineralisation and primary production are all in the water column. 'Other' = gain/loss to the bottom or due to excretion processes of the higher trophic levels (NH4). Positive values = gain, negative = loss. The arrows to the left indicate import from the river, to the right are exchanges with the sea. The small arrows on top indicate waste discharges.

on the speciation. The process occurs whenever aerobic conditions are restored in the estuary (Wollast, 1983). In the Westerschelde model, the zone of highest nitrification is in the upper estuarine part and ammonium is almost completely consumed by this process, especially in summer when nitrification rates are most pronounced. Thus the nitrate concentration increases up to about the Belgian-Dutch border. From then on nitrate concentration gradually decreases due to dilution with sea-water. On average about 47 000 tonnes of nitrate-N are discharged per year into the sea whereas only 15 000 tonnes are imported into the estuary. In contrast to nitrate, the estuary acts as a consumer of ammonium such that only 30% of the total input is discharged into the sea. This net consumption of ammonium is due to the fact that aerobic mineralisation and other ammonium generating processes are not able to meet the losses due to nitrification.

More important in terms of the global nitrogen budget is the process of denitrification. This is the heterotrophic reduction of nitrate to N_2 or N_2O which are lost to the atmosphere (Seitzinger, 1988). The normal site of denitrification is in the bottom, but in the Westerschelde the process can also occur in the water column when oxygen conditions are bad (Billen *et al.*, 1985). In our model, denitrification is negatively influenced by higher oxygen conditions but continues up to the sea in the water column, albeit at very low levels. Due to the high load of organic matter in the water column, small-scale gradients of oxygen will be created in detrital particles of a certain size. Because of the dependence of denitrification both on oxic conditions (which provide nitrate) and anoxic conditions (to reduce nitrate), these microscale gradients could be ideal sites for denitrification (Law & Owens, 1990). Thus we assumed that the process can take place at a very short distance to the nitrification in an otherwise oxygenated water column. The same implementation strategy was adopted in the model of lake Grevelingen (Vries *et al.*, 1988). Evidence that nitrifiers can under anaerobic conditions reduce nitrate to N_2O (Kaplan, 1983; Goreau *et al.*, 1980) further justifies this implementation strategy. Denitrification rates are highest in MOSES compartment 2, which has the best mix of

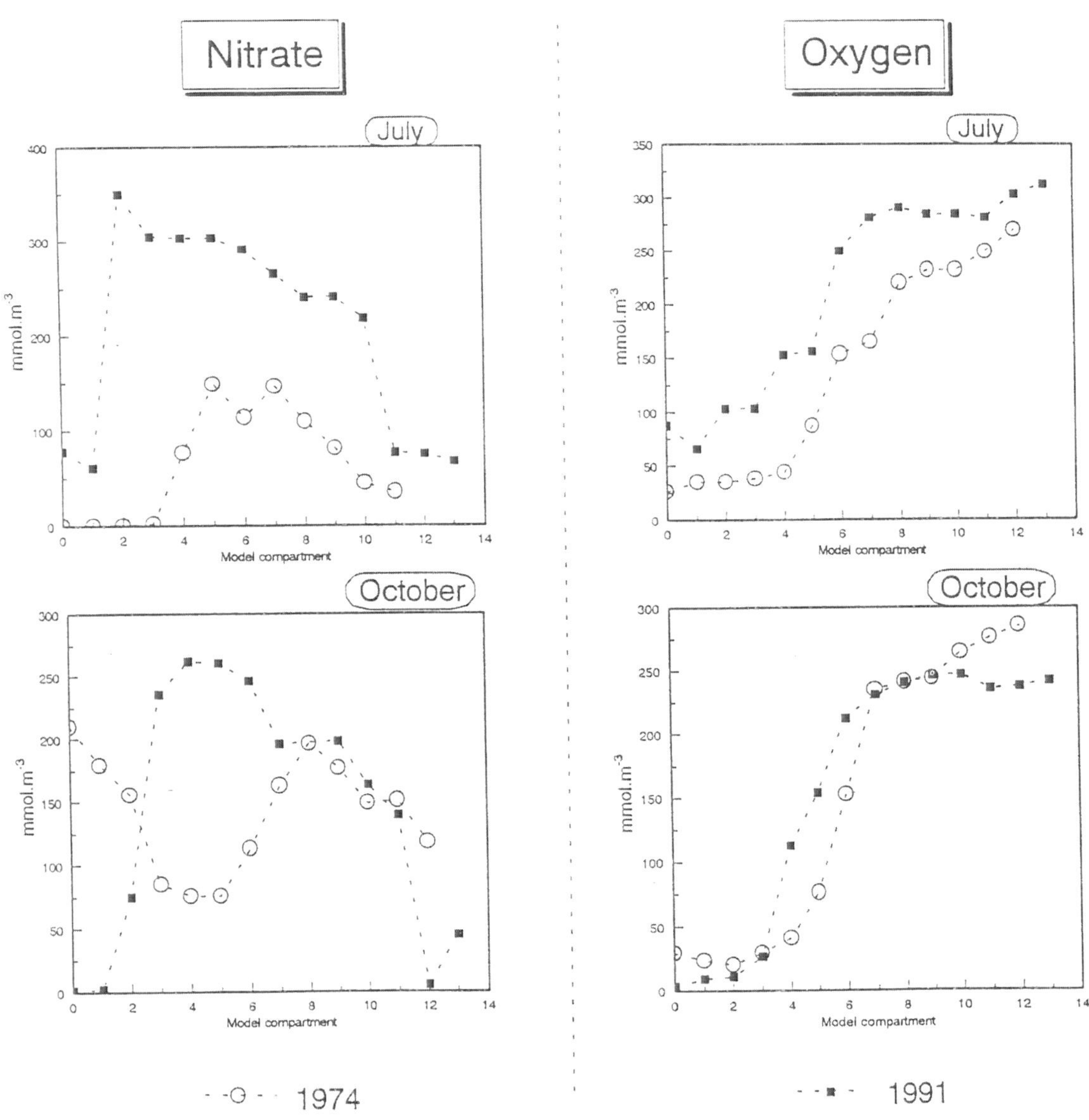

Fig. 10. Comparison of a typical nitrate and oxygen profile in summer and fall 1974 (Billen & Smitz, 1976) and 1991 (data from the JEEP data base). Data from 1974 were transposed to mid-tidal level.

nitrate load and anoxia. This is somewhat more downstream than the peak of nitrification. In the Westerschelde, it has been shown that a clear N_2O peak occurs upstream from the nitrate peak (Deck, 1980 in Nixon & Pilson, 1983), but somewhat downstream from the offset of nitrate production. This is consistent with the position of the denitrification peak in our model.

Part of the denitrification occurs on the intertidal and subtidal flats. The magnitude of this nitrogen loss was determined by means of a diagenetic model proposed by Lancelot & Billen (1985). Thus in the model yearly-averaged denitrification rates of 0.07 to 0.3 gN m^{-2} d^{-1} were calculated in the intertidal area. This range is comparable to denitrification rates of 77 to 1067 μmol N m^{-2} hr^{-1} (0.03–0.36 gN m^{-2} d^{-1}) reported from estuarine sediments in Seitzinger (1988).

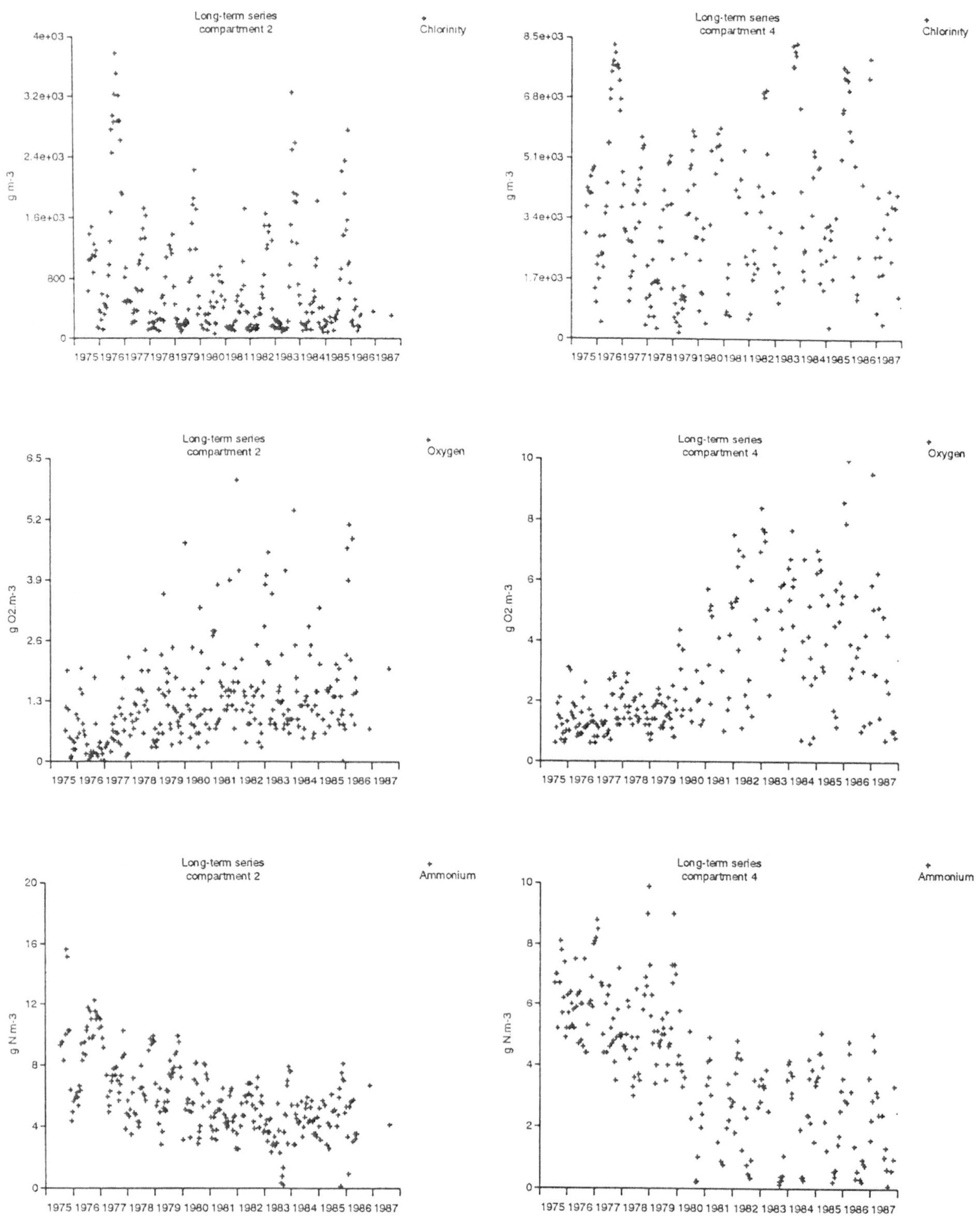

Fig. 11. See p. 243 for legend.

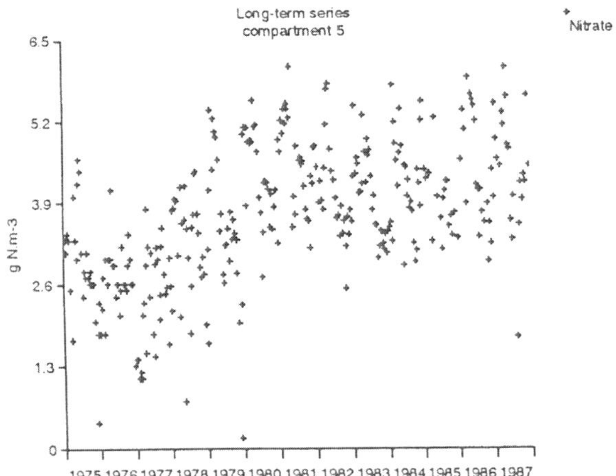

Fig. 11. Long-term series of chlorinity, oxygen, ammonium and nitrate in pelagic spatial compartment 2 and 4 (5 for nitrate).

In the model, the largest amount of nitrogen is lost by pelagic rather than by benthic denitrification. A small calculation shows the extent of benthic denitrification if all nitrogen loss (15 000 ton N yr^{-1}) would be in the Westerschelde bottom. Assuming denitrification occurs equiproportional on the entire Westerschelde bottom (260 km^2) then yearly averaged denitrification rates would be 15 000/260/365 = 0.16 gN m^{-2} d^{-1}. In reality, the main sedimentation occurs on the intertidal flats (59 km^2). Overall intertidal nitrogen loss would necessitate a yearly averaged denitrification rate of 0.7 gN m^{-2} d^{-1}. Both estimates are close to or larger than the maximal denitrification rates, reported in Seitzinger (1988) of 0.36 gN m^{-2} d^{-1} for an area in the Tagus that receives a large amount of sewage. The amount of carbon sedimenting on the Westerschelde bottom was estimated to be 21 10^3 tonnes of carbon per year (Soetaert *et al.*, 1995b), half of which occurs in the downstream part of the estuary. In order to meet the demands of whole-benthic nitrogen loss, it would be necessary that more than 70% of all benthic degradation would occur by means of denitrification (or 60% if phytobenthic production were to be included).

Jørgensen & Sørensen (1985) showed that denitrification was an insignificant pathway of organic carbon oxidation in the sediment of a Danish estuary (3–4%). The main processes were oxygen uptake and sulfate reduction, while nitrate reduction to ammonium was important in the brackish part. More significance of benthic denitrification was suggested by Seitzinger (1988), who reported that the process in marine and estuarine sediments accounted for 20–70% of total sediment-water nitrogen fluxes in estuaries.

Whether benthic or pelagic, all included, a load of about 40 tonnes of nitrogen is lost to the atmosphere per day, which amounts to about 15 000 tonnes per year, i.e. 21% of total nitrogen import, 33% of inorganic nitrogen input. This value is rather low as other estuaries that were examined thus far had a surprisingly narrow range of 40 to 50% removal of inorganic nitrogen input (Seitzinger, 1988). In the Danish estuary examined by Jørgensen & Sørensen (1985) on the other hand, (benthic) denitrification removed only 5% of total nitrate import.

Due to high turbidity in the water column, depth-integrated net phytoplankton production is low (Soetaert *et al.*, 1994) and the nutrient loss due to this process is insignificant in the global estuarine cycle.

A major uncertainty in many budget approaches is a lack of detailed knowledge on the role of organic nitrogen, although it is considered to be a substantial part of the budget (Meybeck, 1982). The pathway of organic matter through the estuary strongly depends on its partitioning over dissolved and particulate forms. The incorporation of particulate matter into a mathematical model is more complicated than for dissolved substances as they exhibit an aberrant transport behavior which is hard to describe in one, or even two dimensions (Odd, 1988).

In the Westerschelde, sediment particles from both the river and the sea enter the estuary and are accumulated in the zone of maximum turbidity (Van Eck & de Rooij, 1990). As these transport and accumulation processes were deemed to be important for the particulate organic matter, we implemented a simple model that represents the horizontal and vertical movement of particulates in the Westerschelde (Soetaert & Herman, 1993). Although crudely modeled, the general gradients (turbidity maximum) and seasonality of suspended matter were adequately represented. Also, the net import of sediment from both the river and the sea (yearly values) and total sedimentation of mud and organic carbon were convincingly represented (see Soetaert & Herman, 1993 for more details).

In the model, the Westerschelde consumes particulate nitrogen both from freshwater, waste and marine origin and about 26 000 tonnes of organic nitrogen is trapped or lost in the estuary per year. As the estuary acts as a net exporter of organic carbon to the sea (Wollast, 1976; Soetaert & Herman, 1995b), the fact that in the model marine organic nitrogen is net imported into the estuary deserves some attention. This imbalance is created as fast decaying detritus (with high nitrogen content) enters the estuary, while more refractory (and nitrogen-poor) detritus net leaves the estuary. However, virtually nothing is known about the degradability and nitrogen content of marine versus estuarine organic matter which makes the direction of both fluxes difficult to estimate based on observations. The exchange in the model could be an artefact due to the fact that a fixed (and maybe too large?) amount of total marine detritus was assumed to be fast decaying. Nevertheless, there is evidence of at least one pool of (nitrogen-rich) carbon that is net imported from the sea into the estuary: each year about 1500 tonnes of marine zooplankton dry weight drifts into the estuary where it decays (Soetaert & Herman, 1994). Several other estuarine environments appear to represent both efficient traps for river-borne particulate organic carbon (Fontugne & Jouanneay, 1987; Lucotte *et al.*, 1991) and sinks for a significant fraction of the carbon metabolized in the coastal area (Tan & Strain, 1978; Gearing *et al.*, 1984).

The oxygen concentration proved to be the most difficult to fit by means of our model. We had to find a trade-off between too high concentrations near the fresh-water boundary and too low concentrations near the sea side. The reason for this discrepancy as yet is unclear, but there is a possibility of an oxygen-demanding process in the upstream part that is unaccounted for in the model, while atmospheric reaeration could be higher than included in the model. In the most upstream compartments, a substantial amount of dredging takes place. The main process of organic matter degradation in the sediments of the brackish and marine part could be by means of sulfate reduction (Middelburg, pers. comm.) and the release of the reduced forms of sulfur in the water column, where they would be oxidized quickly, could provide an additional oxygen demand in the upper estuarine compartments. Similarly, methane produced in the riverine part of the Schelde could oxidize quickly when oxygen conditions improve and hence retard restoration of oxygen concentrations in the water (Middelburg, pers. comm.).

Compared to a previous model of nitrogen cycling in the Westerschelde (Billen *et al.*, 1985, 1986, 1988), we model lower maximal denitrification rates in the water column (0.25 gN m^{-3} d^{-1} compared to 0.7 gN m^{-3} d^{-1} in Billen *et al.*, 1985) while only about 21% of total nitrogen, 33% of inorganic nitrogen, is removed in our model compared to 40–50% as estimated by Billen and co-workers. This discrepancy requires a more thorough analysis. Part of the differences could be methodological: Billen and co-workers defined the estuarine boundaries at ebb tide, as opposed to mid-tide in our model. Moreover, denitrification was not dynamically simulated by Billen *et al.* (1985) but its magnitude was assessed by substraction of export from the total import of nitrogen. Inspection of nitrogen and oxygen profiles along the estuary nevertheless revealed a large difference between typical profiles of the seventies (after transposing them to mid-tidal level, e.g. Billen & Smitz, 1976; Billen *et al.* 1985; Somville, 1980 in Wollast, 1983) with respect to the eighties and early nineties (this paper). This is represented in Fig. 10. In fall 1974, nitrate concentrations declined very rapidly from high values at the fresh-water boundary towards the middle estuarine region. This decline was accompanied by a small decrease in oxygen concentration. Nitrate and oxygen levels were rising from model spatial compartment 5 towards the sea after which, due to dilution with sea water, nitrate concentration declined further downstream. In summer 1974, nitrate was entirely exhausted when entering the estuary and restoration (nitrification) did not start before spatial compartment 4 when oxygen conditions were slightly improving. Maximal nitrate concentrations in the middle estuarine part never exceeded 3 gN m^{-3} in the seventies. Profiles that were used for our model and from the nineties (data from the JEEP database, courtesy J. Kromkamp) showed another nutrient behavior. Nitrate and oxygen concentrations increased almost immediately upon entering the estuary. Moreover, nitrate concentrations in the middle estuarine part attained larger values of up to 8 gN m^{-3} in 1991. Thus it seems as if the oxygen-deficient zone has moved upstream for about 10 kilometres since the late seventies. As pelagic denitrification occurs in these badly aerated water masses, the magnitude of nitrogen loss could have diminished concurrently, which explains the lower values reported in our study. Furthermore, nitrification, which was insignificant in the upper estuary in the seventies (Billen *et al.*, 1985) now proceeds immediately upon entering the estuary. A long-term series of oxygen, nitrate and ammonium covering the years 1975 to 1987 (Fig. 11) indeed confirms that oxygen conditions in the upstream compartments improved. Concurrently the load of ammonium has decreased, while nitrate increased almost abruptly in 1980. Chlorinity data from the same period seem to witness that this improvement could not have been caused by differences in river discharge alone, but also represent real biogeochemical changes. Interestingly, whereas the amount of nitrogen imported into the estuary from the river or waste discharges has not changed a great deal (52 000 tonnes in Billen *et al.*, 1985; 66 000 tonnes in our model), the amount discharged into the sea seems to have doubled (27 000 *vs* 51 000). This is exactly what was put forward by Billen *et al.* (1985).

A yearly export of 51 000 tonnes of nitrogen to the coastal system seems a formidable amount. It amounts to about 3% of the total nitrogen input to the North sea (1.5 10^6 tonnes N yr^{-1} in Law & Owens, 1990), whereas the contribution of fresh water discharge from the Westerschelde to the North sea is about ten times as low (calculated based on total fresh-water fluxes in Otto *et al.*, 1990). Taking an average flood volume of 1030 10^6 m^3 (Van Maldeghem, 1988) and the nitrogen concentration in the sea water, we calculated the degree to which coastal waters are enriched with nitrogen upon mixing with the estuarine waters. Yearly averaged, the nitrogen load of coastal waters that enter and leave the estuary with the tides increased with about 14% per day. As these waters are trapped into a coastal gyre in front of the estuarine mouth (Nihoul & Ronday, 1975), this increase in nutrient load remains apparent, especially in winter. This results in a considerable phytoplankton spring bloom which is higher than in the Channel (Brockmann *et al.*, 1988).

In this paper we presented a set of chemical and biological processes that can explain nutrient behavior observed in the Westerschelde. As for so many modeling studies, the correctness of our results are mainly consolidated by an adequate description of the standing stock of several relevant variables. In the future a further evaluation of the proposed mass balances should be performed by means of *in situ* measured fluxes. Nevertheless, inherent environmental variability – especially in the benthos – will probably make it difficult to verify or falsify the estimates that are presented here.

Acknowledgments

This model was developed as part of the MAST-JEEP project of the C.E.C. and of the ECOLMOD project of Rijkswaterstaat (D.G.W.).

Thanks to Dr Van Eck and A. Schouwenaar for providing us with the data necessary for the construction of the model. This is article number 730 of the NIOO-CEMO. C. Heip, J. Middelburg and G. Billen are acknowledged for critically reading the manuscript. The ecosystem model MOSES is available upon request from the authors. It can only be run under the model environment SENECA, which can be purchased from the same authors (price in 1994 US$100).

References

Billen, G., 1982. Modelling the processes of organic matter degradation and nutrient recycling in sedimentary systems. In D. B. Nedwell & C. M. Brown (eds), Sediment microbiology. Academic Press, New York: 15–52.

Billen, G. & J. Smitz, 1976. Modele mathematique de la qualite de l'eau dans un estuaire partiellement stratifie. In La mecanique des fluides et l'environnement: 1–8.

Billen, G., M. Somville, E. de Becker & P. Servais, 1985. A nitrogen budget of the Scheldt hydrographical basin. Neth. J. Sea Res. 19: 223–230.

Billen, G., C. Lancelot, E. De Becker & P. Servais, 1986. The terrestrial marine interface: modelling nitrogen transformations during its transfer through the Scheldt river system and its estuarine zone. In Nihoul, J. C. J. (ed.), Marine interfaces Ecohydrodynamics. Elsevier: 429–490.

Billen, G. & C. Lancelot, 1988. Modelling benthic nitrogen cycling in temperate coastal ecosystems. In: Blackburn, T. H. & J. Sorensen (eds), Nitrogen cycling in coastal marine environments. SCOPE. John Wiley & Sons, Chichester: 341–378.

Billen, G., C. Lancelot, E. De Becker & P. Servais, 1988. Modelling microbial processes (phyto- and bacterioplankton) in the Schelde Estuary. Hydrobiol. Bull. 22: 43–55.

Billen, G., C. Lancelot & M. Meybeck, 1991. N, P and Si retention along the aquatic continuum from land to ocean. In R. F. C. Mantoura, J.-M. Martin & R. Wollast (eds), Ocean Margin Processes in Global Change. John Wiley & Sons, Chichester: 19–44.

Brock, T. D. & M. T. Madigan, 1991. Biology of microorganisms. 6th edition. Prentice Hall, Englewood Cliffs: 874 pp.

Brockmann, U., G. Billen & W. W. C. Gieskes, 1988. North Sea nutrients and eutrophication. In Salomons, W., B. Bayne, E. Duursma & U. Forstner (eds), Pollution of the North Sea: an assessment. Springer, Berlin: 348–389.

Brockmann, U. H., R. W. P. M. Laane & H. Postma, 1991. Cycling of nutrient elements in the North Sea. Neth. J. Sea Res. 26: 239–264.

Di Toro, D. M., D. J. O'Connor & R. V. Thomann, 1971. A dynamic model of the phytoplankton in the Sacramento-San Joaquin Delta. Adv. chem. Ser. 106: 131–150.

Fontugne, M. R. & J. M. Jouanneau, 1987. Modulation of the particulate organic flux to the ocean by a macrotidal estuary: evidence from measurements of carbon isotopes in organic matter from the Gironde system. Estuar. coast. Shelf Sci. 24: 377–387.

Garber, J. M., 1984. Laboratory study of nitrogen and phosphorus remineralization during the decomposition of coastal plankton and seston. Estuar. coast. Shelf Sci. 18: 685–702.

Gearing, J. N., P. J. Gearing, D. T. Rudnick, A. D. Requejo & M. J. Hutchins, 1984. Isotope variability of organic carbon in a phytoplankton-based temperate estuary. Geochim. Cosmochim. Acta 48: 1089–1098.

Goreau, T. J., W. A. Kaplan, S. C. Wofsy, M. B. McElroy, F. A. Valois & S. W. Watson, 1980. Production of NO_2^- and N_2O by nitrifying bacteria at reduced concentrations of oxygen. Appl. envir. Microbiol 40: 526–532.

Grasshoff, K., M. Ehrhardt & K. Kremling, 1983. Methods of seawater analysis. Verlag Chemie, Weinheim, 419 pp.

de Hoop, B. J., P. M. J. Herman, H. Scholten & K. Soetaert, 1993. SENECA 2.0. A Simulation ENvironment for ECological Application. MANUAL

Heip, C., 1988. Biota and abiotic environment in the Westerschelde estuary. Hydrobiol. Bull. 22: 31–34.

Helder, W. & R. T. P. De Vries, 1983. Estuarine nitrite maxima and nitrifying bacteria (Ems-Dollard estuary). Neth. J. Sea Res. 17: 1–18.

Jorgensen, B. B. & J. Sorensen, 1985. Seasonal cycles of O_2, NO_3, and SO_4^2 reduction in estuarine sediments. The significance of an NO_3^- reduction maximum in spring. Mar. Ecol. Progr. Ser. 24: 65–74.

Kaplan, W. A., 1983. Nitrification. In Carpenter, E. J. & D. G. Capone (eds), Nitrogen in the Marine Environment. Academic Press, New York: 139–190.

Klepper, O., 1989. A Model of Carbon Flows in Relation to Macrobenthic Food Supply in the Oosterschelde Estuary (SW Netherlands). Ph. D. thesis, University of Wageningen, Wageningen: 1–270.

Kromkamp, J., A. van Spaendonk, J. Peene, P. van Rijswijk & N. Goosen, 1992. Light, nutrient and phytoplankton primary production in the eutrophic, turbid Westerschelde estuary (The Netherlands). In JEEP 92. Major biological processes in european tidal estuaries. MAST report: 115–126.

Lancelot, C. & G. Billen, 1984. Activity of heterotrophic bacteria and its coupling to primary production during the spring phytoplankton bloom in the southern bight of the North Sea. Limnol. Oceanogr. 29: 721–730.

Lancelot, C. & G. Billen, 1985. Carbon-nitrogen relationships in nutrient metabolism of coastal marine ecosystems. Adv. aquat. Microbiol. 3: 263–321.

Law, C. S. & N. J. P. Owens, 1990. Denitrification and nitrous oxide in the North Sea. Neth. J. Sea Res. 25: 65–74.

Loder, T. C. & R. P. Reichard, 1981. The dynamics of conservative mixing in estuaries. Estuaries 4: 64–69.

Lucotte, M., C. Hillaire-Marcel & P. Louchouarn, 1991. First-order organic carbon budget in the St Lawrence Estuary from ^{13}C data. Estuar. coast. Shelf Sci. 32: 297–312.

Malcolm, S. J. & S. O. Stanley, 1982. The sediment environment. In D. B. Nedwell & C. M. Brown (eds), Sediment microbiology. Academic Press, New York: 15–52.

Meybeck, M., 1982. Carbon, nitrogen and phosphorus transport by world rivers. Am. J. Sci. 282: 401–450.

Nihoul, J. C. J. & F. C. Ronday, 1975. The influence of tidal stress on the residual circulation. Tellus 27: 484–489.

Nixon, S. W. & M. E. Q. Pilson, 1983. Nitrogen in estuarine and coastal marine ecosystems. In Carpenter, E. J. & D. G. Capone (eds), Nitrogen in the Marine Environment. Academic Press, New York: 565–648.

Odd, N. V. M., 1988. Mathematical modelling of mud transport in estuaries. In Dronkers, D. & W. van Leussen (eds), Physical processes in estuaries. Springer-Verlag, Berlin: 503–531.

O'Kane, J. P., 1980. Estuarine water-quality management. Pitman, Boston.

Otto, L., J. T. F. Zimmerman, G. K. Furnes, M. Mork, R. Saetre & G. Becker, 1990. Review of the physical oceanography of the North Sea. Neth. J. Sea Res. 26: 161–238.

Prosser, J. I., 1990. Mathematical modeling of nitrification. In: Adv. Microb. Ecol. 11: 263–304.

SAWES, 1991. Waterkwaliteitsmodel Westerschelde. WL-rapport T257.

Seitzinger, S. P., 1988. Denitrification in freshwater and coastal marine ecosystems: ecological and geochemical significance. Limnol. Oceanogr. 33: 702–724.

Soetaert, K. & P. M. J. Herman, 1994. One foot in the grave – zooplankton drift in the Westerschelde estuary (The Netherlands). Mar. Ecol. Progr. Ser. 105: 19–25.

Soetaert, K., P. M. J. Herman & J. Kromkamp, 1994. Living in the twilight: estimating net phytoplankton growth in the Westerschelde estuary (The Netherlands) by means of an ecosystem model (MOSES). J. Plankt. Res. 16: 1277–1301.

Soetaert, K. & P. M. J. Herman, 1995a. Estimating estuarine residence times in the Westerschelde (The Netherlands) using a box model with fixed dispersion coefficients. Hydrobiologia 311 (Dev. Hydrobiol. 110): 215–224.

Soetaert, K. & P. M. J. Herman, 1995b. Carbon flows in the Westerschelde estuary (The Netherlands) evaluated by means of an ecosystem model (MOSES). Hydrobiologia 311 (Dev. Hydrobiol. 110): 247–266.

Soetaert, K. & P. M. J. Herman, 1993. MOSES – model of the scheldt estuary – ecosystem model development under SENECA. Report, 89 pp.

Somville, M., G. Billen & J. Smitz, 1982. An ecophysiological model of nitrification in the Scheldt estuary. Mathem. Modelling 3: 523–533.

Spitzy, A. & V. Ittekot, 1991. Dissolved and particulate organic matter in rivers. In R. F. C. Mantoura, J. M. Martin & R. Wollast (eds), Ocean margin processes in global change. Wiley & Sons, Chichester: 5–18.

Streeter, H. W. & E. B. Phelps, 1925. Study of the pollution and natural purification of the Ohio river. III. Factors concerned in the phenomena of oxidation and reaeration. Bull. U.S. Publ. Health Serv. 116.

SYSTAT, 1992. Systat for windows: graphics, version 5. Evanston, IL: SYSTAT inc., 636 pp.

Tan, F. C. & P. M. Strain, 1979. Organic carbon isotope ratios in recent sediments in the St Lawrence estuary and the gulf of St Lawrence. Estuar. coast. mar. Sci. 8: 213–255.

Thomann, R. V. & J. A. Mueller, 1987. Principles of surface water quality modelling and control. New York, Harper & Row.

Van Eck, G. Th. M. & N. M. de Rooij, 1990. Development of a water quality and bio-accumulation model for the Scheldt estuary. In W. Michaelis (ed.), Coastal and Estuarine Studies. Springer Verlag. Berlin, Heidelberg: 95–104.

van Maldegem, D., 1988. Verzeilen van de immissiegegevens van het oppervlaktewater van het Schelde estuarium over de periode 1975 t/m 1986. Internal report RWS. GWAO-88.1267.

Vries, I. de, F. Hopstaken, H. Goossens, M. de Vries, H. de Vries & J. Heringa, 1988. GREWAQ: an ecological model for Lake Grevelingen. Rijkswaterstaat, Tidal Water division report T 0215-03.

Wattel, G. & A. Schouwenaar, 1991. SAWES-nota 19.06. Rijkswaterstaat, D.G.W.

Wollast, R., 1976. Transport et accumulation de polluants dans l'estuaire de l'Escaut. In J. C. Nihoul & R. Wollast (eds). l'Estuaire de l'Escaut. Projet Mer 10: 191–201.

Wollast, R., 1983. Interactions in estuaries and coastal waters. In: Bolin, B. & R. B. Cook (eds), The major biogeochemical cycles and their interactions. SCOPE.

Hydrobiologia **311**: 247–266, 1995.
C. H. R. Heip & P. M. J. Herman (eds), Major Biological Processes in European Tidal Estuaries.

Carbon flows in the Westerschelde estuary (The Netherlands) evaluated by means of an ecosystem model (MOSES)

Karline Soetaert & Peter M. J. Herman
Netherlands Institute of Ecology, Centre for Estuarine and Coastal ecology, Vierstraat 28, NL-4401 EA Yerseke, The Netherlands

Key words: carbon, budget, autotrophy, heterotrophy, Westerschelde, estuary

Abstract

The autotrophic production and heterotrophic consumption of organic matter in the Westerschelde, a highly turbid and eutrophic estuary in the Southwest Netherlands is examined by means of a dynamic simulation model. The model describes the ecologically relevant processes in thirteen spatial compartments and adequately fits most observed data.

Three autotrophic processes are included in the model. Net pelagic photosynthetic production is relatively low (average 41 gC m^{-2} yr^{-1}) and three spatial compartments near the turbidity maximum zone are respiratory sinks of phytoplankton biomass. According to the model, net phytobenthic primary production is more important than pelagic primary production in the upstream half of the Westerschelde. On the scale of the entire estuary, benthic primary production amounts to about 60% of pelagic primary production. Water-column nitrification, which is very important in the nitrogen cycle, is most pronounced near the turbidity zone where it accounts for the major autotrophic fixation of carbon (up to 27 g C m^{-2} yr^{-1}). Viewed on the scale of the total estuary, however, the process is not very important.

Less than 20% of total organic carbon input to the estuary is primary produced, the remainder is imported from waste discharges and from the river.

The degree of heterotrophy of the Westerschelde estuary proved to be one of the highest yet reported. On average 380 g carbon per square metre is net lost per year (range 200–1200 gC m^{-2} yr^{-1}). The yearly community respiration (bacterial mineralization, respiration of higher trophic levels and sedimentation) is 4 to 35 times (estuarine mean of 6) higher than the net production. This degree of heterotrophy is highest near the turbidity maximum and generally decreases from the freshwater to the seaward boundary. About 75% of all carbon losses can be ascribed to pelagic heterotrophic processes; the sediment is only locally important.

Mineralisation rates are highest in the turbidity region, but as only a fraction of total carbon resides here, less than 20% of all organic carbon is lost in this part of the estuary. This result is in contradiction with a previous budget of the estuary, based on data of the early seventies, where more than 80% of all carbon was estimated to be lost in the turbidity zone. Part of this discrepancy is probably caused by changes that have occurred in the estuary since that time.

Due to the high heterotrophic activity, nearly all imported and *in situ* produced carbon is lost in the estuary itself and the Westerschelde is an insignificant source of organic matter to the coastal zone.

The model estuary acts as a trap for reactive organic matter, both from the land, from the sea or *in situ* produced. Internal cycling, mainly in the water column, results in the removal of most of the carbon while the more refractory part is exported to the sea.

Introduction

Estuaries are characterized by a strong diversity in carbon sources. This is because of the existence of an intertidal habitat, the supply of nutrients and organic material from the river and from the sea and the input of matter from human origin. If turbidity remains limited, estuaries can also support a high primary production as nutrients are abundant. Nevertheless, estuaries are usually heterotrophic ecosystems (Billen *et al.*, 1991; Smith & Hollibaugh, 1993) where respiratory processes exceed *in situ* production.

The Westerschelde estuary (260 km^2 in SAWES, 1991) drains about 21580 square kilometres of land in one of the most densely populated and highly industrialised regions of Europe (Wollast, 1988). This estuary is unusual due to the high degree of eutrophication which results from the discharge of untreated wastes (Heip, 1988) and due to the high residence time of the water masses (Soetaert & Herman, 1995a). A turbidity maximum resides in the upper estuarine part which increases the residence time of particulates in the estuary.

According to Wollast (1976, 1983) nearly all the degradable organic matter is mineralized in the upper estuarine region and less than 10% of total land-derived carbon reaches the sea. This clumping of bacterial activity in the turbidity zone seems in contradiction with results from Goosen *et al.* (1992) who found bacterial production in the marine part of the estuary to be substantial, even compared to microbial activity in the turbidity zone.

In what follows we present a tentative annual mass budget of the most important carbon sources and fluxes in the eutrophic Westerschelde estuary of the years 1980–1985. This budget is based on an estuarine ecosystem model that includes all major physical, ecological and biochemical processes. As the model formulations and goodnesses of fit were published elsewhere (Soetaert & Herman, 1995a, b; Soetaert *et al.*, 1994) we restrict ourselves to describing the major carbon fluxes. We illustrate the variability in time and space of the carbon dynamics as put forward by the model. Finally, we examine the degree of heterotrophy along the estuarine length axis and compare this with data from other coastal and estuarine areas.

Table 1. State variables modeled in MOSES.

State variable	Units
Fast-decay detritus	g C m^{-3}
Slow-decay detritus	g C m^{-3}
Freshwater diatoms	g C m^{-3}
Freshwater flagellates	g C m^{-3}
Brackish and marine diatoms	g C m^{-3}
Brackish and marine flagellates	C m^{-3}
Micro-zooplankton	g C m^{-3}
Brackish meso-zooplankton	g C m^{-3}
Marine meso-zooplankton	g C m^{-3}
Hyperbenthos	g C m^{-3}
Detrital silicate	g Si m^{-3}
Dissolved silicate	g Si m^{-3}
Nitrate and nitrite	g N m^{-3}
Ammonia	g N m^{-3}
Oxygen	g O m^{-3}
Chlorides	g Cl m^{-3}
Suspended matter	g m^{-3}
Benthic diatoms	g C m^{-3}
Benthic deposit feeders	g C m^{-3}
Benthic suspension feeders	g C m^{-3}

Material and methods

The MOdel of the Scheldt EStuary (hereafter referred to as MOSES) is a thirteen spatial compartment, tidally averaged model that describes carbon, nitrogen, oxygen and silicate fluxes in the pelagic, subtidal and intertidal area. The model is in essence a carbon model and coupling of all fluxes is done by means of stoichiometric equations. State variables included in MOSES are in Table 1. In the pelagic realm they comprize four phytoplankton and two mesozooplankton pools, the microzooplankton, the hyperbenthos, two detrital fractions (with different degradabilities), a dissolved and particulate silicate fraction, two nitrogen pools, oxygen, chlorides and suspended matter. Bacteria are not modeled separately but included in the detritus fraction. Benthic diatoms, suspension and deposit feeders are modeled in the intertidal bottom.

The spatial compartimentalisation of the pelagic and intertidal area is in Fig. 1. Subtidal areas are defined as that part of a pelagic compartment that is deeper than the intertidal and less deep than the channels. The channel bottom (depth >10 m) is assumed to be biologically and biochemically inert and is not modeled.

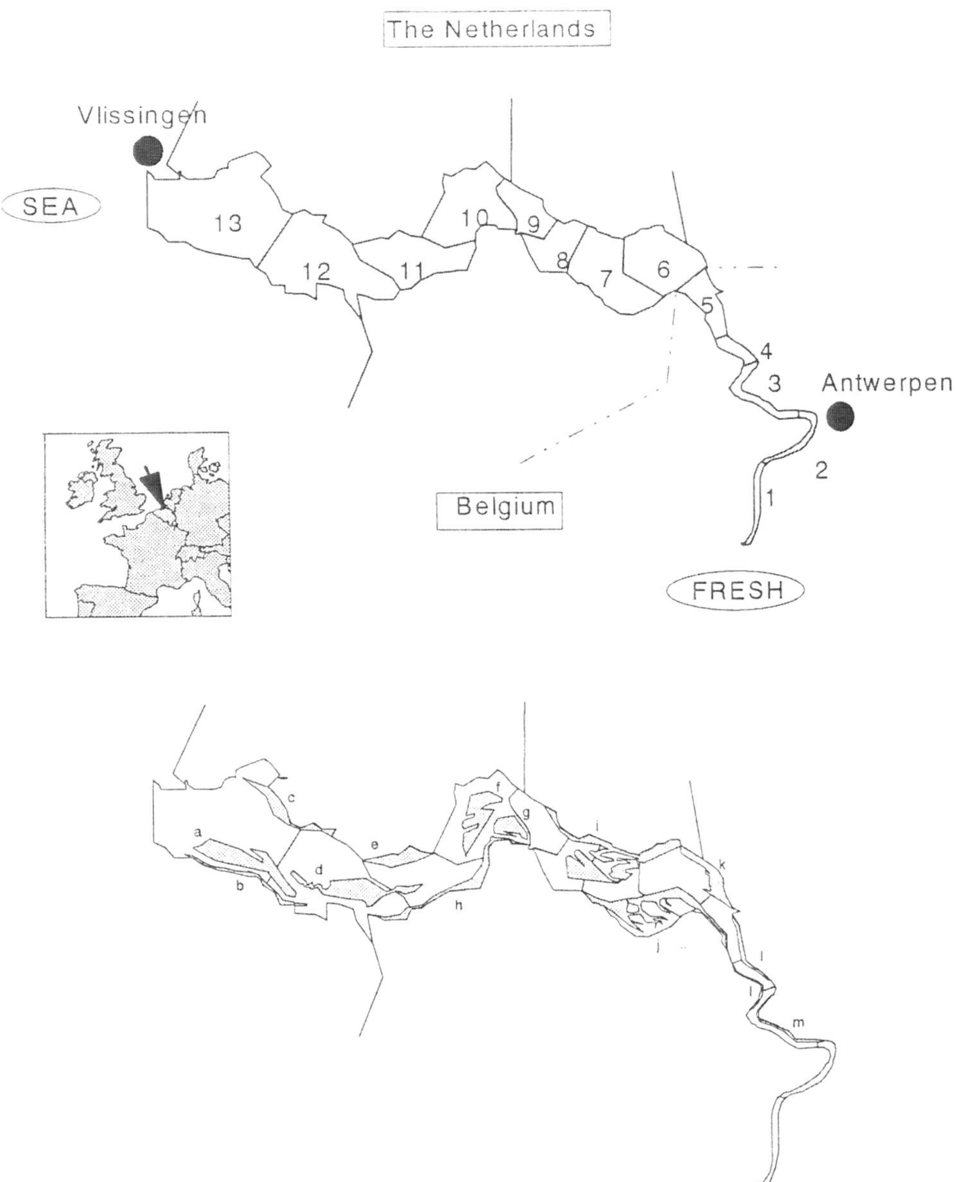

Fig. 1. Compartimentalisation of the pelagic (above) and the intertidal (below) in the Westerschelde model.

Pelagic constituents are moved by means of a tide-averaged, constant volume, advective-diffusive formulation (Soetaert & Herman, 1995a). Both a dissolved and a particulate (mud) transport are included. Resuspension and sedimentation of suspended matter is patterned according to Van Maldeghem *et al.* (1993). It is described as a fixed part of pelagic load that sediments to the bottom per day. The transport and sedimentation behavior of the pelagic detrital and living fractions is inbetween the behavior of dissolved and particulate matter. Organic carbon sedimentation is such that yearly values of benthic C fluxes obtained from Middelburg *et al.* (1995 and pers. comm.) are reproduced (Soetaert & Herman, 1993).

The biological processes described in MOSES are primary production (Soetaert *et al.*, 1994), feeding, respiration, excretion and physiological and predatory mortality. Biotic interactions are summarized in Fig. 2. In most cases the rate of change is expressed as first-rate kinetics with respect to the relevant state variable. Temperature dependence is modeled by means of a Q10 formulation, limitation functions are mostly described as a Monod equation.

Light-limited phytoplankton primary production is based on the model of Eilers & Peeters (1988). The description of algal primary production also includes nutrient limitation. Apart from grazing, dark respiration and salinity-stress mortality are the loss terms of phytoplankton standing stock (Soetaert *et al.*, 1994). Benthic algal production is modeled in much the same way as phytoplankton production, it is limited to the periods the intertidal flats are dry and a CO_2 limitation factor is included.

Higher trophic levels (micro- and mesozooplankton, hyperbenthos, benthic suspension and deposit feeders) are food-limited (threshold formulation). Part of ingested food is lost as faeces, another part is respired. Mortality includes an oxygen-dependent term (higher mortality when oxygen-depleted); the two mesozooplankton groups are also subjected to a salinity-stress mortality. An additional predatory mortality term allows for the inclusion of grazing by predators that are not explicitly modeled (e.g. fish). Benthic larvae (spawning) are included in the microzooplankton.

Aerobic organic matter degradation, denitrification and nitrification are the modeled biochemical processes in the water column (see Soetaert & Herman, 1995b). Oxygen is the limiting resource for aerobic mineralisation and nitrification; denitrification is inhibited at high oxygen concentrations. The availability of detritus is furthermore limiting the denitrification process, while ammonium is a limiting resource for the nitrification process.

Biochemical processes in the bottom are not modeled exhaustively but nutrient and oxygen exchange across the bottom-water interface is represented by means of the diagenetic model of Lancelot & Billen (1985). Organic carbon that sediments to the bottom is considered to be lost to the estuary (mineralized or buried).

Dissolution of particulate silicate and oxygen exchange across the water-air interface are also included in MOSES.

Forcing functions and boundary conditions were either kindly provided by Van Eck (SAWES, 1991) or were obtained from Klepper (1989).

For a detailed description of the model formulations, the obtained parameters and the goodnesses of fit the reader is referred to Soetaert & Herman (1993; 1995a, b) and Soetaert *et al.*, (1994). The model is able to reproduce most observed data that are available (see also Figs 3–4). Data on the benthos and microzooplankton were as yet insufficient to warrant a great calibration effort. Their role in the model is of minor importance compared to the mesozooplankton (Soetaert *et al.*, 1994).

Results

Spatio-temporal plots were produced by averaging the model output of the years 1980–1985 over a year and smoothing this by means of a double-weighted least squares regression (SYSTAT, 1992).

Description of spatial and temporal characteristics of the major organic carbon stocks in the model (Fig. 5)

Phytoplankton carbon in the model ranges from 0.02 gC m^{-3} in winter to 1.9 gC m^{-3} near the freshwater part in summer. Yearly averaged phytoplankton biomass drops from 1 gC m^{-3} near the freshwater part to 0.2 gC m^{-3} in the middle of the estuary after which biomass increases slightly (0.3 gC m^{-3}) towards the sea. The two peaks of phytoplankton biomass correspond with a freshwater and a brackish + marine community. For a detailed description of the factors that are responsible for this spatio-temporal pattern we refer to Soetaert *et al.* (1994).

Total organic carbon in the model ranges from 3 gC m^{-3} to 16 gC m^{-3}. In winter and autumn, organic matter peaks near the turbidity maximum (model spatial compartments 2 and 3). In summer, a maximum is observed very near to the freshwater boundary. This partly reflects the high phytoplankton standing stocks there. In general the organic load in the model decreases towards the sea.

Benthic diatom carbon in the intertidal increases from the seaward boundary (yearly average 4 gC m^{-2}) towards upstream (7 gC m^{-2}). Minimum biomass in the model is 2 gC m^{-2}, maximum biomass is 12 gC m^{-2}.

Description of spatial and temporal patterns of the various processes in the model (Fig. 6)

Aerobic mineralisation processes in the water column (Fig. 6a) are strongly patterned by the load of total organic carbon in the model (Fig. 5) but mineralisation is relatively more pronounced towards the sea

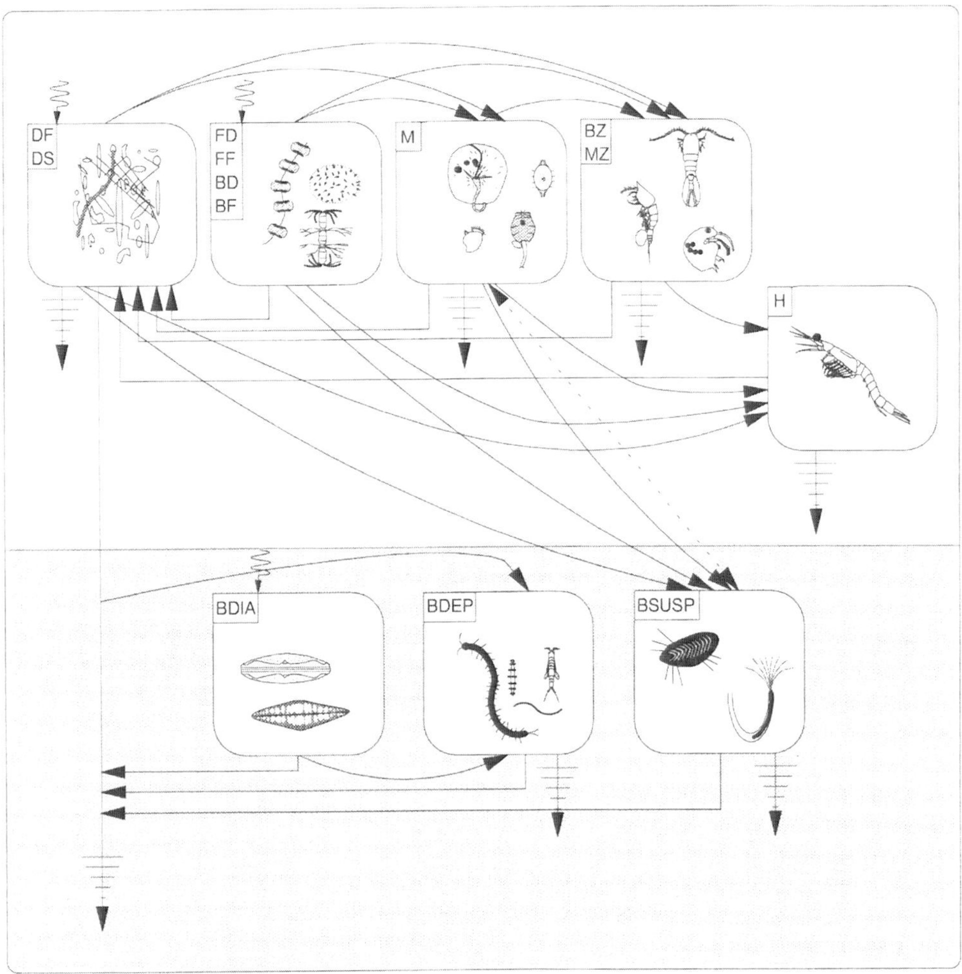

Fig. 2. Biotic interactions in MOSES. Curved arrows denote food relationships; undulated arrows denote primary production; the dotted arrow denotes spawning activity; the vertical arrows that are crossed with small lines indicate carbon loss to the system. DF = fast decaying detritus; DS = slow-decaying detritus; FD = freshwater diatoms; FF = freshwater flagellates; BD = brackish diatoms; BF = brackish flagellates; M = microzooplankton; BZ = brackish meso-zooplankton; MZ = marine meso-zooplankton; H = hyperbenthos; BDIA = benthic diatoms; BDEP = benthic depositfeeders; BSUSP = benthic suspension feeders.

and towards the freshwater boundary. This reflects the higher reactivity of organic matter imported from the river and the sea, compared to the resident organics in the mid-estuary.

Water-column denitrification rates in the model (Fig. 6b) clearly peak towards the fresh-water part of the estuary. They are substantially higher in summer when oxygen content of the water column is lower and temperature higher.

Pelagic carbon loss rates due to sedimentation (Fig. 6c) are most pronounced in the mid-estuary where tidal flats are relatively most abundant. Near the sea and near the river, the sedimentation surface is insignificant compared to total volume and sedimentation rates (expressed per volumetric units) are very small. The seasonality of sedimentation is not very pronounced and somewhat reflects carbon standing stock. When expressed per unit of intertidal surface (Fig. 6j) the amount of carbon lost to the intertidal flats (sedimentation + benthic uptake) decreases towards the sea. The slight increase at the three most downstream intertidal flats reflects the higher benthic biomass there.

The amount of carbon respired by the higher trophic levels (Fig. 6d) is most pronounced in summer in the marine part of the estuary. This reflects mesozooplankton standing stock (Soetaert & Van Rijswijk, 1993;

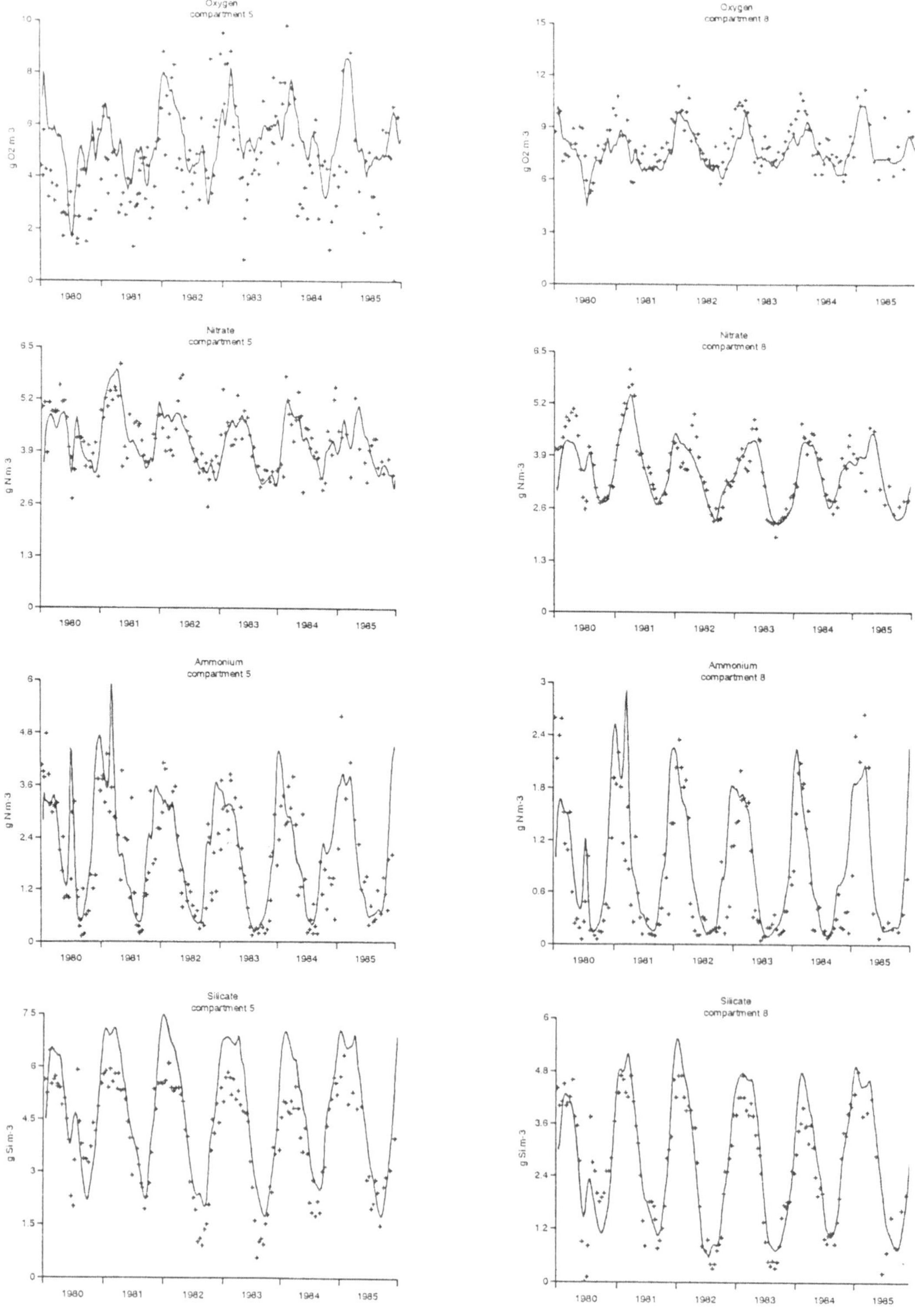

Fig. 3. Goodness of fit of modeled (full line) and observed ('+') variables for model spatial compartments 5 and 8. Variables are Oxygen, Nitrate, Ammonium and silicate.

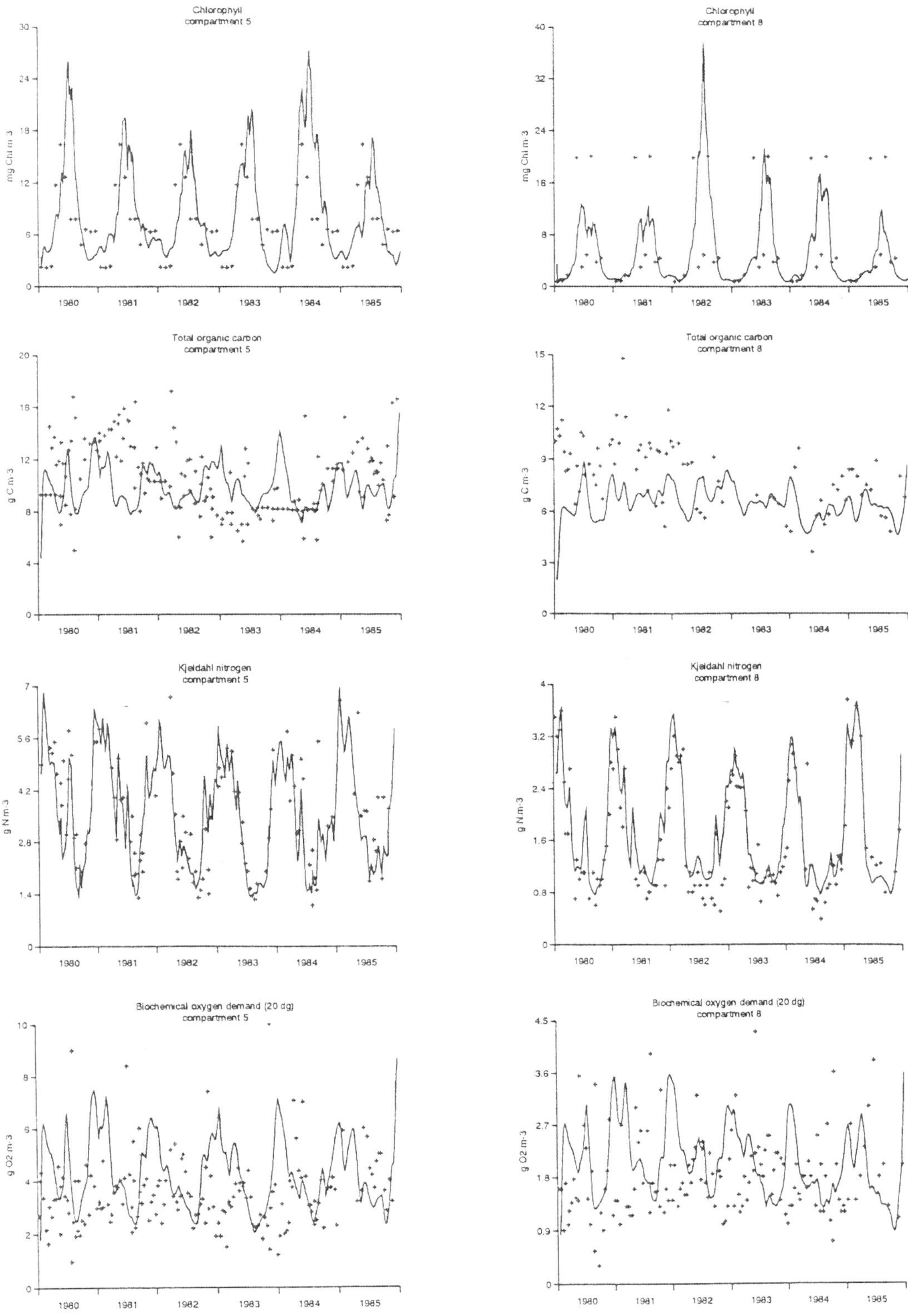

Fig. 4. Goodness of fit of modeled (full line) and observed ('+') variables for model spatial compartments 5 and 8. Variables are Chlorophyll, Total organic carbon, Kjeldahl nitrogen and biochemical oxygen demand.

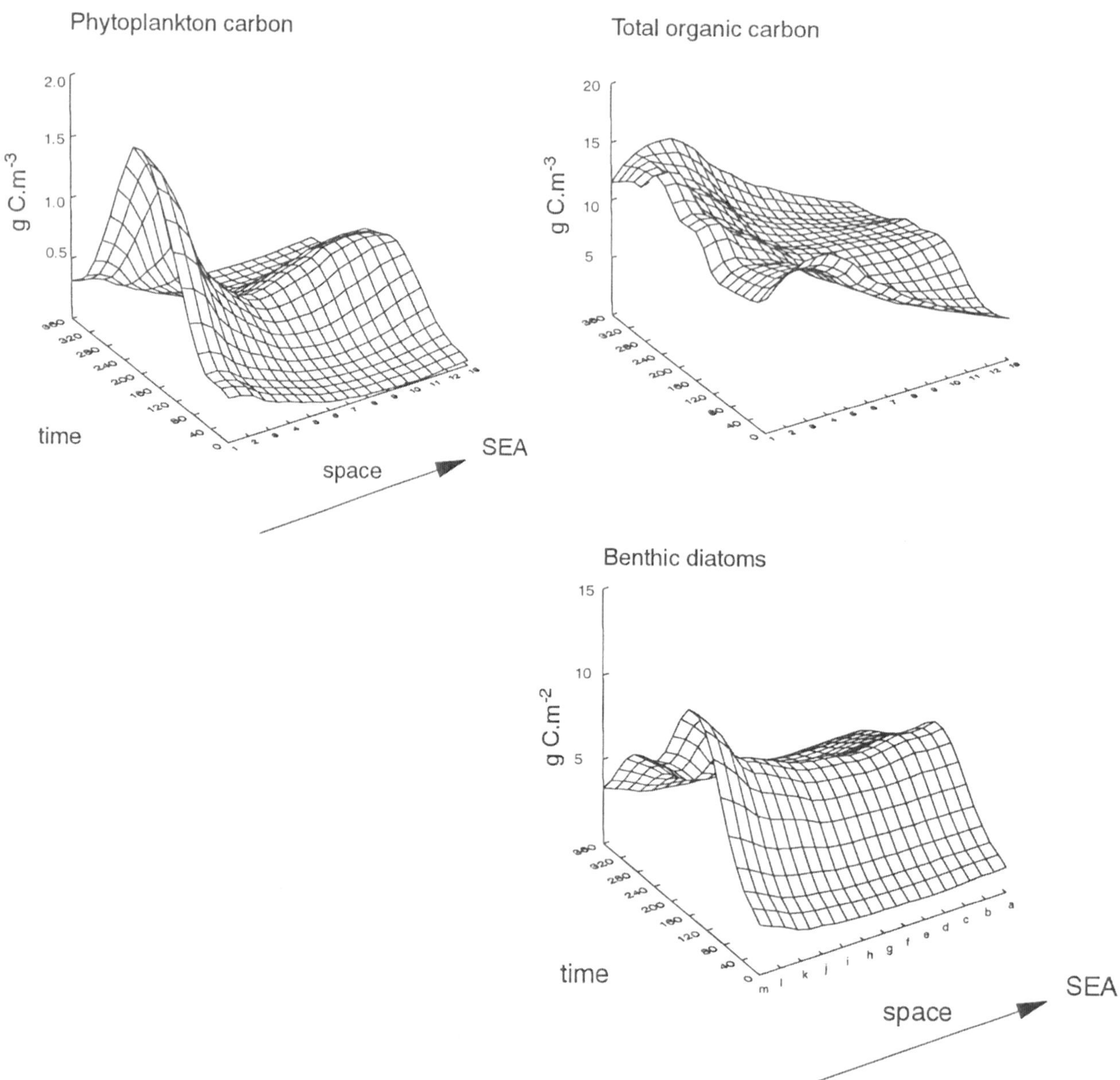

Fig. 5. Temporal and spatial patterns of the major carbon stocks (six years average of model output). The X axis represents the space axis, Y gives the day number (starting on 01-jan, ending on 31-dec). Phytoplankton and total organic carbon are in pelagic compartments (1–13), benthic diatom carbon is in intertidal compartments (a–m).

Soetaert & Herman, 1994). In the brackish part of the estuary, carbon loss by the higher trophic levels is more evenly distributed throughout the year. Winter and spring consumption is mainly caused by mesozooplankton whose biomass peaks at that period (Soetaert & Van Rijswijk, 1993) while in summer microzooplankton consumption is more important in the model. Due to the different spatial and temporal characteristics of its constituents, the distribution of total carbon consumption in time and space (Fig. 6e) is complicated. The general trend reflects pelagic aerobic mineralisation rates, the denitrification has greatest impact upstream, sedimentation processes are most prominent in the mid-estuary and carbon lost by respiration of higher trophic levels is slightly obvious in total C consumption near the sea.

Among the primary production processes, nitrification rates (Fig. 6g) decline steadily towards the sea

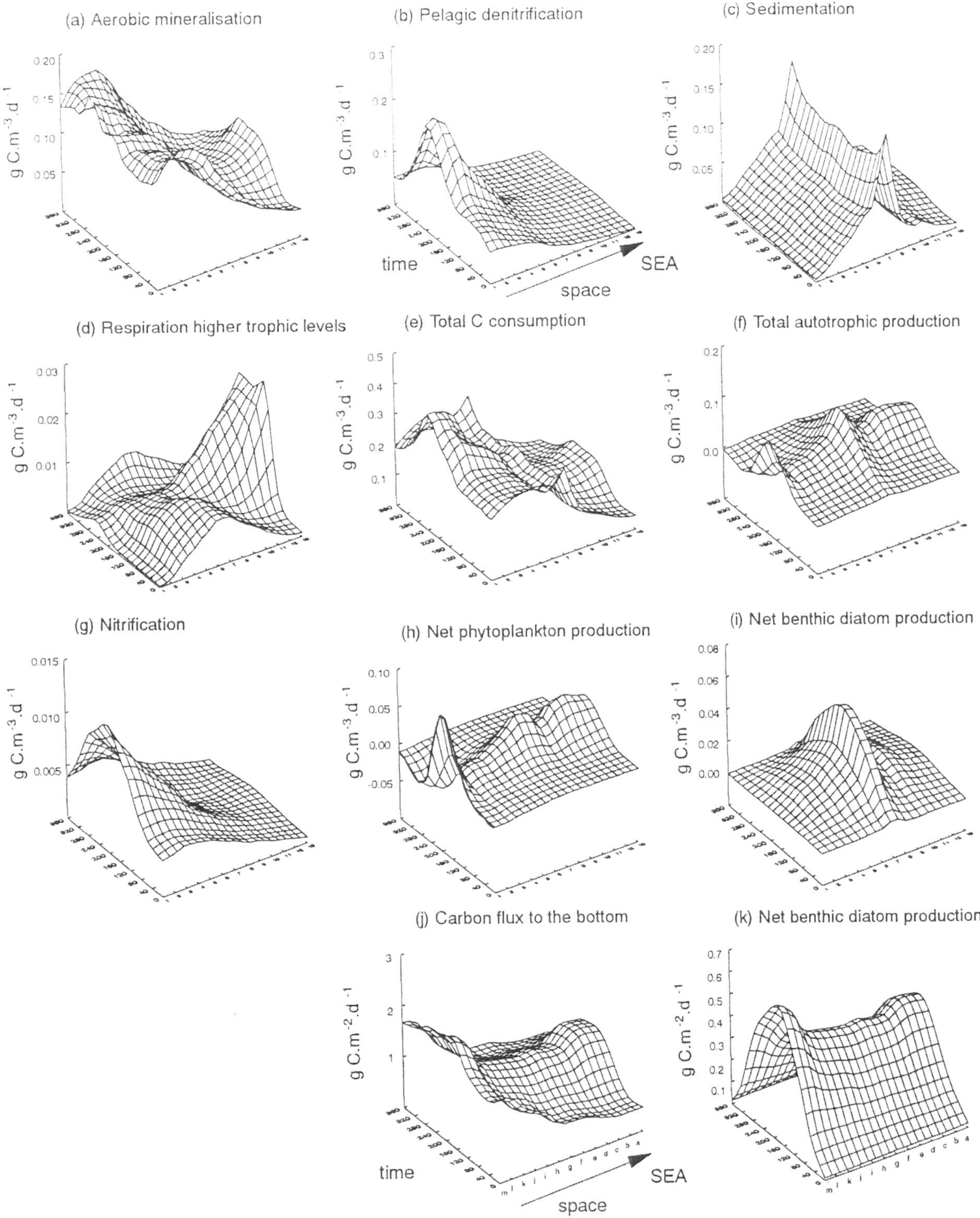

Fig. 6. Temporal and spatial patterns of the various carbon consuming (6a...6e, 6j) and carbon producing processes (6f...6i, 6k) (six years average of model output). Carbon flux to the bottom (Fig. 6j) and benthic diatom production (Fig. 6k) are in the intertidal compartments, the other carbon processes are in the pelagic compartments.

Table 2. Comparison this model-Wollast (1976) in 1000 tonnes carbon per year. 'Upstream' = model compartments 1–4 'Downstream' = model compartments 5–13

10^3 ton C yr^{-1}		Model	Wollast
	Total import	+104	+102
Upstream			
	Degradation	− 13	− 40
	Sedimentation	− 1	− 46
	Production	+ 0.5	-
	Secondary prod.	− 0.3	-
	Other processes	− 0.2	-
	net C-loss	− 14	− 86
Downstream			
	Degradation	− 66	− 8
	Sedimentation	− 20	− 4
	Production	+ 18	+ 4
	Secondary prod.	− 9	-
	Other processes	− 7	-
	net C-loss	− 84	− 8
	Export to the sea	− 6	− 8

to very low levels. They peak in summer. Net phytoplankton production (Fig. 6h) is negative all over the estuary during the darker months, positive production values are attained in summer in all but model spatial compartment 2. Net production is lowest near the turbidity zone (spatial compartments 2–4) and increases both towards the freshwater and marine boundary.

Net benthic diatom production (Fig. 6i), expressed per unit volume is significant in the mid-estuarine region due to the presence of extensive tidal flats there. Benthic diatoms attain positive net production values all over the year. When expressed per unit of intertidal surface (Fig. 6k), net benthic diatom production reflects both the diatom standing stock (increase towards the freshwater zone) and the carbon flux towards the bottom (higher at the three most seaward compartments). The latter stems from the fact that diatom production is mainly CO_2 limited in the model. The sum of all autotrophic processes in the model (Fig. 6f) shows the predominance of benthic diatom production in the mid-estuarine region. The pattern up- and downstream corresponds more closely to phytoplankton primary production but the sharp seasonal gradients of phytoplankton productivity are moderated by the other autotrophic processes.

A yearly averaged carbon budget (Figs 7–9)

Mean net production and consumption rates expressed in pelagic units (g C m^{-3} d^{-1}) as put forward by the model are in Fig. 7. The most upstream part of the estuary (spatial compartment 2–4) is very turbid, causing yearly phytoplankton production to be negative (see Soetaert *et al.*, 1994). As there are almost no intertidal flats there, phytobenthos production is very small. Due to the high load of ammonium in the water column, nitrification is the most important primary production process in MOSES spatial compartments 2 and 3. Towards the middle of the estuary (spatial compartments 5–7), the turbidity of the water column decreases and phytoplankton net production becomes positive. As the intertidal surface/pelagic volume ratio is high in this part, phytobenthos primary production is the main autotrophic process here. Towards the sea, the pelagic gains in importance compared to the intertidal zone. Hence in the downstream part of the estuary (spatial compartments 8–13), phytoplankton production becomes the most important autotrophic process, phytobenthic production ranks second, while nitrification is unimportant here.

Net phytoplankton production in the model varies from −74 to 60 gC m^{-2} (total surface) per year. The phytobenthos produces some 80 to 120 gC m^{-2} (intertidal surface) per year. Total water-column nitrification varies from 4 to 27 gC m^{-2} yr^{-1} (total surface).

Except for model spatial compartment 7, pelagic aerobic mineralisation is the most important heterotrophic process in the estuary. Denitrification rates in the water column are high in the most upstream part of the estuary only. Sedimentation is most important in the mid-estuary. Net carbon consumption by the higher trophic levels is relatively unimportant, except at the most seaward compartments where it comprizes about 15% of all heterotrophic carbon loss.

Due to the funnel shape of the estuary, processes that occur at much lower rates near the sea can nevertheless be more important in absolute terms. The autotrophic and heterotrophic processes, expressed as the total amount of carbon gained or lost per spatial compartment per year are in Fig. 8. The total amount of carbon produced or consumed in the estuary increases almost exponentially with distance downstream. Processes that are linked to the intertidal area have greatest impact in pelagic compartments 6, 7 and 10.

Figure 9 gives the yearly integrated import, export, production and consumption processes for 1980–1985. All in all 104 000 tonnes C yr^{-1} is imported into the

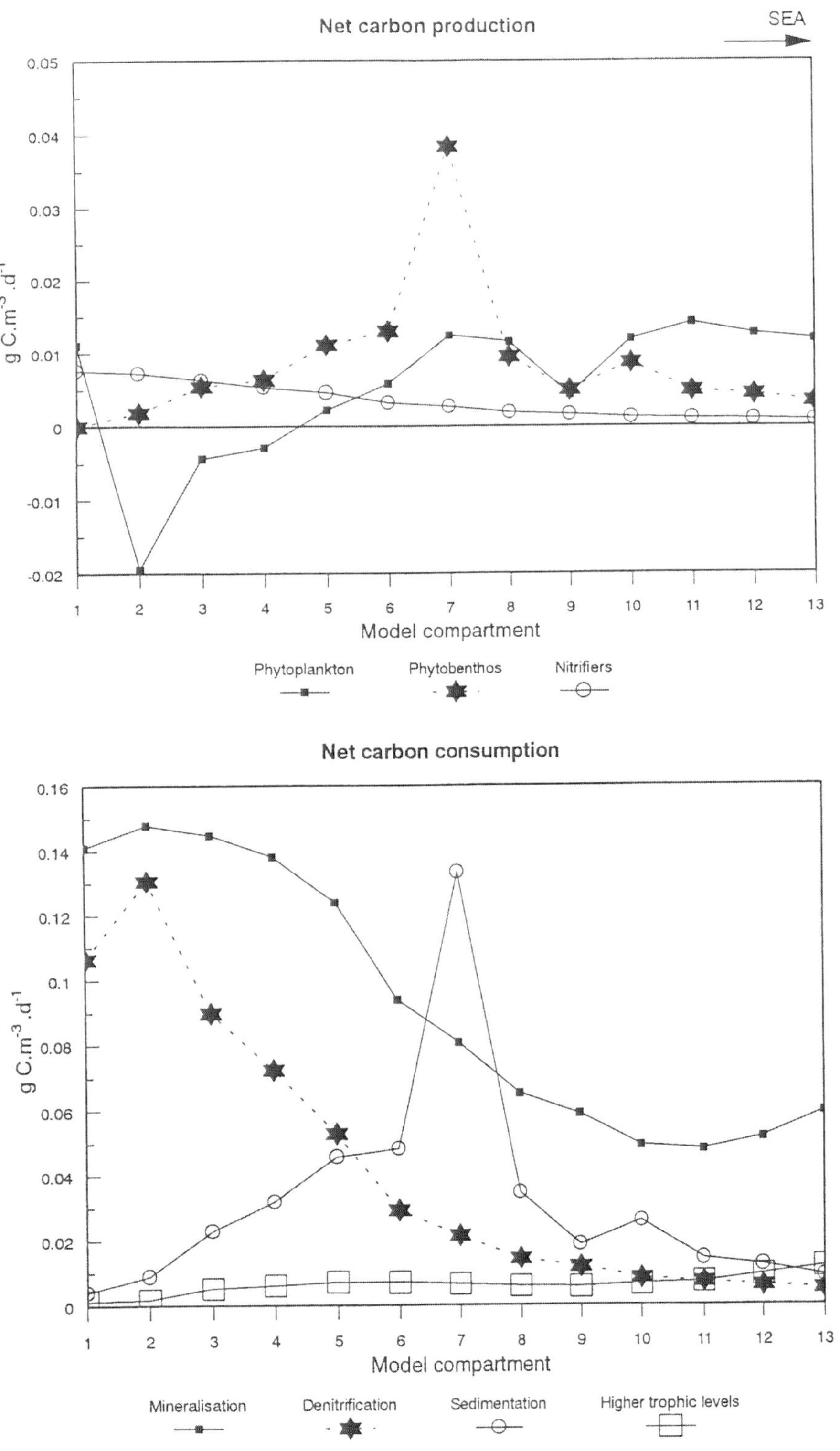

Fig. 7. Yearly averaged process rates (g C. m^{-3} d^{-1}) for the different (pelagic) model compartments.

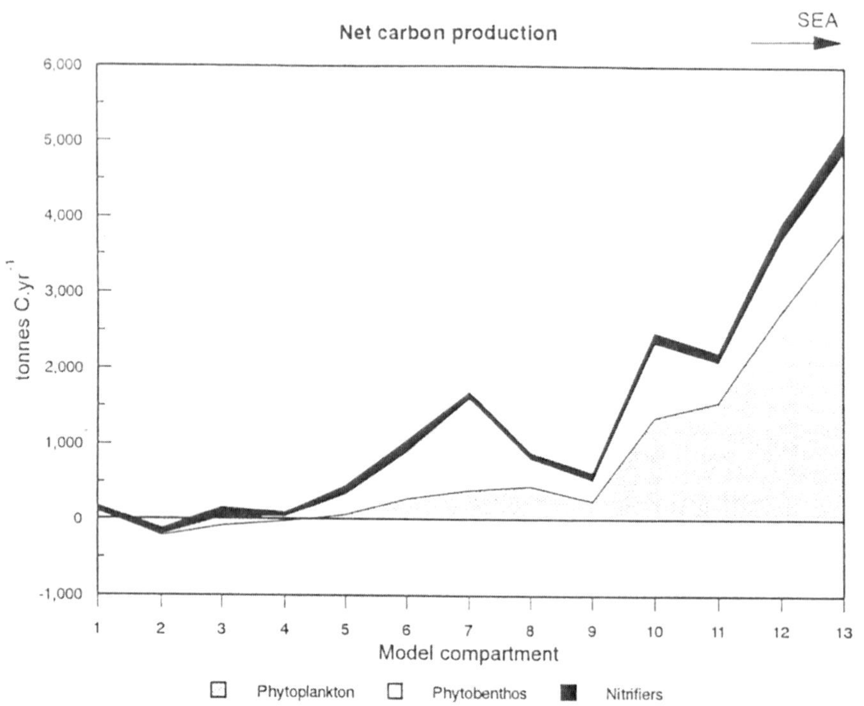

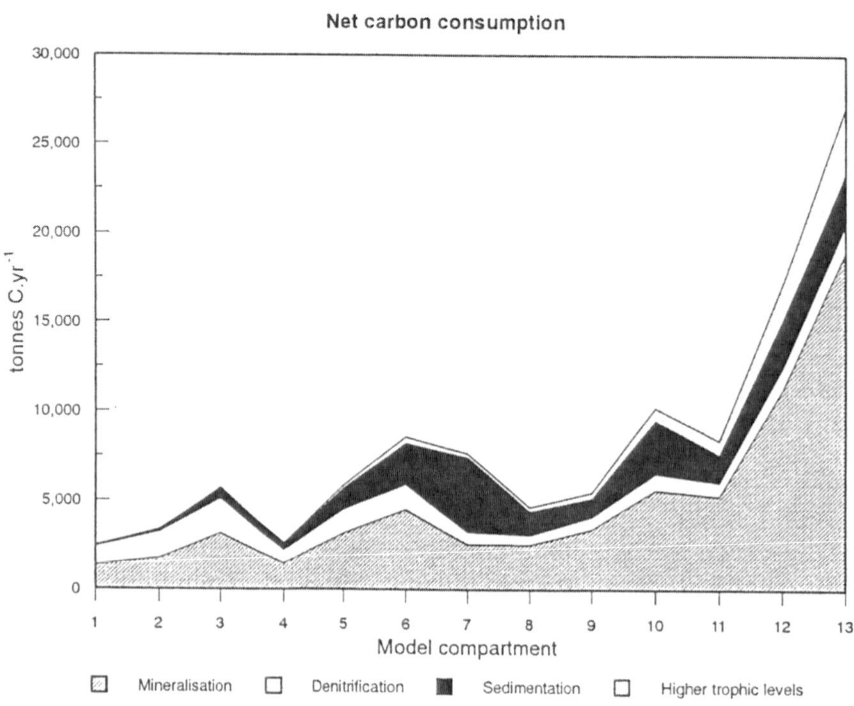

Fig. 8. Yearly carbon budget (tonnes C yr^{-1}) for the different pelagic compartments.

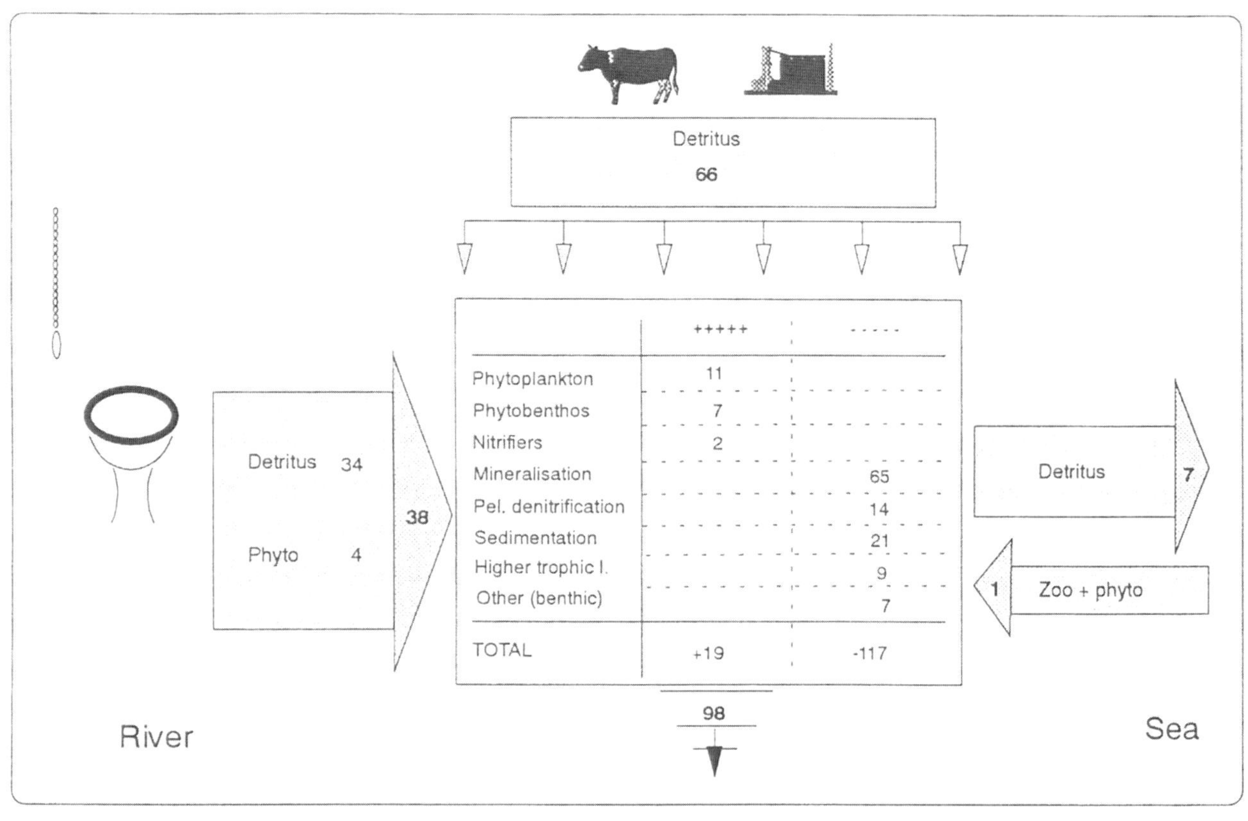

Fig. 9. Global carbon budget of the Westerschelde estuary, in 1000 tonnes C yr^{-1}. Positive values indicate (net) gain, negative values indicate (net) loss. The arrow to the left indicates exchange with the river; the arrows to the right indicate exchange with the sea. The small arrows on top are waste discharges. The bottom arrow indicates net carbon loss in the estuary.

estuary by means of waste discharges or from the river. By far the majority of this carbon consists of detritus. A small amount (1000 tonnes C yr^{-1}) of zoo- and phytoplankton is (net) imported from the sea. The amount of carbon primary produced in the estuary is much less (19 000 tonnes C yr^{-1}). Some sixty percent of this primary carbon is produced by the phytoplankton, thirty percent by phytobenthos, some 10% is formed by nitrifying activity. Amongst the decomposition processes, pelagic aerobic mineralisation is most important in the model (56%) followed by sedimentation (18%) and pelagic denitrification (12%). The higher trophic levels consume some 8% of all carbon in the estuary. Only 7000 tonnes of detrital carbon leaves the estuary at the sea side, while 98 000 tonnes C per year is net lost in the Westerschelde.

Autotrophic/heterotrophic balance (Fig. 10)

The net rate of carbon consumption (g C $m^{-2} yr^{-1}$) and the ratio of respiration versus autotrophic production in the model are in Fig. 10. The Westerschelde estuary is heterotrophic all over and the degree of heterotrophy is very high: carbon respiration is on average six times higher than net primary production and 0.06–0.31 gC m^{-3} is lost in the water column per day. This corresponds to a net carbon loss of 220–1200 gC m^{-2} per year (global estuarine average of 380 gC m^{-2} yr^{-1}). The heterotrophic nature of the estuary is highest near the turbidity maximum (spatial compartment 2) where community respiration is about 35 times higher than net production. The degree of heterotrophy decreases downstream, but respiration/production ratios are overall higher than 4. The slight rise from spatial compartment 11 towards downstream reflects the import of more reactive detritus from the sea, resulting in increased mineralisation rates.

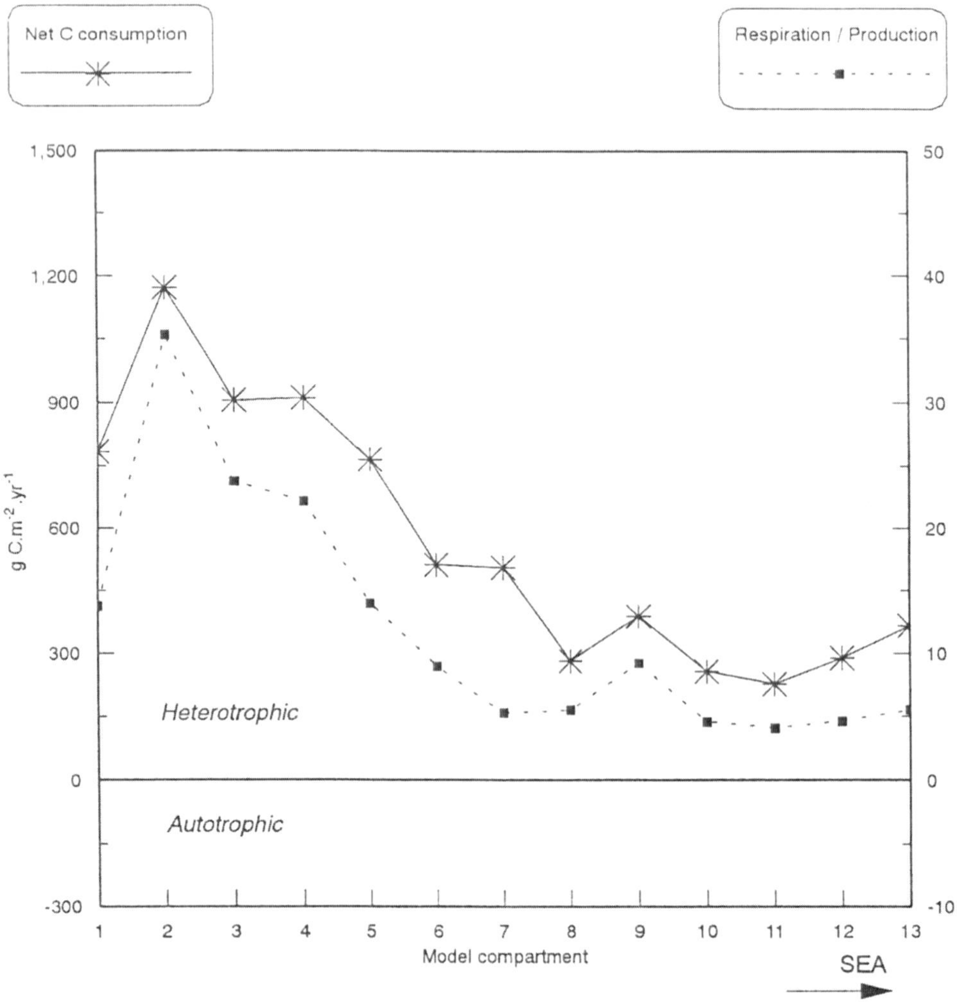

Fig. 10. Yearly net carbon consumption (Respiration – Production) and Respiration/production ratio for the different pelagic model compartments.

Discussion

The distribution of organic carbon in the Westerschelde is affected by physical processes such as residual estuarine circulation and mixing, as well as by a range of biochemical and biological reactions.

Sources of organic carbon

The river Schelde and waste discharges from point sources and tributaries are the main sources of organic matter to the estuary. Each year, they supply about 100 10^3 tonnes of carbon to the ecosystem. Compared to this allochtonous source of carbon, the primary formation of organic matter (in the model) is much less and amounts to 19 10^3 tonnes of carbon per year (some 72 gC m^{-2} yr^{-1}). Some sixty percent of this primary carbon comes from phytoplankton production, the benthic algae account for another 30%, while about 10% is due to pelagic nitrification.

Autotrophic processes

If turbidity remains restricted, the high nutrient load that is typical for many estuaries allows a high primary production at these sites. Hence estuaries are usually considered to be highly productive areas and in some cases they significantly add to coastal enrichment (Peterson *et al.*, 1988). The average phytoplankton production in estuaries was estimated to be about 190 gC m^{-2} yr^{-1} (Boynton *et al.*, 1982; Smith & Hollibaugh, 1993), which is much higher than the average of 41 gC

m^{-2} yr^{-1} that is modeled in the Westerschelde. Reasons for this low phytoplankton productivity are the high turbidity and relatively great depth of the Westerschelde estuary. This results in an adverse environment for the phytoplankton communities which have to spend a large amount of time in darkness (Soetaert *et al.*, 1994). Hence the production potential of the phytoplankton in the Westerschelde is far from being realised.

Due to the high turbidity of the water masses, we assume that benthic primary production is restricted to tidal flats when they are dry (Colijn & de Jonge, 1984). The mean annual phytobenthos biomass in the model of 4–7 gC m^{-2} falls well within ranges reported by de Jonge & Colijn (1992) for the Ems estuary. The model predicts benthic primary production values varying between 80 to 120 g C m^{-2} yr^{-1} (intertidal surface) and the seasonality in primary production is not very pronounced. Colijn & de Jonge (1984) found estuarine benthic primary production to vary between 47 and 314 g C m^{-2} yr^{-1}. Similarly, Jassby *et al.* (1993) found the overall mean productivity for 28 different sites to be 110 g C m^{-2} yr^{-1} and 50% of all values were within 66–180 g C m^{-2} yr^{-1}. Thus phytobenthos productivity in the model appears to be consistently estimated. In a set of preliminary experiments, Kromkamp *et al.* (1995) measured much lower phytobenthos production values in a central intertidal area (roughly estimated about 20 gC m^{-2} yr^{-1}). This could mean that our model estimates are too high. Nevertheless, in view of the difficulties inherent in the calculation of annual production of benthic microflora (Shaffer & Onuf, 1985), an estimate, consistent with literature data, seems to be most appropriate.

Because of the low primary production values in the pelagic, benthic algae are the major primary carbon source in model spatial compartments 4 to 7. Their contribution to total carbon fixation decreases towards the river and the sea as the ratio of intertidal surface versus water volume decreases and as the waters become less turbid towards the sea (allowing higher phytoplankton production). On average about one third of total primary production in the model is derived from phytobenthos. This is consistent with values cited in Smith & Hollibaugh (1993).

Bacterial autotrophy is usually considered to be of minor importance due to its inefficient energy generating system (Helder & de Vries, 1983) and they are mostly ignored in mass balance studies. We use a rather low carbon conversion efficiency of 0.033 gC $(gN)^{-1}$, which is near to efficiencies of 0.045 reported by Helder & de Vries (1983) and of 0.025 gC $(gN)^{-1}$ cited in Jassby *et al.* (1993). Other studies (Indrebø *et al.*, 1979) use higher efficiencies of 0.09 gC $(gN)^{-1}$.

Notwithstanding their low efficiency, in the Westerschelde model, the nitrifying activity in the upper estuarine stretch is so high (Soetaert & Herman, 1995b) that they are the most important carbon fixing organisms in this turbid part (about 25 g C m^{-2} yr^{-1}). On the scale of the entire estuary however, bacterial primary carbon fixation is not very important (about 10%).

High rates of carbon-fixation by nitrification were also observed at the oxic-anoxic interface in a permanently stratified estuary in western Norway (Indrebø *et al.*, 1979) and in a stratified saline lake (Cloern *et al.*, 1983). The latter authors found chemoautotrophic production to be 150 gC m^{-2} yr^{-1}, of which 45 to 80% was contributed by nitrifiers.

In the mid-estuarine region a large salt marsh is situated (Saeftinge), which is characterized by a high primary production (De Leeuw & Buth, 1991). This ecosystem is not included in our model, and according to Hemminga *et al.* (1993), export of vascular plant detritus from the Saeftinge salt marsh in the Westerschelde is insignificant. Hence, salt marshes are not considered to contribute to estuarine enrichment in the model.

Heterotrophic processes

In the model, the allochthonous and autochthonous particulate organic matter in the upstream estuarine part is temporarily retained in the region of high turbidity where it is degraded in an aerobic and anaerobic pathway before being flushed downstream. The more inert organic matter that reaches the middle region of the estuary then joins the carbon of marine origin. A substantial deposition on the extensive tidal flats occurs here in the model. All in all, sedimentation removes some 20% of total carbon in the estuary.

By far the major decomposition occurs by means of aerobic mineralisation (56% of all carbon loss). Water-column denitrification, which is a significant component in the nitrogen cycle (Soetaert & Herman, 1995b) is of less importance in carbon recycling, except in the most upstream compartments, where about 45 percent of total degradation is estimated to be anaerobic. The amount of carbon respired by organisms other than bacteria increases from about 1% in the most upstream compartments to about 14% near the sea; on average

8% of all carbon losses can be attributed to higher trophic levels in the model.

Decomposition rates in the model are most pronounced at the zone of maximal turbidity, where carbon concentrations are highest. As the carbon concentration and reactivity decreases towards the mid-estuary, decomposition rates become lower. With the input of reactive organic matter from the sea, aerobic decomposition rates increase from mid-estuary towards downstream but denitrification, which is inhibited by oxygen, further declines.

If we assume a growth yield of 0.5, our yearly-integrated values of bacterial production are close to production values measured *in situ* by Goosen *et al.* (1992), except for the most upstream compartment (Fig. 11). Bacterial production in spatial compartment 1 as given in Goosen *et al.* (1992) was measured at the freshwater boundary, irrespective of the phase of the tide (and hence some measurements were probably performed in the river rather than in the estuary). As bacterial production is higher in the river compared to the estuary (Goosen *et al.*, 1992) and as our estimates are compartment means, this could explain part of the differences of both estimates. A growth yield of 0.5 is not exceptionally high, although such efficiencies are only found in situations without nutrient limitation (Billen *et al.*, 1990). Thus, although independently derived, bacterial production measured in the field lends validity to our model estimates of bacterial activity.

The rates at which organic matter is formed or degraded are interesting for comparison with other areas, but concerning the estuarine ecosystem as a whole, total amounts are more instructive. The volume of the upstream estuarine compartments are only a fraction of the most seaward compartments. Hence, notwithstanding higher rates near the turbidity zone, by far the major production and consumption of carbon occurs in the most seaward part. Processes that are linked to the benthos are most pronounced in the mid-estuary where tidal flats are most abundant.

All in all, about 117 000 tonnes of carbon is consumed in the model estuary per year.

Impact of the turbidity zone on the estuarine carbon cycle (Fig. 12)

In the model, a small fraction of total carbon input is lost in the turbidity zone (model spatial compartments 1–4), mainly through bacterial degradation. Sedimentation and respiration by other organisms than bacteria are negligible in the turbidity zone. The primary formation of organic matter is almost nil there. Hence, of the 74×10^3 tonnes of carbon imported in this small stretch, some 60×10^3 (80%) is exported to the zone downstream. By far the largest amount of carbon is removed in the area downstream from the turbidity zone (spatial compartments 5–13) and pelagic bacteria are mainly responsible for this (65%). Sedimentation removes some 20% of primary produced and imported carbon, while secondary producers other than bacteria account for 9% of carbon loss in this estuarine stretch. Nearly all primary production occurs in this zone.

The predominance of carbon loss in the downstream zone in our model seems in contradiction with the lower degradation rates there. It is however the consequence of the larger volume residing in this part (95% of the entire Westerschelde volume) and the higher total standing stocks of carbon (85% of total Westerschelde carbon load in 1991, data from the JEEP data base).

Comparison with a previous budget

Another carbon budget of the Westerschelde, based on data of the early seventies was developed by Wollast (1976). He subdivided the Westerschelde into an upstream part near the turbidity maximum (our model spatial compartments 1 to 4) and a downstream part (compartments 5 to 13). There are some striking disagreements between the budget of Wollast and our budget that merit attention (Table 2).

A major difference is in the main centre of degradation. Wollast calculated that more than 80% of total carbon input was lost in the most upstream part of the estuary (compartments 1-4) and about half of this loss was due to sedimentation. Our budget reveals the opposite trend, with 80% of total loss occurring in the more downstream part of the estuary (compartments 5–13), mainly due to pelagic degradation.

Part of the discrepancy between both budgets could be methodological. Bacterial degradation in Wollast's budget was measured by means of dark bicarbonate ^{14}C intake (Billen *et al.*, 1976), which was taken indicative of heterotrophic respiration and the conversion into heterotrophic activity much depends on the use of fixed factors (Billen *et al.*, 1976). Pelagic degradation in the downstream part of the estuary was calculated such as to close the budget and it may have been underestimated: whereas our model and *in situ* measurements by Goosen *et al.* (1992) indicate that bacterial degradation rates are about 5 times lower in

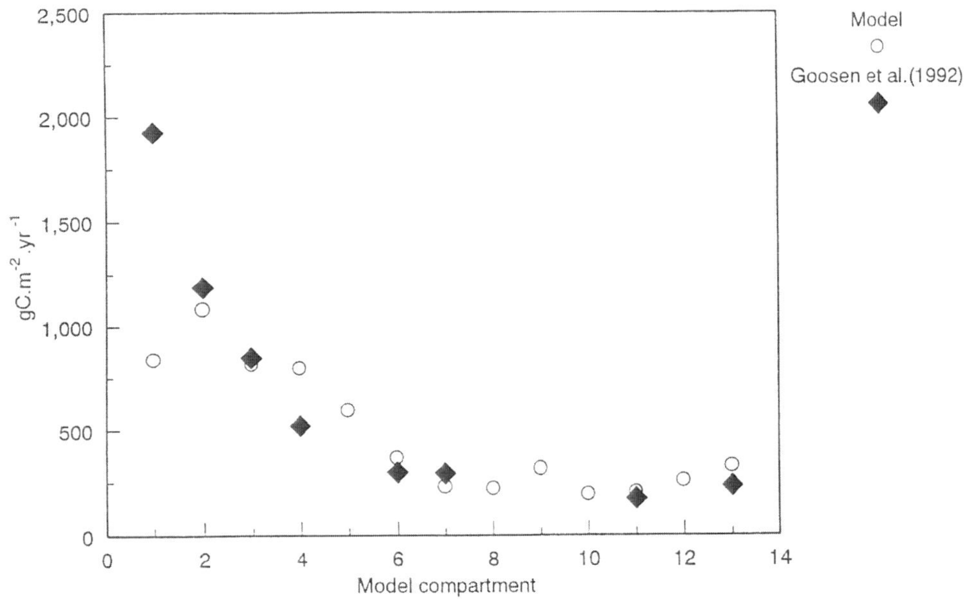

Fig. 11. Modeled versus *in situ* measured yearly bacterial production values (Goosen, 1992), assuming a growth yield of 50%.

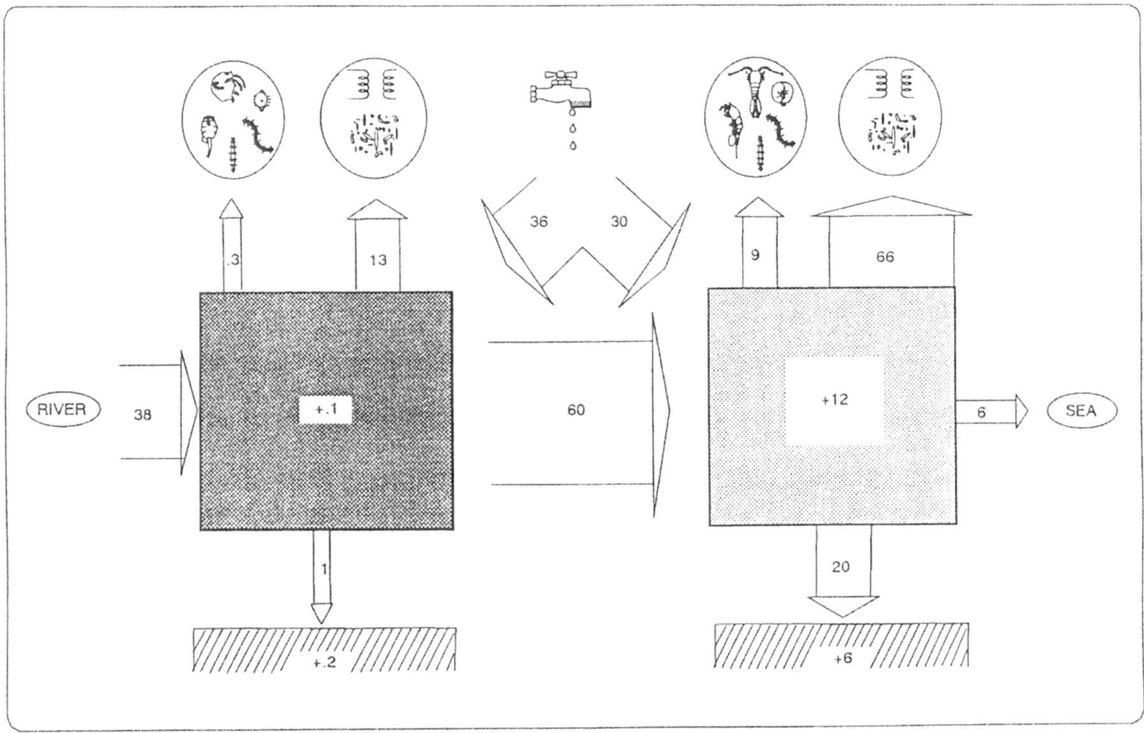

Fig. 12. Yearly carbon budget of the turbidity zone (pelagic compartments 1–4, darkly shaded) and of the zone downstream (compartments 5–13). Arrows directed to the boxes indicate import. The left icon above indicates higher trophic levels, the right icon denotes bacteria. Arrows directed towards the icons denote the amount of carbon respired by these groups. The tap in the middle of the graph denotes waste import.

the downstream part, Wollast (1976) assumed almost a 100-fold difference.

Furthermore, carbon sedimentation was calculated by Wollast by multiplying estimates of sediment accumulation rates with average carbon content in the sediments, while we patterned total organic matter sedimentation to recent estimates of sediment accumulation rates in Van Maldeghem *et al.* (1993) and to annual intertidal C fluxes measured by Middelburg *et al.* (1995, pers. comm.). The sedimentation estimates of Van Maldeghem *et al.* (1993) were based on data from the period 1975 to 1985 and they found net sedimentation to be largest in spatial compartments 6, 7 (Land of Saeftinghe) as well as in model compartment 10 (their compartments 7, 8 and 11). This was reproduced in our model. The upstream part of the estuary (where most sedimentation took place in the budget of Wollast) was even a net exporter of mud in Van Maldeghem *et al.*, (1993) but this is due to dredging activity. Nevertheless, there is evidence that the Westerschelde estuary itself has changed since the mid seventies (Soetaert & Herman, 1995b) and this too could account at least for some of the discrepancies. Based on a comparison of nutrient profiles from 1974 and 1991 and on a long-term series of oxygen, nitrate and ammonium concentrations, it was shown that the oxygen-deficient zone has moved upstream for at least 10 kilometres since 1975 (Soetaert & Herman, 1995b). Due to the larger extent of the anoxic zone in the early seventies, both the processes of denitrification (Soetaert & Herman, 1995b) and sulphate reduction in the water column (Wollast, pers. comm.) were much more important in the estuary then. *In situ* measurements performed in 1991 show that bacterial production immediately upstream from the estuary is some 25% higher than in the upper estuarine part (Goosen *et al.*, 1992). Assuming that the biochemical characteristics that now reside in the river were present in the upper estuarine part in the seventies, part of the budgettary discrepancy can be accounted for. Finally, sedimentation patterns in the estuary are continuously changing and shifts in the main deposition centres may have occurred as well (Wollast, pers. comm.).

Net heterotrophy in the Westerschelde

Many aquatic ecosystems are heterotrophic (respiration $\gg$ autotrophic production) and have food webs based on detritus. In a compilation of available literature, Smith & Hollibaugh (1993) showed most coastal marine communities to be net heterotrophic. With few exceptions, total respiration was between 1 to 2 times higher than autotrophic production, indicating that respiration and autotrophic production were closely linked. Net carbon consumption (respiration-autotrophic production) varied from -43 (net autotrophic) to +250 gC m^{-2} yr^{-1}. Compared to these values, the Westerschelde estuary is much more heterotrophic (average Resp/Prod ratio of 6, maximum of 35; average net C consumption of 380 gC m^{-2} yr^{-1}). Nevertheless, total respiration in the Westerschelde (300–1200, average of 450 gC m^{-2} yr^{-1}) is not so much larger than the estuarine mean of 356 gC m^{-2} yr^{-1} and falls well within estuarine ranges (Smith & Hollibaugh, 1993). Total primary production on the other hand is much lower (72 compared to 300 gC m^{-2} yr^{-1}) than the estuarine mean. Using the regression equation derived by Smith & Hollibaugh (1993), total autotrophic production should be as high as 370 gC m^{-2} yr^{-1} in the Westerschelde. That this is not the case merely stems from the fact that primary production is strongly limited by the high turbidity in the area. Thus, in contrast to the estuaries cited in Smith & Hollibaugh (1993), there is no close coupling of bacterial and primary production in the Westerschelde estuary and bacterial metabolism is mainly fuelled by exogenous organic matter. Similarly in the turbid Hudson River estuary, which receives a significant amount of allochthonous carbon, bacterioplankton production was shown to be about 4 times the value of depth-integrated phytoplankton production (Findlay *et al.*, 1991).

Contribution of the Westerschelde to coastal enrichment

Due to the high heterotrophic rates, the role of the Westerschelde estuary as a subsidizer of organic carbon to the North sea is only minor (a net export of 6000 tonnes of carbon per year in the model). Assuming the same heterotrophic and autotrophic characteristics at the sea boundary compared to the last model compartment, we calculated that if the estuary would be 3 to 4 kilometres longer, all imported and *in situ* produced carbon would be lost in the estuary itself. This probably reflects the scale at which coastal fertilisation by the estuary is tangible. Moreover, not all forms of organic carbon are exported to the sea in the model. Organic detritus is exchanged across the seaward boundary, with more refractory detritus leaving the estuary while more reactive matter (with a higher nitrogen content) enters the estuary. This explains the apparent paradox of our model estuary both exporting organic carbon and importing organic nitrogen (Soetaert & Herman, 1995b). In addition, some 500 tonnes carbon of marine zooplankton biomass is net imported into the estuary per year where they dy (Soetaert & Herman, 1994). A detailed budget of phytoplankton carbon shows that 4000 tonnes of carbon is imported from the river, while a small amount is brought in from the sea (Soetaert *et al.*, 1994). All this imported, relatively reactive organic carbon is incorporated into the estuarine web. Internal cycling, mainly in the watercolumn results in the removal of most organic matter that enters the estuary or is produced *in situ* and the more refractory part escapes to the sea. Because the estuary mineralizes nearly all allochthonous and *in situ* produced carbon, it is a large net nutrient exporter: despite the denitrification process that removes part of the nitrogen, the amount of dissolved nitrogen that is flushed to the sea is some 25% higher than the amount that enters from the river or from wastes (Soetaert & Herman, 1995b).

Acknowledgments

This model was developed as part of the MAST-JEEP project of the C.E.C. and of the ECOLMOD project of Rijkswaterstaat (D.G.W.).

Thanks to Dr Van Eck and A. Schouwenaar for providing us with the data necessary for the construction of the model. This is article number 731 of the NIOO-CEMO. Dr C. Heip is acknowledged for critically reading the manuscript.

The ecosystem model MOSES is available upon request from the authors. It can only be run under the model environment SENECA, which can be purchased from the same authors (price in 1994 US $100).

References

Billen, G., J. Smitz, M. Somville & R. Wollast, 1976. Degradation de la matiere organique et processus d'oxydo-reduction dans l'estuaire de l'Escaut. In: L'Estuaire de l'Escaut. Projet Mer. Rapport final. Bruxelles. Service du Premier Ministre 10: 102–152.

Billen, G., C. Lancelot & M. Meybeck, 1991. N, P and Si retention along the aquatic continuum from land to ocean. In Mantoura, R. F. C., J. M. Martin & R. Wollast (eds), Ocean margin processes is global change. John Wiley, Chichester: 365–381.

Billen, G., C. Joiris, L. Meyer-Reil & H. Lindeboom, 1990. Role of bacteria in the North Sea ecosystem. Neth. J. Sea Res. 26: 265–293.

Boynton, W. R., W. M. Kemp & C. W. Keefe, 1982. A comparative analysis of nutrients and other factors influencing estuarine phytoplankton production. In V. S. Kennedy (ed.), Estuarine comparisons. Academic, San Diego: 69–90.

Cloern, J. E., B. E. Cole & R. S. Oremland, 1983. Autotrophic processes in meromictic Big Soda Lake, Nevada. Limnol. Oceanogr. 28: 1049–1061.

Colijn, F. & V. N. de Jonge, 1984. Primary production of microphytobenthos in the Ems-Dollard estuary. Mar. Ecol. Progr. Ser. 14: 185–196.

de Jonge, V. N. & F. Colijn, 1992. Dynamics of microphytobenthos biomass in the Ems estuary measured as chlorophyll-a and carbon. In Physical processes and dynamics of microphytobenthos in the Ems estuary (the Netherlands). PHD Thesis Groningen: 79–96.

de Leeuw, J. & G. J. Buth, 1991. Spatial and temporal variation in peak standing crop of European tidal marshes. In Elliott, M. & J. P. Ducrotoy (eds), Estuaries and coasts: spatial and temporal inter-comparisons, ECSA 19th symposium. Olsen & Olsen, Copenhagen: 133–137.

Eilers, P. H. C. & J. C. H. Peeters, 1988. A model for the relationship between light intensity and the rate of photosynthesis in phytoplankton. Ecol. Modell. 42: 185–198.

Findlay, S., M. L. Pace, D. Lints, J. J. Cole, N. F. Caraco & B. Peierls, 1991. Week coupling of bacterial and algal production in a heterotrophic ecosystem, the Hudson estuary. Limnol. Oceanogr. 36: 268–278.

Goosen, N., P. van Rijswijk, J. Peene & J. Kromkamp, 1992. Annual patterns of bacterial production in the Scheldt estuary (S.W. Netherlands). In JEEP 92. Major biological processes in european tidal estuaries. MAST report: 109–114.

Heip, C., 1988. Biota and abiotic environment in the Westerschelde estuary. Hydrobiol. Bull. 22: 31–34.

Helder, W. & R. T. P. De vries, 1983. Estuarine nitrite maxima and nitrifying bacteria (Ems-Dollard estuary). Neth. J. Sea Res. 17: 1–18.

Hemminga, M. A., V. A. Klap, J. van Soelen & J. J. Boon, 1993. The effect of salt marsh inundation on estuarine particulate organic matter characteristics. Mar. Ecol. Progr. Ser. 99: 153–161.

Indrebø, G., B. Pengerud & I. Dundas, 1979. Microbial activities in a permanently stratified estuary. II. Microbial activities at the oxic-anoxic interface. Mar. Biol. 51: 305–309.

Jassby, A. D., J. E. Cloern & T. M. Powell, 1993. Organic carbon sources and sinks in San Francisco Bay: variability induced by river flow. Mar. Ecol. Prog. Ser. 95: 39–54.

Klepper, O., 1989. A Model of Carbon Flows in Relation to Macrobenthic Food Supply in the Oosterschelde Estuary (S.W. Netherlands). Ph. D. thesis, University of Wageningen, Wageningen: 1–270.

Kromkamp, J., J. Peene, P. van Rijswijk, A. Sandee & N. Goosen, 1995. Nutrients, light and primary production by phytoplankton and microphytobenthos in the eutrophic, turbid Westerschelde estuary (The Netherlands). Hydrobiologia 311 (Dev. Hydrobiol. 110): 9–19.

Lancelot, C. & G. Billen, 1985. Carbon-nitrogen relationships in nutrient metabolism of coastal marine ecosystems. Adv. aquat. Microbiol. 3: 263–321.

Middelburg, J. J., G. Klaver, J. Nieuwenhuize & T. Vlug, 1995. Carbon and nitrogen cycling in intertidal sediments near Doel, Scheldt Estuary. Hydrobiologia 311 (Dev. Hydrobiol. 110): 57–69.

Peterson, D. H., S. W. Hager & L. E. Schemel, 1988. Riverine C, N, Si and P transport to the coastal ocean: an overview. In B. O. Jansson (ed.), Lecture notes on coastal and estuarine studies vol 22. Coastal-offshore ecosystem interactions. Springer-Verlag, Berlin/Heidelberg: 227–253.

SAWES, 1991. Waterkwaliteitsmodel Westerschelde. WL-rapport T257.

Shaffer, G. P. & C. P. Onuf, 1985. Reducing the error in estimating annual production of benthic microflora: hourly to monthly rates, patchiness in space and time. Mar. Ecol. Progr. Ser. 26: 221–231.

Smith, S. V. & J. T. Hollibaugh, 1993. Coastal metabolism and the oceanic organic carbon balance. Rev. Geophysics 31: 75–89.

Soetaert, K. & P. M. J. Herman, 1993. MOSES – model of the scheldt estuary – ecosystem model development under SENECA. Report, 89 pages.

Soetaert, K. & van Rijswijk, 1993. Spatial and temporal patterns of the zooplankton in the Westerschelde estuary. Mar. Ecol. Progr. ser. 97: 47–59.

Soetaert, K. & P. M. J. Herman, 1994. One foot in the grave: zooplankton drift in the Westerschelde estuary (S.W. Netherlands). Mar. Ecol. Prog. Ser. 105: 19–25.

Soetaert, K., M. J. Herman & J. Kromkamp, 1994. Living in the twilight: estimating net phytoplankton growth in the Westerschelde estuary (The Netherlands) by means of an ecosystem model (MOSES). J. Plankton Res. 16: 1277–1301.

Soetaert, K. & P. M. J. Herman, 1995a. Estimating estuarine residence times in the Westerschelde (The Netherlands) using a box model with fixed dispersion coefficients. Hydrobiologia 311 (Dev. Hydrobiol. 110): 215–224.

Soetaert, K. & P. M. J. Herman, 1995b. Nitrogen dynamics in the Westerschelde estuary (SW Netherlands) estimated by means of the ecosystem model MOSES. Hydrobiologia 311 (Dev. Hydrobiol. 110): 225–246.

SYSTAT, 1992. Systat for windows: graphics, version 5. Evanston, IL: SYSTAT inc., 636 pp.

Van Maldeghem, D. C., H. P. J. Mulder & A. Langerak, 1993. A cohesive sediment balance for the Scheldt estuary. Neth. J. Aquat. Ecol. 27: 247–256.

Wollast, 1988. The Scheldt estuary. In Salomons, W., B. L. Bayne, E. K. Duursma & U. Forstner (eds), Pollution of the North Sea an assessment. Springer-Verlag, Berlin: 183–193.

Wollast, R., 1976. Transport et accumulation de polluants dans l'estuaire de l'Escaut. In J. C. Nihoul & R. Wollast (eds), l'Estuaire de l'Escaut. Projet Mer Rapport final. Bruxelles. Service du Premier Ministre 10: 191–201.

Wollast, R., 1983. Interactions in estuaries and coastal waters. In Bolin, B. & R. B. Cook (eds), The major biogeochemical cycles and their interactions. SCOPE.